Springer Series in Reliability Engineering

Series Editor

Hoang Pham

For further volumes:
http://www.springer.com/series/6917

Halil Karadeniz

Stochastic Analysis of Offshore Steel Structures

An Analytical Appraisal

Springer

Halil Karadeniz
Faculty of Civil Engineering
and Geosciences
Delft University of Technology
Delft
The Netherlands

Vedat Togan
Department of Civil Engineering
Karadeniz Technical University
Trabzon
Turkey

Mehmet Polat Saka
Department of Engineering Sciences
Middle East Technical University
Ankara
Turkey

Additional material to this book can be downloaded from http://extras.springer.com/

ISSN 1614-7839
ISBN 978-1-4471-6094-6 ISBN 978-1-84996-190-5 (eBook)
DOI 10.1007/978-1-84996-190-5
Springer London Heidelberg New York Dordrecht

Softcover reprint of the hardcover 1st edition 2013

Content of Section 1.3 in Chapter one adapted from OMAE-2010, Paper No. OMAE2010-20971, Shanghai, China for use here with kind permission from ASME

Printed on acid-free paper

Springer is part of Springer Science+Business Media (www.springer.com)

Preface

Increasing oil consumption in the world and scarcity of land-oil resources due to political and economical reasons has caused offshore oil exploration and production to become a growing investigation field in the past six decades. The analysis of structures to use energy deposits and other recourses, or for other purposes, in ocean environments requires a special consideration since environmental and loading conditions offshore are very complicated and contain large uncertainties. Offshore structures are continuously subjected to random ocean waves producing stochastic loads that cause mainly fatigue failure in structural components. In tectonic offshore environments, structures are also subjected to earthquake and earthquake-induced hydrodynamic loadings, which are considered to be important as they can cause structural collapse in a short time. Since the ocean environment and random wave phenomenon are highly uncertain, a probabilistic structural analysis needs to be carried out essentially. This requires the knowledge of probability theory and applied probability models, which construct the basis of reliability analysis. For the fatigue damage and fatigue reliability analysis, theoretical knowledge is combined with experimental information to predict correct results. Under these complexities, offshore structures should be designed to give optimal performance within the safe margin. This can be achieved by applying methods of the reliability-based design optimization. This book aims to cover difficult issues encountered in the analysis of offshore steel structures under random wave and earthquake loadings. It provides broad analytical tools for advanced analysis of offshore structures. It serves as a stand-alone reference book for design engineers, researchers, graduate and post graduate students, and for higher education in the field of offshore structural engineering. A corresponding computer program, SAPOS, is also attached to the book via Springer Extras. The program is run at the responsibility of the users. The publisher and the authors of this book are not responsible in any form that may arise from the use of this program. The book contains seven independent chapters of which Chaps. 1–6 have been written by Prof. Dr. H. Karadeniz and Chap. 7 has been written by Prof. Dr. M. P. Saka and Dr. V. Togan. Each chapter is devoted to handle the specific subject as briefly summarized in the following paragraphs.

Chapter 1 explains the mechanics of space frame structures and presents necessary formulations for the finite element analysis of space frames since offshore steel structures are constituted from space frames. In this chapter, the 3D Timoshenko beam theory is presented with emphasis on formulation of partly and eccentrically connected members, as well as formulation of a soil-beam interface element to account for foundation effects in the analysis. This chapter highlights also the eigenvalue problem related to the dynamic analysis of offshore structures.

Chapter 2 presents basic information and essential formulation of random vibration and stochastic analysis that are needed in offshore structural analysis. Having presented commonly used probability models, stochastic processes are summarized. Then, the spectral analysis, transfer functions, and the crossing analysis are highlighted.

Chapter 3 is devoted to ocean wave mechanics and wave forces. Having summarized wave theories in general, the Airy wave theory is explained in detail and formulation for deep-water conditions is presented. The chapter first describes stochastic ocean waves, transfer functions, commonly used short-term sea spectra with directional distribution, wave-current interaction phenomenon, and probabilistic description of sea states in the long term. Then, attention is paid to the calculation of wave and member consistent forces with emphasis on added mass and hydrodynamic damping concept.

Chapter 4 presents spectral analysis of offshore structures under wave and earthquake actions. After describing the problem of spectral analysis and giving general information, formulation of dynamic analysis of offshore structures in the frequency domain, transfer functions of wave and earthquake forces are presented. Then, concentration is focused on response transfer functions of wave and earthquake forces, followed by formulation of the hydrodynamic and inertia forces produced by earthquakes and their combination. This chapter also explains calculation of response spectra of offshore structures under stochastic wave and earthquake forces with earthquake ground motion and its spectral representation including non-uniform earthquake ground motions. Finally, the calculation of response statistical quantities is presented with illustrative examples.

Chapter 5 is devoted to the fatigue phenomenon in structures. First, the fatigue process, source of fatigue, and modeling of fatigue is summarized in general. Then, the calculation of fatigue damages is explained by using the fracture mechanics and *S–N* curve approaches. The cumulative damage is calculated according to the Palmgren–Miner's rule. For non-narrow banded stress processes, the fatigue damage is estimated using probability distribution of random stress ranges obtained from the rain-flow cycle counting algorithm. Finally, calculation of total spectral fatigue damage is presented for a given lifetime by using a multilinear *S–N* fatigue model.

Chapter 6 presents reliability analysis of offshore structures. Having explained uncertainties in general and given information about the reliability methods, basic definitions and structural reliability methods are presented in more detail. This is followed by the calculation of the reliability index β by the FORM and SORM methods for nonlinear failure functions and non-normal correlated design variables. The calculation algorithms and flow diagrams are given. Then, after the Level III reliability methods are outlined, the inverse reliability method and its calculation algorithm are presented. In the final sections, uncertainties in spectral stresses and fatigue damages of offshore structures are explained in a reduced uncertainty space which follows the fatigue reliability calculation and its algorithms.

Chapter 7 is devoted to optimization techniques that are widely applied to determine the optimum solution of structural design problems. First, this chapter introduces the mathematical formulation of optimization problems and their solution techniques among which sequential programming technique and differential evolution algorithm are briefly explained. Second, it presents the mathematical formulation of reliability-based design optimization problems in the uncertainty space regarding load, resistance, and structural response. Then, it summarizes the available solution techniques and explains methods of sensitivity analysis of the reliability-based design optimization of offshore structures.

The first author of this book, Prof. Dr. H. Karadeniz, acknowledges Delft University of Technology (TUDelft), The Netherlands, for giving him the opportunity to carry out research work and to use their facilities for a long period, from 1978 until his retirement at the end of 2010. Especially, he acknowledges Prof. A. L. Bouma of the structural mechanics group in the Civil Engineering Department for accepting him into his research group in 1978. He thanks all the staff of the Structural Mechanics Group in the Civil Engineering Department at TUDelft for the memorable social atmosphere that they created. His special thanks and appreciation are for Prof. Ton Vrouwenvelder of the Civil Engineering Department, TUDelft, for his continuous, stimulating, and invaluable discussions during his stay at TUDelft. He appreciates and acknowledges all his colleagues, friends, and research collaborators from all over the world for their useful discussions and inspiring suggestions. Finally, all the authors of this book thank Springer-Verlag for publishing this book and making it available for interested readers in worldwide.

July 2012

H. Karadeniz
M. P. Saka
V. Togan

Contents

1 Finite Element Analysis of Space Frame Structures 1
1.1 Introduction. 1
1.2 Formulation of a 3D Timoshenko Beam Element. 2
1.2.1 Curvatures of 3D Beams Under Pure Bending. 3
1.2.2 Equilibrium Equations of 3D Beam Elements 5
1.2.3 Contributions of Transverse Shear Forces on the Elastic Curve. 7
1.2.4 Deformation of a Point on a Cross-Section of 3D Beams. 9
1.2.5 Rotation Matrix of a Point on a Cross-Section of 3D Beams and Deformation for Small Rotations . . . 10
1.2.6 Strains and Stresses at a Location on a Cross-Section of a 3D Beam 13
1.2.7 Calculation of Forces and Moments of 3D Beams 14
1.2.8 Differential Equations of the 3D Beam Element 16
1.2.9 Solution of Differential Equations of the Elastic Curve and Shape Functions of the 3D Beam Element 18
1.2.10 Total Potential Energy, Stiffness Matrix, and Static Equilibrium Equation 23
1.2.11 Total Kinetic Energy, Mass Matrix, Damping Matrix, and Dynamic Equilibrium Equation 30
1.2.12 Coordinate Systems and Transformations 37
1.2.13 Transformations of Element Stiffness Matrix, Consistent Load Vector, and Mass Matrix. 41
1.3 Formulation of Member Releases and Partly Connected Members. 43
1.3.1 Representation of a Partly Connected Beam Element. 44

1.3.2 Formulation of Stiffness Matrix, Consistent Load Vector, and Mass Matrix of a Spring-Beam Element 46
1.3.3 Calculation of the Connectivity Matrix 53
1.3.4 Member Releases in a Different Coordinate System . . . 63
1.4 Formulation of Eccentrically Connected Members 65
1.5 An Interface Beam Element for the Soil–Structure Interaction and Deformation of Soil Under R-Wave Propagation 69
1.5.1 Modeling of Soil Medium and Calculation of Interface Loadings 71
1.5.2 Formulation of Interface Element for Soil–Beam Interactions 75
1.5.3 Ground Deformation Under R-Wave Propagation and Calculation of the Exerted Force Vector 79
1.6 Calculation of Natural Frequencies and Mode Shapes, Eigenvalue Solution 87
1.6.1 Eigenvalue Solution 89
1.6.2 Eigenvalue Solution of Deteriorated Structures 94
1.7 Dynamic Response Analysis 97
1.7.1 Time-Domain Solution 98
1.7.2 Frequency Domain Solution 105
1.8 Examples 111
1.8.1 Example of a Portal Frame 111
1.8.2 Example of 2D Offshore Jacket Structure 114
1.8.3 A Simple Beam for Exercise 115
References 115

2 Introduction to Random Vibration and Stochastic Analysis 121
2.1 Introduction 121
2.2 Probability, Random Variables, Stochastic Processes, Probability Distribution and Density Functions 122
2.2.1 Probability Measure 123
2.2.2 Random Variables 125
2.2.3 Stochastic Processes 126
2.2.4 Probability Distribution and Density Functions 127
2.3 Mean Values, Probability Moments, and Variances of Random Variables and Random Functions 134
2.3.1 Functions of Random Variables 136
2.3.2 Some Useful Probability Distributions 141
2.4 Random Processes, Ensemble Averages, Expected Values, Stationary and Ergodic Processes 147
2.4.1 Ensemble Averages and Expected Values 148
2.4.2 Stationary and Ergodic Processes 150
2.4.3 Differentiation of Stochastic Processes 153

2.5 Spectral Analysis of Stochastic Processes ... 154
2.5.1 Spectral Moments, Variances of Derived Processes, Spectral Bandwidth ... 157
2.5.2 Band-Limited, Narrow-Band and Broad-Band Processes ... 158
2.5.3 Crossing Analysis and Probability Distributions of Maxima ... 161
2.6 Input–Output Relations of Stochastic Processes, Transfer Functions ... 167
2.7 Examples ... 170
References ... 175

3 Water Wave Theories and Wave Loads ... 177
3.1 Introduction ... 177
3.2 Introduction to Wave Theories ... 178
3.2.1 Stokes Wave Theory ... 183
3.2.2 Other Wave Theories ... 185
3.3 Linear (Airy) Wave Theory ... 186
3.3.1 Formulation for Deep Water Condition ... 189
3.4 Stochastic Description of Ocean Waves and Short-Term Sea States ... 190
3.4.1 Transfer Functions of a Random Wave in Deep Water ... 193
3.4.2 Statistics and Spectral Functions of the Water Elevation in the Short Term ... 194
3.5 Wave–Current Interaction ... 201
3.6 Probabilistic Description of Sea States in the Long Term ... 205
3.7 Morison's Equation and Wave Forces on Structural Members ... 207
3.7.1 Wave Forces on Large Structural Members ... 209
3.7.2 Wave Forces on Inclined Structural Members ... 214
3.7.3 Calculation of Consistent Wave Forces of Members ... 216
3.8 Linearization of the Morison's Equation ... 220
3.9 Linearization of the Morison's Equation Under Wave–Current and Wave–Structure Interactions ... 227
3.9.1 Calculation of Standard Deviation of the Relative Normal Water Velocity ... 234
3.10 Calculation of Consistent Current Forces of Members, Hydrodynamic Damping Ratio and Added Mass Matrices of Members ... 236
3.10.1 Calculation of Consistent Current Forces of Members ... 237
3.10.2 Calculation of Hydrodynamic Damping Ratio ... 238
3.10.3 Calculation of Added Mass Matrices of Members ... 240

3.11 Examples 242
3.11.1 Example of a Monopod Tower Under Deterministic Wave Loading 242
3.11.2 Example of a Monopod Tower Under Random Wave Loading 245
References 248

4 Spectral Analysis of Offshore Structures Under Wave and Earthquake Loadings 253
4.1 Introduction 253
4.2 Dynamic Analysis of Structures in the Frequency Domain, the Transfer Function Approach 254
4.3 Calculation of Response Transfer Functions of Offshore Structures 256
4.3.1 Response Transfer Functions for Wave Loads 256
4.3.2 Response Transfer Functions for Earthquake Loading 257
4.4 Calculation of Response Spectra of Offshore Structures 262
4.4.1 Response Spectra Under Stochastic Wave Loads 262
4.4.2 Response Spectra Under Stochastic Earthquake Loading 263
4.5 Calculation of Response Statistical Quantities 276
4.6 Example 277
References 281

5 Fatigue Analysis of Offshore Structures 287
5.1 Introduction 287
5.2 Fatigue Phenomenon and Fatigue Damages 288
5.2.1 Fracture Mechanics Approach to Predict Fatigue Damages 289
5.2.2 S–N Curve Approach to Predict Fatigue Damages 293
5.2.3 Cumulative Fatigue Damage Rule 294
5.3 Cycle Counting Procedure for Random Stresses 295
5.3.1 Rainflow Cycle Counting 296
5.4 Probability Distribution of Random Stress Ranges 298
5.4.1 Parameters of the Probability Distribution of Stress Ranges 300
5.5 Spectral Fatigue Damages Based on Multilinear (S–N) Model 305
5.6 Examples 308
References 312

6 Reliability Analysis of Offshore Structures 315
6.1 Introduction 315

6.2 Structural Reliability Methods 316
6.2.1 Design Variables, Limit State Functions, and Failure Probability in the Reliability Analysis 318
6.2.2 First-Order Reliability Methods 321
6.2.3 Second-Order Reliability Method 330
6.2.4 Level-III (Exact) Reliability Methods 335
6.3 Inverse Reliability Method 338
6.4 Uncertainties in Spectral Stresses and Fatigue Damages of Offshore Structures 341
6.4.1 Uncertainties in Stress Statistical Characteristics 342
6.4.2 Uncertainties in Fatigue Model Parameters 348
6.5 Stress Spectrum and Stress Statistical Values in the Uncertainty Space 349
6.6 Fatigue and Stress Based Reliability Calculations of Offshore Structures 353
6.7 Examples 354
6.7.1 Reliability Calculation of a Tubular Steel Member . . . 354
6.7.2 Fatigue Reliability Calculation of an Example Offshore Jacket Structure 358
References 361

7 Optimization of Offshore Structures 365
7.1 Introduction 365
7.2 Structural Optimization 366
7.2.1 Objective Function 367
7.2.2 Design Constraints 368
7.2.3 Design Example 369
7.3 Deterministic and Stochastic Solution Techniques of Optimization 372
7.3.1 Sequential Quadratic Programming 373
7.3.2 Differential Evolution Technique 374
7.4 Mathematical Formulation of the Reliability-Based Design Optimization 376
7.4.1 Reliability Index Approach for the RBDO 378
7.4.2 Performance Measure Approach for the RBDO 379
7.5 Sensitivity Analysis of RBDO of Offshore Structures 380
7.5.1 Finite Difference Method 381
7.5.2 Direct Differentiation Method 382
7.5.3 Adjoint Method 382
7.6 Examples 384
7.6.1 Deterministic Design Optimization 384
7.6.2 Reliability-Based Design Optimization 391
7.6.3 Exercises 404
References 405

Chapter 1
Finite Element Analysis of Space Frame Structures

1.1 Introduction

Frame structures are commonly used in structural engineering applications in different forms as plane (2D) and space (3D) frames, which are made of steel, reinforced and prestressed concrete (RC), or timber. Plane frames are composed of arbitrarily oriented beam elements jointed together in a plane with distributed loading on elements and/or concentrated loads in the same plane. Before the computer technology is developed, they have been mostly used in practice by modeling a structural system and loading in different planes due to analysis simplicity. Numerous investigations have been reported on planar frames for different analysis types and conditions; see i.e., [1–10]. Today, since the capabilities and capacities of computers are at high levels and still increasing, the need of simplification of structural systems is not essential for calculation purposes and more realistic structural models are used by applying 3D beam elements [11–13] from which 2D elements are obtained as a special case. Beams are also categorized as straight or curved beams. Most structural frames are made of straight beams. Curved elements [14–18], which have considerable initial curvatures in the plane of loading, can find limited applications in practice such as arches and hooks. In complex structures, they are used mainly to obtain various structural shapes for either esthetic or load carrying purposes, and curved girders in bridge structures. The structural behavior of curved members differs considerably from the structural behavior of straight members due to the existence of initial curvature [19]. In general, two beam theories are used in the framed structural analyses: (a) classical beam theory, which is also known as the Euler–Bernoulli beam theory, (b) Timoshenko beam theory. The Euler–Bernoulli beam theory [20–22] is based on a simplified linear theory of elasticity [23, 24] and used to calculate the load carrying and deflection characteristics of beams in general [10, 25, 26]. In the development of the Euler–Bernoulli beam theory, the following assumptions are made:

H. Karadeniz, *Stochastic Analysis of Offshore Steel Structures*,
Springer Series in Reliability Engineering, DOI: 10.1007/978-1-84996-190-5_1,

1. The cross-section of the beam is infinitely rigid in its own plane so that no deformations occur in the plane of cross-section.
2. The cross-section of the beam remains plane after deformation.
3. The cross-section of the beam remains normal to the deformed axis of the beam.

These assumptions are valid for long, slender, and thin beams of isotropic materials with solid cross-sections. For short and thick beams and for higher natural frequencies the results of the Euler–Bernoulli beam theory may be incorrect and misleading since the effect of the transverse shear deformation is not included in the formulation of the beam. Timoshenko solved this problem by including the effect of the transverse shear deformation [27, 28] at the first time. Since then many studies have been carried out on the Timoshenko beam theory, see e.g., [29–37]. The aforementioned assumptions of the Euler–Bernoulli beam theory are also valid for the Timoshenko beam theory; except that the cross-section of the beam does not remain normal to the deformed axis of the beam any more due to inclusion of the transverse shear deformation in the beam formulation. Timoshenko beam theory is applicable for both thick and thin beams and the Euler–Bernoulli beam theory is obtained as a special case of the Timoshenko beam theory. As being parallel to the development of computational facilities, classical structural analysis methods [38–40] get replaced by modern matrix and finite element structural analysis methods [41–43]. Today, the finite element method is widely used in almost all analysis disciplines. Its formulations are based on variational principles [44, 45] and explained in general in many text book, see e.g., [42, 43, 46–49].

In this chapter, the finite element formulation of the 3D Timoshenko beam element for linear analysis is presented in detail since it forms the basis of the analysis of frame structural systems. This chapter is organized somewhat elementary that all ingredients of analyses are included in the book for the completeness, which may help the readers to better understand the background information without consulting any theoretical related sources. In this chapter, attention is given particularly to solve problems of some special topics that may occur in practical applications and their programing techniques. This chapter is intended to be a theoretical manual and reference for researchers and postgraduate students, who like to be experts in the linear structural analysis and corresponding calculation algorithms. Since the beam element presented in this chapter is intended to use for the linear analysis of offshore structures, not all types of beam elements and their properties are discussed herein. It is restricted to the Timoshenko beam element with a straight length and a solid cross-section as it is mostly used in the practice.

1.2 Formulation of a 3D Timoshenko Beam Element

A space 3D beam element is a rod oriented arbitrarily in the space, which is defined by (x, y, z) coordinates, and also loaded arbitrarily. It is assumed here that the beam is straight, solid with uniform cross-section and made of homogeneous, isotropic, and elastic material. In order to formulate a 3D Timoshenko beam element we start from the formulation of curvatures under pure bending conditions.

1.2.1 Curvatures of 3D Beams Under Pure Bending

Curvatures of a space beam can be defined in two planes as depending on directions of applied bending moments. It is assumed here that the beam axis is in the x coordinate direction and bending of the beam are in the z and y coordinate directions in the (x–y) and (x–z) planes as shown in Fig. 1.1. Further, we assume that the neutral axis of the beam, which defines the location of zero strains on the cross-section, determines the elastic curve of the beam after the bending deformations. Curvatures are defined in two cases of the bending deformation in (x–y) and (x–z) planes as presented below.

1.2.1.1 Curvature in (x–y) Plane Under Pure Bending

The bending deformation of an infinitesimal length of the beam (dx), in the (x–y) plane is shown in Fig. 1.1a. The applied bending moment is M_z, which is vectorially directed in the positive z direction. After the deformation, the rotation angle of the infinitesimal length (dx), in the (x–y) plane is denoted by $d\theta_z$, the deformation at the center is $\mathrm{d}u_y$ and the radius of the curvature is ρ_z. From Fig. 1.1a the length (ds) and curvature of the infinitesimal elastic curve can be stated as:

$$\begin{aligned} \mathrm{d}s &= \rho_z \mathrm{d}\theta_z = \sqrt{\mathrm{d}x^2 + \mathrm{d}u_y^2} = \left(1 + \left(\frac{\mathrm{d}u_y}{\mathrm{d}x}\right)^2\right)^{1/2} \mathrm{d}x \\ \kappa_z &= \frac{1}{\rho_z} = \frac{\mathrm{d}\theta_z}{\mathrm{d}x}\left(1 + \left(\frac{\mathrm{d}u_y}{\mathrm{d}x}\right)^2\right)^{-1/2} \end{aligned} \tag{1.1}$$

where κ_z is the curvature in the (x–y) plane, which depends on the derivative of the rotation θ_z with respect to x. From the definition of the rotation θ_z it is stated that,

$$\tan\theta_z = \frac{\mathrm{d}u_y}{\mathrm{d}x} \rightarrow \frac{\mathrm{d}^2 u_y}{\mathrm{d}x^2} = \frac{\mathrm{d}(\tan\theta_z)}{\mathrm{d}\theta_z}\frac{\mathrm{d}\theta_z}{\mathrm{d}x} \rightarrow \frac{\mathrm{d}\theta_z}{\mathrm{d}x} = \left(1 + \left(\frac{\mathrm{d}u_y}{\mathrm{d}x}\right)^2\right)^{-1} \frac{\mathrm{d}^2 u_y}{\mathrm{d}x^2} \tag{1.2}$$

Having introduced Eq. (1.2) into Eq. (1.1) the curvature κ_z can be obtained as written by,

$$\kappa_z = \frac{1}{\rho_z} = \left(1 + \left(\frac{\mathrm{d}u_y}{\mathrm{d}x}\right)^2\right)^{-3/2} \frac{\mathrm{d}^2 u_y}{\mathrm{d}x^2} \rightarrow \kappa_z \simeq \frac{\mathrm{d}\theta_z}{\mathrm{d}x} \simeq \frac{\mathrm{d}^2 u_y}{\mathrm{d}x^2} \tag{1.3}$$

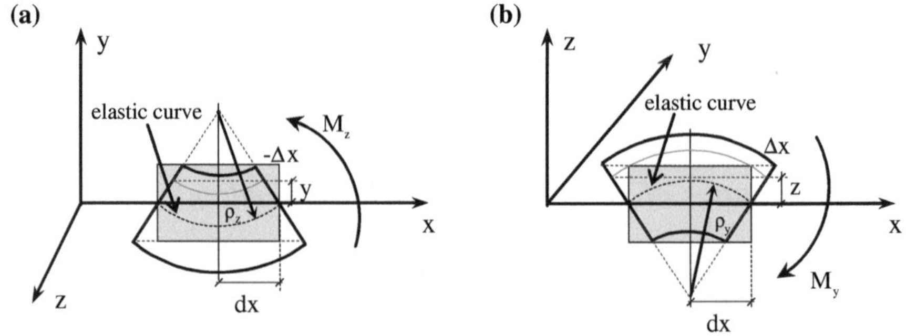

Fig. 1.1 Bending deformations of an infinitesimal beam element **a** Bending in (*x*–*y*) plane **b** Bending in (*x*–*z*) plane

This statements of the curvature κ_z will be used later to formulate differential equation of the elastic curve in the (*x*–*y*) plane.

1.2.1.2 Curvature in (*x*–*z*) Plane Under Pure Bending

As similar to the curvature in the (*x*–*y*) plane, the curvature in the (*x*–*z*) plane is obtained under the deformation of the bending moment M_y, which is shown in Fig. 1.1b with positive y direction vectorially. After the deformation, the rotation angle of the infinitesimal length (d*x*), in the (*x*–*z*) plane is denoted by $\mathrm{d}\theta_y$, the deformation at the center is $\mathrm{d}u_z$ and the radius of the curvature is ρ_y. From Fig. 1.1b the length (ds) and curvature of the infinitesimal elastic curve can be stated as:

$$\begin{aligned} \mathrm{d}s &= \rho_y \mathrm{d}\theta_y = \sqrt{\mathrm{d}x^2 + \mathrm{d}u_z^2} = \left(1 + \left(\frac{\mathrm{d}u_z}{\mathrm{d}x}\right)^2\right)^{1/2} \mathrm{d}x \\ \kappa_y &= \frac{1}{\rho_y} = \frac{\mathrm{d}\theta_y}{\mathrm{d}x}\left(1 + \left(\frac{\mathrm{d}u_z}{\mathrm{d}x}\right)^2\right)^{-1/2} \end{aligned} \tag{1.4}$$

where κ_y is the curvature in (*x*–*z*) plane, which depends on the derivative of the rotation θ_y with respect to *x*. From the definition of the rotation θ_y it is stated that,

$$\tan\theta_y = -\frac{\mathrm{d}u_z}{\mathrm{d}x} \rightarrow -\frac{\mathrm{d}^2u_z}{\mathrm{d}x^2} = \frac{\mathrm{d}(\tan\theta_y)}{\mathrm{d}\theta_y}\frac{\mathrm{d}\theta_y}{\mathrm{d}x} \rightarrow \frac{\mathrm{d}\theta_y}{\mathrm{d}x} = -\left(1 + \left(\frac{\mathrm{d}u_z}{\mathrm{d}x}\right)^2\right)^{-1}\frac{\mathrm{d}^2u_z}{\mathrm{d}x^2} \tag{1.5}$$

Having introduced Eq. (1.5) into Eq. (1.4) the curvature κ_y can be obtained as written by,

$$\kappa_y = \frac{1}{\rho_y} = -\left(1 + \left(\frac{\mathrm{d}u_z}{\mathrm{d}x}\right)^2\right)^{-3/2} \frac{\mathrm{d}^2 u_z}{\mathrm{d}x^2} \rightarrow \kappa_y \simeq \frac{d\theta_y}{\mathrm{d}x} \simeq -\frac{\mathrm{d}^2 u_z}{\mathrm{d}x^2} \tag{1.6}$$

This statements of the curvature κ_y will be used later to formulate differential equation of the elastic curve in the (x–z) plane. To find the differential equations of the elastic curve the equilibrium equations of a space beam element are also required. These are presented briefly in the following section.

1.2.2 Equilibrium Equations of 3D Beam Elements

The internal forces and moments of an infinitesimal element (dx) in a deformed state are as shown in Fig. 1.2 in the (x–y) and (x–z) planes. For small deformations, the horizontal and vertical components of the forces are written as:

$$\begin{aligned} H &= N - Q_y\theta_z + Q_z\theta_y \\ V_y &= Q_y + N\,\theta_z \\ V_z &= Q_z - N\,\theta_y \end{aligned} \tag{1.7}$$

where N is the axial force, Q_y and Q_z are the transverse shear forces, θ_y and θ_z are small rotations about y and z coordinate axes, respectively, as shown in Fig. 1.2. From the equilibriums of forces in the horizontal and vertical directions the following relations can be obtained:

$$\begin{aligned} \frac{\mathrm{d}H}{\mathrm{d}x} &= -q_x \rightarrow \frac{\mathrm{d}N}{\mathrm{d}x} = -q_x + \frac{\mathrm{d}Q_y}{\mathrm{d}x}\theta_z + Q_y\frac{d\theta_z}{\mathrm{d}x} - \frac{\mathrm{d}Q_z}{\mathrm{d}x}\theta_y - Q_z\frac{d\theta_y}{\mathrm{d}x} \\ \frac{\mathrm{d}V_y}{\mathrm{d}x} &= -q_y \rightarrow \frac{\mathrm{d}Q_y}{\mathrm{d}x} = -q_y - \left(\frac{\mathrm{d}N}{\mathrm{d}x}\theta_z + N\frac{d\theta_z}{\mathrm{d}x}\right) \\ \frac{\mathrm{d}V_z}{\mathrm{d}x} &= -q_z \rightarrow \frac{\mathrm{d}Q_z}{\mathrm{d}x} = -q_z + \left(\frac{\mathrm{d}N}{\mathrm{d}x}\theta_y + N\frac{d\theta_y}{\mathrm{d}x}\right) \end{aligned} \tag{1.8a}$$

where q_x, q_y and q_z are respectively applied distributed loadings on the element in the horizontal (x) and transverse (y and z) directions as shown in Fig. 1.2. Having neglected second-order small quantities, the force–loading relations can be stated as:

$$\begin{aligned} &\frac{\mathrm{d}N}{\mathrm{d}x} = -q_x - q_y\theta_z + q_z\theta_y + Q_y\kappa_z - Q_z\kappa_y \\ &\frac{\mathrm{d}Q_y}{\mathrm{d}x} = -q_y + q_x\theta_z - N\kappa_z, \quad \frac{\mathrm{d}Q_z}{\mathrm{d}x} = -q_z - q_x\theta_y + N\kappa_y \end{aligned} \tag{1.8b}$$

From the equilibriums of moments in the (x–y) and (x–z) planes the following moment–force relations can be obtained:

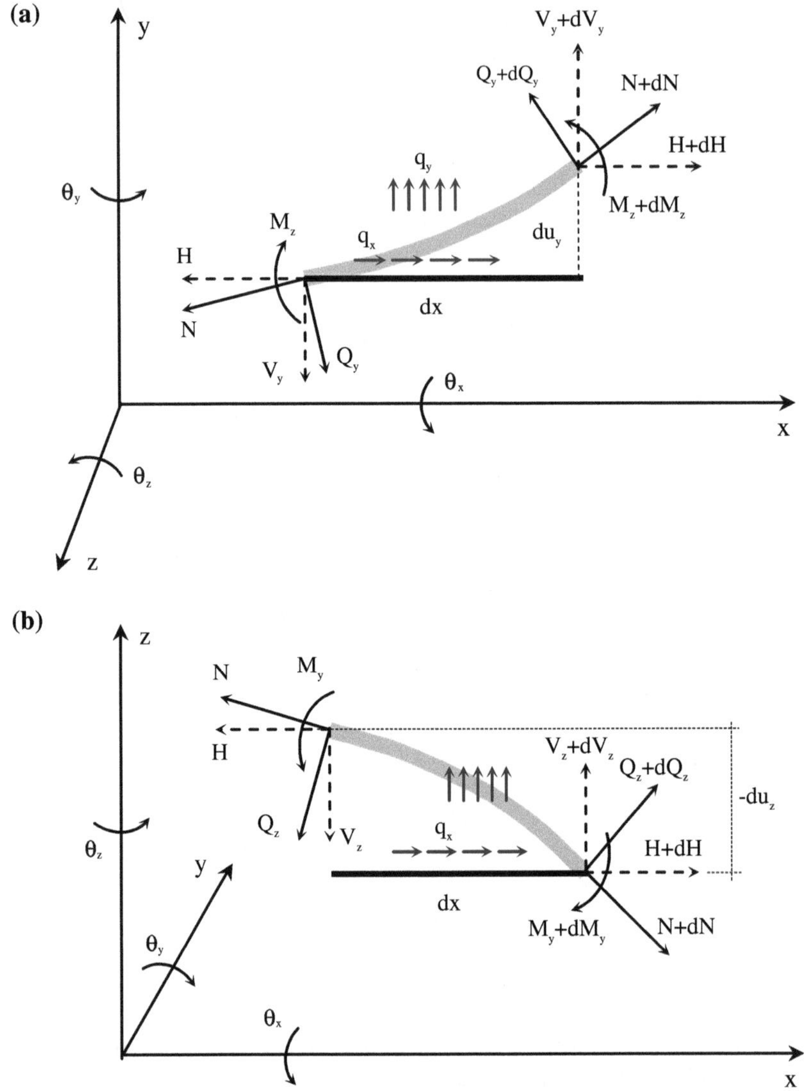

Fig. 1.2 Force and moment components of an infinitesimal beam element **a** Projection on (*x*–*y*) plane **b** Projection on (*x*–*z*) plane

$$\frac{\mathrm{d}M_z}{\mathrm{d}x} = \left(-V_y + H\frac{\mathrm{d}u_y}{\mathrm{d}x}\right) \rightarrow \frac{\mathrm{d}M_z}{\mathrm{d}x} = \left[-Q_y - N\,\theta_z + (N - Q_y\theta_z + Q_z\theta_y)\frac{\mathrm{d}u_y}{\mathrm{d}x}\right]$$
$$\frac{\mathrm{d}M_y}{\mathrm{d}x} = \left(V_z - H\frac{\mathrm{d}u_z}{\mathrm{d}x}\right) \rightarrow \frac{\mathrm{d}M_y}{\mathrm{d}x} = \left[Q_z - N\,\theta_y - (N - Q_y\theta_z + Q_z\theta_y)\frac{\mathrm{d}u_z}{\mathrm{d}x}\right] \tag{1.9a}$$

Having neglected second order small quantities these relations can be simplified as written by,

$$\begin{aligned} \frac{\mathrm{d}M_z}{\mathrm{d}x} &= -Q_y + N\left(\frac{\mathrm{d}u_y}{\mathrm{d}x} - \theta_z\right) \\ \frac{\mathrm{d}M_y}{\mathrm{d}x} &= Q_z - N\left(\frac{\mathrm{d}u_z}{\mathrm{d}x} + \theta_y\right) \end{aligned} \tag{1.9b}$$

The moment–force relations will be used later in the determination of differential equations of the elastic curve. So far, curvatures of Euler–Bernoulli beams and equilibrium equations are presented through Eqs. (1.1) and (1.9b). In the following section the effects (contributions) of transverse shear forces on the elastic curve are presented.

1.2.3 Contributions of Transverse Shear Forces on the Elastic Curve

Deformation of a beam under transverse shear forces and bending moments (Timoshenko beam) are somewhat different from the deformation under pure bending moments (Euler–Bernoulli beam). As mentioned in the introduction, in the Timoshenko beam theory, due to large shear deformations a cross-section of the beam does not remain normal to the deformed axis of the beam unlike the Euler–Bernoulli beam theory, in which the cross-section of the beam remains normal to the deformed axis. In the Timoshenko beam theory, the transverse shear deformations are also taken into account in the formulation of the differential equations of the elastic curve. As it is shown in Fig. 1.3, the total infinitesimal displacements, $\mathrm{d}u_y$ and $\mathrm{d}u_z$, can be considered in two parts as: (a) contributions of the bending moments and (b) contributions of the transverse shear forces. From Fig. 1.3 the total infinitesimal displacements and their derivatives can be written as

$$\begin{aligned} \mathrm{d}u_y &= \theta_z\mathrm{d}x + \gamma_y\mathrm{d}x \rightarrow \frac{\mathrm{d}u_y}{\mathrm{d}x} = \gamma_y + \theta_z \\ \mathrm{d}u_z &= -\theta_y\mathrm{d}x + \gamma_z\mathrm{d}x \rightarrow \frac{\mathrm{d}u_z}{\mathrm{d}x} = \gamma_z - \theta_y \end{aligned} \tag{1.10a}$$

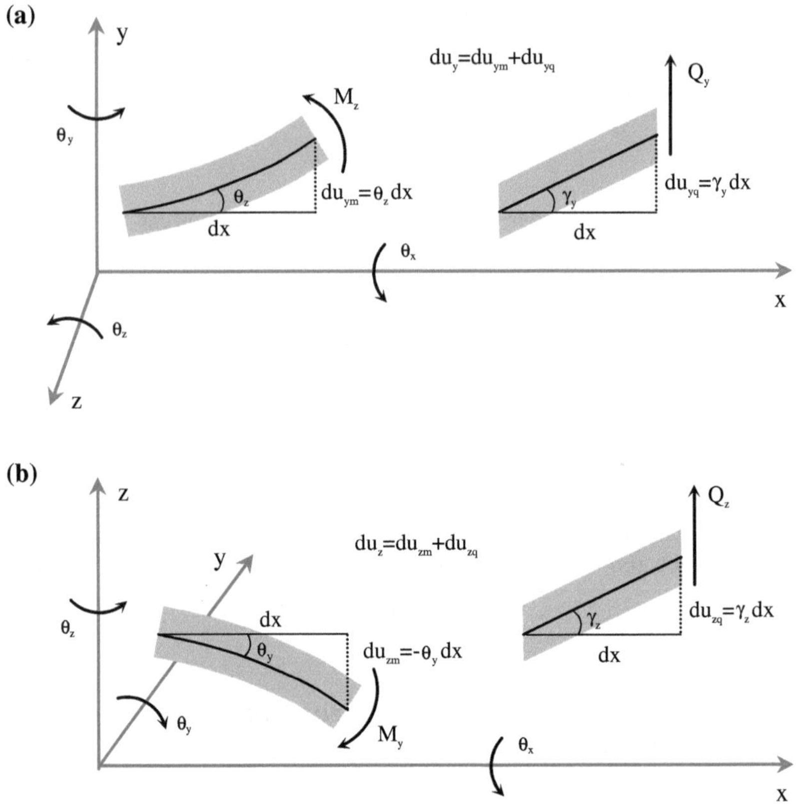

Fig. 1.3 Bending and shear deformations of an infinitesimal beam element **a** Bending and shear deformations in (*x*–*y*) plane **b** Bending and shear deformations in (*x*–*z*) plane

in which γ_y and γ_z are the average transverse shear strains in the (*y*) and (*z*) coordinate directions, which can be written [44] as,

$$\gamma_y = \frac{Q_y}{A_y G} \quad \text{and} \quad \gamma_z = \frac{Q_z}{A_z G} \tag{1.10b}$$

in which A_y and A_z are the cross-sectional areas for the shear forces in the (*y*) and (*z*) directions, respectively and G is the shear modulus of the beam material. The areas of the shear forces, A_y and A_z, are defined as $\left(A_y = k_y A\right)$ and $(A_z = k_z A)$ where A is the cross-sectional area. The coefficients k_y and k_z are dimensionless shear correction factors [44] that depend on the cross-sectional shape. Having introduced Eqs. (1.10a, b) into Eq. (1.9b) the differential equations of the bending moments can be obtained as written by,

$$\left.\begin{aligned} \frac{\mathrm{d}M_z}{\mathrm{d}x} &= -Q_y + \gamma_y N \\ \frac{\mathrm{d}M_y}{\mathrm{d}x} &= Q_z - \gamma_z N \end{aligned}\right\} \rightarrow \left\{\begin{aligned} \frac{\mathrm{d}M_z}{\mathrm{d}x} &= -Q_y\left(1 - \frac{N}{A_y G}\right) \\ \frac{\mathrm{d}M_y}{\mathrm{d}x} &= Q_z\left(1 - \frac{N}{A_z G}\right) \end{aligned}\right. \tag{1.11a}$$

or having neglected the effects of the axial force N the differential equations of the bending moments simplify as written by,

$$\frac{\mathrm{d}M_z}{\mathrm{d}x} = -Q_y \quad \text{and} \quad \frac{\mathrm{d}M_y}{\mathrm{d}x} = Q_z \tag{1.11b}$$

which will be used for the formulation of the differential equations of the elastic curve of the beam. In order to calculate strains at a point on a cross-section of the beam, the deformation and rotation of the point are presented in the following two sections.

1.2.4 Deformation of a Point on a Cross-Section of 3D Beams

A point on a cross-section of a space beam element is denoted by A with the location vector $\{r\}$ before deformation as shown in Fig. 1.4. The point A moves to the point A' with the location vector $\{r'\}$ after the deformation. The displacement of the point A, the distance of $(A - A')$, is denoted by the vector $\{v\}$ and the displacement vector of the coordinates of the center is denoted by $\{u\}$. The displacement vector of the point A can be stated as,

$$\{v\} = \{u\} + (\{r'\} - \{r\}) \tag{1.12}$$

which consists of two parts as being translational and rotational deformations. The location vector, $\{r'\}$, on the deformed cross-section, can be obtained from the rotational transformation of the original location vector $\{r\}$ in the space. This transformation can be written as,

$$\{r'\} = [R]\{r\} \tag{1.13a}$$

where $[R]$ is the rotation matrix, which will be presented in the next section. Having introduced Eq. (1.13a) into Eq. (1.12) the deformation vector of the point A can be stated in terms of the deformation vector $\{u\}$ of the central point C, at which the origin of the coordinate system is located, and the rotation of the location vector $\{r\}$ of the point A. Thus, the deformation vector $\{v\}$ of the location A is written as,

$$\{v\} = \{u\} + ([R] - I_3)\{r\} \tag{1.13b}$$

where I_3 is a (3×3) unit matrix.

Fig. 1.4 Deformation of a point on a cross-section

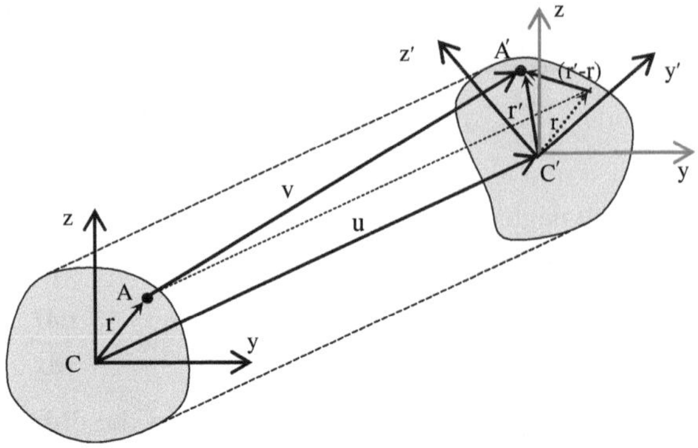

1.2.5 Rotation Matrix of a Point on a Cross-Section of 3D Beams and Deformation for Small Rotations

The rotation matrix $[R]$ in Eqs. (1.13a, b) is calculated from the Rodriguez rotation formula [50–53]. It is stated in the series solution as,

$$[R] = \exp(\tilde{\Psi}) = I_3 + \tilde{\Psi} + \frac{1}{2!}\tilde{\Psi}^2 + \frac{1}{3!}\tilde{\Psi}^3 + \cdots + \frac{1}{n!}\tilde{\Psi}^n \tag{1.14}$$

in which I_3 is a (3×3) unit matrix and $\tilde{\Psi}$ denotes a skew-symmetric matrix defined as,

$$\tilde{\Psi} = \begin{bmatrix} 0 & -\theta_z & \theta_y \\ \theta_z & 0 & -\theta_x \\ -\theta_y & \theta_x & 0 \end{bmatrix} \tag{1.15}$$

Here θ_x, θ_y, and θ_z denote small angles of rotations about the local (x, y, z) coordinate axes respectively. They are the components of the rotation $\left(\vartheta = \sqrt{\theta_x^2 + \theta_y^2 + \theta_z^2}\right)$ in the coordinate axes. The rotation matrix $[R]$ for a finite rotation can be obtained by infinitesimal rotations successively, i.e., the rotations θ_x, θ_y and θ_z are divided into n equal infinitesimal angles. Then, the matrix $[R]$ can be obtained from the repetition of rotations with infinitesimal angles, i.e.,:

$$\begin{aligned} [R]_{1/n} &= \left(I_3 + \frac{1}{n}\tilde{\Psi}\right) \\ \{r_1\} &= [R]_{1/n}\{r\},\ \{r_2\} = [R]_{1/n}\{r_1\} = [R]^2_{1/n}\{r\},\ \ldots, \\ &\ldots,\ \{r_n\} = [R]^n_{1/n}\{r\} = \left(I_3 + \frac{1}{n}\tilde{\Psi}\right)^n\{r\} \end{aligned} \tag{1.16a}$$

In the limit case, when (n) approaches infinity, the final vector $\{r_n\}$ will be the vector $\{r'\}$ and the rotation matrix $[R]$ will be calculated from:

$$[R] = \lim_{n\to\infty}\left(I_3 + \frac{1}{n}\tilde{\Psi}\right)^n \tag{1.16b}$$

The series solution of $[R]$, given by Eq. (1.14) can be proved by using the binomial theorem as stated below.

$$\begin{aligned}
&(1+y)^n = \sum_{k=0}^{n}\frac{n!}{k!(n-k)!}y^k \\
&[R] = \lim_{n\to\infty}\left(I_3 + \frac{1}{n}\tilde{\Psi}\right)^n \to [R] = \lim_{n\to\infty}\sum_{k=0}^{n}\frac{\tilde{\Psi}^k}{k!}\frac{n!}{(n-k)!\,n^k} \\
&[R] = \lim_{n\to\infty}\sum_{k=0}^{n}\frac{\tilde{\Psi}^k}{k!}\left(1-\frac{k-1}{n}\right)\left(1-\frac{k-2}{n}\right)\left(1-\frac{k-3}{n}\right)\ldots \to [R] = \sum_{k=0}^{n}\frac{\tilde{\Psi}^k}{k!}
\end{aligned} \tag{1.17}$$

which results in the same statement as given by Eq. (1.14). If we define a rotation vector as,

$$\{\theta\}^T = \{\theta_x, \theta_y, \theta_z\} \tag{1.18}$$

then the following relation can be obtained.

$$\tilde{\Psi}^2 = \tilde{\Psi}\tilde{\Psi} = \begin{bmatrix} (\theta_x^2 - \vartheta^2) & \theta_x\theta_y & \theta_x\theta_z \\ \theta_x\theta_y & \left(\theta_y^2 - \vartheta^2\right) & \theta_y\theta_z \\ \theta_x\theta_z & \theta_y\theta_z & (\theta_z^2 - \vartheta^2) \end{bmatrix} = \left(\{\theta\}\{\theta\}^T - \vartheta^2 I_3\right) \tag{1.19}$$

Further, the skew-symmetric matrix $\tilde{\Psi}$ possesses the properties:

$$\begin{aligned}
&\tilde{\Psi}^3 = -\vartheta^2\tilde{\Psi} \\
&\tilde{\Psi}^4 = -\vartheta^2\tilde{\Psi}^2,\quad \tilde{\Psi}^5 = \vartheta^4\tilde{\Psi},\quad \tilde{\Psi}^6 = \vartheta^4\tilde{\Psi}^2,\quad \tilde{\Psi}^7 = -\vartheta^6\tilde{\Psi},\quad \tilde{\Psi}^8 = -\vartheta^6\tilde{\Psi}^2
\end{aligned} \tag{1.20}$$

Having introduced these properties of $\tilde{\Psi}$ into Eq. (1.14) the rotation matrix $[R]$ becomes as written by,

$$[R] = I_3 + \tilde{\Psi} + \frac{1}{2!}\tilde{\Psi}^2 - \frac{\vartheta^2}{3!}\tilde{\Psi} - \frac{\vartheta^2}{4!}\tilde{\Psi}^2 + \frac{\vartheta^4}{5!}\tilde{\Psi} + \frac{\vartheta^4}{6!}\tilde{\Psi}^2 - \ldots \tag{1.21a}$$

and having reordered Eq. (1.21a) the rotation matrix $[R]$ will be:

$$[R] = I_3 + \left(1 - \frac{\vartheta^2}{3!} + \frac{\vartheta^4}{5!} - \frac{\vartheta^6}{7!} + \cdots\right)\tilde{\Psi} + \left(\frac{1}{2} - \frac{\vartheta^2}{4!} + \frac{\vartheta^4}{6!} - \frac{\vartheta^6}{8!} + \cdots\right)\tilde{\Psi}^2 \tag{1.21b}$$

Or

$$\begin{aligned}[R] = I_3 &+ \frac{1}{\vartheta}\left(\vartheta - \frac{\vartheta^3}{3!} + \frac{\vartheta^5}{5!} - \frac{\vartheta^7}{7!} + \cdots\right)\tilde{\Psi} \\ &+ \frac{1}{\vartheta^2}\left[1 - \left(1 - \frac{\vartheta^2}{2} + \frac{\vartheta^4}{4!} - \frac{\vartheta^6}{6!} + \frac{\vartheta^8}{8!} - \cdots\right)\right]\tilde{\Psi}^2\end{aligned} \tag{1.21c}$$

Since the series expansion in the brackets (.) are sin and cos functions, i.e.,:

$$\sin\vartheta = \vartheta - \frac{\vartheta^3}{3!} + \frac{\vartheta^5}{5!} - \frac{\vartheta^7}{7!} + \cdots \quad \text{and} \quad \cos\vartheta = 1 - \frac{\vartheta^2}{2} + \frac{\vartheta^4}{4!} - \frac{\vartheta^6}{6!} + \frac{\vartheta^8}{8!} + \cdots \tag{1.22a}$$

the final statement of the rotation matrix can be written as,

$$[R] = I_3 + \frac{\sin\vartheta}{\vartheta}\tilde{\Psi} + \frac{(1-\cos\vartheta)}{\vartheta^2}\tilde{\Psi}^2 \tag{1.22b}$$

Since we assume the linear beam theory, the rotations will be small. Thus, the second and higher order terms in the rotation matrix $[R]$ are neglected, which results in the small rotation matrix as:

$$[R] \simeq I_3 + \tilde{\Psi} \tag{1.23}$$

Having introduced Eq. (1.23) into Eq. (1.13b), the displacement vector of the location A on a cross-section of the beam shown in Fig. 1.4 can be written as:

$$\{v\} = \{u\} + \tilde{\Psi}\,\{r\} \quad \text{where} \quad \{r\}^T = \{0,\ y,\ z\} \tag{1.24}$$

The deformation vector $\{v\}$ of the location A can now be stated explicitly as,

$$\begin{aligned}\begin{Bmatrix} v_x \\ v_y \\ v_z \end{Bmatrix} &= \begin{Bmatrix} u_x \\ u_y \\ u_z \end{Bmatrix} + \begin{bmatrix} 0 & -\theta_z & \theta_y \\ \theta_z & 0 & -\theta_x \\ -\theta_y & \theta_x & 0 \end{bmatrix} \begin{Bmatrix} 0 \\ y \\ z \end{Bmatrix} \\ \begin{Bmatrix} v_x \\ v_y \\ v_z \end{Bmatrix} &= \begin{Bmatrix} u_x - \theta_z y + \theta_y z \\ u_y - \theta_x z \\ u_z + \theta_x y \end{Bmatrix}\end{aligned} \tag{1.25a}$$

In Eq. (1.25a), the axial displacement v_x does not include warping effect of the torsion. To complete the displacement field in the axial direction the warping contribution is also included by using the Saint–Venant torsion theory [44, 54]. This contribution is defined as $\kappa_x f_x(y,z)$ where κ_x is a twist along the beam axis

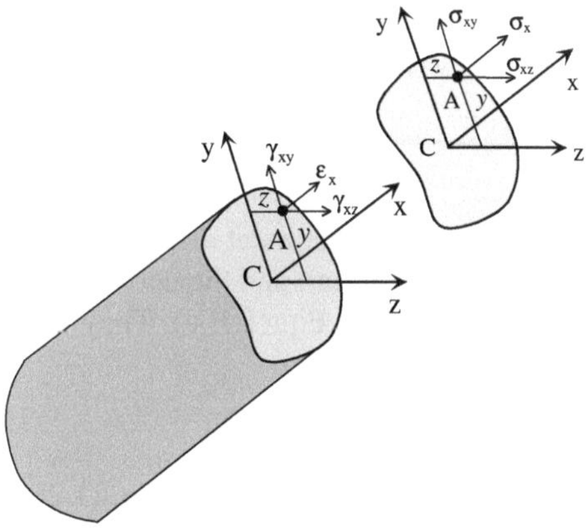

Fig. 1.5 Strains and stresses at a point of a cross-section of a beam element

(a unit axial rotation), which is assumed to be constant, and $f_x(y, z)$ is an unknown warping function of y and z coordinates, which is calculated from the solution of a stress boundary value problem. Since κ_x is assumed constant, a linear rotation angle θ_x is obtained, i.e., $(\theta_x = \kappa_x x)$. With the warping contribution, the axial displacement at point on a cross-section is stated:

$$v_x = \left(u_x - \theta_z y + \theta_y z\right) + \kappa_x f_x(y,z) \tag{1.25b}$$

This displacements v_x from Eq. (1.25b), and v_y and v_z from Eq. (1.25a) will be used to calculate strains at a location on a cross-section of the beam presented in the next section.

1.2.6 Strains and Stresses at a Location on a Cross-Section of a 3D Beam

From the definition of the linear strains in the theory of elasticity [24], using Eqs. (1.25a, b), the strain components at the location A of a beam shown in Fig. 1.5 can be expressed:

$$\left.\begin{aligned} \varepsilon_{xx} &= \frac{\partial v_x}{\partial x} \\ \gamma_{xy} &= \frac{\partial v_y}{\partial x} + \frac{\partial v_x}{\partial y} \\ \gamma_{xz} &= \frac{\partial v_z}{\partial x} + \frac{\partial v_x}{\partial z} \end{aligned}\right\} \rightarrow \left\{\begin{aligned} \varepsilon_{xx} &= \frac{\partial u_x}{\partial x} - \frac{\partial \theta_z}{\partial x} y + \frac{\partial \theta_y}{\partial x} z \\ \gamma_{xy} &= \frac{\partial u_y}{\partial x} - \frac{\partial \theta_x}{\partial x} z - \theta_z + \kappa_x \frac{\partial f_x(y,z)}{\partial y} \\ \gamma_{xz} &= \frac{\partial u_z}{\partial x} + \frac{\partial \theta_x}{\partial x} y + \theta_y + \kappa_x \frac{\partial f_x(y,z)}{\partial z} \end{aligned}\right. \tag{1.26}$$

In these equations, κ_x is assumed as a constant twist along the beam axis (a unit axial rotation) which is obtained from the derivative of the axial rotation θ_x, and the derivatives of the rotations θ_z and θ_y are the curvatures in the (*x*–*y*) and (*x*–*z*) planes, which are given in Eqs. (1.3) and (1.6), respectively. They are written as:

$$\frac{\partial\theta_x}{\partial x} = \kappa_x, \ \frac{\partial\theta_z}{\partial x} = \kappa_z \quad \text{and} \quad \frac{\partial\theta_y}{\partial x} = \kappa_y \tag{1.27a}$$

The derivative of the axial displacement of the elastic curve u_x is simply the axial strain (normal strain), and the derivatives of transverse displacements of the elastic curve are given in Eq. (1.10a). They are stated:

$$\frac{\partial u_x}{\partial x} = \varepsilon_x, \ \frac{\partial u_y}{\partial x} = \left(\gamma_y + \theta_z\right) \quad \text{and} \quad \frac{\partial u_z}{\partial x} = \left(\gamma_z - \theta_y\right) \tag{1.27b}$$

Having used Eqs. (1.27a, b) in Eq. (1.26), the strain and stress components of the beam at the location A can be obtained as written:

$$\begin{aligned} \varepsilon_{xx} &= \varepsilon_x - \kappa_z y + \kappa_y z & \quad & \sigma_{xx} = E\left(\varepsilon_x - \kappa_z y + \kappa_y z\right) \\ \gamma_{xy} &= \gamma_y + \kappa_x\left(\frac{\partial f_x(y,z)}{\partial y} - z\right) & \quad \text{and} \quad & \tau_{xy} = G\left(\gamma_y + \kappa_x\left(\frac{\partial f_x(y,z)}{\partial y} - z\right)\right) \\ \gamma_{xz} &= \gamma_z + \kappa_x\left(\frac{\partial f_x(y,z)}{\partial z} + y\right) & \quad & \tau_{xz} = G\left(\gamma_z + \kappa_x\left(\frac{\partial f_x(y,z)}{\partial z} + y\right)\right) \end{aligned} \tag{1.28}$$

in which E is the Young's modulus (elasticity modulus) and G is the shear modulus of the beam material. The warping function $f_x(y, z)$ is determined from the solution of the boundary value problem [44, 54]:

$$\frac{\partial\tau_{xy}}{\partial y} + \frac{\partial\tau_{xz}}{\partial z} = 0 \ \rightarrow \ \left(\frac{\partial^2 f_x(y,z)}{\partial y^2} + \frac{\partial^2 f_x(y,z)}{\partial z^2}\right) = 0 \tag{1.29}$$

with zero shear stresses at the boundary of the cross-section. It is seen from Eq. (1.28) that second terms in the statements of the shear strains/stresses are due to pure torsion (twisting) and they do not produce any resultant shear forces in the *y* and *z* directions, except a resultant torsional moment. This feature will be taken into account when calculating the shear forces presented in the next section.

1.2.7 Calculation of Forces and Moments of 3D Beams

The forces in a 3D beam are calculated from the following integrations:

$$
\begin{aligned}
N &= \int \sigma_{xx}\,\mathrm{d}A = E\varepsilon_x \int \mathrm{d}A - E\kappa_z \int y\,\mathrm{d}A + E\kappa_y \int z\,\mathrm{d}A \\
Q_y &= \int \tau_{xy}\,\mathrm{d}A = G\left[\int \gamma_y\,\mathrm{d}A + \kappa_x \int \left(\frac{\partial f_x(y,z)}{\partial y} - z\right)\mathrm{d}A\right] \\
Q_z &= \int \tau_{xz}\,\mathrm{d}A = G\left[\int \gamma_z\,\mathrm{d}A + \kappa_x \int \left(\frac{\partial f_x(y,z)}{\partial z} + y\right)\mathrm{d}A\right]
\end{aligned}
\tag{1.30a}
$$

in which N is the axial force, Q_y and Q_z are the transverse shear forces in the (x–y) and (x–z) planes, respectively. The moments are calculated from:

$$
\left.
\begin{aligned}
M_y &= \int z\,\sigma_{xx}\,\mathrm{d}A \\
M_z &= -\int y\,\sigma_{xx}\,\mathrm{d}A
\end{aligned}
\right\} \rightarrow
\left\{
\begin{aligned}
M_y &= E\varepsilon_x \int z\,\mathrm{d}A - E\kappa_z \int z\,y\,\mathrm{d}A + E\kappa_y \int z^2\,\mathrm{d}A \\
M_z &= -E\varepsilon_x \int y\,\mathrm{d}A + E\kappa_z \int y^2\,\mathrm{d}A - E\kappa_y \int y\,z\,\mathrm{d}A
\end{aligned}
\right.
\tag{1.30b}
$$

$$
\begin{aligned}
M_x &= \int \left(y\,\tau_{xz} - z\,\tau_{xy}\right)\mathrm{d}A \qquad \text{or} \\
M_x &= G\int \left[y\gamma_z - z\gamma_y + \kappa_x\left(y^2 + z^2 - z\frac{\partial f_x(y,z)}{\partial y} + y\frac{\partial f_x(y,z)}{\partial z}\right)\right]\mathrm{d}A
\end{aligned}
\tag{1.30c}
$$

in which M_x is the torsional moment, M_z and M_y are the bending moments in the (x–y) and (x–z) planes, respectively. It is assumed that local coordinates (x, y, z) are principal axes and the origin is at the gravity center of the cross-section from which the following sectional properties can be obtained:

$$
\begin{aligned}
&\int \mathrm{d}A = A, \quad A_y\gamma_y = \int \gamma_y\mathrm{d}A, \quad A_z\gamma_z = \int \gamma_z\mathrm{d}A \\
&\int y\,\mathrm{d}A = 0, \quad \int z^2\,\mathrm{d}A = I_y \quad \text{and} \quad \int y^2\,\mathrm{d}A = I_z \\
&\int z\,\mathrm{d}A = 0 \qquad\qquad\qquad \text{and} \quad \int y\,z\,\mathrm{d}A = 0
\end{aligned}
\tag{1.31a}
$$

In these statements, A_y and A_z are the effective areas of shear forces shown symbolically. Their calculation follows energy principles or solving stress boundary value problems [55, 56]. Here, energy principles are used to define these cross-sectional values. The shear stresses produced by the shear force Q_y on the cross-section in the y and z directions are denoted by $(\tau_y)_y$ and $(\tau_z)_y$, and those produced by $\mathrm{Q_z}$ are denoted by $(\tau_y)_z$ and $(\tau_z)_z$ as defined:

$$
\begin{aligned}
(\tau_y)_y &= \frac{Q_y}{A}\frac{\partial\psi_y}{\partial y} & & & (\tau_y)_z &= \frac{Q_z}{A}\frac{\partial\psi_z}{\partial y} \\
(\tau_z)_y &= \frac{Q_y}{A}\frac{\partial\psi_y}{\partial z} & &\text{and} & (\tau_z)_z &= \frac{Q_z}{A}\frac{\partial\psi_z}{\partial z}
\end{aligned}
\tag{1.31b}
$$

in which ψ_y and ψ_z are potentials of shear stress distributions due to Q_y and Q_z, respectively. Using energy equivalences the shear areas A_y and A_z are obtained from:

$$\frac{A^2}{A_y} = \int \left[\left(\frac{\partial \psi_y}{\partial y} \right)^2 + \left(\frac{\partial \psi_y}{\partial z} \right)^2 \right] dA \quad \text{and} \quad \frac{A^2}{A_z} = \int \left[\left(\frac{\partial \psi_z}{\partial y} \right)^2 + \left(\frac{\partial \psi_z}{\partial z} \right)^2 \right] dA \tag{1.31c}$$

As mentioned above the shear stresses produced by a pure torsional (twisting) moment do not produce resultant shear forces. Thus,

$$\int \left(\frac{\partial f_x(y,z)}{\partial y} - z \right) dA = 0 \quad \text{and} \quad \int \left(\frac{\partial f_x(y,z)}{\partial z} + y \right) dA = 0 \tag{1.31d}$$

Using the cross-sectional properties in Eq. (1.31a) and the conditions in Eq. (1.31d) the forces can be stated as:

$$\begin{aligned} N &= \mathrm{EA}\varepsilon_x \\ Q_y &= \mathrm{GA}_y\gamma_y \\ Q_z &= \mathrm{GA}_z\gamma_z \end{aligned} \tag{1.32a}$$

and the moments are:

$$\begin{aligned} M_x &= \mathrm{GJ}\,\kappa_x \;\rightarrow\; \mathrm{J} = I_p - \int \left(z\frac{\partial f_x(y,z)}{\partial y} - y\frac{\partial f_x(y,z)}{\partial z} \right) dA \\ M_y &= \mathrm{EI}_y\kappa_y \\ M_z &= \mathrm{EI}_z\kappa_z \end{aligned} \tag{1.32b}$$

where GJ is the torsional rigidity of the beam and $I_p = (I_y + I_z)$ is the polar inertia moment and J is the torsional constant of the cross-section. Equations (1.32a, b) will be used in the formulation of the elastic curve and calculation of the strain energy of the beam.

1.2.8 Differential Equations of the 3D Beam Element

For a zero loading case, the derivatives of the normal force (N) and torsional moment (M_x) are equal to zero, i.e.,:

$$\frac{dN}{dx} = 0 \quad \text{and} \quad \frac{dM_x}{dx} = 0 \tag{1.33a}$$

Using N and M_x from Eqs. (1.32a, b) in Eq. (1.22a) it can be written as,

$$\mathrm{EA}\frac{\mathrm{d}\varepsilon_x}{\mathrm{d}x}=0 \quad \text{and} \quad \mathrm{GJ}\frac{\mathrm{d}\kappa_x}{\mathrm{d}x}=0 \tag{1.33b}$$

Having introduced ε_x and κ_x from Eqs. (1.27b, a) into Eq. (1.33b) the differential equations of the axial displacement and torsional rotation can be obtained as written:

$$\frac{\mathrm{d}^2u_x}{\mathrm{d}x^2}=0 \quad \text{and} \quad \frac{\mathrm{d}^2\theta_x}{\mathrm{d}x^2}=0 \tag{1.33c}$$

In order to find the differential equations of the elastic curve in the (x–y) and (x–z) planes (transverse displacements and rotations) of a Timoshenko beam, Eqs. (1.11b), (1.10b) and (1.10a) will be used. It is stated as:

$$\begin{aligned}
&\frac{\mathrm{d}M_z}{\mathrm{d}x}=-Q_y \rightarrow \frac{\mathrm{d}M_z}{\mathrm{d}x}=-\mathrm{GA}_y\gamma_y \rightarrow \frac{\mathrm{d}M_z}{\mathrm{d}x}=-\mathrm{GA}_y\left(\frac{\mathrm{d}u_y}{\mathrm{d}x}-\theta_z\right)\\
&\frac{\mathrm{d}M_y}{\mathrm{d}x}=Q_z \rightarrow \frac{\mathrm{d}M_y}{\mathrm{d}x}=\mathrm{GA}_z\gamma_z \rightarrow \frac{\mathrm{d}M_y}{\mathrm{d}x}=\mathrm{GA}_z\left(\frac{\mathrm{d}u_z}{\mathrm{d}x}+\theta_y\right)
\end{aligned} \tag{1.34}$$

The differential equations can be stated in two alternative forms as:

- In the first alternative, the curvatures κ_z and κ_y defined in terms of second derivatives of displacements in Eqs. (1.3) and (1.6) are used in Eq. (1.32b), and then using Eq. (1.34) the differential equations of the elastic curve can be obtained in terms of displacements as stated by,

$$\begin{aligned}
&\frac{\mathrm{d}M_y}{\mathrm{d}x}=-\mathrm{EI}_y\frac{\mathrm{d}^3u_z}{\mathrm{d}x^3}=\mathrm{GA}_z\left(\frac{\mathrm{d}u_z}{\mathrm{d}x}+\theta_y\right) \rightarrow \frac{\mathrm{EI}_y}{\mathrm{GA}_z}\frac{\mathrm{d}^3u_z}{\mathrm{d}x^3}+\frac{\mathrm{d}u_z}{\mathrm{d}x}+\theta_y=0\\
&\frac{\mathrm{d}M_z}{\mathrm{d}x}=\mathrm{EI}_z\frac{\mathrm{d}^3u_y}{\mathrm{d}x^3}=-\mathrm{GA}_y\left(\frac{\mathrm{d}u_y}{\mathrm{d}x}-\theta_z\right) \rightarrow \frac{\mathrm{EI}_z}{\mathrm{GA}_y}\frac{\mathrm{d}^3u_y}{\mathrm{d}x^3}+\frac{\mathrm{d}u_y}{\mathrm{d}x}-\theta_z=0
\end{aligned} \tag{1.35}$$

- In the second alternative, the moment–curvature relations given in Eq. (1.32b) are used in Eq. (1.34), and the differential equations of the elastic curve can be obtained in terms of rotations as written by,

$$\begin{aligned}
&\frac{\mathrm{d}M_y}{\mathrm{d}x}=\mathrm{EI}_y\frac{\mathrm{d}^2\theta_y}{\mathrm{d}x^2}=\mathrm{GA}_z\left(\frac{\mathrm{d}u_z}{\mathrm{d}x}+\theta_y\right) \rightarrow \frac{\mathrm{EI}_y}{\mathrm{GA}_z}\frac{\mathrm{d}^2\theta_y}{\mathrm{d}x^2}-\frac{\mathrm{d}u_z}{\mathrm{d}x}-\theta_y=0\\
&\frac{\mathrm{d}M_z}{\mathrm{d}x}=\mathrm{EI}_z\frac{\mathrm{d}^2\theta_z}{\mathrm{d}x^2}=-\mathrm{GA}_y\left(\frac{\mathrm{d}u_y}{\mathrm{d}x}-\theta_z\right) \rightarrow \frac{\mathrm{EI}_z}{\mathrm{GA}_y}\frac{\mathrm{d}^2\theta_z}{\mathrm{d}x^2}+\frac{\mathrm{d}u_y}{\mathrm{d}x}-\theta_z=0
\end{aligned} \tag{1.36}$$

One of these differential equations can be used to find the elastic curve of the spatial beam element. The solutions of the differential equations are presented in the next section for both alternatives.

1.2.9 Solution of Differential Equations of the Elastic Curve and Shape Functions of the 3D Beam Element

The solutions of the differential equations of the axial displacement and torsional rotation are obtained directly from Eq. (1.33c). They are obtained to be linear functions of the axial coordinate (x) as stated by,

$$\begin{aligned} u_x &= A_0 + A_1 x \\ \theta_x &= A_2 + A_3 x \end{aligned} \tag{1.37}$$

From the solutions of differential equations given in Eqs. (1.35) and (1.36), the transverse displacements and rotations of the elastic curve can be obtained respectively to be cubic and quadratic functions of the axial coordinate (x) as explained below for two cases:

- In the first alternative, the differential equations given in Eq. (1.35) are used. The displacements are chosen to be cubic functions of (x) as written by:

$$\begin{aligned} u_y &= C_0 + C_1 x + C_2 x^2 + C_3 x^3 \\ u_z &= C_4 + C_5 x + C_6 x^2 + C_7 x^3 \end{aligned} \tag{1.38}$$

  Using Eqs. (1.35) and (1.38) the rotations θ_y and θ_z are obtained as stated by,

$$\begin{aligned} \theta_y &= -C_5 - 2C_6 x - \left(\frac{6\mathrm{EI}_y}{\mathrm{GA}_z} + 3x^2\right) C_7 \\ \theta_z &= C_1 + 2C_2 x + \left(\frac{6\mathrm{EI}_z}{\mathrm{GA}_y} + 3x^2\right) C_3 \end{aligned} \tag{1.39}$$

- In the second alternative, the differential equations given in Eq. (1.36) are used. The rotations θ_y and θ_z are chosen to be quadratic functions of (x) as stated by,

$$\begin{aligned} \theta_y &= B_5 + B_6 x + B_7 x^2 \\ \theta_z &= B_1 + B_2 x + B_3 x^2 \end{aligned} \tag{1.40}$$

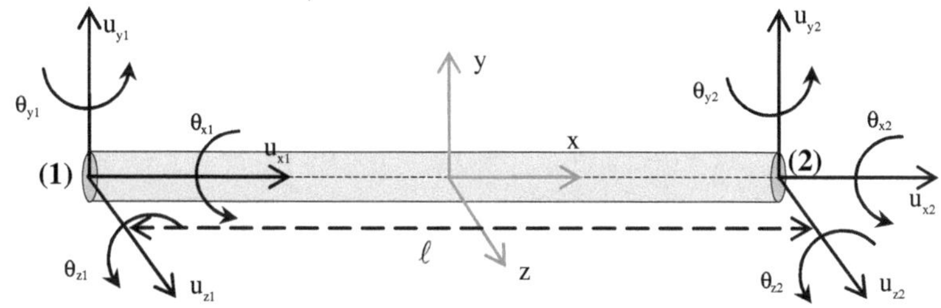

Fig. 1.6 A spatial beam element with nodal displacements and rotations

Using Eqs. (1.36) and (1.40) the displacements u_y and u_z are obtained as stated by,

$$
\begin{aligned}
u_y &= B_0 + B_1 x + \frac{1}{2} B_2 x^2 + \left(\frac{1}{3} x^3 - \frac{2\mathrm{EI}_z}{\mathrm{GA}_y} x \right) B_3 \\
u_z &= B_4 - B_5 x - \frac{1}{2} B_6 x^2 - \left(\frac{1}{3} x^3 - \frac{2\mathrm{EI}_y}{\mathrm{GA}_z} x \right) B_7
\end{aligned}
\tag{1.41}
$$

The constants, A_i ($i = 0$–3) in Eq. (1.37), C_i ($i = 0$–7) in Eqs. (1.38) and (1.39), and B_i ($i = 0$–7) in Eqs. (1.40) and (1.41) are calculated by using the kinematic boundary conditions at the member ends (1) and (2), i.e., for $(x = 0)$, and $(x = \ell)$. These conditions are imposed as:

$$
\text{for } (x = 0) \rightarrow \begin{aligned} u_x &= u_{x1} \\ u_y &= u_{y1} \\ u_z &= u_{z1} \\ \theta_x &= \theta_{x1} \\ \theta_y &= \theta_{y1} \\ \theta_z &= \theta_{z1} \end{aligned} \quad \text{and} \quad \text{for } (x = \ell) \rightarrow \begin{aligned} u_x &= u_{x2} \\ u_y &= u_{y2} \\ u_z &= u_{z2} \\ \theta_x &= \theta_{x2} \\ \theta_y &= \theta_{y2} \\ \theta_z &= \theta_{z2} \end{aligned}
\tag{1.42}
$$

where ℓ denotes the length of the beam element and the definitions of displacements and rotations at the member ends (1) and (2) are shown in Fig. 1.6. By using these boundary conditions the constants are obtained as given below.

For the constants, A_i ($i = 0$–3):

$$
\begin{aligned} A_0 &= u_{x1} \\ A_1 &= (u_{x2} - u_{x1})/\ell \end{aligned} \quad \text{and} \quad \begin{aligned} A_2 &= \theta_{x1} \\ A_3 &= (\theta_{x2} - \theta_{x1})/\ell \end{aligned}
\tag{1.43}
$$

For the constants, C_i $(i = 0\text{–}7)$:

$$\begin{aligned}
C_0 &= u_{y1} \\
C_1 &= \theta_{z1} + \frac{1}{2}\Phi_y\mu_y\left[\frac{2}{\ell}\left(u_{y2} - u_{y1}\right) - \theta_{z1} - \theta_{z2}\right] \\
C_2 &= \frac{1}{2\ell}\left[\theta_{z2} - \theta_{z1} + 3\mu_y\left[\frac{2}{\ell}\left(u_{y2} - u_{y1}\right) - \theta_{z1} - \theta_{z2}\right]\right] \\
C_3 &= -\frac{1}{\ell^2}\mu_y\left[\frac{2}{\ell}\left(u_{y2} - u_{y1}\right) - \theta_{z1} - \theta_{z2}\right]
\end{aligned} \tag{1.44a}$$

$$\begin{aligned}
C_4 &= u_{z1} \\
C_5 &= -\theta_{y1} + \frac{1}{2}\Phi_z\mu_z\left[\frac{2}{\ell}\left(u_{z2} - u_{z1}\right) + \theta_{y2} + \theta_{y1}\right] \\
C_6 &= -\frac{1}{2\ell}\left(\theta_{y2} - \theta_{y1}\right) + \frac{3}{2\ell}\mu_z\left[\frac{2}{\ell}\left(u_{z2} - u_{z1}\right) + \theta_{y2} + \theta_{y1}\right] \\
C_7 &= -\frac{1}{\ell^2}\mu_z\left[\frac{2}{\ell}\left(u_{z2} - u_{z1}\right) + \theta_{y2} + \theta_{y1}\right]
\end{aligned} \tag{1.44b}$$

For the constants, B_i $(i = 0\text{–}7)$:

$$\begin{aligned}
B_0 &= u_{y1}, \quad B_1 = \theta_{z1} \\
B_2 &= \frac{3}{\ell}\mu_y\left[\frac{2}{\ell}\left(u_{y2} - u_{y1}\right) - \theta_{z2} - \theta_{z1}\right] + \frac{1}{\ell}\left(\theta_{z2} - \theta_{z1}\right) \\
B_3 &= \frac{3}{\ell^2}\mu_y\left[-\frac{2}{\ell}\left(u_{y2} - u_{y1}\right) + \theta_{z1} + \theta_{z2}\right]
\end{aligned} \tag{1.45a}$$

$$\begin{aligned}
B_4 &= u_{z1}, \quad B_5 = \theta_{y1} \\
B_6 &= \frac{1}{\ell}\left(\theta_{y2} - \theta_{y1}\right) - \frac{3}{\ell}\mu_z\left[\frac{2}{\ell}\left(u_{z2} - u_{z1}\right) + \theta_{y1} + \theta_{y2}\right] \\
B_7 &= \frac{3}{\ell^2}\mu_z\left[\frac{2}{\ell}\left(u_{z2} - u_{z1}\right) + \theta_{y1} + \theta_{y2}\right]
\end{aligned} \tag{1.45b}$$

In these constants, the parameters (Φ_y, μ_y) and (Φ_z, μ_z) are the transverse shear force parameters, which are defined as,

$$\Phi_y = 12\mathrm{EI}_z/\ell^2\mathrm{GA}_y, \quad \mu_y = 1/\left(1 + \Phi_y\right) \tag{1.46a}$$

$$\Phi_z = 12\mathrm{EI}_y/\ell^2\mathrm{GA}_z, \quad \mu_z = 1/\left(1 + \Phi_z\right) \tag{1.46b}$$

If the effect of transverse shear forces on the elastic curve is not considered, then the parameters Φ_y and Φ_z will be zero. This special case produces formulations of the Euler–Bernoulli beam theory. Since the Timoshenko beam theory is more general than the Euler–Bernoulli beam theory, it is preferably used in the analysis of frame structures as, in the limit case, it is equivalent to the Euler–Bernoulli beam theory. It can be verified from Eq. (1.46) as $(\ell \to \infty)$ the parameters Φ_y and

Φ_z approach zero. The shape functions of a 3D beam element can be obtained by introducing the constants of displacements and rotations into their corresponding functions, i.e., A_i ($i = 0$–3) into Eq. (1.43), C_i ($i = 0$–7) into Eqs. (1.44a, b), or B_i ($i = 0$–7) into Eqs. (1.45a, b). The solutions of the differential equations, which are given by Eqs. (1.38) and (1.39), and Eqs. (1.41) and (1.40), produce the same shape functions for transverse displacements and rotations. Having used the dimensionless variable:

$$\xi = x/\ell \tag{1.47}$$

the functions of displacements and rotations can be stated in terms of their nodal values as written by,

$$\{u\} = [N_u]\{d\} \quad \text{and} \quad \{\theta\} = [N_\theta]\{d\} \tag{1.48a}$$

where the definitions of the vectors $\{u\}$, $\{\theta\}$ are:

$$\{u\}^T = \{u_x, u_y, u_z\} \quad \text{and} \quad \{\theta\}^T = \{\theta_x, \theta_y, \theta_z\} \tag{1.48b}$$

and $\{d\}$ is the vector of nodal values of the displacements and rotations defined as,

$$\{d\}^T = \{u_{x1},\ u_{y1},\ u_{z1},\ \theta_{x1},\ \theta_{y1},\ \theta_{z1},\ u_{x2},\ u_{y2},\ u_{z2},\ \theta_{x2},\ \theta_{y2},\ \theta_{z2}\} \tag{1.49}$$

where $\{.\}^T$ denotes the transpose of a vector. The matrices, $[N_u]$ and $[N_\theta]$, are the shape functions matrices for displacements and rotations, respectively, which are obtained as written below in Eqs. (1.50a, b).

$$[N_u]^T = \begin{bmatrix} (1-\xi) & 0 & 0 \\ 0 & [1-\mu_y(\Phi_y\xi+3\xi^2-2\xi^3)] & 0 \\ 0 & 0 & [1-\mu_z(\Phi_z\xi+3\xi^2-2\xi^3)] \\ 0 & 0 & 0 \\ 0 & 0 & -\frac{\mu_z\ell}{2}[(2+\Phi_z)\xi-(4+\Phi_z)\xi^2+2\xi^3] \\ 0 & \frac{\mu_y\ell}{2}[(2+\Phi_y)\xi-(4+\Phi_y)\xi^2+2\xi^3] & 0 \\ \xi & 0 & 0 \\ 0 & \mu_y(\Phi_y\xi+3\xi^2-2\xi^3) & 0 \\ 0 & 0 & \mu_z(\Phi_z\xi+3\xi^2-2\xi^3) \\ 0 & 0 & 0 \\ 0 & 0 & -\frac{\mu_z\ell}{2}[-\Phi_z\xi+(\Phi_z-2)\xi^2+2\xi^3] \\ 0 & \frac{\mu_y\ell}{2}[-\Phi_y\xi+(\Phi_y-2)\xi^2+2\xi^3] & 0 \end{bmatrix} \tag{1.50a}$$

and

$$[N_\theta]^T = \begin{bmatrix} 0 & 0 & 0 \\ 0 & 0 & 6\mu_y(-\xi+\xi^2)/\ell \\ 0 & 6\mu_z(\xi-\xi^2)/\ell & 0 \\ (1-\xi) & 0 & 0 \\ 0 & [1-\mu_z((4+\Phi_z)\xi-3\xi^2)] & 0 \\ 0 & 0 & [1-\mu_y((4+\Phi_y)\xi-3\xi^2)] \\ 0 & 0 & 0 \\ 0 & 0 & 6\mu_y(\xi-\xi^2)/\ell \\ 0 & 6\mu_z(-\xi+\xi^2)/\ell & 0 \\ \xi & 0 & 0 \\ 0 & \mu_z[(\Phi_z-2)\xi+3\xi^2] & 0 \\ 0 & 0 & \mu_y[(\Phi_y-2)\xi+3\xi^2] \end{bmatrix} \tag{1.50b}$$

where $[.]^T$ denotes a matrix transposition. By using Eqs. (1.27a–1.28) the strains at the center of principal coordinates and the curvatures of the beam can be obtained in terms of nodal displacements and rotations as stated in vector notation by,

$$\{\delta\} = [B]\{d\} \quad \text{where} \quad \{\delta\}^T = \{\varepsilon_x,\ \gamma_y,\ \gamma_z,\ \kappa_x,\ \kappa_y,\ \kappa_z\} \tag{1.51a}$$

in which $\{\delta\}$ is the deformation vector and $[B]$ is a matrix of functions that obtained as written by,

$$[B]^T = \begin{bmatrix} -\frac{1}{\ell} & 0 & 0 & 0 & 0 & 0 \\ 0 & -\frac{1}{\ell}\Phi_y\mu_y & 0 & 0 & 0 & -\frac{6}{\ell^2}\mu_y(1-2\xi) \\ 0 & 0 & -\frac{1}{\ell}\Phi_z\mu_z & 0 & \frac{6}{\ell^2}\mu_z(1-2\xi) & 0 \\ 0 & 0 & 0 & -\frac{1}{\ell} & 0 & 0 \\ 0 & 0 & \frac{1}{2}\Phi_z\mu_z & 0 & \frac{1}{\ell}\mu_z[-(4+\Phi_z)+6\xi] & 0 \\ 0 & -\frac{1}{2}\Phi_y\mu_y & 0 & 0 & 0 & \frac{1}{\ell}\mu_y[-(4+\Phi_y)+6\xi] \\ \frac{1}{\ell} & 0 & 0 & 0 & 0 & 0 \\ 0 & \frac{1}{\ell}\Phi_y\mu_y & 0 & 0 & 0 & \frac{6}{\ell^2}\mu_y(1-2\xi) \\ 0 & 0 & \frac{1}{\ell}\Phi_z\mu_z & 0 & -\frac{6}{\ell^2}\mu_z(1-2\xi) & 0 \\ 0 & 0 & 0 & \frac{1}{\ell} & 0 & 0 \\ 0 & 0 & \frac{1}{2}\Phi_z\mu_z & 0 & \frac{1}{\ell}\mu_z[(\Phi_z-2)+6\xi] & 0 \\ 0 & -\frac{1}{2}\Phi_y\mu_y & 0 & 0 & 0 & \frac{1}{\ell}\mu_y[(\Phi_y-2)+6\xi] \end{bmatrix} \tag{1.51b}$$

The aforementioned statements will be used to calculate the total potential energy and stiffness matrix of a Timoshenko beam element presented in the next section.

1.2.10 Total Potential Energy, Stiffness Matrix, and Static Equilibrium Equation

The total potential energy of a beam is stated as,

$$\Pi = U - W_p \tag{1.52}$$

where U is the total strain energy and W_p is the total work done by all external loads. Since the total potential energy is stationary its variation will be zero, i.e.,:

$$\delta\Pi = \delta(U - W_p) = 0 \tag{1.53}$$

The total strain energy is calculated from the integration:

$$U = \frac{1}{2}\int_0^{\ell}\int \{\varepsilon\}^T\{\sigma\}\,\mathrm{d}A\,\mathrm{d}x \tag{1.54}$$

where $\{\varepsilon\}$ and $\{\sigma\}$ are the vectors of strains and stresses at a point on the cross-section of a beam element, which are given by Eqs. (1.28) and (1.29), respectively. These vectors are defined as:

$$\{\varepsilon\} = \begin{Bmatrix} \varepsilon_{xx} \\ \gamma_{xy} \\ \gamma_{xz} \end{Bmatrix} = \begin{Bmatrix} \varepsilon_x - \kappa_z y + \kappa_y z \\ \gamma_y + \kappa_x\left(\frac{\partial f_x}{\partial y} - z\right) \\ \gamma_z + \kappa_x\left(\frac{\partial f_x}{\partial z} + y\right) \end{Bmatrix}$$
$$\{\sigma\} = \begin{Bmatrix} \sigma_{xx} \\ \tau_{xy} \\ \tau_{xz} \end{Bmatrix} = \begin{Bmatrix} E(\varepsilon_x - \kappa_z y + \kappa_y z) \\ G\left(\gamma_y + \kappa_x\left(\frac{\partial f_x}{\partial y} - z\right)\right) \\ G\left(\gamma_z + \kappa_x\left(\frac{\partial f_x}{\partial z} + y\right)\right) \end{Bmatrix} \tag{1.55}$$

Having introduced Eq. (1.55) into Eq. (1.54), the total strain energy can be written:

$$U = \frac{1}{2}\int_0^{\ell} \left[\begin{array}{l} E\int \left(\varepsilon_x - \kappa_z y + \kappa_y z\right)^2 \mathrm{d}A + G\int \left(\gamma_y + \kappa_x\left(\dfrac{\partial f_x}{\partial y} - z\right)\right)^2 \mathrm{d}A + \cdots \\ \cdots + G\int \left(\gamma_z + \kappa_x\left(\dfrac{\partial f_x}{\partial z} + y\right)\right)^2 \mathrm{d}A \end{array} \right] \mathrm{d}x \tag{1.56}$$

Having carried out integrations in Eq. (1.56), and using section properties in Eq. (1.31a) and the conditions in Eq. (1.31d), the total strain energy can be obtained as:

$$U = \frac{1}{2}\int_0^{\ell} E\left(A\varepsilon_x^2 + I_z\kappa_z^2 + I_y\kappa_y^2\right)\mathrm{d}x + \frac{1}{2}\int_0^{\ell} G\left(A_y\gamma_y^2 + A_z\gamma_z^2 + J\kappa_x^2\right)\mathrm{d}x \tag{1.57}$$

Using Eqs. (1.32a, b) in Eq. (1.57), the total strain energy can be written alternatively in terms of member forces and moments as:

$$\begin{aligned} U &= \frac{1}{2}\int_0^{\ell} \left(N\varepsilon_x + Q_y\gamma_y + Q_z\gamma_z + M_x\kappa_x + M_y\kappa_y + M_z\kappa_z\right)\mathrm{d}x, \quad \text{or} \\ U &= \frac{1}{2}\int_0^{\ell} \left(\frac{N^2}{\mathrm{EA}} + \frac{Q_y^2}{\mathrm{GA}_y} + \frac{Q_z^2}{\mathrm{GA}_z} + \frac{M_x^2}{\mathrm{GJ}} + \frac{M_y^2}{\mathrm{EI}_y} + \frac{M_z^2}{\mathrm{EI}_z}\right)\mathrm{d}x \end{aligned} \tag{1.58}$$

In vector notations, it can be written,

$$\left.\begin{array}{l} U = \dfrac{1}{2}\displaystyle\int_0^{\ell} \{\delta\}^T\{F\}\,\mathrm{d}x \\ \{F\} = [G]\{\delta\} \end{array}\right\} \rightarrow U = \frac{1}{2}\int_0^{\ell} \{\delta\}^T[G]\{\delta\}\mathrm{d}x \tag{1.59}$$

In Eq. (1.59), $\{\delta\}$ is the deformation vector given in Eq. (1.51a), $\{F\}$ is the force vector defied as,

$$\{F\}^T = \{N,\ Q_y,\ Q_z,\ M_x,\ M_y,\ M_z\} \tag{1.60}$$

and $[G]$ is the rigidity matrix defined as,

$$[G] = \begin{bmatrix} \mathrm{EA} & 0 & 0 & 0 & 0 & 0 \\ 0 & \mathrm{GA}_y & 0 & 0 & 0 & 0 \\ 0 & 0 & \mathrm{GA}_z & 0 & 0 & 0 \\ 0 & 0 & 0 & \mathrm{GJ} & 0 & 0 \\ 0 & 0 & 0 & 0 & \mathrm{EI}_y & 0 \\ 0 & 0 & 0 & 0 & 0 & \mathrm{EI}_z \end{bmatrix} \tag{1.61}$$

Having introduced the deformation vector $\{\delta\}$ from Eq. (1.51a) into Eq. (1.59) the total strain energy of the beam can be stated as written by,

$$U = \frac{1}{2}\{d\}^T\left(\int_0^{\ell} [B]^T[G][B]\mathrm{d}x\right)\{d\}, \quad \text{or} \quad U = \frac{1}{2}\{d\}^T[k]\{d\} \tag{1.62}$$

where $[k]$ is the stiffness matrix of the beam element in local coordinates defined as:

$$[k] = \int_0^{\ell} [B]^T[G][B]\mathrm{d}x \tag{1.63}$$

The total work done by external distributed and concentrated loads as well as member-end forces, which are all specified in the direction of displacements and rotations, can be written as,

$$\begin{aligned} W_p = \int_0^{\ell} \{u\}^T\{q\}\,\mathrm{d}x + \int_0^{\ell} \{\theta\}^T\{\mathrm{m}\}\,dx + \sum_{i=1}^{n} \{u(\xi_i)\}^T\{P_i\} + \cdots \\ \cdots + \sum_{j=1}^{m} \{\theta(\xi_j)\}^T\{M_j\} + \{d\}^T\{f\} \end{aligned} \tag{1.64}$$

in which $\{q\}$ and $\{\mathrm{m}\}$ are respectively distributed load and moment vectors, $\{u\}$ and $\{\theta\}$ are displacements and rotation vectors given in Eq. (1.48b), $\{\mathrm{P_i}\}$ and $\{\mathrm{M_j}\}$ are concentrated load and moment vectors at the location j of the element, $\{u(\xi_i)\}$ and $\{\theta(\xi_j)\}$ are the displacement and rotation vectors at the location j, and $\{f\}$ is the vector of internal forces and moments at the element ends (1) and (2). These applied loads and moments are shown in Fig. 1.7. Having introduced the displacements and rotations, $\{u\}$ and $\{\theta\}$, from Eq. (1.48b) into Eq. (1.64), the total work of external loads and member-end forces can be expressed as:

$$W_p = \{d\}^T\left[\begin{aligned} \int_0^{\ell} ([N_u]^T\{q\} + [N_\theta]^T\{\mathrm{m}\})\mathrm{d}x + \sum_{i=1}^{n} [N_u(\xi_i)]^T\{\mathrm{P_i}\} + \cdots \\ \cdots + \sum_{j=1}^{m} [N_\theta(\xi_j)]^T\{\mathrm{M_j}\} + \{f\} \end{aligned}\right] \tag{1.65a}$$

or using an equivalent load vector of the element, the total work can be simplified as written by,

$$W_p = \{d\}^T(\{p\} + \{f\}) \tag{1.65b}$$

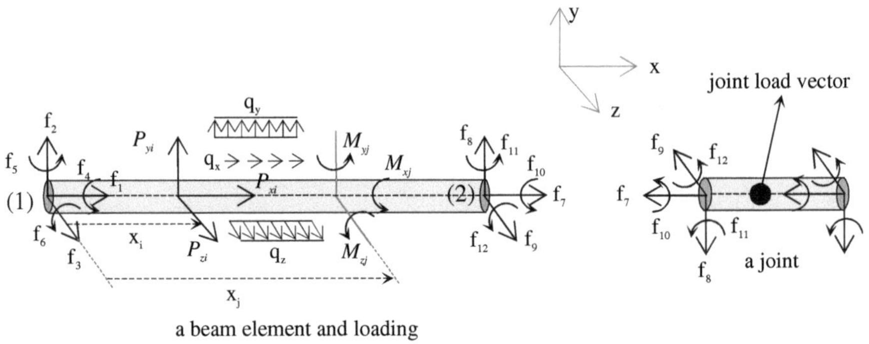

Fig. 1.7 Distributed and concentrated applied loads and member-end forces of a beam element

The equivalent, or in other words the consistent, load vector is simply defined from Eq. (1.65a) as written by,

$$\{p\} = \int_0^{\ell} \left([N_u]^T\{q\} + [N_\theta]^T\{\mathrm{m}\}\right)\mathrm{d}x + \sum_{i=1}^{n} [N_u(\xi_i)]^T\{\mathrm{P_i}\} + \cdots$$
$$\cdots + \sum_{j=1}^{m} \left[N_\theta(\xi_j)\right]^T\{\mathrm{M_j}\} \tag{1.66}$$

With these definitions of the total strain energy and external work, the variation of the total potential energy given by Eq. (1.53) is stated as:

$$\begin{aligned} \delta\Pi &= \delta\left(\frac{1}{2}\{d\}^T[k]\{d\} - \{d\}^T(\{p\} + \{f\})\right) = 0 \\ &\delta\{d\}^T([k]\{d\} - \{p\} - \{f\}) = 0 \end{aligned} \tag{1.67}$$

Since $\delta\{d\}$ is arbitrary, the stiffness equation of the beam element under static loads can be obtained from Eq. (1.67) in local coordinates as:

$$\{f\} = [k]\{d\} - \{p\} \tag{1.68}$$

From this equation it is seen that, when the member-end displacements and rotations are zero, the internal force vector will only be dependent on the applied forces and moments. This case of the internal forces and moments is known as the fixed-end member forces and moments. They are defined from Eq. (1.68) in the vector notation as:

$$\{f_0\} = -\{p\} \tag{1.69}$$

so that with member fixed-end forces, the stiffness equilibrium equation of the element can be written as:

$$\{f\} = [k]\{d\} + \{f_0\} \tag{1.70}$$

In Eq. (1.70), the first term on the right hand side is the contribution of the stiffness forces (forces due to member-end displacements) and the second term is contribution of the applied external loads. The stiffness matrix, $[k]$, is calculated by using Eq. (1.63) and the consistent load vector is calculated from Eq. (1.66). Equation (1.68) or (1.70) states the equilibrium condition of the element. In a similar way, we can also write the equilibrium condition of a joint (a connection point of different elements) in the system, e.g., like the one shown in Fig. 1.7. This is achieved by superimposing member-end forces of all elements connected at the joint and applied loads, which are all stated in global coordinates, and then equalizing them to zero. This superposition is indicated symbolically by the summation of all member-end forces in global coordinates as written by,

$$\sum_{e=1}^{n_e} \{f_{Gj}\}_e - \{Q_{Gj}\} = 0 \rightarrow \sum_{e=1}^{n_e} ([k_G]\{d_G\})_e = \sum_{e=1}^{n_e} \{p_G\}_e + \{Q_{Gj}\} \tag{1.71}$$

where (n_e) indicates total number of elements joining at the joint (j), the subscript (G) denotes global coordinates, and $\{Q_{Gj}\}$ is the external load vector applied at the joint (j) in the global coordinates. To obtain an equilibrium condition of the system (equilibriums of all joints in the system), an assembly process of the element stiffness matrix and consistent load vector as written symbolically in Eq. (1.71) is carried out. For this assembly process, the stiffness matrices and consistent load vectors in local coordinates of elements must be transformed first to the global coordinates, and then the assemblage is processed. The coordinate transformation process will be explained later in the Sect. 1.2.12. After the assemblage, the equilibrium equation of the system under static loadings can be stated as:

$$[K]\{D\} = \{P\} \tag{1.72}$$

where $[K]$ is the system stiffness matrix and $\{P\}$ is the system load vector, which are all in global coordinate directions. The vector $\{D\}$ is the system displacement vector in the global coordinates, which is calculated from the solution of Eq. (1.72). In order to form the stiffness matrix and load vector of the system, the element stiffness matrix and load vector in the local coordinates must be calculated first as presented below.

1.2.10.1 Element Stiffness Matrix in Local Coordinates

The stiffness matrix of an element will be calculated in the local principal coordinates by using Eq. (1.63), in which the rigidity matrix $[G]$ and the deformation matrix $[B]$ are defined respectively by Eqs. (1.61) and (1.51b). Having carried out the integration of Eq. (1.63), the stiffness matrix can be obtained in the local principal coordinates as written by,

$$[k] = \begin{bmatrix} [k]_{11} & [k]_{12} & -[k]_{11} & [k]_{12} \\ & [k]_{22} & -[k]_{12}^T & [k]_{24} \\ & & [k]_{11} & -[k]_{12} \\ & & & [k]_{22} \end{bmatrix} \tag{1.73a}$$

where the submatrices are as defined below.

$$[k]_{11}= \begin{bmatrix} \dfrac{EA}{\ell} & 0 & 0 \\ 0 & \dfrac{12EI_z\mu_y}{\ell^3} & 0 \\ 0 & 0 & \dfrac{12EI_y\mu_z}{\ell^3} \end{bmatrix}, \quad [k]_{12}= \begin{bmatrix} 0 & 0 & 0 \\ 0 & 0 & \dfrac{6EI_z\mu_y}{\ell^2} \\ 0 & -\dfrac{6EI_y\mu_z}{\ell^2} & 0 \end{bmatrix} \tag{1.73b}$$

$$[k]_{22} = \begin{bmatrix} \dfrac{GJ}{\ell} & 0 & 0 \\ 0 & \dfrac{EI_y\mu_{4z}}{\ell} & 0 \\ 0 & 0 & \dfrac{EI_z\mu_{4y}}{\ell} \end{bmatrix}, \quad [k]_{24}= \begin{bmatrix} -\dfrac{GJ}{\ell} & 0 & 0 \\ 0 & \dfrac{EI_y\mu_{2z}}{\ell} & 0 \\ 0 & 0 & \dfrac{EI_z\mu_{2y}}{\ell} \end{bmatrix}$$

$$\mu_{4z} = (3\mu_z + 1),\ \mu_{4y} = (3\mu_y + 1) \quad \text{and} \quad \mu_{2z} = (3\mu_z - 1),\ \mu_{2y} = (3\mu_y - 1) \tag{1.73c}$$

where the parameters, μ_y and μ_z, are given in Eq. (1.46).

1.2.10.2 Element Consistent Load Vector in Local Coordinates

The consistent load vector of an element will be calculated in the local principal coordinates by using Eq. (1.66), which can be expressed as:

$$\{p\} = \{p_q\} + \sum_{i=1}^{n} \{p_p\}_i + \sum_{j=1}^{m} \{p_m\}_j \Bigg\} \rightarrow \begin{cases} \{p_q\} = \displaystyle\int_0^\ell \left([N_u]^T\{q\} + [N_\theta]^T\{\text{m}\}\right) dx \\ \{p_p\}_i = [N_u(\xi_i)]^T\{\text{P}_\text{i}\} \\ \{p_m\}_j = [N_\theta(\xi_j)]^T\{\text{M}_\text{j}\} \end{cases} \tag{1.74}$$

where $\{p_q\}$ is the contribution of the distributed loads, $\{p_p\}_i$ is the contribution of the concentrated forces and $\{p_m\}_j$ is the contribution of concentrated moments. For fully distributed constant loads and moments, $\{q\}$ and $\{m\}$, and for concentrated forces $\{P_i\}$, these contributions $\{p_q\}$ and $\{p_p\}_i$ can be obtained as written by,

$$\{p_q\} = \begin{Bmatrix} q_x\ell/2 \\ (q_y\ell/2) - m_z\mu_y \\ (q_z\ell/2) + m_y\mu_z \\ m_x\ell/2 \\ -(q_z\ell^2/12) + m_y\ell(1-\mu_z)/2 \\ (q_y\ell^2/12) + m_z\ell(1-\mu_y)/2 \\ q_x\ell/2 \\ (q_y\ell/2) + m_z\mu_y \\ (q_z\ell/2) - m_y\mu_z \\ m_x\ell/2 \\ (q_z\ell^2/12) + m_y\ell(1-\mu_z)/2 \\ -(q_y\ell^2/12) + m_z\ell(1-\mu_y)/2 \end{Bmatrix} \tag{1.75a}$$

and

$$\{p_p\}_i = \begin{Bmatrix} P_x\beta_{px} \\ P_y\beta_{py}\left[1+\mu_y\alpha_{py}(\beta_{py}-\alpha_{py})\right] \\ P_z\beta_{pz}\left[1+\mu_z\alpha_{pz}(\beta_{pz}-\alpha_{pz})\right] \\ 0 \\ -P_z\ell\mu_z\alpha_{pz}\beta_{pz}(\Phi_z/2+\beta_{pz}) \\ P_y\ell\mu_y\alpha_{py}\beta_{py}(\Phi_y/2+\beta_{py}) \\ P_x\alpha_{px} \\ P_y\alpha_{py}\left[1-\mu_y\beta_{py}(\beta_{py}-\alpha_{py})\right] \\ P_z\alpha_{pz}\left[1-\mu_z\beta_{pz}(\beta_{pz}-\alpha_{pz})\right] \\ 0 \\ P_z\ell\mu_z\alpha_{pz}\beta_{pz}(\Phi_z/2+\alpha_{pz}) \\ -P_y\ell\mu_y\alpha_{py}\beta_{py}(\Phi_y/2+\alpha_{py}) \end{Bmatrix} \rightarrow \begin{cases} \beta_{px} = (1-\alpha_{px}) \\ \beta_{py} = (1-\alpha_{py}) \\ \beta_{pz} = (1-\alpha_{pz}) \\ \\ \alpha_{px} = \xi_i \rightarrow \text{for } P_x \\ \alpha_{py} = \xi_i \rightarrow \text{for } P_y \\ \alpha_{pz} = \xi_i \rightarrow \text{for } P_z \end{cases} \tag{1.75b}$$

The contribution of concentrated moments $\{M_j\}$ to the consistent load vector, $\{p_m\}_j$, is obtained as written by,

$$\{p_m\}_j = \begin{Bmatrix} 0 \\ -6M_z\mu_y\alpha_{mz}\beta_{mz}/\ell \\ 6M_y\mu_z\alpha_{my}\beta_{my}/\ell \\ M_x\beta_{mx} \\ M_y\beta_{my}(1-3\mu_z\alpha_{my}) \\ M_z\beta_{mz}(1-3\mu_y\alpha_{mz}) \\ 0 \\ 6M_z\mu_y\alpha_{mz}\beta_{mz}/\ell \\ -6M_y\mu_z\alpha_{my}\beta_{my}/\ell \\ M_x\alpha_{mx} \\ M_y\alpha_{my}(1-3\mu_z\beta_{my}) \\ M_z\alpha_{mz}(1-3\mu_y\beta_{mz}) \end{Bmatrix} \rightarrow \begin{cases} \beta_{mx} = (1-\alpha_{mx}) \\ \beta_{my} = (1-\alpha_{my}) \\ \beta_{mz} = (1-\alpha_{mz}) \\ \\ \alpha_{mx} = \xi_j \rightarrow \text{for } M_x \\ \alpha_{my} = \xi_j \rightarrow \text{for } M_y \\ \alpha_{mz} = \xi_j \rightarrow \text{for } M_z \end{cases} \tag{1.75c}$$

For other types of loadings, the consistent load vector can be calculated by using Eq. (1.74). For a dynamic analysis, the mass and damping matrices of the element are also required. These quantities and the dynamic equilibrium equation of a structural system are presented in the following section.

1.2.11 Total Kinetic Energy, Mass Matrix, Damping Matrix, and Dynamic Equilibrium Equation

When the applied loads are time dependent, such as earthquake and wave loadings, the static equilibrium equation given by Eq. (1.72) is not valid any more. In this case, the dynamic equilibrium equation will be used to calculate response displacements and element internal forces as being time functions. The dynamic equilibrium equation of an element can be obtained by using the fundamental form of the Lagrange's equation [57–59], which is stated in terms of the generalized coordinates (here, the displacements) as written in the matrix form by,

$$\frac{\mathrm{d}}{\mathrm{d}t}\left\{\frac{\partial T}{\partial \dot{d}_i}\right\} - \left\{\frac{\partial T}{\partial d_i}\right\} + \left\{\frac{\partial\left(U - W_p\right)}{\partial d_i}\right\} + \left\{\frac{\partial D}{\partial \dot{d}_i}\right\} = 0 \tag{1.76}$$

in which d_i is the ith term of the displacement vector, T is the total kinetic energy of the element, U is the strain energy given by Eq. (1.62), W_p is the total work done by all external loads, which is given by Eq. (1.65b) in the static case, D is the dissipation energy in the element due to internal friction, in other words due to structural damping, and a dot means a time derivative. Since U and W_p are previously determined, attention is paid here on the kinetic and dissipation energies, T and D.

The total kinetic energy of a beam is obtained from the integration of the kinetic energy of an infinitesimal volume in the element, dV. It is stated as,

$$T = \frac{1}{2}\int \mathrm{v}^2 \mathrm{d}m \;\rightarrow\; T = \frac{1}{2}\int \rho_s \mathrm{v}^2 \mathrm{d}V \tag{1.77}$$

in which ρ_s is the mass density of the structural material and v is the velocity of the mass of the infinitesimal volume dV. Its square is written as:

$$\mathrm{v}^2 = \dot{v}_x^2 + \dot{v}_y^2 + \dot{v}_z^2 \tag{1.78a}$$

where $\dot{v}_x$, $\dot{v}_y$ and $\dot{v}_z$ are the velocity components at a point on the cross-section of the beam in the coordinate directions, x, y, and z, respectively. Having neglected the warping effect and using Eq. (1.25a) in Eq. (1.78a) the velocity square can be obtained as written:

$$\begin{aligned}v^2 = &\dot{u}_x^2 + \dot{u}_y^2 + \dot{u}_z^2 + \dot{\theta}_x^2(y^2 + z^2) + \dot{\theta}_z^2 y^2 + \dot{\theta}_y^2 z^2 - 2\dot{u}_x\dot{\theta}_z y + \cdots \\ &\cdots + 2\dot{u}_x\dot{\theta}_y z - 2\dot{\theta}_z\dot{\theta}_y yz - 2\dot{u}_y\dot{\theta}_x z + 2\dot{u}_z\dot{\theta}_x y\end{aligned} \tag{1.78b}$$

Having introduced Eq. (1.78b) into Eq. (1.77) and using the cross-sectional properties given by Eq. (1.31a), the total kinetic energy in the local principal coordinates can be obtained as:

$$T = \frac{1}{2}\int \rho_s\left[A\left(\dot{u}_x^2 + \dot{u}_y^2 + \dot{u}_z^2\right) + I_p\dot{\theta}_x^2 + I_z\dot{\theta}_z^2 + I_y\dot{\theta}_y^2\right]\mathrm{d}x \tag{1.79}$$

where I_p is the polar inertia moment. Using Eqs. (1.48a, b) in Eq. (1.79), the kinetic energy can be expressed in the matrix form as:

$$T = \frac{1}{2}\int \rho_s\left(A\{\dot{u}\}^T\{\dot{u}\} + \left\{\dot{\theta}\right\}^T[J]\left\{\dot{\theta}\right\}\right)\mathrm{d}x \rightarrow [J] = \begin{bmatrix} I_p & & \\ & I_y & \\ & & I_z\end{bmatrix} \tag{1.80a}$$

or

$$T = \frac{1}{2}\{\dot{d}\}^T[m]\{\dot{d}\} \tag{1.80b}$$

in which $[m]$ is the consistent mass matrix of the element in the local principal coordinates. It is defined as:

$$[m] = \int_0^\ell \rho_s\left(A[N_u]^T[N_u] + [N_\theta]^T[J][N_\theta]\right)\mathrm{d}x \tag{1.81}$$

The dissipation energy is the work done by viscous forces due to internal friction in the element, and it can be stated in a similar form of the kinetic energy [60] as:

$$D = \frac{1}{2}\{\dot{d}\}^T[c]\{\dot{d}\} \tag{1.82}$$

where $[c]$ is the damping matrix of the element, which can be obtained as being proportional to the mass matrix and the stiffness matrix for a linear viscous damping. It is also referred to as the Rayleigh damping. In general, it is stated [61] as:

$$[c] = \alpha[m] + \beta[k] \tag{1.83}$$

where α and β are the proportionality factors. Having substituted the strain energy U from Eq. (1.62), the kinetic energy T from Eq. (1.80b), the dissipation energy D from Eq. (1.82), and the external work W_p from Eq. (1.65b) into the Lagrange's equation in Eq. (1.76), the dynamic equilibrium equation of the beam element can be obtained as,

$$[k]\{d(t)\} + [c]\left\{\dot{d}(t)\right\} + [m]\left\{\ddot{d}(t)\right\} - \{p(t)\} = \{f(t)\} \tag{1.84}$$

where $\{f(t)\}$ is the vector of time dependent internal forces at the element ends, $\{p(t)\}$ is the vector of time dependent distributed loads on the element. Similar to the static analysis, the dynamic equilibrium of a joint (j) in a structural system can be stated in the global coordinates as:

$$\sum_{e=1}^{n_e} \{f_{\mathrm{Gj}}(t)\}_e - \{Q_{\mathrm{Gj}}(t)\} = 0, \qquad \text{or}$$

$$\sum_{e=1}^{n_e} \left([k_G]\{d_G(t)\} + [c_G]\left\{\dot{d}_G(t)\right\} + [m_G]\left\{\ddot{d}_G(t)\right\}\right)_e = \sum_{e=1}^{n_e} \{p_G(t)\}_e + \{Q_{\mathrm{Gj}}(t)\} \tag{1.85}$$

in which the subscript (G) denotes global coordinates, $\{Q_{Gj}(t)\}$ is the vector of time dependent applied loads at the joint in the global coordinates. For all joints of a structural system, this equation can be stated similarly to the static analysis in the global coordinates as written:

$$[K]\{D(t)\} + [C]\left\{\dot{D}(t)\right\} + [M]\left\{\ddot{D}(t)\right\} = \{P(t)\} \tag{1.86}$$

which defines the dynamic equilibrium equation of the system. In Eq. (1.86), $[M]$ is the mass matrix of the system obtained from the assembly process of element mass matrices and $[C]$ is the Rayleigh damping matrix of the system. The mass matrix of an element is calculated in two manners as presented in the next sections.

1.2.11.1 Consistent Mass Matrix in Local Coordinates

The consistent mass matrix will be calculated in the local principal coordinates by using Eq. (1.81), in which the shape functions matrices, $[N_u]$ and $[N_\theta]$, are as given by Eqs. (1.50a, b), and the matrix $[J]$ is given in Eq. (1.80a). Having carried out the integration of Eq. (1.81), the consistent mass matrix can be stated as:

$$[m] = \rho A \ell \begin{bmatrix} [m]_{11} & [m]_{12} & [m]_{13} & [m]_{14} \\ & [m]_{22} & [m]_{23} & [m]_{24} \\ & & [m]_{11} & [m]_{34} \\ & & & [m]_{22} \end{bmatrix} \tag{1.87}$$

where the submatrices are defined as:

$$[m]_{11} = \begin{bmatrix} 1/3 & 0 & 0 \\ 0 & m_{22} & 0 \\ 0 & 0 & m_{33} \end{bmatrix}, \ [m]_{12} = \begin{bmatrix} 0 & 0 & 0 \\ 0 & 0 & m_{26} \\ 0 & -m_{35} & 0 \end{bmatrix} \tag{1.88a}$$

$$[m]_{13} = \begin{bmatrix} 1/6 & 0 & 0 \\ 0 & m_{28} & 0 \\ 0 & 0 & m_{39} \end{bmatrix}, \ [m]_{14} = \begin{bmatrix} 0 & 0 & 0 \\ 0 & 0 & -m_{212} \\ 0 & m_{311} & 0 \end{bmatrix} \tag{1.88b}$$

$$[m]_{22} = \begin{bmatrix} m_{44} & 0 & 0 \\ 0 & m_{55} & 0 \\ 0 & 0 & m_{66} \end{bmatrix}, \ [m]_{23} = \begin{bmatrix} 0 & 0 & 0 \\ 0 & 0 & -m_{59} \\ 0 & m_{68} & 0 \end{bmatrix} \tag{1.88c}$$

$$[m]_{24} = \begin{bmatrix} m_{410} & 0 & 0 \\ 0 & -m_{511} & 0 \\ 0 & 0 & -m_{612} \end{bmatrix}, \ [m]_{34} = \begin{bmatrix} 0 & 0 & 0 \\ 0 & 0 & -m_{812} \\ 0 & m_{911} & 0 \end{bmatrix} \tag{1.88d}$$

The elements of the mass matrix (m_{ij}) are obtained as presented:

$$\begin{aligned}
m_{22} &= \mu_y^2\left(\frac{13}{35} + \frac{7}{10}\Phi_y + \frac{\Phi_y^2}{3} + \frac{6}{5}\frac{I_z}{A\ell^2}\right) \\
m_{26} &= \mu_y^2\ell\left[\frac{11}{210} + \frac{11}{120}\Phi_y + \frac{\Phi_y^2}{24} + \frac{I_z}{A\ell^2}\left(\frac{1}{10} - \frac{3}{2}\Phi_y - \Phi_y^2\right)\right] \\
m_{28} &= \mu_y^2\left(\frac{9}{70} + \frac{3}{10}\Phi_y + \frac{\Phi_y^2}{6} - \frac{6}{5}\frac{I_z}{A\ell^2}\right) \\
m_{212} &= \mu_y^2\ell\left[\frac{13}{420} + \frac{3}{40}\Phi_y + \frac{\Phi_y^2}{24} - \frac{I_z}{A\ell^2}\left(\frac{1}{10} - \frac{\Phi_y}{2}\right)\right]
\end{aligned} \tag{1.89a}$$

$$\begin{aligned}
m_{33} &= \mu_z^2\left(\frac{13}{35} + \frac{7}{10}\Phi_z + \frac{\Phi_z^2}{3} + \frac{6}{5}\frac{I_y}{A\ell^2}\right) \\
m_{35} &= \mu_z^2\ell\left[\frac{11}{210} + \frac{11}{120}\Phi_z + \frac{\Phi_z^2}{24} + \frac{I_y}{A\ell^2}\left(\frac{1}{10} - \frac{3}{2}\Phi_z - \Phi_z^2\right)\right] \\
m_{39} &= \mu_z^2\left(\frac{9}{70} + \frac{3}{10}\Phi_z + \frac{\Phi_z^2}{6} - \frac{6}{5}\frac{I_y}{A\ell^2}\right) \\
m_{311} &= \mu_z^2\ell\left[\frac{13}{420} + \frac{3}{40}\Phi_z + \frac{\Phi_z^2}{24} - \frac{I_y}{A\ell^2}\left(\frac{1}{10} - \frac{\Phi_z}{2}\right)\right] \\
m_{44} &= I_p/(3A), \qquad m_{410} = I_p/(6A)
\end{aligned} \tag{1.89b}$$

$$\begin{aligned}
m_{55} &= \mu_z^2\ell^2\left[\frac{1}{105}+\frac{\Phi_z}{60}+\frac{\Phi_z^2}{120}+\frac{I_y}{A\ell^2}\left(\frac{2}{15}+\frac{\Phi_z}{6}+\frac{\Phi_z^2}{3}\right)\right]\\
m_{59} &= \mu_z^2\ell\left[\frac{13}{420}+\frac{3}{40}\Phi_z+\frac{\Phi_z^2}{24}-\frac{I_y}{A\ell^2}\left(\frac{1}{10}-\frac{3}{2}\Phi_z-\Phi_z^2\right)\right]\\
m_{511} &= \mu_z^2\ell^2\left[\frac{1}{140}+\frac{\Phi_z}{60}+\frac{\Phi_z^2}{120}+\frac{I_y}{A\ell^2}\left(\frac{1}{30}+\frac{\Phi_z}{6}-\frac{\Phi_z^2}{6}\right)\right]
\end{aligned} \tag{1.89c}$$

$$\begin{aligned}
m_{66} &= \mu_y^2\ell^2\left[\frac{1}{105}+\frac{\Phi_y}{60}+\frac{\Phi_y^2}{120}+\frac{I_z}{A\ell^2}\left(\frac{2}{15}+\frac{\Phi_y}{6}+\frac{\Phi_y^2}{3}\right)\right]\\
m_{68} &= \mu_y^2\ell\left[\frac{13}{420}+\frac{3}{40}\Phi_y+\frac{\Phi_y^2}{24}-\frac{I_z}{A\ell^2}\left(\frac{1}{10}-\frac{3}{2}\Phi_y-\Phi_y^2\right)\right]\\
m_{612} &= \mu_y^2\ell^2\left[\frac{1}{140}+\frac{1}{60}\Phi_y+\frac{1}{120}\Phi_y^2+\frac{I_z}{A\ell^2}\left(\frac{1}{30}+\frac{\Phi_y}{6}-\frac{\Phi_y^2}{6}\right)\right]\\
m_{812} &= \mu_y^2\ell\left[\frac{11}{210}+\frac{11}{120}\Phi_y+\frac{1}{24}\Phi_y^2+\frac{I_z}{A\ell^2}\left(\frac{1}{10}-\frac{\Phi_y}{2}\right)\right]\\
m_{911} &= \mu_z^2\ell\left[\frac{11}{210}+\frac{11}{120}\Phi_z+\frac{\Phi_z^2}{24}+\frac{I_y}{A\ell^2}\left(\frac{1}{10}-\frac{\Phi_z}{2}\right)\right]
\end{aligned} \tag{1.89d}$$

The parameters, Φ_y, μ_y, and Φ_z, μ_z, are given in Eq. (1.46). This consistent mass matrix will be transformed to the global coordinates and then assembly process will be performed to obtain the system mass matrix. An alternative choice of using mass matrix is the lumped mass matrix as it is explained in the next section.

1.2.11.2 Lumped Mass Matrix in Local Coordinates

The consistent mass matrix presented above is the general formulation of the mass matrix of an element in the local principal coordinates, which produces more accurate results. Since it is a symmetric full matrix, in practical applications, using the consistent mass matrix is relatively costly in terms of computation time. An alternative choice may be to use a diagonal (lumped) mass matrix, which offers computational and storage advantages in certain cases, notably in explicit time integration, within acceptable precision bounds of the results for dynamic sensitive structures. The construction of the consistent mass matrix is fully defined by the choice of kinetic energy functional and shape functions, whereas the construction of a diagonally lumped mass matrix is not a unique process. Once the consistent mass matrix is calculated, the lumped mass matrix can be formed in different ways. One of the following methods can be widely used in the practice:

1. The lumped mass matrix can be obtained by using a rigid body motion in a selected coordinate direction, i.e., in one of u_x, u_y, u_z, θ_x, θ_y, and θ_z directions at a time.
2. The HRZ lumping [62]. The lumped mass matrix can be obtained by a heuristic procedure as it follows.

   - For each coordinate direction, select the degrees of freedom (DOF) that contribute to motion in that direction. From this set, separate translational DOF and rotational DOF subsets.
   - Add up diagonal entries of the consistent mass matrix pertaining to the translational DOF subset only. This summation is denoted by **S**.
   - Find the terms of the lumped mass matrix of both subsets by dividing the diagonal entries of the consistent mass matrix by the sum **S**.
   - Repeat this process for all coordinate directions.

These two methods of mass lumping have three advantages: (a) easy to explain and implement, (b) applicable to any element as long as the consistent mass matrix is available and (c) retaining non-negativity. The last property is particularly important as it means that the lumped mass matrix is physically admissible, preventing numerical instability. As a general assessment, it gives reasonable results if the element has only translational freedoms. The lumped mass matrices, which are obtained from the consistent mass matrix using above methods, are presented in the local principal coordinates in Eqs. (1.90a, b) with and without including shear deformation effects. In these equations, the nonzero diagonal terms of the lumped mass matrix are shown in vector notations. As it is seen from Eqs. (1.90a, b), both methods produce the same results in the translational directions and satisfy the mass conservation of the element as the total mass is equally concentrated at both ends of the element. In the rotational degrees of freedom, there is no unique lumped mass. However, in the light of comparatively small contributions of rotational degrees of freedom to the total kinetic energy, lumped masses at rotational directions can also be taken as being zero with an acceptable precision of the results. However, in this case, the lumped mass matrix becomes singular and therefore it produces numerical difficulties. In order to prevent such difficulties, the lumped masses in rotational DOF can be assumed a small quantity in practice, e.g., $(m_{\text{rot}} = m\ell^2/\alpha)$ with α is an assumed large number.

With shear deformation:

Rigid body motion (a) HRZ (b)

$$[m]_{\text{lump}} = \rho A\ell \left\{ \begin{array}{c} 1/2 \\ 1/2 \\ 1/2 \\ I_p/(2A) \\ \ell^2\mu_z^2\left[1 + I_y\left(42 + 210\Phi_z^2\right)/(A\ell^2)\right]/420 \\ \ell^2\mu_y^2\left[1 + I_z\left(42 + 210\Phi_y^2\right)/(A\ell^2)\right]/420 \\ 1/2 \\ 1/2 \\ 1/2 \\ I_p/(2A) \\ \ell^2\mu_z^2\left[1 + I_y\left(42 + 210\Phi_z^2\right)/(A\ell^2)\right]/420 \\ \ell^2\mu_y^2\left[1 + I_z\left(42 + 210\Phi_y^2\right)/(A\ell^2)\right]/420 \end{array} \right\} \left\{ \begin{array}{c} 1/2 \\ 1/2 \\ 1/2 \\ I_p/(2A) \\ m_{55}/(2m_{33}) \\ m_{66}/(2m_{22}) \\ 1/2 \\ 1/2 \\ 1/2 \\ I_p/(2A) \\ m_{55}/(2m_{33}) \\ m_{66}/(2m_{22}) \end{array} \right\} \tag{1.90a}$$

Without shear deformation:

Rigid body motion (a) HRZ (b)

$$[m]_{\text{lump}} = \rho A\ell \left\{ \begin{array}{c} 1/2 \\ 1/2 \\ 1/2 \\ I_p/(2A) \\ \ell^2\left[1 + 42I_y/(A\ell^2)\right]/420 \\ \ell^2\left[1 + 42I_z/(A\ell^2)\right]/420 \\ 1/2 \\ 1/2 \\ 1/2 \\ I_p/(2A) \\ \ell^2\left[1 + 42I_y/(A\ell^2)\right]/420 \\ \ell^2\left[1 + 42I_z/(A\ell^2)\right]/420 \end{array} \right\} \left\{ \begin{array}{c} 1/2 \\ 1/2 \\ 1/2 \\ I_p/(2A) \\ \ell^2\left[1 + 140I_y/(13A\ell^2)\right]/78 \\ \ell^2\left[1 + 140I_z/(13A\ell^2)\right]/78 \\ 1/2 \\ 1/2 \\ 1/2 \\ I_p/(2A) \\ \ell^2\left[1 + 140I_y/(13A\ell^2)\right]/78 \\ \ell^2\left[1 + 140I_z/(13A\ell^2)\right]/78 \end{array} \right\} \tag{1.90b}$$

The lumped mass matrix produces accurate results for small natural frequencies. For higher natural frequencies, it produces approximate results, and for more correct results, the consistent mass matrix should be used.

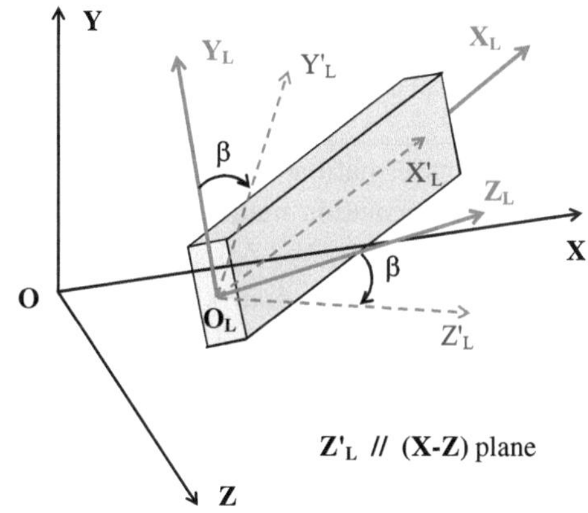

Fig. 1.8 Element coordinate systems, **a** global (*X*, *Y*, *Z*), **b** local principal (X_L, Y_L, Z_L), and **c** local auxiliary (X'_L, Y'_L, Z'_L)

1.2.12 Coordinate Systems and Transformations

For a spatial beam element, three coordinate systems are involved as: (a) global coordinate system, (*X*, *Y*, *Z*), (b) local principal coordinate system (X_L, Y_L, Z_L), and (c) local auxiliary coordinate system $\left(X'_L, Y'_L, Z'_L\right)$, as shown in Fig. 1.8. The stiffness, load, mass, and cross-sectional properties of the element are formulated in the local principal coordinate system (X_L, Y_L, Z_L) as presented above. In order to form the stiffness and mass matrices, and the load vector, of the system using the assembly process, the element stiffness and mass matrices, and the load vector, must be transformed to the global coordinate system (*X*, *Y*, *Z*). This transformation can be done by using a local auxiliary coordinate system $\left(X'_L, Y'_L, Z'_L\right)$. As it is shown in Fig. 1.8, the coordinates, X_L and X'_L, of the principal and auxiliary coordinate systems are assumed in the axial direction of the element. The other axes $\left(Y'_L, Z'_L\right)$ of the auxiliary coordinate systems are obtained by rotating the principal coordinate axes (Y_L, Z_L) about the axial coordinate X_L until the axis Z_L becomes parallel to the (*X*–*Z*) plane of the global coordinate system. This position of the rotation of (Y_L, Z_L) is assumed to be the auxiliary coordinates $\left(Y'_L, Z'_L\right)$. Thus, the condition of the auxiliary coordinate system is $\left(Z'_L//(X-Z)\text{ plane}\right)$. The rotation angle satisfying this condition is denoted by β as shown in Fig. 1.8. The clockwise rotation of (Y_L, Z_L) is assumed to have a (+) sign, which produces a vector in the (+) axial coordinate direction, X_L. For further development, the following definitions are made:

- $\{u\}$ Displacement vector in the local principal coordinate system (X_L, Y_L, Z_L), given in Eq. (1.48b)
- $\{u'\}$ Displacement vector in the local auxiliary coordinate system $\left(X'_L, Y'_L, Z'_L\right)$
- $\{u_G\}$ Displacement vector in the global coordinate system (*X*, *Y*, *Z*)

The displacement vectors in these coordinate systems are calculated from the following transformation:

$$\left.\begin{array}{l}\{u\} = [t_1]\{u'\} \\ \{u'\} = [t_2]\{u_G\}\end{array}\right\} \rightarrow \{u\} = [t_1][t_2]\{u_G\} \rightarrow \left\{\begin{array}{l}\{u\} = [t]\{u_G\} \\ [t] = [t_1][t_2]\end{array}\right. \tag{1.91}$$

The transformation matrices are defined as:

$[t_1]$ Transformation matrix between the local principal and auxiliary coordinate systems, (X_L, Y_L, Z_L) and $\left(X'_L, Y'_L, Z'_L\right)$.

$[t_2]$ Transformation matrix between the auxiliary and global coordinate systems, $\left(X'_L, Y'_L, Z'_L\right)$ and (X, Y, Z)

$[t]$ Transformation matrix between the local principal and global coordinate systems, (X_L, Y_L, Z_L) and (X, Y, Z).

The transformation matrix between the local principal and auxiliary coordinate systems, $[t_1]$, can be easily written from Fig. 1.8 in terms of the rotation angle β as:

$$[t_1] = \begin{bmatrix} 1 & 0 & 0 \\ 0 & \cos\beta & -\sin\beta \\ 0 & \sin\beta & \cos\beta \end{bmatrix} \tag{1.92}$$

The transformation matrix between the local auxiliary and global coordinate systems, $[t_2]$, can be defined in general as:

$$[t_2] = \begin{bmatrix} c_x & c_y & c_z \\ \ell_x & \ell_y & \ell_z \\ m_x & m_y & m_z \end{bmatrix} \tag{1.93}$$

where the parameters are:

$$c_x, c_y, c: \text{ cosine directions of the axis } X'_L\ (X_L)$$
$$\ell_x, \ell_y, \ell_z: \text{ cosine directions of the axis } Y'_L$$
$$m_x, m_y, m_z: \text{ cosine directions of the axis } Z'_L$$

The element orientation in the global coordinate system, which is defined by the cosine directions (c_x, c_y, c_z), is given. The cosine directions of the axes Y'_L and Z'_L, $\left(\ell_x, \ell_y, \ell_z\right)$ and (m_x, m_y, m_z), will be calculated in terms of (c_x, c_y, c_z) using normality and orthogonality properties of orthogonal transformations, and also the condition that $\left(Z'_L//(X - Z) \text{ plane}\right)$. The normality and orthogonality properties are satisfied by the condition:

$$[t_2]^T[t_2] = I_3 \rightarrow I_3: \text{ a unit matrix with } (3 \times 3) \tag{1.94}$$

The condition $\left(Z'_L//(X - Z) \text{ plane}\right)$ implies that ($m_y = 0$). Using these conditions the unknowns $\left(\ell_x, \ell_y, \ell_z\right)$ and (m_x, m_y, m_z) can be obtained as written:

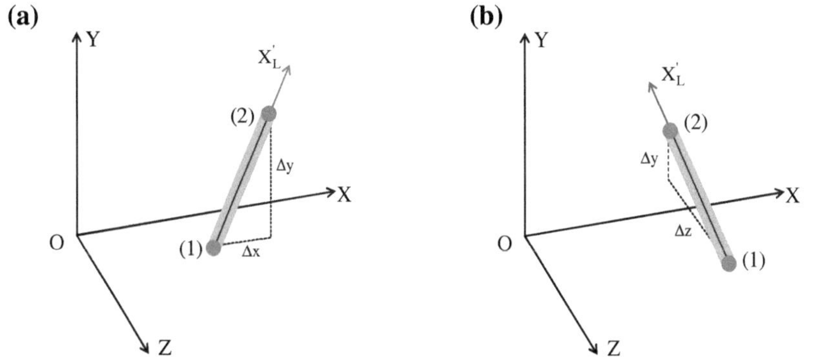

Fig. 1.9 Special cases of the element orientation, ($c_z = 0$) and ($c_x = 0$). **a** The case of $c_z = 0$. **b** The case of $c_x = 0$

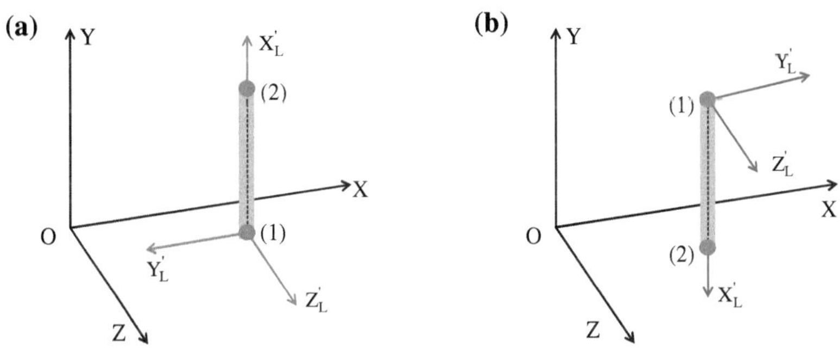

Fig. 1.10 Element orientations in the global Y direction. **a** Positive orientation $c_y = +1$. **b** Negative orientation $c_y = -1$

$$
\begin{aligned}
\ell_x &= -c_x c_y \Big/ \sqrt{1 - c_y^2} & \qquad & m_x = -c_z \Big/ \sqrt{1 - c_y^2} \\
\ell_y &= \sqrt{1 - c_y^2} & \text{and} \qquad & m_y = 0 \\
\ell_z &= -c_y c_z \Big/ \sqrt{1 - c_y^2} & \qquad & m_z = c_x \Big/ \sqrt{1 - c_y^2}
\end{aligned}
\tag{1.95}
$$

With these cosine directions the transformation matrix $[t_2]$ given by Eq. (1.93) can be written as:

$$[t_2] = \begin{bmatrix} c_x & c_y & c_z \\ -\dfrac{c_x c_y}{\sqrt{1-c_y^2}} & \sqrt{1-c_y^2} & -\dfrac{c_y c_z}{\sqrt{1-c_y^2}} \\ -\dfrac{c_z}{\sqrt{1-c_y^2}} & 0 & \dfrac{c_x}{\sqrt{1-c_y^2}} \end{bmatrix} \tag{1.96a}$$

or using the property of cosine directions that $\left(c_x^2 + c_y^2 + c_z^2 = 1\right)$ Eq. (1.96a) can be stated as:

$$[t_2] = \begin{bmatrix} c_x & c_y & c_z \\ -\dfrac{c_y}{\sqrt{1+\lambda^2}} & \sqrt{1-c_y^2} & -c_y\dfrac{\lambda}{\sqrt{1+\lambda^2}} \\ -\dfrac{\lambda}{\sqrt{1+\lambda^2}} & 0 & \dfrac{1}{\sqrt{1+\lambda^2}} \end{bmatrix} \quad \text{with} \quad \left(\lambda = \frac{c_z}{c_x}\right) \tag{1.96b}$$

This matrix covers all positions of element orientations, except in the global Y direction. For the orientation in the global Y direction, i.e.,$\left(c_y = \pm 1\right)$ and $(c_x = 0,\ c_z = 0)$, the parameter λ in Eq. (1.96b) becomes indefinite, which introduces a numerical instability. In order to prevent this problem, a special treatment is required. This can be done in different ways. In the first way, it is assumed that c_z is approaching, or equal to, zero as shown in Fig. 1.9a. In the second way, it is assumed that c_x is approaching, or equal to, zero as shown in Fig. 1.9b. A third way may be diagonal approach, for which $(\lambda = 1)$ is assumed in Eq. (1.96b). However, this option is not considered here. Instead, the first two ways can be applied since they are relatively simpler. In this case, Eq. (1.96b) becomes as written:

$$[t_2]_{c_z\to 0} = \begin{bmatrix} c_x & c_y & 0 \\ -c_y & c_x & 0 \\ 0 & 0 & 1 \end{bmatrix} \quad \text{and} \quad [t_2]_{c_x\to 0} = \begin{bmatrix} 0 & c_y & c_z \\ 0 & c_z & -c_y \\ -1 & 0 & 0 \end{bmatrix} \tag{1.97a}$$

For the critical element orientation $\left(c_y = \pm 1\right)$ as shown in Fig. 1.10, if the first way is used, c_x will be zero $(c_x = 0)$ in the matrix $[t_2]_{c_z\to 0}$ and, if the second way is used, c_z will be zero $(c_z = 0)$ in the matrix $[t_2]_{c_x\to 0}$, which are both written in Eq. (1.97a). The transformation matrices for these two ways for $(c_y = \pm 1)$ are stated as

$$[t_2]_{c_z\to 0} = \begin{bmatrix} 0 & c_y & 0 \\ -c_y & 0 & 0 \\ 0 & 0 & 1 \end{bmatrix},\ [t_2]_{c_x\to 0} = \begin{bmatrix} 0 & c_y & 0 \\ 0 & 0 & -c_y \\ -1 & 0 & 0 \end{bmatrix} \quad \text{where} \ \left(c_y = \pm 1\right) \tag{1.97b}$$

The solutions of both ways are correct and one of them must be adopted. We adopt the first way, i.e., $[t_2]_{c_z\to 0}$ leading to the transformation matrices for positive and negative orientations of the element as written by,

$$[t_2]_{(c_y=+1)} = \begin{bmatrix} 0 & 1 & 0 \\ -1 & 0 & 0 \\ 0 & 0 & 1 \end{bmatrix} \quad \text{and} \quad [t_2]_{(c_y=-1)} = \begin{bmatrix} 0 & -1 & 0 \\ 1 & 0 & 0 \\ 0 & 0 & 1 \end{bmatrix} \tag{1.97c}$$

Having determined the transformation matrices, $[t_1]$ and $[t_2]$, the transformation matrix $[t]$ between the local principal and global coordinates will be calculated using Eq. (1.91). It is stated for the general case and for the special case of $(c_y = \pm 1)$ as written:

$$[t] = \begin{bmatrix} c_x & c_y & c_z \\ \dfrac{(c_z \sin\beta - c_x c_y \cos\beta)}{\sqrt{1-c_y^2}} & \cos\beta\sqrt{1-c_y^2} & -\dfrac{(c_x \sin\beta + c_y c_z \cos\beta)}{\sqrt{1-c_y^2}} \\ -\dfrac{(c_z \cos\beta + c_x c_y \sin\beta)}{\sqrt{1-c_y^2}} & \sin\beta\sqrt{1-c_y^2} & \dfrac{(c_x \cos\beta - c_y c_z \sin\beta)}{\sqrt{1-c_y^2}} \end{bmatrix} \tag{1.98a}$$

$$[t]_{(c_y=\pm 1)} = \begin{bmatrix} 0 & c_y & 0 \\ -c_y\cos\beta & 0 & -\sin\beta \\ -c_y \sin\beta & 0 & \cos\beta \end{bmatrix} \text{ for } (c_y = \pm 1) \tag{1.98b}$$

As similar to the coordinate transformations the transformation of displacements and rotations can now be constructed as written by,

$$\left.\begin{aligned} \{d\} &= [T]\{d_G\} \\ \{d_G\} &= [T]^T\{d\} \end{aligned}\right\} \rightarrow [T] = \begin{bmatrix} [t] & 0 & 0 & 0 \\ 0 & [t] & 0 & 0 \\ 0 & 0 & [t] & 0 \\ 0 & 0 & 0 & [t] \end{bmatrix} \tag{1.99}$$

where $\{d\}$ and $\{d_G\}$ are respectively displacement vectors in the local principal and global coordinate systems. The transformation matrix $[T]$ is used to calculate element properties in the global coordinates as explained in the following section.

1.2.13 Transformations of Element Stiffness Matrix, Consistent Load Vector, and Mass Matrix

The element stiffness matrix given by Eq. (1.73a), consistent load vector given by Eq. (1.74), and mass matrix given by Eq. (1.87) are defined and calculated in the element local principal coordinates. For the equilibrium of the system, all elements must be transformed to a common coordinate system, which is known as the global coordinate system. The transformations from element local to the global coordinates systems are carried out using the energy conservations, i.e., the total strain energy, the total work of external forces, and the total kinetic energy are invariant. These quantities are defined in the local principal coordinates by Eq. (1.62) for the strain energy, Eq. (1.65b) for the external work and Eq. (1.80b) for the kinetic energy. They are equalized to values stated in the global coordinates as written by:

$$\begin{aligned} U &= \frac{1}{2}\{d_{\mathrm{G}}\}^T[k_{\mathrm{G}}]\{d_{\mathrm{G}}\} = \frac{1}{2}\{d\}^T[k]\{d\} \\ W_p &= \{d_{\mathrm{G}}\}^T(\{p_{\mathrm{G}}\} + \{f_{\mathrm{G}}\}) = \{d\}^T(\{p\} + \{f\}) \\ T &= \frac{1}{2}\{\dot{d}_{\mathrm{G}}\}^T[m_{\mathrm{G}}]\{\dot{d}_{\mathrm{G}}\} = \frac{1}{2}\{\dot{d}\}^T[m]\{\dot{d}\} \end{aligned} \tag{1.100}$$

Having substituted $\{d\}$ from Eq. (1.99) to these statements they can be obtained as written:

$$\begin{aligned} U &= \frac{1}{2}\{d_{\mathrm{G}}\}^T[k_{\mathrm{G}}]\{d_{\mathrm{G}}\} = \frac{1}{2}\{d_{\mathrm{G}}\}^T\left([T]^T[k][T]\right)\{d_{\mathrm{G}}\} \\ W_p &= \{d_{\mathrm{G}}\}^T(\{p_{\mathrm{G}}\} + \{f_{\mathrm{G}}\}) = \{d_{\mathrm{G}}\}^T\left[[T]^T(\{p\} + \{f\})\right] \\ T &= \frac{1}{2}\{\dot{d}_{\mathrm{G}}\}^T[m_{\mathrm{G}}]\{\dot{d}_{\mathrm{G}}\} = \frac{1}{2}\{\dot{d}_{\mathrm{G}}\}^T\left([T]^T[m][T]\right)\{\dot{d}_{\mathrm{G}}\} \end{aligned} \tag{1.101}$$

From these statements of energy equivalences the stiffness matrix, mass matrix, consistent load vector, and internal force vector can be expressed in the global coordinates as:

$$\left.\begin{aligned} [k_{\mathrm{G}}] &= [T]^T[k][T] \rightarrow \text{ stiffness matrix} \\ [m_{\mathrm{G}}] &= [T]^T[m][T] \rightarrow \text{ mass matrix} \\ \{p_{\mathrm{G}}\} &= [T]^T\{p\} \cdots \rightarrow \text{ consistent load vector} \\ \{f_{\mathrm{G}}\} &= [T]^T\{f\} \cdots \rightarrow \text{ internal force vector} \end{aligned}\right\} \text{in GLOBAL coordinates} \tag{1.102}$$

in which the transformation matrix $[T]$ is given in Eq. (1.99). Since $[T]$ contains only diagonal submatrices with (3×3) dimensions, the transformations can be carried out easily.

1.3 Formulation of Member Releases and Partly Connected Members[1]

In structural analysis, it is mostly assumed or constructed that structural elements are rigidly connected to each other at joints, such as in the case of reinforced concrete frames. Under cyclic or ultimate loadings, allowable damages and deteriorations of elements at some joints can happen due to some stress concentrations. Such occurrences result in unsatisfactory response performances of the structural system since functionalities of the damaged elements are reduced considerably as depending on the degree of the damage rate. For a correct analysis, the damaged elements should be modeled to allow the damages in the element formulation. Steel structures are largely used in offshore structural industry because of their topological varieties, constructional and building flexibilities, well-known material properties, easy reparability, etc. These structures consist of a large number of tubular elements with various dimensions (diagonals and legs), which are joined to each other by welding that makes connections to be rigid. Diagonal members (braces) have relatively small dimensions and legs or chords have larger dimensions in general. Although the connections at joints are made by weld, the actual joint behavior under dynamic loadings, such as wave and earthquake loadings, is not fully rigid in the vicinity of connections due to local deformations of elements having large diameters [63, 64], or due to fatigue damages in the long term and also due to plastic deformations under ultimate loadings in the short term, which are schematically shown in Fig. 1.11. The phenomenon of the deterioration of elements can be taken into account in the analysis by using a computational model that allows flexibility at joints. It is assumed here that all deteriorations of an element are represented by massless spring systems, which allow flexibilities at the element ends. This subject has been studied by several investigators, see e.g., [65–71]. Most of these works deal with investigation of local flexibility of tubular members rather than addressing a full structural analysis procedure taking into account joint flexibilities. A fictitious element at the deteriorated joint [70] can be used to solve this problem, which considers local flexibilities in the system. This fictitious element may be derived as depending on actual member dimensions and joint configurations. However, the technique of using fictitious members introduces additional degrees of freedom that are not desirable in the analysis. A procedure which uses modified stiffness and mass matrices for flexibly connected elements is more practical and attractive [72, 73] since:

1. no additional degrees of freedom are introduced,
2. element-release and fixed-connection conditions can be directly obtained,
3. a general element-end condition in any direction can be easily specified,
4. a failure mechanism can be easily determined,

[1] The materials through pages (43–62) in Sect. 1.3 are taken partly from OMAE-2010 [73] and the publisher ASME is greatly acknowledged for granting permission.

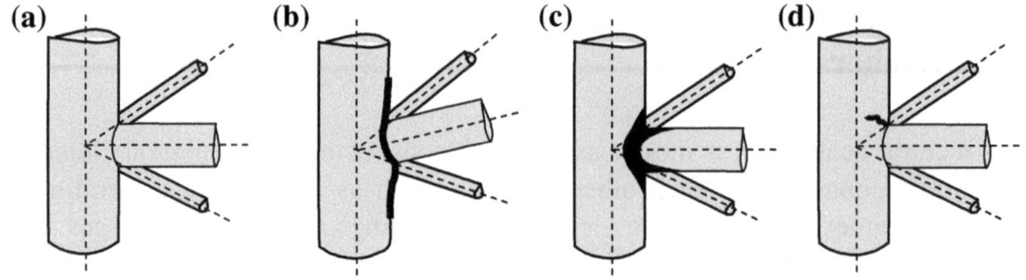

Fig. 1.11 Examples of deteriorations of an element at a joint. **a** An original joint. **b** Local deformation. **c** Plastic deformation. **d** Cracked deformation

5. in a reliability analysis, the influence of local flexibilities can be easily considered,
6. in the fatigue damage calculation the load carrying capacity of the element can be used until the whole cross-section of the element is damaged,
7. natural frequencies and mode shapes of damaged structural system can be estimated in terms of the natural frequencies and mode shapes of the undamaged structural system.

In this chapter, formulations of stiffness and mass matrices and consistent load vector of partly connected members, which are taken from [73], are presented. Parameters of the local flexibilities can be determined experimentally or analytically using a detailed FE analysis of related joints. It is assumed that member connectivity conditions are known or determined a priori.

1.3.1 Representation of a Partly Connected Beam Element

As it is mentioned in the previous section, all deteriorations in an element are represented by massless spring systems, and therefore they do not carry inertia forces, at the element ends as shown in Fig. 1.12. These spring systems are denoted by $[r_i]$ and $[r_j]$ at the element ends (i) and (j), which are assumed to be uncoupled, i.e., they include only diagonal terms and known a priori. The stiffness matrix $[k]$, mass matrix $[m]$ and the consistent load vector $\{p\}$ of the element (i–j) are known as explained in previous sections. In the local principal coordinates, the joint displacement vectors are respectively $\{d_i\}$ and $\{d_j\}$ at joints (i) and (j), $\{d'_i\}$ and $\left\{d'_j\right\}$ at joints (i') and (j') as shown in Fig. 1.12. Our purpose here is to find the stiffness matrix, consistent load vector, and mass matrix of the spring-beam element (i'–j') in the local principal coordinates, representing a deteriorated element in a structural system. As it is indicated in Fig. 1.12, the joints (i') and (j') are the nodal joints in the system, and the joints (i) and (j) are the internal joints of the element. The internal joints of the element will be eliminated from the system equilibrium equations. Relative displacement vectors of the spring systems at both

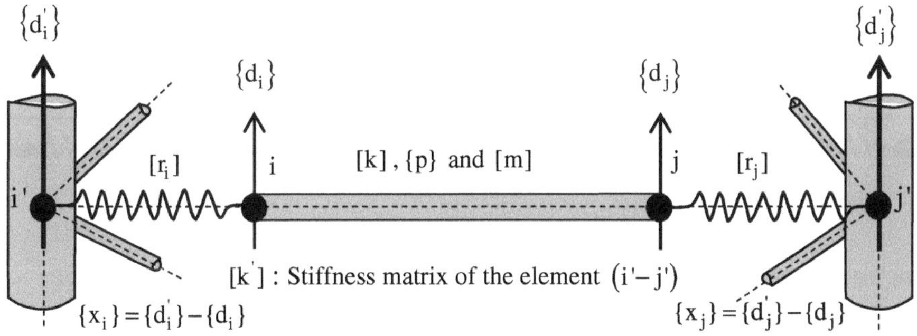

Fig. 1.12 Representation of a partly connected element in a structural system

ends of the structural element (*i–j*) are also defined as being $\{x_i\}$ and $\{x_j\}$ in the local principal coordinates, which are:

$$\begin{aligned} \{x_i\} &= \left(\{d_i'\} - \{d_i\}\right) \\ \{x_j\} &= \left(\{d_j'\} - \{d_j\}\right) \end{aligned} \tag{1.103}$$

For the whole spring-beam element the displacement vectors are:

$$\{d\} = \begin{Bmatrix} \{d_i\} \\ \{d_j\} \end{Bmatrix}, \{d'\} = \begin{Bmatrix} \{d_i'\} \\ \{d_j'\} \end{Bmatrix} \quad \text{and} \quad \{x\} = \begin{Bmatrix} \{x_i\} \\ \{x_j\} \end{Bmatrix} \tag{1.104}$$

Under a zero external loading condition, a relation between the displacement vectors, $\{d'\}$ and $\{d\}$, will be constructed. For this purpose, equilibrium equations at the internal joints, (*i*) and (*j*), are used. From Fig. 1.13 it can be written that,

$$\left.\begin{aligned} [r_i]\{x_i\} &= \{f_i\} \\ [r_j]\{x_j\} &= \{f_j\} \end{aligned}\right\} \rightarrow \left\{\begin{aligned} [r]\{x\} &= [k]\{d\} \\ \{x\} &= (\{d'\} - \{d\}) \end{aligned}\right. \quad \text{where} \quad [r] = \begin{bmatrix} [r_i] & 0 \\ 0 & [r_j] \end{bmatrix} \tag{1.105}$$

The total equilibrium equation of the spring-beam element can be stated as:

$$\begin{aligned} [r](\{d'\} - \{d\}) = [k]\{d\} \;\rightarrow\; [r]\{d'\} = ([k] + [r])\{d\} \text{ from which} \\ \{d'\} = \left(I_{12} + [r]^{-1}[k]\right)\{d\} \end{aligned} \tag{1.106}$$

where I_{12} is a unit matrix with the dimension of (12 × 12), and $[.]^{-1}$ denotes the inverse of a matrix. The transformation matrix between displacement vectors, $\{d'\}$ and $\{d\}$, can readily be written from Eq. (1.106) as:

$$\{d\} = [T]\{d'\} \quad \text{where} \quad [T] = \left(I_{12} + [r]^{-1}[k]\right)^{-1} \tag{1.107}$$

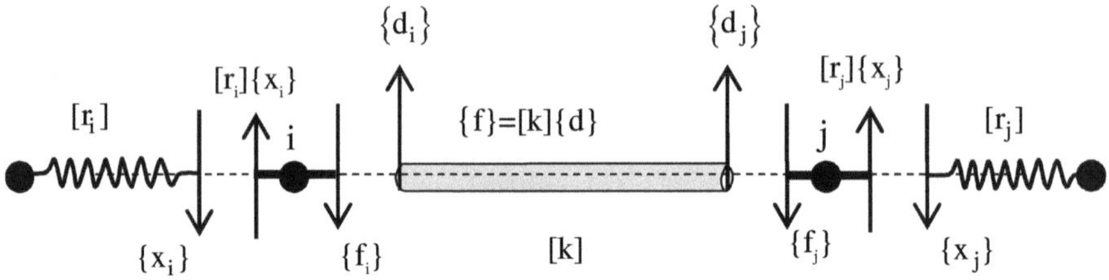

Fig. 1.13 Forces at internal joints of a spring-beam element under zero external loadings

which will be used in the formulation of the spring-beam element explained in the next section.

1.3.2 Formulation of Stiffness Matrix, Consistent Load Vector, and Mass Matrix of a Spring-Beam Element

Formulation of the stiffness matrix of a spring-beam element can be done in two different ways: (a) using equivalent forces of the spring-beam element at the joints (i') and (j'), (b) using variation of the total potential energy. These two formulation ways are explained below.

1. In the first alternative, the stiffness matrix of the spring-beam element is formulated by using equivalent forces at the element ends, (i') and (j'), i.e., at the element ends the spring and stiffness forces must be equal. Thus,

$$[r]\{x\} = [k']\{d'\} \tag{1.108a}$$

Having used the relative displacement vector $\{x\}$ from Eq. (1.105) and the displacement vector $\{d\}$ from Eq. (1.107) in Eq. (1.108a) the following relation can be obtained:

$$[r](I_{12} - [T])\{d'\} = [k']\{d'\} \tag{1.108b}$$

from which the stiffness matrix of a spring-beam element can be readily written as:

$$[k'] = [r](I_{12} - [T]) \to \left(\begin{array}{l} [k'][T]^{-1} = [r]\left([T]^{-1} - I_{12}\right) \\ [T]^{-1} = \left(I_{12} + [r]^{-1}[k]\right) \end{array} \right) \to \left\{ \begin{array}{l} [k] = [k'][T]^{-1} \\ [k'] = [k][T] \end{array} \right. \tag{1.109}$$

The consistent load vector is formulated using equivalent external works done by loads on the beam element (i–j) and the consistent load vector of the spring-beam element (i'–j'). Thus,

$$\left.\begin{array}{l}\{d'\}^T\{p'\} = \{d\}^T\{p\} \\ \{d\} = [T]\{d'\}\end{array}\right\} \rightarrow \{d'\}^T\{p'\} = \{d'\}^T[T]^T\{p\} \rightarrow \{p'\} = [T]^T\{p\} \tag{1.110}$$

2. In the second alternative, the stiffness matrix and consistent load vector of the spring-beam element can be formulated by using variation of the total potential energy. The total potential energy of the spring-beam element contains the strain energy, the energy stored in the spring system, and the work of external loads. It is expressed as:

$$\Pi = \frac{1}{2}\{d\}^T[k]\{d\} + \frac{1}{2}\{x\}^T[r]\{x\} - \{d\}^T(\{p\} + \{f\}) \tag{1.111}$$

Having used $\{d\}$ from Eq. (1.107) and $\{x\}$ from Eq. (1.105) in Eq. (1.111) the total potential energy becomes as:

$$\begin{aligned}\Pi =& \frac{1}{2}\{d'\}^T[T]^T[k][T]\{d'\} + \frac{1}{2}\{d'\}^T(I_{12} - [T])^T[r](I_{12} - [T])\{d'\} - \cdots \\ & \cdots - \{d'\}^T[T]^T(\{p\} + \{f\})\end{aligned} \tag{1.112a}$$

or

$$\begin{aligned}\Pi =& \frac{1}{2}\{d'\}^T\left[[T]^T[k][T] + (I_{12} - [T])^T[r](I_{12} - [T])\right]\{d'\} - \cdots \\ & \cdots - \{d'\}^T[T]^T(\{p\} + \{f\})\end{aligned} \tag{1.112b}$$

Since the total potential energy is stationary, its variation will be zero, i.e., $(\delta\Pi = 0)$ which leads to:

$$\begin{aligned}&\delta\{d'\}^T\begin{bmatrix}\left([T]^T[k][T] + (I_{12} - [T])^T[r](I_{12} - [T])\right)\{d'\} - \cdots \\ \cdots - [T]^T(\{p\} + \{f\})\end{bmatrix} = 0, \text{ or} \\ &\delta\{d'\}^T[[k']\{d'\} - (\{p'\} + \{f'\})] = 0\end{aligned} \tag{1.113}$$

from which the stiffness matrix $[k']$, the consistent load vector $\{p'\}$ and the vector of member internal forces $\{f'\}$ can be stated as:

$$\begin{aligned}[k'] &= [T]^T[k][T] + (I_{12} - [T])^T[r](I_{12} - [T]) \\ \{p'\} &= [T]^T\{p\} \\ \{f'\} &= [T]^T\{f\}\end{aligned} \tag{1.114}$$

Since $[r]$ is a diagonal matrix, the stiffness matrix $[k']$ will be symmetric. From the multiplication of $[k']$ by $[T]^{-1}$ from the right hand side it is obtained that,

$$[k'][T]^{-1} = [T]^T[k] + (I_{12} - [T])^T[r]\left([T]^{-1} - I_{12}\right) \tag{1.115a}$$

and using $[T]^{-1}$ from Eq. (1.109) in Eq. (1.115a) it is obtained that,

$$\begin{aligned} [k] &= [k'][T]^{-1} \\ [k'] &= [k][T] \end{aligned} \tag{1.115b}$$

which is the same as that given in Eq. (1.109).

The mass matrix of the spring-beam element is obtained from the total kinetic energy as similar to the stiffness matrix. Since the spring system is assumed to be massless, the total kinetic energy of the spring-beam element will be equal to that of the beam element given by Eq. (1.80b). Thus, using Eqs. (1.80b) and (1.107) the mass matrix of the spring-beam element can be obtained as written:

$$\left.\begin{aligned} T &= \frac{1}{2}\{\dot{d}'\}^T[T]^T[m]\,[T]\,\{\dot{d}'\} \\ T &= \frac{1}{2}\{\dot{d}'\}^T[m']\,\{\dot{d}'\} \end{aligned}\right\} \rightarrow [m'] = [T]^T[m]\,[T] \tag{1.116}$$

If the spring matrix $[r]$ is known, the transformation, or connectivity, matrix $[T]$ is calculated from Eq. (1.107). If, however, the connectivity matrix $[T]$ is provided directly rather than providing the spring matrix $[r]$, an equivalent spring matrix can also be calculated from Eq. (1.107). A unit value in a diagonal term of $[T]$, i.e., $(r_i = \infty)$, means that a rigid connection is made in this direction while a zero value, i.e., $(r_i = 0)$, indicates that a free connection is made, which produces zero member-end force accordingly.

The formulation of the spring-beam element can be summarized as follows:

$$\left.\begin{aligned} \{d\} &= [T]\{d'\} \\ \{x\} &= (I_{12} - [T])\{d'\} \\ [T] &= \left(I_{12} + [r]^{-1}[k]\right)^{-1} \end{aligned}\right\} \rightarrow \text{Connectivity relation} \tag{1.117a}$$

$$\left.\begin{aligned} [k] &= [k'][T]^{-1} \\ [k'] &= [k][T] \end{aligned}\right\} \rightarrow \text{stiffness relations} \tag{1.117b}$$

$$\left.\begin{aligned} \{p'\} &= [T]^T\{p\} \\ \{f'\} &= [T]^T\{f\} \end{aligned}\right\} \rightarrow \text{load and force relations} \tag{1.117c}$$

$$[m'] = [T]^T[m]\,[T]\} \rightarrow \text{mass relation} \tag{1.117d}$$

In these formulations, the same coordinate system for springs and the structural element must be maintained. If different coordinate systems are used, i.e., if the spring matrix is defined in a different coordinate system from the member coordinates, then they must be transformed to the same coordinates before the

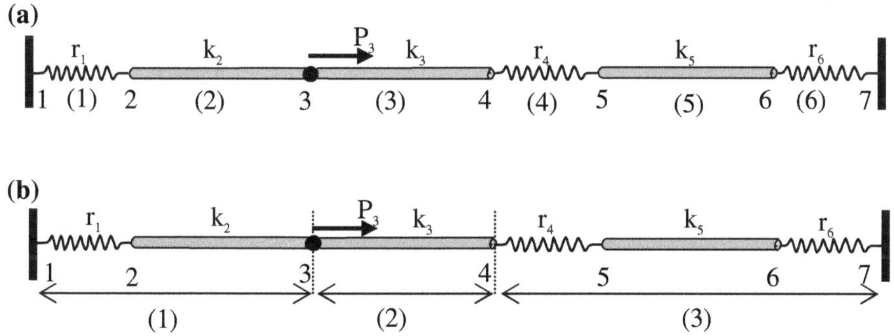

Fig. 1.14 An example of a simple bar structural system, **a** standard FE idealization, **b** spring-beam element idealization

aforementioned transformations are carried out. The calculation procedure is demonstrated by an example in the following section.

1.3.2.1 Spring-Beam Element Idealization of a Bar Structural System

As a demonstration of the spring-beam element, a simple bar structural system shown in Fig. 1.14, is analyzed. The system allows only axial deformation. It is assumed that joints 1 and 7 are fixed, i.e., displacements at these joints are zero. Displacements at joints (2–6) will be calculated under a point (concentrated) load applied at the joint 3 as shown in Fig. 1.14. There are two approaches for the solution, (a) using standard FE idealization, (b) using spring-beam element idealization, which are both presented below.

Standard FE Idealization

In this solution, all spring and solid elements are taken as being parts of the standard FE idealization with unknown displacements at the joints (2–6) as shown in Fig. 1.14a. For this simple system, the stiffness matrix of a spring element, $[r]$, and the stiffness matrix of a solid element, $[k]$, are expressed as,

$$[r] = \begin{bmatrix} r & -r \\ -r & r \end{bmatrix} \quad \text{and} \quad [k] = \begin{bmatrix} k & -k \\ -k & k \end{bmatrix} \tag{1.118}$$

Using the boundary conditions, i.e., ($d_1 = 0$) and ($d_7 = 0$), the system equation can be obtained as written by,

$$\begin{bmatrix} r_1+k_2 & -k_2 & 0 & 0 & 0 \\ -k_2 & k_2+k_3 & -k_3 & 0 & 0 \\ 0 & -k_3 & r_4+k_3 & -r_4 & 0 \\ 0 & 0 & -r_4 & r_4+k_5 & -k_5 \\ 0 & 0 & 0 & -k_5 & r_6+k_5 \end{bmatrix} \begin{Bmatrix} d_2 \\ d_3 \\ d_4 \\ d_5 \\ d_6 \end{Bmatrix} = \begin{Bmatrix} 0 \\ P_3 \\ 0 \\ 0 \\ 0 \end{Bmatrix} \tag{1.119}$$

From the solution of this equation the unknown displacements can be obtained in terms of the stiffness and spring constants as written by,

$$\begin{aligned} d_2 &= \frac{k_2\left(r_4+r_6+\dfrac{r_4r_6}{k_3}+\dfrac{r_4r_6}{k_5}\right)P_3}{\left[r_1k_2\left(r_4+r_6+\dfrac{r_4r_6}{k_3}+\dfrac{r_4r_6}{k_5}\right)+r_4r_6(r_1+k_2)\right]} \\ d_3 &= \frac{(r_1+k_2)\left(r_4+r_6+\dfrac{r_4r_6}{k_3}+\dfrac{r_4r_6}{k_5}\right)P_3}{\left[r_1k_2\left(r_4+r_6+\dfrac{r_4r_6}{k_3}+\dfrac{r_4r_6}{k_5}\right)+r_4r_6(r_1+k_2)\right]} \end{aligned} \tag{1.120a}$$

$$\begin{aligned} d_4 &= \frac{(r_1+k_2)\left(r_4+r_6+\dfrac{r_4r_6}{k_5}\right)P_3}{\left[r_1k_2\left(r_4+r_6+\dfrac{r_4r_6}{k_3}+\dfrac{r_4r_6}{k_5}\right)+r_4r_6(r_1+k_2)\right]} \\ d_5 &= \frac{r_4(r_1+k_2)(k_5+r_6)P_3}{k_5\left[r_1k_2\left(r_4+r_6+\dfrac{r_4r_6}{k_3}+\dfrac{r_4r_6}{k_5}\right)+r_4r_6(r_1+k_2)\right]} \\ d_6 &= \frac{r_4(r_1+k_2)P_3}{\left[r_1k_2\left(r_4+r_6+\dfrac{r_4r_6}{k_3}+\dfrac{r_4r_6}{k_5}\right)+r_4r_6(r_1+k_2)\right]} \end{aligned} \tag{1.120b}$$

In these statements, all special cases of the spring systems can be produced by varying the spring coefficients, r_1, r_4, and r_6. For example, for rigid connections, i.e., $(r_1=\infty)$, $(r_4=\infty)$ and $(r_6=\infty)$, the displacements will be:

$$\begin{matrix} d_2=0 \\ d_6=0 \\ d_4=d_5 \end{matrix} \quad \text{and} \quad \begin{aligned} d_3 &= \frac{(k_3+k_5)P_3}{[k_2(k_3+k_5)+k_3k_5]} \\ d_5 &= \frac{k_3P_3}{[k_2(k_3+k_5)+k_3k_5]} \end{aligned} \tag{1.120c}$$

Spring-Beam Element Idealization

In this solution, the system is considered as being consisted of three elements, (1–3), (3–4), and (4–7), as shown in Fig. 1.14b. The unknown displacements are d_3 and d_4. The rest, d_2, d_5, and d_6, will be calculated in terms of these unknown displacements. A general case of the spring-beam element for this system, which

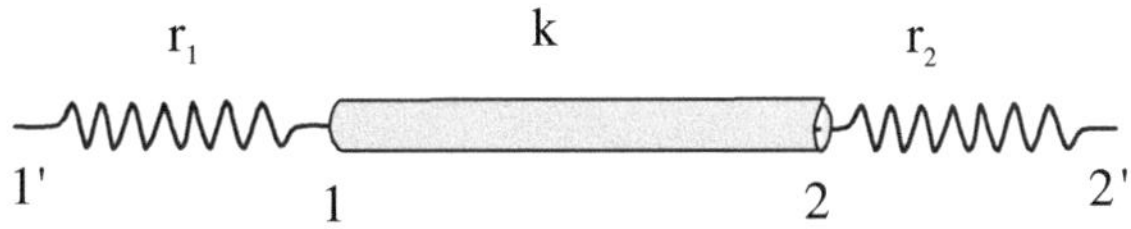

Fig. 1.15 A simple spring-beam element

allows only axial deformation, is shown in Fig. 1.15. The stiffness matrices of the solid and spring parts are written as,

$$[r] = \begin{bmatrix} r_1 & 0 \\ 0 & r_2 \end{bmatrix} \quad \text{and} \quad [k] = \begin{bmatrix} k & -k \\ -k & k \end{bmatrix} \tag{1.121}$$

Using Eq. (1.109) the inverse of the connectivity matrix, $[T]^{-1}$, and consequently the connectivity matrix, $[T]$ can be obtained as written:

$$\begin{aligned} [T]^{-1} &= \begin{bmatrix} \left(1+\dfrac{k}{r_1}\right) & -\dfrac{k}{r_1} \\ -\dfrac{k}{r_2} & \left(1+\dfrac{k}{r_2}\right) \end{bmatrix} \\ [T] &= \frac{1}{\left[1+k\left(\dfrac{1}{r_1}+\dfrac{1}{r_2}\right)\right]} \begin{bmatrix} \left(1+\dfrac{k}{r_2}\right) & \dfrac{k}{r_1} \\ \dfrac{k}{r_2} & \left(1+\dfrac{k}{r_1}\right) \end{bmatrix} \end{aligned} \tag{1.122}$$

Using Eq. (1.117b) the stiffness matrix of a general spring-beam element, $(1' - 2')$, for this simple system can be expressed as:

$$[k'] = \frac{1}{[1+k(1/r_1+1/r_2)]} \begin{bmatrix} k & -k \\ -k & k \end{bmatrix} \tag{1.123}$$

The connectivity matrices of the spring-beam elements are stated as,

$$\text{element (1)}: \; [T]_1 = \frac{1}{(1+k_2/r_1)} \begin{bmatrix} 1 & k_2/r_1 \\ 0 & (1+k_2/r_1) \end{bmatrix} \tag{1.124a}$$

$$\text{element (2)}: \; [T]_2 = \begin{bmatrix} 1 & 0 \\ 0 & 1 \end{bmatrix} \tag{1.124b}$$

$$\text{element (3)}: \; [T]_3 = \frac{1}{[1+k_5(1/r_4+1/r_6)]} \begin{bmatrix} (1+k_5/r_6) & k_5/r_4 \\ k_5/r_6 & (1+k_5/r_4) \end{bmatrix} \tag{1.124c}$$

The stiffness matrices of the spring-beam elements of the FE idealization of the system shown in Fig. 1.14b are calculated as written:

$$\text{element (1)}: \; [k']_1 = \frac{1}{(1+k_2/r_1)} \begin{bmatrix} k_2 & -k_2 \\ k_2 & k_2 \end{bmatrix} \tag{1.125a}$$

$$\text{element (2)}: \ [k']_2 = \begin{bmatrix} k_3 & -k_3 \\ -k_3 & k_3 \end{bmatrix} \tag{1.125b}$$

$$\text{element (3)}: \ [k']_3 = \frac{1}{[1 + k_5(1/r_4 + 1/r_6)]} \begin{bmatrix} k_5 & -k_5 \\ -k_5 & k_5 \end{bmatrix} \tag{1.125c}$$

After the assembly process of the elements (1), (2), and (3) and using the boundary conditions of ($d_1 = 0$) and ($d_7 = 0$), the system equilibrium equation can be obtained as stated:

$$[K]\{D\} = \{P\} \ \rightarrow \begin{bmatrix} k_3 + \dfrac{k_2}{\left(1 + \dfrac{k_2}{r_1}\right)} & -k_3 \\ -k_3 & k_3 + \dfrac{k_5}{\left[1 + k_5\left(\dfrac{1}{r_4} + \dfrac{1}{r_6}\right)\right]} \end{bmatrix} \begin{Bmatrix} d_3 \\ d_4 \end{Bmatrix} = \begin{Bmatrix} P_3 \\ 0 \end{Bmatrix} \tag{1.126}$$

from which the displacements d_3 and d_4 are calculated to be:

$$d_3 = \frac{(r_1 + k_2)(r_4 + r_6 + r_4 r_6/k_3 + r_4 r_6/k_5)P_3}{[r_1 k_2(r_4 + r_6 + r_4 r_6/k_3 + r_4 r_6/k_5) + r_4 r_6(r_1 + k_2)]} \tag{1.127a}$$

$$d_4 = \frac{(r_1 + k_2)[r_4 + r_6 + r_4 r_6/k_5]P_3}{[r_1 k_2(r_4 + r_6 + r_4 r_6/k_3 + r_4 r_6/k_5) + r_4 r_6(r_1 + k_2)]} \tag{1.127b}$$

The displacements (d_2), (d_5), and (d_6) are calculated from the connectivity relations in terms of (d_3) and (d_4), i.e., from:

$$\begin{Bmatrix} d_2 \\ d_3 \end{Bmatrix} = [T]_1 \begin{Bmatrix} d_1' \\ d_3' \end{Bmatrix} = \frac{1}{(1 + k_2/r_1)} \begin{bmatrix} 1 & k_2/r_1 \\ 0 & (1 + k_2/r_1) \end{bmatrix} \begin{Bmatrix} 0 \\ d_3 \end{Bmatrix} \tag{1.128a}$$

$$d_2 = k_2 \, d_3/(r_1 + k_2)$$

$$\begin{Bmatrix} d_5 \\ d_6 \end{Bmatrix} = [T]_3 \begin{Bmatrix} d_5' \\ d_6' \end{Bmatrix} = \frac{1}{[1 + k_5(1/r_4 + 1/r_6)]} \begin{bmatrix} (1 + k_5/r_6) & k_5/r_4 \\ k_5/r_6 & (1 + k_5/r_4) \end{bmatrix} \begin{Bmatrix} d_4 \\ 0 \end{Bmatrix}$$

$$d_5 = \frac{r_4(r_6 + k_5)}{[r_4 r_6 + k_5(r_4 + r_6)]} d_4$$

$$d_6 = \frac{r_4 k_5}{[r_4 r_6 + k_5(r_4 + r_6)]} d_4 \tag{1.128b}$$

which produce the same results obtained in the first solution by the standard FE idealization.

1.3.3 Calculation of the Connectivity Matrix

For a given or assumed spring system, the connectivity matrix of a partly connected element, $[T]$, which is given in Eq. (1.117a), will be calculated. For convenience it is rewritten below.

$$[T] = \left(I_{12} + [r]^{-1}[k]\right)^{-1} \rightarrow [T]^{-1} = \left(I_{12} + [r]^{-1}[k]\right) \tag{1.129}$$

As it can be realized from this equation, its inverse $[T]^{-1}$ will be formed first, and then using the reverse inversion the matrix $[T]$ will be calculated. The stiffness matrix $[k]$ is given by Eq. (1.73a) in the local coordinates and the nonzero diagonal terms of the spring matrix $[r]$ are stored in a vector, say $\{r\}$, as written:

$$\{r\}^T = \{r_1,\ r_2,\ r_3,\ r_4,\ r_5,\ r_6,\ r_7,\ r_8,\ r_9,\ r_{10},\ r_{11},\ r_{12}\} \tag{1.130}$$

If the release conditions are specified in the principal local coordinates, the stiffness matrix $[k]$ can also be stated in a different form for the simplicity by rearranging the order of nodal degrees of freedom as written:

$$[k_*] = \begin{bmatrix} [k_*]_1 & 0 & 0 \\ 0 & [k_*]_2 & 0 \\ 0 & 0 & [k_*]_3 \end{bmatrix} \tag{1.131}$$

where $[k_*]$ indicates reordered form of $[k]$. The submatrices in Eq. (1.131) are expressed below with the corresponding degrees of freedom (DOF) of the element.

$$[k_*]_1 = \begin{array}{c} \\ 1 \\ 7 \\ 4 \\ 10 \end{array} \begin{array}{c} \begin{array}{cccc} 1 & 7 & 4 & 10 \end{array} \\ \begin{bmatrix} k_{11} & -k_{11} & 0 & 0 \\ -k_{11} & k_{11} & 0 & 0 \\ 0 & 0 & k_{44} & -k_{44} \\ 0 & 0 & -k_{44} & k_{44} \end{bmatrix} \end{array} \tag{1.132a}$$

$$[k_*]_2 = \begin{array}{c} \\ 3 \\ 5 \\ 9 \\ 11 \end{array} \begin{array}{c} \begin{array}{cccc} 3 & 5 & 9 & 11 \end{array} \\ \begin{bmatrix} k_{33} & k_{35} & -k_{33} & k_{35} \\ k_{35} & k_{55} & -k_{35} & k_{511} \\ -k_{33} & -k_{35} & k_{33} & -k_{35} \\ k_{35} & k_{511} & -k_{35} & k_{55} \end{bmatrix} \end{array} \tag{1.132b}$$

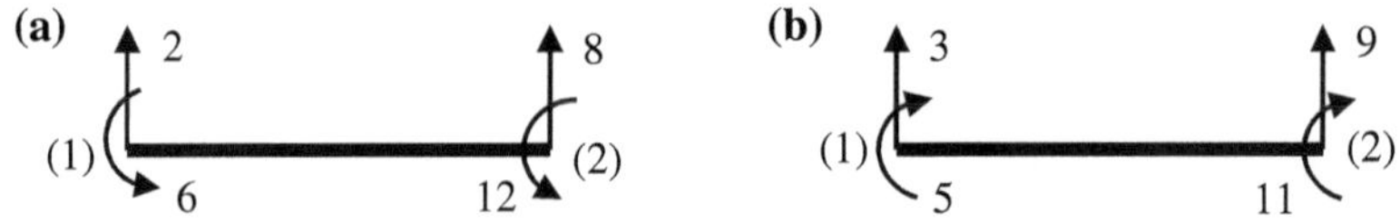

Fig. 1.16 Degrees of freedom of the submatrices $[k_*]_3$ and $[k_*]_2$ defined by Eqs. (1.132c, b). **a** DOF for $[k_*]_3$ in (*x*–*y*) plane. **b** DOF for $[k_*]_2$ in (*x*–*z*) plane

$$[k_*]_3 = \begin{array}{c} \\ 2 \\ 6 \\ 8 \\ 12 \end{array} \begin{array}{c} \begin{array}{cccc} 2 & 6 & 8 & 12 \end{array} \\ \begin{bmatrix} k_{22} & k_{26} & -k_{22} & k_{26} \\ k_{26} & k_{66} & -k_{26} & k_{612} \\ -k_{22} & -k_{26} & k_{22} & -k_{26} \\ k_{26} & k_{612} & -k_{26} & k_{66} \end{bmatrix} \end{array} \quad (1.132c)$$

The submatrix $[k_*]_1$ contains axial and torsional degrees of freedom (DOF) only. The submatrices $[k_*]_3$and $[k_*]_2$ contain translational and rotational DOF in transverse directions as shown in Fig. 1.16. In the reordered form of the element DOF, the connectivity matrix, which is indicated by $[T_*]$, and its inverse $[T_*]^{-1}$ can be expressed as:

$$[T_*] = \begin{bmatrix} [T_*]_1 & 0 & 0 \\ 0 & [T_*]_2 & 0 \\ 0 & 0 & [T_*]_3 \end{bmatrix} \rightarrow [T_*]^{-1} = \begin{bmatrix} [T_*]_1^{-1} & 0 & 0 \\ 0 & [T_*]_2^{-1} & 0 \\ 0 & 0 & [T_*]_3^{-1} \end{bmatrix} \quad (1.133)$$

where the submatrices, $[T_*]_1^{-1}$, $[T_*]_2^{-1}$ and $[T_*]_3^{-1}$ can be obtained from Eq. (1.129) as written by,

$$[T_*]_1^{-1} = \begin{array}{c} \\ 1 \\ 7 \\ 4 \\ 10 \end{array} \begin{array}{c} \begin{array}{cccc} 1 & 7 & 4 & 10 \end{array} \\ \begin{bmatrix} (1 + k_{11}/r_1) & -k_{11}/r_1 & 0 & 0 \\ -k_{11}/r_7 & (1 + k_{11}/r_7) & 0 & 0 \\ 0 & 0 & (1 + k_{44}/r_4) & -k_{44}/r_4 \\ 0 & 0 & -k_{44}/r_{10} & (1 + k_{44}/r_{10}) \end{bmatrix} \end{array} \quad (1.134a)$$

$$[T_*]_2^{-1} = \begin{array}{c} \\ 3 \\ 5 \\ 9 \\ 11 \end{array} \begin{array}{c} \begin{array}{cccc} 3 & 5 & 9 & 11 \end{array} \\ \begin{bmatrix} (1 + k_{33}/r_3) & k_{35}/r_3 & -k_{33}/r_3 & k_{35}/r_3 \\ k_{35}/r_5 & (1 + k_{55}/r_5) & -k_{35}/r_5 & k_{511}/r_5 \\ -k_{33}/r_9 & -k_{35}/r_9 & (1 + k_{33}/r_9) & -k_{35}/r_9 \\ k_{35}/r_{11} & k_{511}/r_{11} & -k_{35}/r_{11} & (1 + k_{55}/r_{11}) \end{bmatrix} \end{array} \quad (1.134b)$$

$$[T_*]_3^{-1} = \begin{array}{c} \\ 2 \\ 6 \\ 8 \\ 12 \end{array} \begin{array}{c} \begin{array}{cccc} 2 & 6 & 8 & 12 \end{array} \\ \begin{bmatrix} (1+k_{22}/r_2) & k_{26}/r_2 & -k_{22}/r_2 & k_{26}/r_2 \\ k_{26}/r_6 & (1+k_{66}/r_6) & -k_{26}/r_6 & k_{612}/r_6 \\ -k_{22}/r_8 & -k_{26}/r_8 & (1+k_{22}/r_8) & -k_{26}/r_8 \\ k_{26}/r_{12} & k_{612}/r_{12} & -k_{26}/r_{12} & (1+k_{66}/r_{12}) \end{bmatrix} \end{array} \tag{1.134c}$$

Since calculations of $[T_*]_1$, $[T_*]_2$ and $[T_*]_3$ are more efficient and simpler than the calculation of $[T]$ using Eq. (1.129), the reordered form of DOF is preferably used. If, however, the release conditions are specified in a different coordinate system than the principal local coordinates, then the transformation matrix $[T]$ will be calculated using Eq. (1.129). Here, we assume that the release conditions are specified in the local principal coordinates and the reordered form of $[T]$ is used. In this case, the calculation of $[T_*]_1$ can easily be carried out analytically using Eq. (1.134a). The result is written as:

$$[T_*]_1 = \begin{array}{c} \\ 1 \\ 7 \\ 4 \\ 10 \end{array} \begin{array}{c} \begin{array}{cccc} 1 & 7 & 4 & 10 \end{array} \\ \begin{bmatrix} t_{11} & t_{17} & 0 & 0 \\ t_{71} & t_{77} & 0 & 0 \\ 0 & 0 & t_{44} & t_{410} \\ 0 & 0 & t_{104} & t_{1010} \end{bmatrix} \end{array} \tag{1.135a}$$

where the elements of the matrix are obtained as stated below.

$$\begin{aligned} t_{11} &= \frac{r_1(r_7+k_{11})}{[r_1r_7+k_{11}(r_1+r_7)]} & & t_{44} = \frac{r_4(r_{10}+k_{44})}{[r_4r_{10}+k_{44}(r_4+r_{10})]} \\ t_{17} &= \frac{r_7k_{11}}{[r_1r_7+k_{11}(r_1+r_7)]} & & t_{410} = \frac{r_{10}k_{44}}{[r_4r_{10}+k_{44}(r_4+r_{10})]} \\ t_{71} &= \frac{r_1k_{11}}{[r_1r_7+k_{11}(r_1+r_7)]} & \text{and} \quad & t_{104} = \frac{r_4k_{44}}{[r_4r_{10}+k_{44}(r_4+r_{10})]} \\ t_{77} &= \frac{r_7(r_1+k_{11})}{[r_1r_7+k_{11}(r_1+r_7)]} & & t_{1010} = \frac{r_{10}(r_4+k_{44})}{[r_4r_{10}+k_{44}(r_4+r_{10})]} \end{aligned} \tag{1.135b}$$

For nonzero spring coefficients, the submatrices $[T_*]_2$ and $[T_*]_3$ will be numerically calculated from the reverse inversion of $[T_*]_2^{-1}$ and $[T_*]_3^{-1}$ using Eqs. (1.134b, c). In this numerical calculation, some special conditions of the spring coefficients may occur as pointed out next. In the structural analysis, most elements are rigidly connected and also there may be some elements that partly connected in some directions with assumed spring coefficients. A rigid connection can be made by using an infinitely large spring coefficient in the corresponding direction. This produces a unit diagonal term and zero off-diagonal terms of the related matrix,

$[T_*]_2^{-1}$ or $[T_*]_3^{-1}$ so that the numerical calculation of the reverse inversion can be carried out without any difficulty. However, for a zero-spring condition, i.e., a loss or no-connection case (fully released), the numerical inversion of $[T_*]_2^{-1}$ or $[T_*]_3^{-1}$ cannot be carried out simply since, in the fully released direction, infinite values in $[T_*]_2^{-1}$ or $[T_*]_3^{-1}$ are obtained. The solution of this special case is presented in the following section.

1.3.3.1 The Case of Zero-Spring (Fully Released) Conditions

As mentioned above, in the case of zero-spring values in directions of some DOF, the inversion of the matrix $[T_*]_2^{-1}$ or $[T_*]_3^{-1}$ cannot be carried out numerically. This problem can be solved in an alternative way. It is assumed for the generality that a combination of nonzero spring and zero-spring conditions are considered. The solution procedure is performed in two steps as explained below.

1. First, it is assumed that the element is rigidly connected in loss or fully released directions (directions with zero springs) and it is partly connected in directions of nonzero springs in the reordered form of DOF. Using Eqs. (1.117a, b) the related transformations of this step are given as:

$$\left.\begin{aligned} \{d_*\}_i &= [T_*]_{i\,(r=\infty)}\{d'_*\}_{i(r=\infty)} \\ [k'_*]_{i\,(r=\infty)} &= [k_*]_i[T_*]_{i\,(r=\infty)} \end{aligned}\right\} \rightarrow \ (i = 2 \text{ or } 3) \qquad (1.136a)$$

The matrix $[T_*]_{i\,(r=\infty)}$ is calculated from the inversion of the matrix $[T_*]_2^{-1}$ or $[T_*]_3^{-1}$ whichever is applicable, with unity in diagonal and zero values in off-diagonal terms in fully released (with zero-spring values) directions, $[k_*]_i$ denotes $[k_*]_2$ or $[k_*]_3$, whichever is applicable, given by Eq. (1.132b) or (c).

2. In the second step, it is assumed that the element is rigidly connected in all directions, except fully released or loss directions since all springs are included in the first step. Loss directions indicate zero connectivities in associated directions and they have to be released accordingly to obtain the original release conditions. From this release operation, another transformation matrix is obtained with the related transformations written as:

$$\left.\begin{aligned} \{d'_*\}_{i(r=\infty)} &= [T_*]_{i\,(r=0)}\{d'_*\}_i \\ [k'_*]_i &= [T_*]^T_{i\,(r=0)}[k'_*]_{i\,(r=\infty)}[T_*]_{i\,(r=0)} \end{aligned}\right\} \rightarrow \ (i = 2 \text{ or } 3) \qquad (1.136b)$$

where the stiffness matrix $[k'_*]_{i\,(r=\infty)}$ is the same as calculated above in step (a). The transformation matrix $[T_*]_{i\,(r=0)}$ will be calculated by using the criterion that stiffness forces are all zero in fully released directions. In order to calculate this matrix easily, the DOF of the displacement vector $\{d'_*\}_{i(r=\infty)}$ are rearranged in the order of released directions first and then rigidly connected directions. The displacement vector with rearranged DOF is denoted by $\{d'_{**}\}_{i(r=\infty)}$ and the corresponding stiffness matrix is denoted by $[k'_{**}]_{i\,(r=\infty)}$ as stated by,

$$\left.\begin{aligned} \{d'_{**}\}_{i(r=\infty)} &= \begin{Bmatrix} \{d'_{**}\}_1 \\ \{d'_{**}\}_2 \end{Bmatrix}_{i\,(r=\infty)} \\ [k'_{**}]_{i\,(r=\infty)} &= \begin{bmatrix} [k'_{**}]_{11} & [k'_{**}]_{12} \\ [k'_{**}]^T_{12} & [k'_{**}]_{22} \end{bmatrix}_{i\,(r=\infty)} \end{aligned}\right\} \rightarrow (i = 2 \text{ or } 3) \tag{1.136c}$$

where $\{d'_{**}\}_1$ is the displacement vector in the released directions and $\{d'_{**}\}_2$ is that in the rigidly connected directions. The stiffness forces of this system are calculated from the following equation:

$$\begin{bmatrix} [k'_{**}]_{11} & [k'_{**}]_{12} \\ [k'_{**}]^T_{12} & [k'_{**}]_{22} \end{bmatrix}_{i\,(r=\infty)} \begin{Bmatrix} \{d'_{**}\}_1 \\ \{d'_{**}\}_2 \end{Bmatrix}_{i\,(r=\infty)} = \begin{Bmatrix} 0 \\ \{f'_{**}\}_2 \end{Bmatrix}_{i\,(r=\infty)} \rightarrow (i = 2, 3) \tag{1.136d}$$

The displacement vector $\{d'_{**}\}_1$ in the released directions can be calculated in terms of the displacements vector $\{d'_{**}\}_2$ in rigidly connected directions as written from Eq. (1.136d):

$$[k'_{**}]_{11}\{d'_{**}\}_1 + [k'_{**}]_{12}\{d'_{**}\}_2 = 0 \rightarrow \{d'_{**}\}_1 = -[k'_{**}]^{-1}_{11}\,[k'_{**}]_{12}\,\{d'_{**}\}_2 \tag{1.136e}$$

Having introduced $\{d'_{**}\}_1$ from Eq. (1.136e) into Eq. (1.136c) the displacement vector $\{d'_{**}\}_{i(r=\infty)}$ can be expressed as written:

$$\left.\begin{aligned} &\begin{Bmatrix} \{d'_{**}\}_1 \\ \{d'_{**}\}_2 \end{Bmatrix}_{i,\,(r=\infty)} = \begin{bmatrix} 0 & -[k'_{**}]^{-1}_{11}[k'_{**}]_{12} \\ 0 & [I] \end{bmatrix}_{i\,(r=0)} \begin{Bmatrix} \{d'_{**}\}_1 \\ \{d'_{**}\}_2 \end{Bmatrix}_i \\ &\text{or} \\ &\{d'_{**}\}_{i(r=\infty)} = [T_{**}]_{i\,(r=0)}\{d'_{**}\}_i \end{aligned}\right\} \rightarrow (i = 2, 3) \tag{1.136f}$$

from which the transformation matrix $[T_{**}]_{i\,(r-0)}$ is defined as:

$$[T_{**}]_{i\,(r=0)} = \begin{bmatrix} 0 & -[k'_{**}]^{-1}_{11}[k'_{**}]_{12} \\ 0 & [I] \end{bmatrix}_{i\,(r=0)} \tag{1.136g}$$

Next step is to rearrange the DOF to obtain the previous sequence of displacements, i.e., $\{d'_*\}_{i(r=\infty)}$, and accordingly to reorder $[T_{**}]_{i\,(r=0)}$ to obtain the required transformation matrix $[T_*]_{i\,(r=0)}$, which will be used in Eq. (1.136b). Unlike the stiffness matrix, in order to calculate the concentrated load vector and the mass matrix for this special case, the transformation matrix $[T_*]_i$ between the displacement vectors, $\{d_*\}_i$ and $\{d'_*\}_i$, must be constructed. For this purpose, Eqs. (1.136a, b) are used. The results are as written:

$$\left.\begin{aligned} \{d_*\}_i &= [T_*]_i\{d'_*\}_i \\ [T_*]_i &= [T_*]_{i\,(r=\infty)}[T_*]_{i\,(r=0)} \end{aligned}\right\} \rightarrow (i = 2 \text{ or } 3) \tag{1.137}$$

The calculation procedure of this special case is summarized below.

$$\left.\begin{aligned} \{d_*\}_i &= [T_*]_i\{d'_*\}_i \\ [T_*]_i &= [T_*]_{i\,(r=\infty)}[T_*]_{i\,(r=0)} \end{aligned}\right\} \rightarrow \text{Connectivity relation} \tag{1.138a}$$

$$[k'_*]_i = [T_*]^T_{i\,(r=0)}[k'_*]_{i\,(r=\infty)}[T_*]_{i\,(r=0)} \rightarrow \text{Stiffness matrix} \tag{1.138b}$$

$$\left.\begin{aligned} \{p'_*\}_i &= [T_*]_i^T\{p_*\} \\ \{f'_*\}_i &= [T_*]_i^T\{f_*\} \end{aligned}\right\} \rightarrow \text{Load and force vectors} \tag{1.138c}$$

$$[m'_*] = [T_*]_i^T[m_*]\,[T_*]_i \rightarrow \text{Mass matrix} \tag{1.138d}$$

where (i = 2 or 3). The calculation procedure is explained by an example in the next section.

1.3.3.2 Example of Transformation and Stiffness Matrices of a Spring-Beam Element with Fully Released and Partly Connected DOF

The calculation procedure of the spring-beam element with fully released and partly connected conditions is explained in this example. For this purpose, it is assumed that the DOF, 6 and 12, in Fig. 1.16a are fully released and the DOF, 2 and 8, are partly connected with spring coefficients, r_2 and r_8. The transformation matrix, $[T_*]_3$, and the stiffness matrix, $[k'_*]_3$, will be calculated analytically step by step to explain the aforementioned calculation procedure.

Step 1: The transformation matrix, $[T_*]_{3\,(r=\infty)}$, which is given in Eq. (1.136a), will be calculated using $[T_*]_3^{-1}$ provided that the spring coefficients in the released directions are infinite, i.e., r_6 and r_{12} are infinite (rigid connection). For this special case, $[T_*]^{-1}_{3(r=\infty)}$ can be stated using Eq. (1.134c) as written by,

$$[T_*]^{-1}_{3(r=\infty)} = \begin{matrix} & 2 & 6 & 8 & 12 \\ 2 & (1+k_{22}/r_2) & k_{26}/r_2 & -k_{22}/r_2 & k_{26}/r_2 \\ 6 & 0 & 1 & 0 & 0 \\ 8 & -k_{22}/r_8 & -k_{26}/r_8 & (1+k_{22}/r_8) & -k_{26}/r_8 \\ 12 & 0 & 0 & 0 & 1 \end{matrix} \tag{1.139a}$$

From the reverse inversion of $[T_*]^{-1}_{3(r=\infty)}$, the transformation matrix $[T_*]_{3\,(r=\infty)}$ can be found as written:

$$[T_*]_{3(r=\infty)} = \Delta \begin{matrix} & 2 & 6 & 8 & 12 \\ 2 & (1+k_{22}/r_8) & -k_{26}/r_2 & k_{22}/r_2 & -k_{26}/r_2 \\ 6 & 0 & 1/\Delta & 0 & 0 \\ 8 & k_{22}/r_8 & k_{26}/r_8 & (1+k_{22}/r_2) & k_{26}/r_8 \\ 12 & 0 & 0 & 0 & 1/\Delta \end{matrix}$$

$$\text{where} \quad \Delta = r_2 r_8/[r_2 r_8 + k_{22}(r_2+r_8)] \tag{1.139b}$$

The stiffness matrix $[k'_*]_{3\,(r=\infty)}$ will be calculated using Eq. (1.136a), in which $(i=3)$ and $[k_*]_3$ is as given by Eq. (1.132c). The result is:

$$[k'_*]_{3\,(r=\infty)} = \Delta \begin{matrix} & 2 & 6 & 8 & 12 \\ 2 & k_{22} & k_{26} & -k_{22} & k_{26} \\ 6 & k_{26} & k_{*11} & -k_{26} & k_{*12} \\ 8 & -k_{22} & -k_{26} & k_{22} & -k_{26} \\ 12 & k_{26} & k_{*12} & -k_{26} & k_{*11} \end{matrix} \tag{1.140a}$$

in which the stiffness terms, k_{*11} and k_{*12}, are defined as:

$$\begin{aligned} k_{*11} &= \left(\frac{1}{r_2}+\frac{1}{r_8}\right)\left(k_{22}k_{66}-k_{26}^2\right)+k_{66} \\ k_{*12} &= \left(\frac{1}{r_2}+\frac{1}{r_8}\right)\left(k_{22}k_{612}-k_{26}^2\right)+k_{612} \end{aligned} \tag{1.140b}$$

Step 2: The element possessing the stiffness matrix $[k'_*]_{3\,(r=\infty)}$ is now released. For this operation, the element DOF are rearranged so that DOF in the released directions (6 and 12) are replaced in the first order and DOF in the rigidly connected directions (2 and 8) are in the second order. The associated stiffness matrix is denoted by $[k'_{**}]_{3\,(r=\infty)}$ as stated:

$$[k'_{**}]_{3\,(r=\infty)} = \begin{matrix} & 6 & 12 & 2 & 8 \\ 6 \\ 12 \\ 2 \\ 8 \end{matrix} \Delta \begin{bmatrix} k_{*11} & k_{*12} & k_{26} & -k_{26} \\ k_{*12} & k_{*11} & k_{26} & -k_{26} \\ k_{26} & k_{26} & k_{22} & -k_{22} \\ -k_{26} & -k_{26} & -k_{22} & k_{22} \end{bmatrix} \tag{1.141a}$$

From this stiffness matrix, the submatrices $[k'_{**}]_{11}$, $[k'_{**}]_{12}$ and $[k'_{**}]_{22}$, which are given in Eq. (1.136c), can be easily expressed as:

$$\begin{aligned} &[k'_{**}]_{11} = \Delta \begin{bmatrix} k_{*11} & k_{*12} \\ k_{*12} & k_{*11} \end{bmatrix}, \quad [k'_{**}]_{12} = k_{26}\,\Delta \begin{bmatrix} 1 & -1 \\ 1 & -1 \end{bmatrix} \\ &[k'_{**}]_{22} = k_{22}\,\Delta \begin{bmatrix} 1 & -1 \\ 1 & -1 \end{bmatrix} \end{aligned} \tag{1.141b}$$

The inverse of $[k'_{**}]_{11}$ can be expressed as written by,

$$[k'_{**}]_{11}^{-1} = \frac{1}{\Delta\Delta_*} \begin{bmatrix} k_{*11} & -k_{*12} \\ -k_{*12} & k_{*11} \end{bmatrix} \rightarrow \begin{cases} \Delta = \dfrac{r_2 r_8}{[r_2 r_8 + k_{22}(r_2 + r_8)]} \\ \Delta_* = k_{*11}^2 - k_{*12}^2 \end{cases} \tag{1.141c}$$

Having introduced $[k'_{**}]_{11}^{-1}$ from Eq. (1.141c) and $[k'_{**}]_{12}$ from Eq. (1.141b) into Eq. (1.136g) the transformation matrix, $[T_{**}]_{3\,(r=0)}$ can be obtained as written by,

$$[T_{**}]_{3\,(r=0)} = \begin{matrix} & 6 & 12 & 2 & 8 \\ 6 \\ 12 \\ 2 \\ 8 \end{matrix} \begin{bmatrix} 0 & 0 & -k_{26}/(k_{*11}+k_{*12}) & k_{26}/(k_{*11}+k_{*12}) \\ 0 & 0 & -k_{26}/(k_{*11}+k_{*12}) & k_{26}/(k_{*11}+k_{*12}) \\ 0 & 0 & 1 & 0 \\ 0 & 0 & 0 & 1 \end{bmatrix} \tag{1.142a}$$

This transformation matrix is now rearranged according to the order of DOF used above in step (a). This new form is denoted by $[T_*]_{3\,(r=0)}$ and stated as:

$$[T_*]_{3\,(r=0)} = \begin{matrix} & 2 & 6 & 8 & 12 \\ 2 \\ 6 \\ 8 \\ 12 \end{matrix} \begin{bmatrix} 1 & 0 & 0 & 0 \\ -k_{26}/(k_{*11}+k_{*12}) & 0 & k_{26}/(k_{*11}+k_{*12}) & 0 \\ 0 & 0 & 1 & 0 \\ -k_{26}/(k_{*11}+k_{*12}) & 0 & k_{26}/(k_{*11}+k_{*12}) & 0 \end{bmatrix} \tag{1.142b}$$

The corresponding stiffness matrix is calculated using Eq. (1.138b) as written by,

$$[k'_*]_3 = \begin{array}{c} \\ 2 \\ 6 \\ 8 \\ 12 \end{array} \begin{array}{c} \begin{array}{cccc} 2 & 6 & 8 & 12 \end{array} \\ \begin{bmatrix} k'_{*11} & 0 & -k'_{*11} & 0 \\ 0 & 0 & 0 & 0 \\ -k'_{*11} & 0 & k'_{*11} & 0 \\ 0 & 0 & 0 & 0 \end{bmatrix} \end{array} \rightarrow k'_{*11} = \Delta\left[k_{22} - \frac{2k_{26}^2}{(k_{*11} + k_{*12})}\right] \tag{1.143}$$

The transformation matrix, which is used to calculate the consistent load vector and mass matrix, is calculated using Eq. (1.138a) as written by,

$$[T_*]_3 = \begin{array}{c} \\ 2 \\ 6 \\ 8 \\ 12 \end{array} \Delta \begin{array}{c} \begin{array}{cccc} 2 & 6 & 8 & 12 \end{array} \\ \begin{bmatrix} 1 + \dfrac{k_{22}}{r_8} + \dfrac{2k_{26}^2}{r_2(k_{*11} + k_{*12})} & 0 & \dfrac{k_{22}}{r_2} - \dfrac{2k_{26}^2}{r_2(k_{*11} + k_{*12})} & 0 \\ -\dfrac{k_{26}}{\Delta(k_{*11} + k_{*12})} & 0 & \dfrac{k_{26}}{\Delta(k_{*11} + k_{*12})} & 0 \\ \dfrac{k_{22}}{r_8} - \dfrac{2k_{26}^2}{r_8(k_{*11} + k_{*12})} & 0 & 1 + \dfrac{k_{22}}{r_2} + \dfrac{2k_{26}^2}{r_8(k_{*11} + k_{*12})} & 0 \\ -\dfrac{k_{26}}{\Delta(k_{*11} + k_{*12})} & 0 & \dfrac{k_{26}}{\Delta(k_{*11} + k_{*12})} & 0 \end{bmatrix} \end{array} \tag{1.144}$$

As a special case, when r_2 and r_8 approach infinity, i.e.,($r_2 \rightarrow \infty$ and $r_8 \rightarrow \infty$), the beam shown in Fig. 1.16a becomes a simply supported beam, i.e., the beam is hinged at both ends. In this case, the transformation matrix $[T_*]_3$ and the stiffness matrix $[k_*]_3$ become as written by,

$$[T_*]_3 = \begin{array}{c} \\ 2 \\ 6 \\ 8 \\ 12 \end{array} \begin{array}{c} \begin{array}{cccc} 2 & 6 & 8 & 12 \end{array} \\ \begin{bmatrix} 1 & 0 & 0 & 0 \\ -k_{26}/(k_{66} + k_{612}) & 0 & k_{26}/(k_{66} + k_{612}) & 0 \\ 0 & 0 & 1 & 0 \\ -k_{26}/(k_{66} + k_{612}) & 0 & k_{26}/(k_{66} + k_{612}) & 0 \end{bmatrix} \end{array} \tag{1.145a}$$

$$[k'_*]_3 = \begin{array}{c} \\ 2 \\ 6 \\ 8 \\ 12 \end{array} \begin{array}{c} \begin{array}{cccc} 2 & 6 & 8 & 12 \end{array} \\ \begin{bmatrix} k'_{*11} & 0 & -k'_{*11} & 0 \\ 0 & 0 & 0 & 0 \\ -k'_{*11} & 0 & k'_{*11} & 0 \\ 0 & 0 & 0 & 0 \end{bmatrix} \end{array} \rightarrow k'_{*11} = \left[k_{22} - \frac{2k_{26}^2}{(k_{66} + k_{612})}\right] \tag{1.145b}$$

The stiffness terms k_{22}, k_{26}, k_{66} and k_{612} are extracted from Eqs. (1.73b, c) as written by,

$$\begin{matrix} k_{22} = 12EI_z\mu_y/\ell^3 \\ k_{26} = 6EI_z\mu_y/\ell^2 \end{matrix} \quad \text{and} \quad \begin{matrix} k_{66} = EI_z(3\mu_y + 1)/\ell \\ k_{612} = EI_z(3\mu_y - 1)/\ell \end{matrix} \tag{1.145c}$$

Having used Eq. (1.145c) in Eq. (1.145b) it is obtained that $\left(k'_{*11} = 0\right)$, which results in a zero stiffness matrix as expected. For these values of k_{22}, k_{26}, k_{66} and k_{612} the transformation matrix $[T_*]_3$ is obtained as written:

$$[T_*]_3 = \begin{matrix} & 2 & 6 & 8 & 12 \\ 2 & 1 & 0 & 0 & 0 \\ 6 & -1/\ell & 0 & 1/\ell & 0 \\ 8 & 0 & 0 & 1 & 0 \\ 12 & -1/\ell & 0 & 1/\ell & 0 \end{matrix} \tag{1.145d}$$

Since a simply supported beam is statically determinate, its stiffness forces will be zero and only the external loads produce member internal forces. If it is assumed that the beam is subject to a constant distributed load q_y, then the consistent load vector is obtained as:

$$\{p_*\}_3^T = \begin{matrix} 2 & 6 & 8 & 12 \\ \Big\{ \dfrac{q_y\ell}{2}, & \dfrac{q_y\ell^2}{12}, & \dfrac{q_y\ell}{2}, & -\dfrac{q_y\ell^2}{12} \Big\} \end{matrix} \tag{1.146a}$$

The consistent load vector of the released element $\{p'_*\}_3$ is now calculated using Eq. (1.138c) as written by,

$$\{p'_*\}_3 = \begin{bmatrix} 1 & -1/\ell & 0 & -1/\ell \\ 0 & 0 & 0 & 0 \\ 0 & 1/\ell & 1 & 1/\ell \\ 0 & 0 & 0 & 0 \end{bmatrix} \begin{Bmatrix} q_y\ell/2 \\ q_y\ell^2/12 \\ q_y\ell/2 \\ -q_y\ell^2/12 \end{Bmatrix} = \begin{matrix} 2 \\ 6 \\ 8 \\ 12 \end{matrix} \begin{Bmatrix} q_y\ell/2 \\ 0 \\ q_y\ell/2 \\ 0 \end{Bmatrix} \tag{1.146b}$$

which is equivalent to support reactions of a simply supported beam under a constant distributed loading.

The calculation of the connectivity matrix $[T]$ explained above looks like a complicated task analytically. But, since it is calculated numerically in a structural analysis program, it can be programed easily and systematically as explained.

1.3.4 Member Releases in a Different Coordinate System

Sometimes it is also possible that the released directions do not coincide with the principal coordinate directions. In this case, the DOF, stiffness matrix, mass matrix and the consistent load vector of the element must be transformed to the coordinate system of the released directions. In Fig. 1.17, the coordinates of the released directions are denoted by (X_R, Y_R, Z_R), the local principal coordinates are denoted by (X_L, Y_L, Z_L) and the global coordinates are (X, Y, Z). It is given that member releases are in the (X_R, Y_R, Z_R) coordinate directions, which are fully defined by the rotations (θ_X, θ_R, θ_Y) shown in Fig. 1.17. The coordinate transformation between global and released directions is stated as:

$$\{X_R\} = [t_R]\{X\} \quad \rightarrow \quad \{X_R\}^T = \{X_R, Y_R, Z_R\} \text{ and } \{X\}^T = \{X, Y, Z\} \tag{1.147a}$$

where $[t_R]$ is as defined:

$$[t_R] = \begin{bmatrix} c_x & c_y & c_z \\ \ell_x & \ell_y & \ell_z \\ m_x & m_y & m_z \end{bmatrix} \rightarrow \begin{cases} c_x,\ c_y,\ c_z : \text{ cosine directions of the axis } X_R \\ \ell_x,\ \ell_y,\ \ell_z : \text{ cosine directions of the axis } Y_R \\ m_x,\ m_y,\ m_z : \text{ cosine directions of the axis } Z_R \end{cases} \tag{1.147b}$$

The vectors of released directions are stated in terms of the cosine directions as can be written:

$$\left.\begin{aligned} \vec{X}_R &= c_x\vec{i} + c_y\vec{j} + c_z\vec{k} \\ \vec{Y}_R &= \ell_x\vec{i} + \ell_y\vec{j} + \ell_z\vec{k} \\ \vec{Z}_R &= m_x\vec{i} + m_y\vec{j} + m_z\vec{k} \end{aligned}\right\} \rightarrow \begin{Bmatrix} c_x = \cos\theta_X \cos\theta_R \\ c_y = \sin\theta_X \cos\theta_R \\ c_z = \sin\theta_R \end{Bmatrix} \text{ and } \begin{Bmatrix} \ell_x = \sin\varphi \\ \ell_y = \cos\varphi \cos\theta_Y \\ \ell_z = \cos\varphi \sin\theta_Y \end{Bmatrix} \tag{1.148}$$

in which (θ_x, θ_R and θ_y) are given or known, (φ, m_x, m_y, m_z) are unknown and to be calculated from the dot (*scalar*) and cross (*vector*) products of the vectors $\vec{X}_R$ and $\vec{Y}_R$, which result in the following relations.

$$\begin{aligned} &(\vec{X}_R.\vec{Y}_R = 0) \rightarrow \tan\varphi = -\frac{(c_y \cos\theta_Y + c_z \sin\theta_Y)}{c_x} \\ &(\vec{Z}_R = \vec{X}_R \mathrm{x} \vec{Y}_R) \rightarrow \begin{cases} m_x = c_y\ell_z - c_z\ell_y \\ m_y = c_z\ell_x - c_x\ell_z \\ m_z = c_x\ell_y - c_y\ell_x \end{cases} \end{aligned} \tag{1.149}$$

The transformation relations between the displacements in the global and released coordinates are stated below.

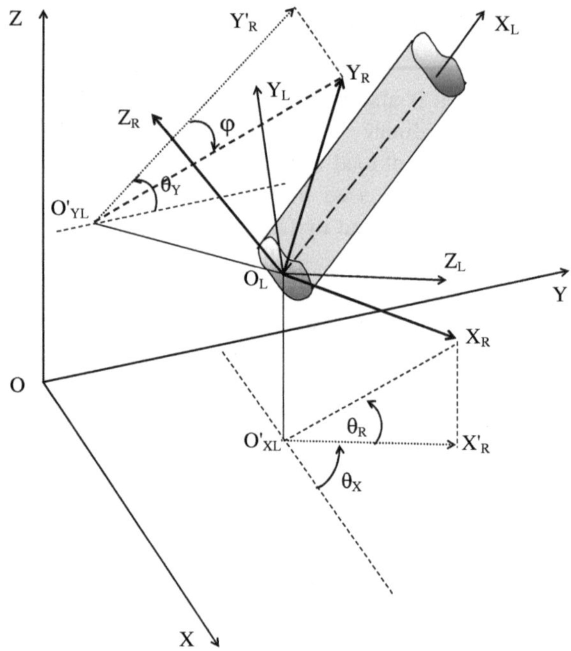

Fig. 1.17 Definition of rotation angles (θ_X, θ_R, and θ_Y) of member release directions. θ_X: Angle between global X and projection of X_R on (X–Y) plane, measured from the global X to the direction of global Y as positive. θ_R: Angle between projection of X_R on (X–Y) plane and X_R, measured from the projection to the direction of X_R as positive. θ_Y: Angle between global Y and projection of Y_R on (Y–Z) plane, measured from the global Y to the direction of global Z as positive. φ: Angle between projection of Y_R on (Y–Z) plane and Y_R, measured from the projection to the direction of Y_R as positive. This angle must be calculated

$$\left.\begin{aligned}\{d_R\} &= [T_R]\{d_G\} \;\rightarrow \text{Displacements in released coordinates}\\ \{d_G\} &= [T_R]^T\{d_R\} \;\rightarrow \text{Displacements in global coordinates}\end{aligned}\right\} \tag{1.150a}$$

where the transformation matrix $[T_R]$ is:

$$[T_R] = \begin{bmatrix} [t_R] & 0 & 0 & 0 \\ 0 & [t_R] & 0 & 0 \\ 0 & 0 & [t_R] & 0 \\ 0 & 0 & 0 & [t_R] \end{bmatrix} \tag{1.150b}$$

As it can be realized from Eq. (1.150a) transformations of the element must be made to the global coordinates first as presented in Eq. (1.102). Then, the transformations from the global coordinates to the released coordinates are made using the relations given in Eq. (1.150a). The results of these transformations are written as:

$$\left.\begin{aligned}[k_R] &= [T_R][k_G][T_R]^T\\ \{p_R\} &= [T_R]\{p_G\}\\ [m_R] &= [T_R][m_G][T_R]^T\end{aligned}\right\} \text{ in RELEASED coordinates} \tag{1.151}$$

where $[k_R]$ is the stiffness matrix, $\{p_R\}$ is the consistent load vector, and $[m_R]$ is the mass matrix in the released coordinates. Having obtained the stiffness matrix, consistent load vector and mass matrix in the released coordinates, the release process will be carried out using Eqs. (1.117b, c, d). At the end of the release process, the stiffness matrix, consistent load vector and mass matrix are denoted by the following notations.

$$\left.\begin{array}{ll} [k_{\mathrm{R}}^{'}] : & \text{stiffness matrix of released element} \\ \{p_{\mathrm{R}}^{'}\} : & \text{consistent load vector of released element} \\ [m_{\mathrm{R}}^{'}] : & \text{mass matrix of released element} \end{array}\right\} \text{in RELEASED coordinates}$$

These quantities must be transformed to the global coordinates again for the assembly process of the system as written below.

$$\left.\begin{array}{l} \left[k_{\mathrm{G}}^{'}\right] = [T_{\mathrm{R}}]^{T}[k_{\mathrm{R}}^{'}]\,[T_{\mathrm{R}}] \\ \left\{p_{\mathrm{G}}^{'}\right\} = [T_{\mathrm{R}}]^{T}\{p_{\mathrm{R}}^{'}\} \\ [m_{\mathrm{G}}^{'}] = [T_{\mathrm{R}}]^{T}[m_{\mathrm{R}}^{'}][T_{\mathrm{R}}] \end{array}\right\} \text{in GLOBAL coordinates} \quad (1.152)$$

where $[k_{\mathrm{G}}^{'}]$, $\{p_{\mathrm{G}}^{'}\}$ and $[m_{\mathrm{G}}^{'}]$ are respectively the stiffness matrix, consistent load vector, and mass matrix of the released element in the global coordinates. If the released coordinates coincide with the global coordinates, then the transformation matrix $[T_{\mathrm{R}}]$ becomes a unit matrix, which simplifies the calculations.

1.4 Formulation of Eccentrically Connected Members

In the calculation model of a structural system, it is assumed that center lines (theoretical axes) of members are connected to each other at points of member intersections, which are defined as theoretical joints or nodes. Although it does not occur usually, sometimes member connections at a theoretical joint can be possible only through some eccentricities, or theoretical axes of some members do not pass through the idealized nodal points. More commonly, theoretical joints are not always located at member ends and lie outside the actual member connections. Distances between the theoretical (idealized) joints and the actual member ends can be considered as rigid blocks. This is demonstrated by a simple example shown in Fig. 1.18. The theoretical axis of the member (2–3) does not cross the theoretical axis of the member (4–5) as shown in Fig. 1.18a. The connection of these two members can be made only using a rigid block, member (3–4) in Fig. 1.18a. In the calculation model, the theoretical joints used as structural nodes are (1, 2, 3, 4) as shown in Fig. 1.18b. For the analysis, there may be two possibilities as (a) member (2–*A*) is assumed to be eccentrically connected to the member (3–4) at the member end (*A*) with the rigid block (*A*–3), (b) member (*B*–4) is assumed to be eccentrically connected to the member (2–3) at the beginning of

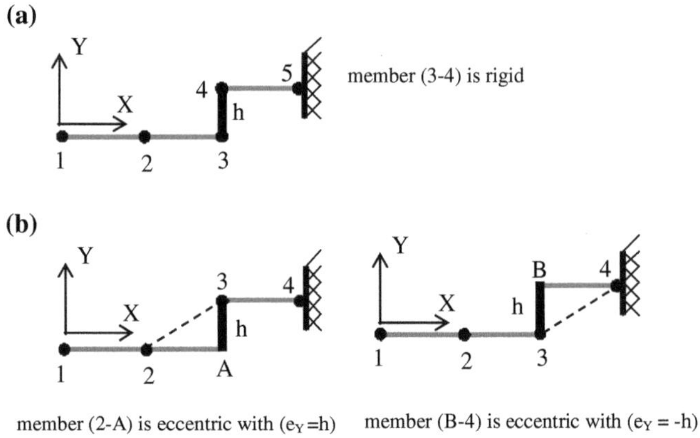

Fig. 1.18 Demonstration of member eccentricities. **a** Original members. **b** Eccentrically connected member

the member (*B*) with the rigid block (*B*–3). For the analysis, the stiffness, load, and mass properties at the end (*A*) of the member (2–*A*), or at the beginning (*B*) of the member (*B*–4), must be transformed to the joint (3). Then, the assembly process will be carried out to find the system stiffness and mass matrices and the load vector. The transformation of eccentrically connected member properties to associated theoretical joints of the system can be made by using rigid body kinematics [74] as explained in the following paragraph.

In order to explain the transformation of eccentrically connected member, a member (*i*–*j*) with eccentricity at the member end (*i*) is assumed as shown in Fig. 1.19 in global coordinates (*X*, *Y*, *Z*). The member is (*A*–*i*–*j*) before deformation and (A'–i'–j') after deformation. The element (*A*–*i*) of the undeformed state is a rigid block and (*i*–*j*) is the actual flexible member. Under the deformation the rigid block translates and rotates with rigid body motion while the flexible part (*i*–*j*) deforms. The rigid block is fully determined by the location vector ($\vec{e}$) with projections (e_X, e_Y, e_Z) on the global coordinates (*X*, *Y*, *Z*), respectively. Further, the following definitions are made:

$\vec{e}'$: location vector of the rigid block after the deformation.
$\vec{u}_i$: displacement vector of the joint (*i*) of the member (*i*–*j*).
$\vec{u}_A$: displacement vector of the joint (*A*) of the rigid block (*A*–*i*)

The location vector ($\vec{e}'$) can be stated from Eq. (1.13a) as:

$$\{e'\} = [R]\{e\} \tag{1.153}$$

where [*R*] is the rotation matrix given by Eq. (1.23). The displacement vectors can be written from Fig. 1.19 as:

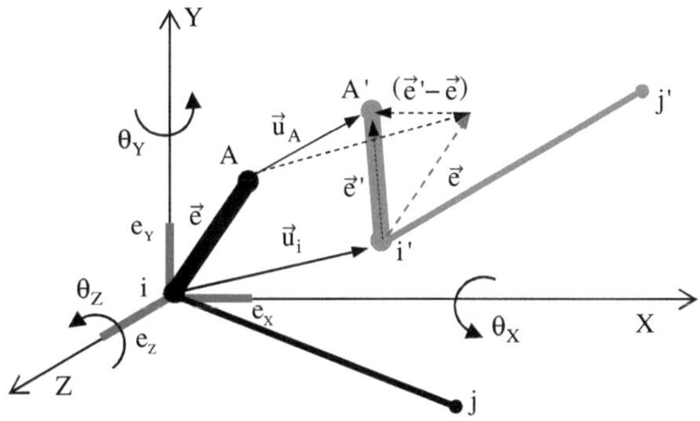

Fig. 1.19 Deformation of an eccentrically connected member

$$\begin{aligned}\vec{u}_i &= \vec{u}_A - (\vec{e}' - \vec{e}) \\ \vec{u}_A &= \vec{u}_i + (\vec{e}' - \vec{e})\end{aligned} \quad \text{or} \quad \begin{aligned}\{u\}_i &= \{u\}_A - (\{e'\} - \{e\}) \\ \{u\}_A &= \{u\}_i + (\{e'\} - \{e\})\end{aligned} \tag{1.154a}$$

from which it can be stated as similar to Eq. (1.24):

$$\begin{aligned}\{u\}_i &= \{u\}_A - \tilde{\Psi}\{e\} \\ \{u\}_A &= \{u\}_i + \tilde{\Psi}\{e\}\end{aligned} \tag{1.154b}$$

where the matrix $\tilde{\Psi}$ is given by Eq. (1.15) provided that the small rotation angles $(\theta_X, \theta_Y, \theta_Z)$ are about the global coordinates as shown in Fig. 1.19 and $\{e\}^T = \{e_X, e_Y, e_Z\}$. Eq. (1.154b) can also be written in an alternative way as:

$$\begin{aligned}\{u\}_i &= \{u\}_A + [E]_i\{\theta\}_i \\ \{u\}_A &= \{u\}_i - [E]_i\{\theta\}_i\end{aligned} \tag{1.155}$$

in which a subscript (i or A) denotes the corresponding member end, $[E]_i$ is a skew-symmetric eccentricity matrix, and $\{\theta\}_i$ is the rotation vector defined by,

$$\{\theta\}_i = \begin{Bmatrix} \theta_X \\ \theta_Y \\ \theta_Z \end{Bmatrix}_i \quad \text{and} \quad [E]_i = \begin{bmatrix} 0 & -e_Z & e_Y \\ e_Z & 0 & -e_X \\ -e_Y & e_X & 0 \end{bmatrix}_i \tag{1.156}$$

From the rigid body mechanics, it is stated that the rotation vectors at both ends of the rigid block (A–i) are the same, i.e., $(\{\theta\}_i = \{\theta\}_A)$. Thus, the FE displacement vector of the joints (i) and (A) are stated as:

$$\{d_G\}_A = \begin{Bmatrix} \{u\}_A \\ \{\theta\}_i \end{Bmatrix} \text{ and } \{d_G\}_i = \begin{Bmatrix} \{u\}_i \\ \{\theta\}_i \end{Bmatrix} \rightarrow \{d_G\}_i = \begin{Bmatrix} \{u\}_A + [E]_i\{\theta\}_i \\ \{\theta\}_i \end{Bmatrix} \tag{1.157}$$

from which the displacement vectors $\{d\}_i$ and $\{d\}_A$ are linked together by a transformation matrix $[T_e]_i$ as written by,

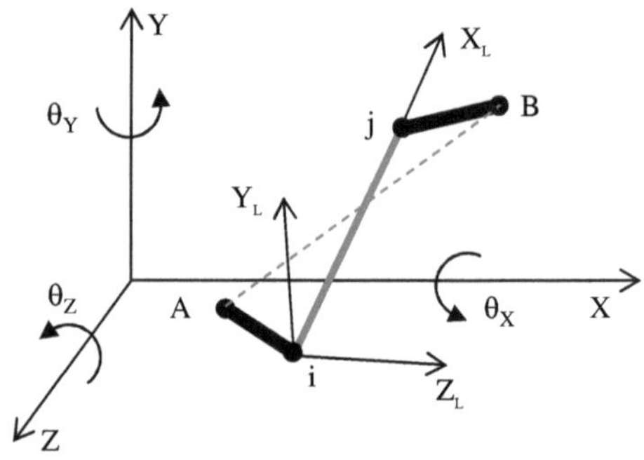

Fig. 1.20 An eccentrically connected member from both ends

$$\{d_G\}_i = [T_e]_i\{d_G\}_A \rightarrow [T_e]_i = \begin{bmatrix} I_3 & [E]_i \\ 0 & I_3 \end{bmatrix} \tag{1.158a}$$

in which I_3 is a unit matrix with (3 × 3) dimensions. The subscript (G) denotes global coordinates. Here, the eccentricities are measured from the flexible member ends in the global coordinates. For the generality, now suppose that the member is eccentrically connected from both ends as shown in Fig. 1.20. The flexible part is (*i*–*j*), the rigid blocks are (*A*–*i*) at the member end (*i*), and (*j*–*B*) at the member end (*j*). The joints (*A*) and (*B*) are the theoretical nodal points (system joints). For the assembly process, the stiffness, mass, and loading properties of the flexible member (*i*–*j*) must be transferred to the joints (*A*) and (*B*) of the rigid blocks. The transformation of the FE displacement vector at the member end (*i*) to the joint (*A*) of the rigid block (*A*–*i*) is given by Eq. (1.158a). Similarly, the transformation of the displacement vector at the member end (*j*) to the joint (*B*) of the rigid block (*j*–*B*) can be written as:

$$\{d_G\}_j = [T_e]_j\{d_G\}_B \rightarrow [T_e]_j = \begin{bmatrix} I_3 & [E]_j \\ 0 & I_3 \end{bmatrix} \tag{1.158b}$$

where $[E]_j$ is the eccentricity matrix defined at the end (*j*) of the member (*i*–*j*) as similar to Eq. (1.156). The displacement vector of the element (*i*–*j*) can now be transferred to the joints (*A*) and (*B*) of the fictitious element (*A*–*B*). This is stated as:

$$\{d_G\} = \begin{Bmatrix} \{d_G\}_i \\ \{d_G\}_j \end{Bmatrix} \text{ and } \{d_G\}_e = \begin{Bmatrix} \{d_G\}_A \\ \{d_G\}_B \end{Bmatrix} \rightarrow \{d_G\} = [T_e]\{d_G\}_e \tag{1.159a}$$

where $\{d_G\}_e$ denotes the displacement vector of the eccentric (fictitious) member (*A*–*B*). The transformation matrix $[T_e]$ is defined as:

$$[T_{\rm e}] = \begin{bmatrix} [T_{\rm e}]_i & 0 \\ 0 & [T_{\rm e}]_j \end{bmatrix} \text{ or } [T_{\rm e}] = \begin{bmatrix} I_3 & [E]_i & 0 & 0 \\ 0 & I_3 & 0 & 0 \\ 0 & 0 & I_3 & [E]_j \\ 0 & 0 & 0 & I_3 \end{bmatrix} \qquad (1.159\text{b})$$

The back transformation of the displacements can be stated from Eq. (1.159a) as written by,

$$\{d_{\rm G}\}_{\rm e} = [T_{\rm e}]^{-1}\{d_{\rm G}\} \rightarrow [T_{\rm e}]^{-1} = \begin{bmatrix} I_3 & -[E]_i & 0 & 0 \\ 0 & I_3 & 0 & 0 \\ 0 & 0 & I_3 & -[E]_j \\ 0 & 0 & 0 & I_3 \end{bmatrix} \qquad (1.160)$$

By using energy equivalences, the stiffness, mass, and loading properties of the eccentric element can be stated in global coordinates as similar to Eq. (1.102), which are:

$$\left.\begin{aligned} [k_{\rm G}]_{\rm e} &= [T_{\rm e}]^T[k_{\rm G}][T_{\rm e}] && \rightarrow \text{stiffness matrix} \\ [m_{\rm G}]_{\rm e} &= [T_{\rm e}]^T[m_{\rm G}][T_{\rm e}] && \rightarrow \text{mass matrix} \\ \{p_{\rm G}\}_{\rm e} &= [T_{\rm e}]^T\{p_{\rm G}\}\cdots && \rightarrow \text{consistent load vector} \end{aligned}\right\} \qquad (1.161)$$

For eccentric members, these equations are used in the assembly process to find the system matrices and vectors. After the solution of the system equation, the displacement vector $\{d_{\rm G}\}_{\rm e}$ is extracted from the system displacement vector and using Eq. (1.159a) the displacement vector of the flexible element [(i–j) in Fig. 1.20] $\{d_{\rm G}\}$ is calculated. Accordingly, member internal forces are calculated once the displacements, $\{d_{\rm G}\}$ in vector notation, are known.

1.5 An Interface Beam Element for the Soil–Structure Interaction and Deformation of Soil Under R-Wave Propagation

Structures are supported on ground either by piles or on a foundation basis, such as offshore platforms and building structures, so that soil–structure interaction phenomenon becomes an inevitable reality. Some other structural types are completely buried in, or rested on the ground such as pipelines, tunnels, and underground structures. The analysis of such structures may be carried out in practice by using conventional finite element methods assuming that the phenomenon of the soil–structure interactions is represented by some viscoelastic models (massless spring and dashpot systems) at the interface nodes (supports). This procedure may produce reasonable results if a fine structural mesh is used. In this section, an alternative 3D soil–beam element is presented as it is more precise and general than representing soil effects by some spring-dashpot models at the supports. For an accurate analysis of the soil-structure interaction phenomenon a continuum finite, or boundary, element modeling is required [75–77], which is more general and straightforward. But, it is costly in terms of data preparation and calculation time, especially in nonlinear and dynamic analyses. Therefore, a simple finite element formulation for the soil–structure interaction problem seems to be more attractive and practical [78]. In the literature, there are numerous publications on 2D beams resting on Winkler type elastic foundations, see i.e., [79–82], without including effects of shear forces in the soil whereas they are included in some works, i.e., [78, 83, 84], for linear static analyses. For dynamic analyses, these models are approximate since radiation and hysteretic damping terms of the soil are not included. In the formulation of the interface element for the soil–beam interactions, which is presented here, it is assumed that the soil medium is represented by a linear viscoelastic continuum including the shear force effects. For this purpose, the differential equations of the beam in the transverse directions (y, z) will be formulated in a different way under external distributed loadings as explained below.

Having used Eqs. (1.10a, b) the following equations in the (y) and (z) directions can be obtained as:

$$\left.\begin{aligned} \frac{du_y}{dx} &= \gamma_y + \theta_z \qquad & \gamma_y &= \frac{Q_y}{A_yG} \\ & \text{and} & & \\ \frac{du_z}{dx} &= \gamma_z - \theta_y \qquad & \gamma_z &= \frac{Q_z}{A_zG} \end{aligned}\right\} \rightarrow \left\{\begin{aligned} \mathrm{EI}_z\frac{d^2u_y}{dx^2} &= \frac{\mathrm{EI}_z}{A_yG}\frac{dQ_y}{dx} + M_z \\ \mathrm{EI}_y\frac{d^2u_z}{dx^2} &= \frac{\mathrm{EI}_y}{A_zG}\frac{dQ_z}{dx} - M_y \end{aligned}\right. \tag{1.162a}$$

Having introduced (dQ_y/dx) and (dQ_z/dx) from Eq. (1.8b) into Eq. (1.162a), and then taking the differentiation with respect to (x) under a constant axial force (N) it is obtained that,

$$\left.\begin{aligned} \mathrm{EI}_z\frac{d^3u_y}{dx^3} &= -\alpha_y\left(\frac{dq_y}{dx} + N\frac{d\kappa_z}{dx}\right) + \frac{dM_z}{dx} \\ \mathrm{EI}_y\frac{d^3u_z}{dx^3} &= \alpha_z\left(-\frac{dq_z}{dx} + N\frac{d\kappa_y}{dx}\right) - \frac{dM_y}{dx} \end{aligned}\right\} \text{where} \left\{\begin{aligned} \alpha_y &= \frac{\mathrm{EI}_z}{A_yG} \\ \alpha_z &= \frac{\mathrm{EI}_y}{A_zG} \end{aligned}\right. \tag{1.162b}$$

Now, we use κ_z and κ_y from Eqs. (1.3) and (1.6), and (dM_z/dx) and (dM_y/dx) from Eq. (1.11a) in Eq. (1.162b), and then taking the differentiation with respect to (x) (keeping in mind that the fourth derivatives of displacements on the right-hand side of Eq.(1.162b) are zero since they are cubic functions) it is obtained that,

$$\begin{aligned} \mathrm{EI}_z\frac{d^4u_y}{dx^4} &= -\alpha_y\frac{d^2q_y}{dx^2} - \frac{dQ_y}{dx} \\ \mathrm{EI}_y\frac{d^4u_z}{dx^4} &= -\alpha_z\frac{d^2q_z}{dx^2} - \frac{dQ_z}{dx} \end{aligned} \tag{1.162c}$$

Having introduced (dQ_y/dx) and (dQ_z/dx) from Eq. (1.8b) into Eq. (1.162c), the required differential equations can be obtained in terms of external loadings as written by,

$$\left.\begin{aligned} EI_z\frac{d^4u_y}{dx^4} &= q_y - \alpha_y\frac{d^2q_y}{dx^2} + N\frac{d^2u_y}{dx^2} \\ EI_y\frac{d^4u_z}{dx^4} &= q_z - \alpha_z\frac{d^2q_z}{dx^2} + N\frac{d^2u_z}{dx^2} \end{aligned}\right\} \rightarrow \left\{\begin{aligned} EI_z\frac{d^4u_y}{dx^4} &= \bar{q}_y \\ EI_y\frac{d^4u_z}{dx^4} &= \bar{q}_z \end{aligned}\right. \tag{1.163}$$

where $\bar{q}_y$ and $\bar{q}_z$ can be considered as pseudo-distributed loadings applied on the beam, which include shear effects (α_y and α_z) and effects of the axial force (N) of the beam. They are defined as:

$$\begin{aligned} \bar{q}_y &= q_y - \alpha_y\frac{d^2q_y}{dx^2} + N\frac{d^2u_y}{dx^2} \\ \bar{q}_z &= q_z - \alpha_z\frac{d^2q_z}{dx^2} + N\frac{d^2u_z}{dx^2} \end{aligned} \tag{1.164}$$

Using these applied pseudo-distributed loadings on the beam the external work will be calculated. But, before doing that some extra distributed loadings at the soil–beam interfaces, which are defined as interface loadings arising from the soil due to deformations of the beam, must be added to these pseudo-distributed loadings. Calculation of the interface loadings and updating the pseudo-distributed loadings are explained in the following subsection.

1.5.1 Modeling of Soil Medium and Calculation of Interface Loadings

In the dynamic analysis of ground or soil-structures based on the theory of wave propagation or on the finite element principle, an important issue is to represent the cyclic behavior of soils in a form of material model to construct a relation between shear stress and shear strain [85]. Modeling of soil behavior under dynamic loading conditions must be made so that the model can represent deformation characteristics in the range of strains under consideration. When the soil behavior is in the range of small strains, an elastic soil model can be used and the shear modulus will be a key parameter in properly modeling the soil behavior. When the soil behavior is in the range of medium strains, the soil behavior becomes approximately elastoplastic and the shear modulus decreases as the shear strain increases. In this case, energy dissipation occurs during application of load cycles in soils. The energy dissipation is mostly rate-independent and of hysteretic nature. A damping ratio can be used to represent the energy dissipation in soils. Since the strain level is still small the shear modulus and damping ratio do not change with cyclic load application. This kind of behavior can be represented by using the linear viscoelastic theory to an acceptable degree of accuracy. The shear modulus and damping ratio are the key parameters to represent soil properties in this medium strain range and the useful analytical tool is the equivalent linear method based on the viscoelastic concept. In this model, the stress–strain relation is assumed linear with energy dissipation in the soil. This model has usually been used to represent soil behavior even in the slightly nonlinear range where damping has important effects. This model of the soil is also assumed in this book. Among others the Kelvin–Voigt model, which is also called Voigt model, is widely used to represent soil behavior. In this model, the elastic property of the soil is represented by a purely elastic spring and the damping characteristics are expressed by a purely viscous damper (dashpot), which are connected in parallel. In this model, the shear stress of the soil, e.g., in the z direction, is expressed [85] as:

$$\tau_{sz} = \left(G_{sz}\gamma_{sz} + \mu_{sz}\dot{\gamma}_{sz}\right) \;\rightarrow\; \tau_{sz} = \left(G_{sz} + i\omega\mu_{sz}\right)\gamma_{sz} \tag{1.165a}$$

or in terms of complex shear modulus it is written as:

$$\tau_{sz} = G^{*}_{sz}\gamma_{sz} \;\rightarrow\; G^{*}_{sz} = \left(G_{sz} + i\omega\mu_{sz}\right) \tag{1.165b}$$

in which G_{sz} and μ_{sz} are respectively the shear modulus and coefficient of viscosity of the soil in the z direction, ω is the angular frequency. Equation (1.165b) will be used to calculate the shear force of the soil, Q_{sz}, in the z direction.

The soil medium around the beam element is assumed to be homogenous, isotropic half-space. For the linear analysis, it can be idealized in different ways as to be elastic and time dependent models among which Winkler, elastic continuum, and two-parameter elastic models [80] may be used largely in practice. The dynamic Winkler model, which is defined by a frequency dependent complex-subgrade-modulus, is commonly used to represent the soil behavior during the response process. This model assumes that the soil reaction is related to the surface deflection only, whereas the two-parameter soil model takes into account not only the surface deflection but also the shear deformation of the soil. In this book, it is assumed that the Pasternak model [80] represents the soil medium to calculate an interface loading of a soil–beam element. This model consists of two parts, one is for purely shear deformation (viscoelastic Shear layers with unit depths) and the other one is for the deflection (dynamic Winkler model) as shown in Fig. 1.21 on the (x–z) plane of member local coordinates. In this figure, k_{sz} and c_{sz} denote the spring and radiation damping coefficients of the soil, respectively, and u_{zr} is the relative vertical displacement of the beam, which is the difference between the absolute beam and ground displacements ($u_z - u_{gz}$). The shear force Q_{sz} due to the shear deformation of a viscoelastic soil with a unit area (unit depths) under the relative deformation u_{zr} of the beam can be stated as:

$$Q_{sz} = G^{*}_{sz}\frac{\mathrm{d}u_{zr}}{\mathrm{d}x} \tag{1.166}$$

From the equilibrium of an infinitesimal soil element (dx), the complex interface loading in the z direction of the member local coordinates can be obtained as:

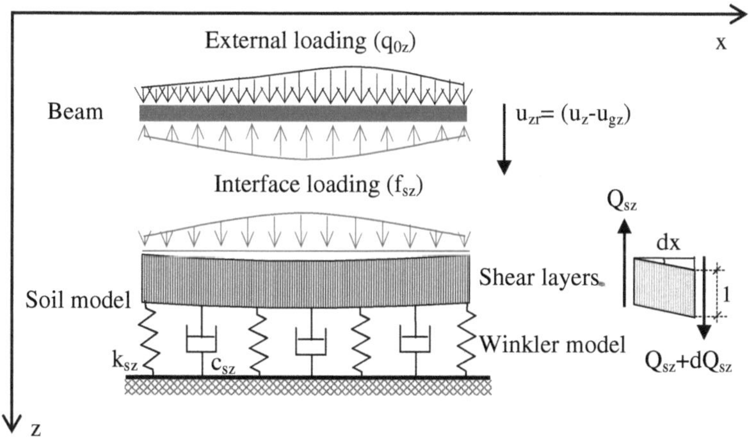

Fig. 1.21 Representation of the soil medium and soil–beam interface loading

$$f^*_{sz} = k^*_{sz}\, u_{zr} - \frac{\mathrm{d}Q_z}{\mathrm{d}x} \rightarrow f^*_{sz} = k^*_{sz}\, u_{zr} - G^*_{sz}\frac{\mathrm{d}^2 u_{zr}}{\mathrm{d}x^2} \tag{1.167a}$$

in which the complex shear modulus G^*_{sz} is defined in Eq. (1.165b) and the first term on the right hand side is due to the dynamic Winkler model with a complex-subgrade-modulus, k^*_{sz} defined by,

$$k^*_{sz} = k_{sz} + i\omega\, c_{sz} \tag{1.167b}$$

Here, k_{sz} is the spring coefficient and c_{sz} is the radiation damping coefficient in the z direction of member local coordinates. This simple model of the interface loading, which is given by Eq. (1.167a), can be extended in other (x and y) coordinate directions so that complete interface loadings may be conveniently stated by vector notation in the member local coordinates as:

$$\{f_s\}^* = [k_s]^*\{u_r\} - [G_s]^*\frac{\mathrm{d}^2\{u_r\}}{\mathrm{d}x^2} \tag{1.168}$$

where $\{u_\mathrm{r}\}$ is the relative displacement vector of the beam in the local coordinates of the beam, which is stated as:

$$\{u_\mathrm{r}\} = \{u\} - \{u_\mathrm{g}\} \tag{1.169a}$$

In Eq. (1.169a), $\{u\}$ is the displacement vector of the beam and $\{u_\mathrm{g}\}$ is the displacement vector of the soil in the beam local coordinates due to excitation in the ground such as earthquake motion. These are defined as:

$$\begin{aligned} \{u\}^T &= \{u_x,\, u_y,\, u_z,\, \theta_x\,,\, \theta_y\,,\, \theta_z\} \\ \{u_\mathrm{g}\}^T &= \{u_{gx},\, u_{gy},\, u_{gz},\, \theta_{gx}\,,\, \theta_{gy}\,,\, \theta_{gz}\} \end{aligned} \tag{1.169b}$$

The matrices, $[k_s]^*$ and $[G_s]^*$ in Eq. (1.168) are the complex spring- and shear-rigidity matrices of the soil. These matrices are diagonal as defined:

$$[k_s]^* = \begin{bmatrix} k^*_{sx} & & & & & 0 \\ & k^*_{sy} & & & & \\ & & k^*_{sz} & & & \\ & & & r^*_{sx} & & \\ & & & & r^*_{sy} & \\ 0 & & & & & r^*_{sz} \end{bmatrix} \quad \text{for a diagonal term, see Eq. (1.167b)} \tag{1.170a}$$

$$[G_s]^* = \begin{bmatrix} G^*_{sx} & & & & & 0 \\ & G^*_{sy} & & & & \\ & & G^*_{sz} & & & \\ & & & 0 & & \\ & & & & 0 & \\ 0 & & & & & 0 \end{bmatrix} \quad \text{for a diagonal term, see Eq. (1.165b)} \tag{1.170b}$$

In Eq. (1.170a), the subgrade modulus of the soil in the axial direction of the beam, k^*_{sx}, represents an axial skin friction and the end bearing resistance of the member if the member end is free in the soil such as a pile end. The subgrade moduli, k^*_{sy} and k^*_{sz}, are due to lateral deformations, whereas the modulus r^*_{sx} represents the torsional skin friction, r^*_{sy} and r^*_{sz} represent rotational subgrade moduli. In Eq. (1.170b), G^*_{sx}, G^*_{sy} and G^*_{sz}, are the complex shear moduli of the soil in the beam local coordinate directions (x, y, z), respectively. These soil properties are time and loading dependable in practice [85], so that average design values may be used in the analysis. A rapid loading rate increases the stiffness of the soil, which is a favorable case unlike the soil-degradation effects under a cyclic loading. One other issue in the soil–structure interaction phenomenon during strong motion earthquakes is that separation between the soil and structure can occur [86], especially at soil layers close to the mudline of offshore pile foundations. However, due to the soil liquefaction after a number of stress cycles, a perfect bond between the soil and structure may occur as it is assumed in this book.

Having introduced Eq. (1.169a) into Eq. (1.168) the soil–beam interface load vector $\{f_s\}^*$ can be stated in terms of the displacement vectors of the beam and soil as written:

$$\{f_s\}^* = \left([k_s]^*\{u\} - [G_s]^*\frac{d^2\{u\}}{dx^2}\right) - \left([k_s]^*\{u_g\} - [G_s]^*\frac{d^2\{u_g\}}{dx^2}\right) \tag{1.171}$$

This interface load vector is defined in the member local coordinates. Due to spatial variations of soil properties, if the soil properties are defined in a different coordinate system, e.g., (x', y', z') as shown in Fig. 1.22, transformations are used to find the interface load vector in the member local coordinates (x, y, z). For this

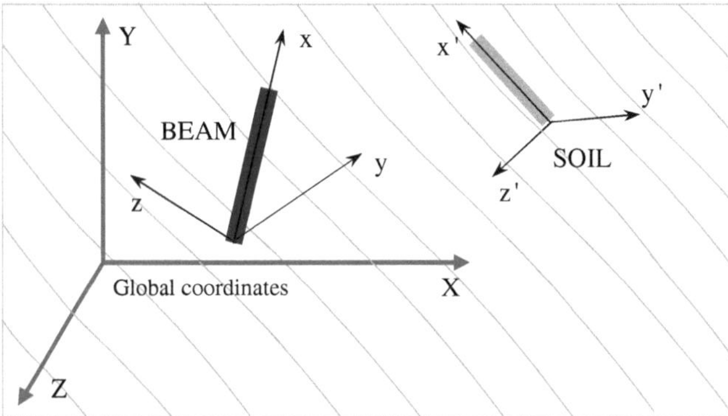

Fig. 1.22 Coordinate systems of the interface beam and soil properties

purpose, in the soil coordinate system (x', y', z'), $\{u'\}$ and $\{u'_g\}$ denote respectively displacement vectors of the beam and soil, $\{u'_r\}$ denotes the relative displacement vector, $[k'_s]^*$ and $[G'_s]^*$ are the matrices of soil properties, and $\{f'_s\}^*$ is the interface load vector. Further, it is assumed that the transformation between the relative displacement vectors $\{u_r\}$ and $\{u'_r\}$ is known as stated:

$$\{u'_r\} = [T']\{u_r\} \tag{1.172}$$

By using equivalent external works of interface loadings in the soil and beam coordinate systems, i.e., $\left(\{u'_r\}^T\{f'_s\}^* = \{u_r\}^T\{f_s\}^*\right)$, the soil property matrices $[k_s]^*$ and $[G_s]^*$ in the beam local coordinates (x, y, z) are obtained as stated:

$$[k_s]^* = [T']^T[k'_s]^*[T'] \text{ and } [G_s]^* = [T']^T[G'_s]^*[T'] \tag{1.173}$$

which are used in Eq. (1.171) to calculate the required interface load vector $\{f_s\}^*$. It is worth mentioning that, due to transformation, the matrices $[k_s]^*$ and $[G_s]^*$ calculated from Eq. (1.173) are not diagonal any more although $[k'_s]^*$ and $[G'_s]^*$ are diagonal, but remain symmetric. Having calculated the interface load vector, formulation of the soil–beam interface element can be carried out as explained in the following section.

1.5.2 Formulation of Interface Element for Soil–Beam Interactions

In the derivation of an interface element for soil–beam interactions the total potential energy, which is stated in the complex domain, is used. It is assumed here that the classical beam theory is applied using pseudo-distributed loadings as stated in Eq. (1.164). For convenience, the components of the pseudo-distributed loadings are stated by vector notation in the complex domain as:

$$\{\bar{q}\}^* = \{q\}^* - [L_\alpha]\frac{\mathrm{d}^2\{q\}^*}{\mathrm{d}x^2} + [L_n]\frac{\mathrm{d}^2\{u\}}{\mathrm{d}x^2} \tag{1.174a}$$

where $\{q\}^*$ is the applied effective complex load vector of the beam, $[L_\alpha]$ and $[L_n]$ are diagonal matrices representing the shear effects, α_y and α_z, and the effect of the axial force N. They are defined as:

$$[L_\alpha] = \begin{bmatrix} 0 & & & & & 0 \\ & \alpha_y & & & & \\ & & \alpha_z & & & \\ & & & 0 & & \\ & & & & 0 & \\ 0 & & & & & 0 \end{bmatrix} \text{ and } [L_n] = \begin{bmatrix} 0 & & & & & 0 \\ & N & & & & \\ & & N & & & \\ & & & 0 & & \\ & & & & 0 & \\ 0 & & & & & 0 \end{bmatrix} \tag{1.174b}$$

The applied effective complex load vector $\{q\}^*$ of the beam is expressed from Fig. 1.21 as written:

$$\{q\}^* = \{q_0\} - \{f_s\}^* \tag{1.174c}$$

in which $\{q_0\}$ is the externally applied distributed load vector as shown in Fig. 1.21 for the component in the z direction, q_{0z}, and $\{f_s\}^*$ is the complex interface load vector given by Eq. (1.171). Having introduced Eq. (1.174c) into Eq. (1.174a) the complex pseudo-distributed load vector $\{\bar{q}\}^*$ is obtained as stated:

$$\{\bar{q}\}^* = \{q_0\} - \{f_s\}^* + [L_\alpha]\frac{\mathrm{d}^2\{f_s\}^*}{\mathrm{d}x^2} + [L_n]\frac{\mathrm{d}^2\{u\}}{\mathrm{d}x^2} \tag{1.175a}$$

assuming that the second derivative of $\{q_0\}$ is zero, which indicates a constant or linear distribution. Now introducing Eq. (1.171) into Eq. (1.175a) the complex pseudo-distributed load vector $\{\bar{q}\}^*$ can be expressed in terms of displacements and their derivatives in vector notations as written:

$$\begin{aligned}\{\bar{q}\}^* =& \{q_0\} + \left([k_s]^*\{u_g\} - [L]^*\frac{\mathrm{d}^2\{u_g\}}{\mathrm{d}x^2} + [L_\alpha][G_s]^*\frac{\mathrm{d}^4\{u_g\}}{\mathrm{d}x^4}\right) + \cdots \\ & \cdots + ([L]^* + [L_n])\frac{\mathrm{d}^2\{u\}}{\mathrm{d}x^2} - [k_s]^*\{u\}\end{aligned} \tag{1.175b}$$

where the complex matrix $[L]^*$ is defined as:

$$[L]^* = [G_s]^* + [L_\alpha][k_s]^* \tag{1.175c}$$

The total external work of the beam done by the complex pseudo-distributed load vector $\{\bar{q}\}^*$ is a complex scalar and can be calculated from $\left(\mathrm{d}W_q^* = d\{u\}^T\{\bar{q}\}^*\right)$

or $\left(\mathrm{d}W_q^* = \{\bar{q}\}^{*T} d\{u\}\right)$. In order to maintain symmetry property of stiffness matrices induced by the soil, we define the external work as written:

$$W_q^* = \frac{1}{2}\int_0^{\ell} \left(\{u\}^T\{\bar{q}\}^* + \{\bar{q}\}^{*T}\{u\}\right) \mathrm{d}x \tag{1.176a}$$

or using $\{\bar{q}\}^*$ from Eq. (1.175b) the total external work W_q^* can be expressed as:

$$W_q^* = \int_0^{\ell} \{u\}^T\{q_0\}\,\mathrm{d}x + \int_0^{\ell} \{u\}^T\{q_g\}^*\mathrm{d}x - \frac{1}{2}\int_0^{\ell} \{u\}^T[k_s]^*\{u\}\,\mathrm{d}x - \cdots$$
$$\cdots - \frac{1}{2}\left(\int_0^{\ell} \frac{d\{u\}^T}{\mathrm{d}x}[L_n]\frac{d\{u\}}{\mathrm{d}x}\,\mathrm{d}x\right) - \frac{1}{2}\left(\frac{1}{2}\int_0^{\ell} \frac{d\{u\}^T}{\mathrm{d}x}\left([L]^* + [L]^{*T}\right)\frac{d\{u\}}{\mathrm{d}x}\,\mathrm{d}x\right) \tag{1.176b}$$

in which $\{q_g\}^*$ is the complex distributed load vector due to ground motion depending on the soil properties and shear constants of the beam α_y and α_z. It is defined as:

$$\{q_g\}^* = [k_s]^*\{u_g\} - [L]^*\frac{\mathrm{d}^2\{u_g\}}{\mathrm{d}x^2} + [L_\alpha][G_s]^*\frac{\mathrm{d}^4\{u_g\}}{\mathrm{d}x^4} \tag{1.176c}$$

From Eq. (1.48a) the displacement vector of the beam can be sated as:

$$\{u\} = \begin{bmatrix}[N_u]\\ [N_\theta]\end{bmatrix}\{d\} \;\rightarrow\; \{u\} = [N]\{d\} \text{ with } [N] = \begin{bmatrix}[N_u]\\ [N_\theta]\end{bmatrix} \tag{1.177}$$

where the matrices $[N_u\}$ and $[N_\theta]$ are given by Eqs. (1.50a, b). Having introduced $\{u\}$ from Eq. (1.177) into Eq. (1.176b) the total external work W_q^* can be expressed as:

$$W_q^* = \{d\}^T\left(\{p_q\} + \{p_g\}^*\right) - \frac{1}{2}\{d\}^T\left([k_{bn}] + [k_{bw}]^* + [k_{bs}]^*\right)\{d\} \tag{1.178}$$

In this equation, $\{p_q\}$ is the consistent load vector of the beam due to external applied loads as given by Eq. (1.66), $\{p_g\}^*$ is the consistent load vector due to ground motion, $[k_{bn}]$ is the stiffness matrix introduced by the axial force N, $[k_{bw}]^*$ and $[k_{bs}]^*$ are respectively stiffness matrices of the beam due to soil Winkler and shear properties. These quantities of the soil–beam interface element, which are introduced by the ground motion and soil properties, are defined as:

Consistent load vector due to ground motion:

$$\{p_g\}^* = \int_0^{\ell} [N]^T\{q_g\}^*\mathrm{d}x \tag{1.179a}$$

Stiffness matrix due to axial force N:

$$[k_{bn}] = \int_0^{\ell} \frac{d[N]^T}{\mathrm{d}x}[L_n]\frac{d[N]}{\mathrm{d}x}\,\mathrm{d}x \tag{1.179b}$$

Stiffness matrix due to soil shear model:

$$[k_{bs}]^* = \int_0^{\ell} \frac{d[N]^T}{\mathrm{d}x}[G_s]^*\frac{d[N]}{\mathrm{d}x}\,\mathrm{d}x \tag{1.179c}$$

Stiffness matrix due to soil Winkler model:

$$\begin{aligned}[k_{bw}]^* = &\int_0^{\ell} [N]^T[k_s]^*[N]\,\mathrm{d}x + \cdots \\ &\cdots + \frac{1}{2}\int_0^{\ell} \frac{d[N]^T}{\mathrm{d}x}\left([L_\alpha][k_s]^* + [k_s]^{*T}[L_\alpha]^T\right)\frac{d[N]}{\mathrm{d}x}\,\mathrm{d}x\end{aligned} \tag{1.179d}$$

By using the stationary property of the total potential energy, given by Eq. (1.53), with the total complex external work in Eq. (1.178), the stiffness equation of the soil–beam interface element can be obtained in local coordinates as written in complex domain:

$$\{f\}^* = [k]^*\{d\} - \{p\}^* \tag{1.180}$$

where $\{f\}^*$ is the complex member-end force vector, $[k]^*$ is the complex member stiffness matrix and $\{p\}^*$ is the complex member consistent force vector, which are defined as:

$$\begin{aligned}\{p\}^* &= \{p_q\} + \{p_g\}^* \\ [k]^* &= [k_b] + [k_{bn}] + [k_{bw}]^* + [k_{bs}]^*\end{aligned} \tag{1.181}$$

In Eq. (1.181), $[k_b]$ denotes the member stiffness matrix defined in Eq. (1.63) and given by Eq. (1.73a). The real and imaginary parts of the complex stiffness matrices, $[k_{bw}]^*$ and $[k_{bs}]^*$, can be separated as stated:

$$[k_{bw}]^* = ([k_{bw}] + i\omega[c_{bw}]) \quad \text{and} \quad [k_{bs}]^* = ([k_{bs}] + i\omega[c_{bs}]) \tag{1.182}$$

in which the real parts are the stiffness contribution and the imaginary parts are the damping contributions as written by,

$$\begin{aligned}[k] &= [k_b] + ([k_{bn}] + [k_{bw}] + [k_{bs}]) \\ [c] &= [c_b] + ([c_{bw}] + [c_{bs}])\end{aligned} \tag{1.183}$$

In Eq. (1.183), the subscripts, b, w, and s, denote respectively the beam, soil Winkler model and soil shear model, $[c_b]$ is the material damping matrix of the beam as given by Eq. (1.83) and the terms in brackets denote the contributions of the soil. The stiffness and damping matrices, $[k_{bw}]$ and $[c_{bw}]$, are calculated from Eq. (1.179d) by using spring coefficients and dashpot constants of the dynamic soil Winkler model given in Eq. (1.170a). The stiffness and damping matrices, $[k_{bs}]$ and $[c_{bs}]$, are calculated from Eq. (1.179c) by using shear modulli and coefficients of viscosity of the soil given in Eq. (1.170b). In order to calculate the consistent load vector of the beam due to a ground motion using Eqs. (1.176c) and (1.179a) the displacement vector of the soil $\{u_g\}$ is required. For this purpose, earthquake ground motion based on the Rayleigh wave propagation is used in this book as explained in the next section.

1.5.3 Ground Deformation Under R-Wave Propagation and Calculation of the Exerted Force Vector

Seismic waves are related to ground vibration caused by an earthquake which releases huge energy in the Earth. Seismic waves originate from the source (focus) of the earthquake, which lies underground at a shallow, intermediate, or deep depth, and they travel in all directions through the body of the Earth. These waves, traveling through Earth, are called body waves. Seismic waves also travel along the surface of Earth, which are called surface waves. There are two main types of body waves as P-waves (primary) and S-waves (secondary). P-waves compress and expand the material as they propagate through solids, liquids, and gases. They have the greatest velocity and reach first to the Earth's surface. S-waves travel through materials by shearing them. They propagate only through solids since liquids and gases do not posses particular shapes. These waves travel slower through Earth and reach to Earth's surface as being the second, see e.g., [87–90].

Surface waves differ from body waves as they do not travel through the Earth, but instead, they travel along or parallel to the surface of the Earth. Surface waves are the most destructive and they cause the most damage, which behave like S-waves. They cause up and down and side to side movements as they travel. They are slower than S-waves. There are two types of S-waves as Love and Rayleigh waves. Love waves cause horizontal shearing of the ground and vibrate in a perpendicular direction to wave motion as they propagate. Rayleigh waves cause both horizontal and vertical movement within the ground. They vibrate in a rolling motion in the same direction as wave motion. Most of the shaking of an earthquake is due to these waves. Since Rayleigh waves are the most destructive and cause the most damage [91], they are used in this book as the source of ground deformations due to an earthquake motion.

Rayleigh wave propagation is a special case of wave equations that satisfy a couple of boundary conditions as being a zero-stress state at the ground surface.

Fig. 1.23 An infinitesimal element in 2D isotropic, homogenous elastic soil medium

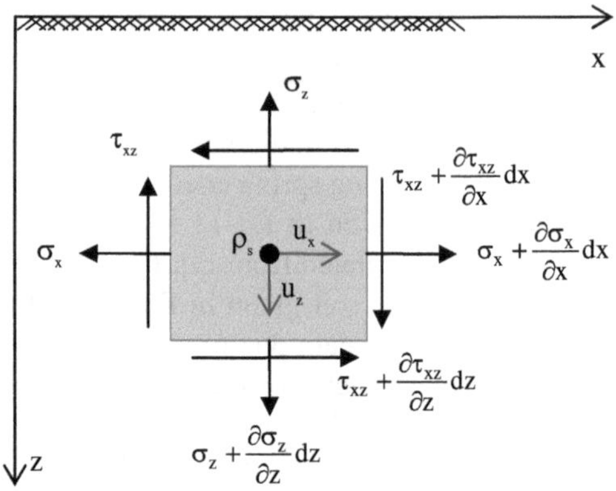

Rayleigh wave is an interaction phenomenon of the P-wave and S-wave, traveling on the surface of the Earth. Wave equations can be found elsewhere, see e.g., [89], but for the completeness, the Rayleigh wave equation is presented here first, and then the ground deformations are derived. It is assumed that the soil medium is represented by an isotropic, homogenous elastic half-space. The dynamic equilibrium equations of a 2D-soil element, which is shown in Fig. 1.23, are stated as:

$$\begin{aligned} \rho_s \frac{\partial^2 u_x}{\partial t^2} &= \frac{\partial \sigma_x}{\partial x} + \frac{\partial \tau_{xz}}{\partial z} \\ \rho_s \frac{\partial^2 u_z}{\partial t^2} &= \frac{\partial \sigma_z}{\partial z} + \frac{\partial \tau_{xz}}{\partial x} \end{aligned} \tag{1.184}$$

in which ρ_s is the mass density of the soil. From the theory of elasticity [23, 24] the stress and strain components are written as:

$$\begin{aligned} \sigma_x &= \frac{E_s}{(1+\nu_s)(1-2\nu_s)}[(1-\nu_s)\varepsilon_x + \nu_s\varepsilon_z] & \qquad \varepsilon_x &= \frac{\partial u_x}{\partial x} \\ \sigma_z &= \frac{E_s}{(1+\nu_s)(1-2\nu_s)}[\nu_s\varepsilon_x + (1-\nu_s)\varepsilon_z] \quad \text{and} & \qquad \varepsilon_z &= \frac{\partial u_z}{\partial z} \\ \tau_{xz} &= G_s\,\gamma_{xz} \;\rightarrow\; G_s = \frac{E_s}{2(1+\nu_s)} & \qquad \gamma_{xz} &= \frac{\partial u_x}{\partial z} + \frac{\partial u_z}{\partial x} \end{aligned} \tag{1.185}$$

where E_s, G_s, and ν_s are respectively the elasticity modulus, shear modulus and Poisson's ratio of the soil. Having defined $\mathrm{v_p}$ and $\mathrm{v_s}$ as being velocities of the P-wave and S-wave respectively by:

$$\begin{aligned} \mathrm{v}_p^2 &= \frac{E_s}{\rho_s}\,\frac{(1-\nu_s)}{(1+\nu_s)(1-2\nu_s)} \\ \mathrm{v}_s^2 &= \frac{G_s}{\rho_s} \end{aligned} \tag{1.186}$$

and using Eq. (1.185) into Eq. (1.184), the dynamic equilibrium equations of the soil element can be stated in terms of derivatives of the displacements as written:

$$\begin{aligned} \frac{\partial^2 u_x}{\partial t^2} &= \mathrm{v}_\mathrm{p}^2 \frac{\partial}{\partial x}\left(\frac{\partial u_x}{\partial x} + \frac{\partial u_z}{\partial z}\right) + \mathrm{v}_\mathrm{s}^2 \frac{\partial}{\partial z}\left(\frac{\partial u_x}{\partial z} - \frac{\partial u_z}{\partial x}\right) \\ \frac{\partial^2 u_z}{\partial t^2} &= \mathrm{v}_\mathrm{p}^2 \frac{\partial}{\partial z}\left(\frac{\partial u_x}{\partial x} + \frac{\partial u_z}{\partial z}\right) - \mathrm{v}_\mathrm{s}^2 \frac{\partial}{\partial x}\left(\frac{\partial u_x}{\partial z} - \frac{\partial u_z}{\partial x}\right) \end{aligned} \tag{1.187}$$

Now, let us define two displacement potentials as ϕ and ψ such that:

$$u_x = \frac{\partial \phi}{\partial x} - \frac{\partial \psi}{\partial z} \quad \text{and} \quad u_z = \frac{\partial \phi}{\partial z} + \frac{\partial \psi}{\partial x} \tag{1.188}$$

and having used these potentials in Eq. (1.187) the following differential equations can be obtained.

$$\begin{aligned} \frac{\partial^2 \phi}{\partial t^2} &= \mathrm{v}_\mathrm{p}^2 \left(\frac{\partial^2 \phi}{\partial x^2} + \frac{\partial^2 \phi}{\partial z^2}\right) \\ \frac{\partial^2 \psi}{\partial t^2} &- \mathrm{v}_\mathrm{s}^2 \left(\frac{\partial^2 \psi}{\partial x^2} + \frac{\partial^2 \psi}{\partial z^2}\right) \end{aligned} \tag{1.189}$$

The solution of these differential equations yields the potential functions, ϕ and ψ, as written:

$$\left.\begin{aligned} \phi &= A\, e^{-\alpha_\mathrm{p} z}\, e^{i(\omega t - kx)} \\ \psi &= B\, e^{-\alpha_\mathrm{s} z}\, e^{i(\omega t - kx)} \end{aligned}\right\} \quad \text{where} \quad \alpha_\mathrm{p}^2 = k^2\left(1 - \frac{c^2}{\mathrm{v}_\mathrm{p}^2}\right),\ \alpha_\mathrm{s}^2 = k^2\left(1 - \frac{c^2}{\mathrm{v}_\mathrm{s}^2}\right) \tag{1.190a}$$

in which A and B are constants to be determined from boundary conditions, k is the wave number and c is the wave velocity as defined by:

$$\left. k = \frac{2\pi}{\lambda},\ c = \frac{\lambda}{T},\ \omega = \frac{2\pi}{T} \right\} \rightarrow \left(k = \frac{\omega}{c}\right) \tag{1.190b}$$

In Eq. (1.190b), λ is the wave length, T and ω are respectively the period and frequency of the wave. As it is seen from Eq. (1.190a), the potential ϕ is characterized by the P-wave and ψ is characterized by the S-wave. Having introduced these potentials from Eq. (1.190a) into Eq. (1.188), the absolute displacements u_x and u_z of the soil can be obtained as:

$$\begin{aligned} u_x &= \left(-ikA\, e^{-\alpha_\mathrm{p} z} + B\alpha_\mathrm{s} e^{-\alpha_\mathrm{s} z}\right) e^{i(\omega t - kx)} \\ u_z &= \left(-A\, \alpha_\mathrm{p} e^{-\alpha_\mathrm{p} z} - ikB\, e^{-\alpha_\mathrm{s} z}\right) e^{i(\omega t - kx)} \end{aligned} \quad \text{with} \quad \left(i = \sqrt{-1}\right) \tag{1.191}$$

Equation (1.191) is general statements of displacements due to plane waves. In order to obtain the R-wave propagation the zero-stress condition ($\sigma_z = 0$ and $\tau_{xz} = 0$) at the ground surface ($z = 0$) is imposed. Having carried out this imposition from Eq. (1.185), the following equations are obtained.

$$\left(-\frac{\nu_s}{(1-\nu_s)}k^2+\alpha_p^2\right)A+ik\alpha_s\left(\frac{1-2\nu_s}{1-\nu_s}\right)B=0$$
$$2ik\alpha_p A-\left(\alpha_s^2+k^2\right)B=0 \tag{1.192a}$$

or using the relations:

$$\frac{(1-2\nu_s)}{(1-\nu_s)}=2\frac{\mathrm{v}_s^2}{\mathrm{v}_p^2} \quad \text{and} \quad \left(\alpha_s^2+k^2\right)=k^2\left(2-\frac{c^2}{\mathrm{v}_s^2}\right) \tag{1.192b}$$

it is written as:

$$k^2\left(2-\frac{c^2}{\mathrm{v}_s^2}\right)A+2ik\alpha_s B=0$$
$$2ik\alpha_p A-k^2\left(2-\frac{c^2}{\mathrm{v}_s^2}\right)B=0 \tag{1.192c}$$

For the nontrivial solution of Eq. (1.192c) its determinate must be zero, which leads to the wave dispersion relation as:

$$\left(2-\frac{c^2}{\mathrm{v}_s^2}\right)^2=\frac{4}{k^2}\alpha_s\alpha_p \;\rightarrow\; c=\mathrm{v}_R \tag{1.193a}$$

where v_R is the velocity of the R-wave propagation as obtained from the solution of the dispersion relation, which leads to:

$$\left(\eta^3-8\eta^2+8\left(\frac{2-\nu_s}{1-\nu_s}\right)\eta-\frac{8}{(1-\nu_s)}=0\right) \quad \text{where} \quad \eta=\left(\frac{\mathrm{v}_R}{\mathrm{v}_s}\right)^2 \tag{1.193b}$$

From Eq. (1.192c) the constant B can be obtained in terms of the constant A as written by:

$$B=\frac{i}{2\alpha_s k}\left(k^2+\alpha_s^2\right)A \quad \text{or} \quad B=i\sqrt{\frac{\alpha_p}{\alpha_s}}A \tag{1.194}$$

Having introduced B from Eq. (1.194) into Eq. (1.191) the displacements u_x and u_z, which will be denoted by u_{gx} and u_{gz} henceforth (the subscript g indicates ground), can be stated as:

$$\left.\begin{array}{l} u_{gx}=-i\,\hat{u}_{g0}\,H_{gx}(z)\,e^{i(\omega t-kx)} \\ u_{gz}=\hat{u}_{g0}\,H_{gz}(z)\,e^{i(\omega t-kx)} \end{array}\right\}\rightarrow\left\{\begin{array}{l} H_{gx}(z)=A\left(-\frac{\alpha_s}{k}\sqrt{\frac{\alpha_p}{\alpha_s}}e^{-\alpha_s z}+e^{-\alpha_p z}\right) \\ H_{gz}(z)=A\left(\sqrt{\frac{\alpha_p}{\alpha_s}}\,e^{-\alpha_s z}-\frac{\alpha_p}{k}e^{-\alpha_p z}\right) \end{array}\right. \tag{1.195a}$$

where $\hat{u}_{g0}$ is the maximum amplitude of the ground displacement at the surface. Since the vertical displacement u_z is greater than the horizontal displacement u_x,

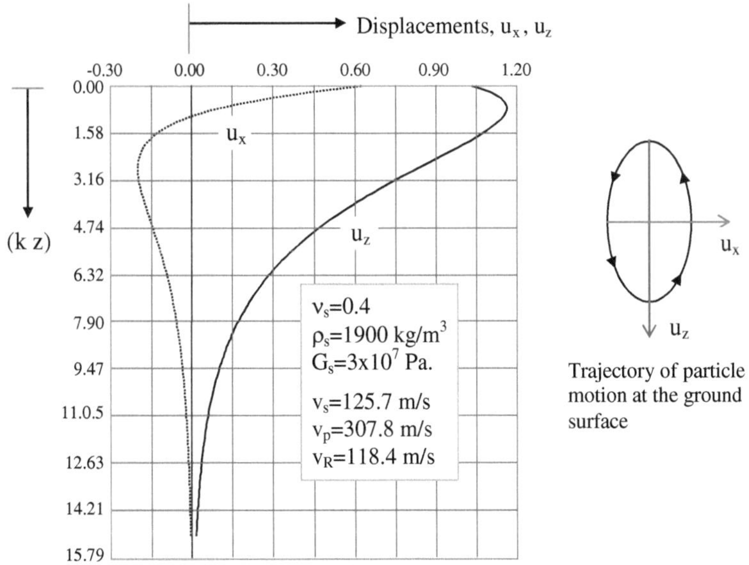

Fig. 1.24 Ground displacements under R-wave propagation and trajectory of particle motion

the amplitude $\hat{u}_{g0}$ is taken as the amplitude of u_z, i.e., $\hat{u}_{g0} = \hat{u}_{zg}(0)$, at the ground surface. The ratio between amplitudes of u_z and u_x at the surface can be obtained as written:

$$\beta = \frac{\hat{u}_{gz}(0)}{\hat{u}_{gx}(0)} \rightarrow \beta = \frac{(k^2 + \alpha_s^2)}{2k\alpha_s} \quad \text{or} \quad \beta = \sqrt{\frac{\alpha_p}{\alpha_s}} \tag{1.195b}$$

The coefficient A may be considered as an amplification factor and determined using $\hat{u}_{g0}$ at the ground surface as written:

$$A = \frac{k}{\alpha_p}\frac{(k^2 + \alpha_s^2)}{(k^2 - \alpha_s^2)} \quad \text{or} \quad A = \frac{2}{\beta}\left(\frac{v_s}{v_R}\right)^2 \tag{1.195c}$$

In the reel domain, the trajectory of the soil particle motion at a depth of z is an elliptical form with the equation expressed as:

$$\left(\frac{\mathrm{Re}.u_{gx}}{H_{gx}(z)}\right)^2 + \left(\frac{\mathrm{Re}.u_{gz}}{H_{gz}(z)}\right)^2 = \hat{u}_{g0}^2 \tag{1.196}$$

The absolute displacement functions, $u_x(z)$ and $u_z(z)$, decay with increasing depth of the ground as shown in Fig. 1.24 for a unit amplitude of $\hat{u}_{zg}(0)$ and soil properties given in the figure. The velocities of waves are calculated to be, $v_P = 307.8$ m/s, $v_s = 125.7$ m/s, and $v_R = 118.4$ m/s. The velocity of the R-wave is somewhat less than that of the S-wave.

Having determined the ground displacements as given in Eq. (1.195a) in the wave propagation coordinates, the rotation of the soil will also be calculated. Since a 2D wave propagation is considered in the (x–z) plane, shown in Fig. 1.23, there is only one rotation in this plane as $\theta_{gy}(z)$, which is vectorially in the y coordinate direction normal to the (x–z) plane. Since a small rectangular element of the soil becomes a parallelogram after the deformation, the rotation can be expressed as:

$$\theta_{gy}(z) = \left(\frac{\partial u_{gx}}{\partial z} - \frac{\partial u_{gz}}{\partial x}\right) \tag{1.197a}$$

Having introduced u_{gx} and u_{gz} from Eq. (1.195a) into Eq. (1.197a) the absolute rotation $\theta_{gy}(z)$ is obtained as stated:

$$\theta_{gy}(z) = i\,\hat{u}_{g0} H_{g\theta}(z) e^{i(\omega t - kx)} \;\rightarrow\; H_{g\theta}(z) = A\left(k\,\beta\left(\frac{V_R}{V_s}\right)^2 e^{-\alpha_s z}\right) \tag{1.197b}$$

As mentioned above other rotations (θ_{gx} and θ_{gz}) are zero in the coordinates of wave propagation. The displacement and rotation vectors of the absolute ground deformation $\{u_{gR}\}_a$ and $\{\theta_{gR}\}_a$ under the R-wave propagation can be written in the coordinates of wave propagation as:

$$\{u_{gR}\}_a = \widehat{u}_{g0} \begin{Bmatrix} -\,i H_{gx}(z) \\ 0 \\ H_{gz}(z) \end{Bmatrix} e^{i(\omega t - kx)} \quad \text{and} \quad \{\theta_{gR}\}_a = \widehat{u}_{g0} \begin{Bmatrix} 0 \\ i H_{g\theta}(z) \\ 0 \end{Bmatrix} e^{i(\omega t - kx)} \tag{1.198}$$

Having determined displacements and rotations of the ground under the R-wave propagation, the exerted consistent load vector of the beam can be calculated using Eqs. (1.176c) and (1.179a) as presented in the next section.

1.5.3.1 Member Consistent Load Vector Due to Ground Deformation Under R-Wave Propagation

Formulation of the member consistent load vector exerted by ground deformations is presented in Sect. 1.5.2. It is calculated from Eq. (1.179a), in which the complex distributed load vector of ground motion $\{q_g\}^*$ is defined by Eq. (1.176c) in member local coordinates as depending on ground displacements and their derivatives. The ground displacements are formulated in the R-wave propagation coordinates as given by Eq. (1.198). To calculate the associated consistent load vector of the member the R-wave ground displacements will be transformed to the member local coordinates. For this purpose, the R-wave propagation coordinates (x_r, y_r, z_r) are defined in the system global coordinates (X, Y, Z), and then transformations are made to obtain ground displacements in the member local coordinates. The R-wave propagates in the x_r direction making an angle φ_r from the

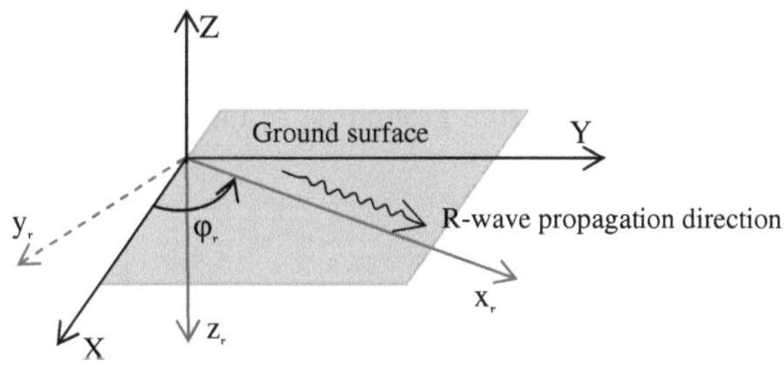

Fig. 1.25 R-wave propagation with respect to global coordinates (*X*, *Y*, *Z*)

global *X* axis as shown in Fig. 1.25 assuming that the ground surface is on the (*X*, *Y*) plane and the *Z* axis is in upward position. The absolute ground displacements and rotations in the global coordinates, $\{u_{GR}\}_a$ and $\{\theta_{GR}\}_a$, can be expressed as:

$$\{u_{GR}\}_a = [t_R]\{u_{gR}\}_a \quad \text{and} \quad \{\theta_{GR}\}_a = [t_R]\{\theta_{gR}\}_a \tag{1.199a}$$

where $[t_R]$ is the transformation matrix defined from Fig. 1.25 as:

$$[t_R] = \begin{bmatrix} \cos\varphi_r & \sin\varphi_r & 0 \\ \sin\varphi_r & -\cos\varphi_r & 0 \\ 0 & 0 & -1 \end{bmatrix} \tag{1.199b}$$

Next step is to transform these displacement and rotation vectors to the member local coordinates, in which they are denoted by $\{u_R\}_a$ and $\{\theta_R\}_a$. Using Eqs. (1.91) and (1.199a) these vectors are written:

$$\begin{aligned} \{u_R\}_a &= [t]\{u_{GR}\}_a \quad & \text{and} \quad \{\theta_R\}_a &= [t]\{\theta_{GR}\}_a \\ \{u_R\}_a &= [t][t_R]\{u_{gR}\}_a \quad & \text{and} \quad \{\theta_R\}_a &= [t][t_R]\{\theta_{gR}\}_a \end{aligned} \tag{1.200}$$

In Eq. (1.200), the transformation matrix $[t]$ is given by Eq. (1.98a). With these definitions the absolute ground deformation vector can be stated:

$$\{u_g\}_a = \begin{Bmatrix} [t][t_R]\{u_{gR}\}_a \\ [t][t_R]\{\theta_{gR}\}_a \end{Bmatrix} \rightarrow \{u_g\}_a = [T_R]\begin{Bmatrix} \{u_{gR}\}_a \\ \{\theta_{gR}\}_a \end{Bmatrix} \tag{1.201a}$$

where the transformation matrix $[T_R]$ is defined:

$$[T_R] = \begin{bmatrix} [t][t_R] & 0 \\ 0 & [t][t_R] \end{bmatrix} \tag{1.201b}$$

In the earthquake analysis, structural deformations are defined as being relative to the ground surface so that absolute deformations of the structure consist of rigid body motion of the ground surface and the relative structural deformations. For the consistency, the ground deformation vector $\{u_g\}$ in the load vector $\{q_g\}^*$ must also be defined relative to the ground surface as written:

$$\{u_g\} = \{u_g(z_r)\}_a - \{u_g(0)\}_a \tag{1.202}$$

This vector of relative ground deformations is stated explicitly as:

$$\{u_g\} = \hat{u}_{g0}\{H_g\}^* e^{i\omega t} \rightarrow \{H_g\}^* = [T_R]\begin{Bmatrix} -i\left(H_{gx}(z_r) - H_{gx}(0)\right) \\ 0 \\ \left(H_{gz}(z_r) - H_{gz}(0)\right) \\ 0 \\ \left(H_{g\theta}(z_r) - H_{g\theta}(0)\right) \\ 0 \end{Bmatrix} e^{-ikx_r} \tag{1.203}$$

where the transformation matrix $[T_R]$ is defined in Eq. (1.201b). In the load vector $\{q_g\}^*$, the derivatives of $\{u_g\}$ are calculated from Eq. (1.203). Having introduced these derivatives into Eq. (1.176c) the load vector becomes:

$$\{q_g\}^* = [k_g]^*\{u_g\} \rightarrow [k_g]^* = [k_s]^* + k^2[L]^* + k^4[L_\alpha][G_s]^* \tag{1.204}$$

in which $[k_s]^*$, $[L]^*$, $[L_\alpha]$ and $[G_s]^*$ are defined, respectively in Eqs. (1.170a), (1.175c), (1.174b) and (1.170b). Since the wave number k is a small quantity, the second and third terms in Eq. (1.204) will be small and may be neglected. Having introduced Eq. (1.203) into Eq. (1.204) the load vector $\{q_g\}^*$ in the member local coordinates can be stated as:

$$\{q_g\}^* = \hat{u}_{g0}[k_g]^*\{H_g(x_r, z_r)\}^* e^{i\omega t} \tag{1.205}$$

in which the vector $\{H_g(x_r, z_r)\}^*$ is a function of the R-wave propagation coordinates x_r and z_r. Due to transformations, these coordinates will depend on the member axial coordinate x, which is the variable of shape functions of the member. Having carried out these transformations the coordinates of R-wave propagation can be obtained in terms of member axial coordinate x as written:

$$\begin{aligned} x_r &= (\cos\varphi_r X_1 + \sin\varphi_r Y_1) + (\cos\varphi_r c_x + \sin\varphi_r c_y)x \\ z_r &= -Z_1 - c_z x \end{aligned} \tag{1.206}$$

in which (X_1, Y_1, Z_1) are the global coordinates of the member end (1) shown in Fig. 1.6, (c_x, c_y, c_z) are the cosine directions of the member axis as defined in Eq. (1.93). By using Eq. (1.205) in Eq. (1.179a) the complex consistent load vector $\{p_g\}^*$ due to ground deformations under the R-wave propagation can be expressed as:

$$\{p_g\}^* = \hat{u}_{g0}\{H_{gp}\}^* e^{i\omega t} \rightarrow \{H_{gp}\}^* = \int_0^\ell [N]^T [k_g]^* \{H_g(x_r, z_r)\}^* dx \tag{1.207}$$

The vector $\{H_{gp}\}^*$ is numerically calculated using the shape functions matrix $[N]$ from Eq. (1.177), $\{H_g(x_r, z_r)\}^*$ from Eq. (1.203), and x_r and z_r from Eq. (1.206). With the calculation of the consistent load vector due to ground deformation, the interface element for the soil–beam interaction is fully determined since the stiffness and damping matrices introduced by the soil have been already explained as presented in Eq. (1.183). So far, at element level, major ingredients are explained to make static and dynamic analyses of a structural system. Since the static analysis is straightforward, attention is paid further to the dynamic analysis. As being preliminary to dynamic response analysis, calculation of natural frequencies and mode shapes of a structural system is outlined in the following section.

1.6 Calculation of Natural Frequencies and Mode Shapes, Eigenvalue Solution

Natural frequencies and mode shapes of structures are important characteristics in structural vibrations. To make a safe design under dynamic excitations, it is useful to know these characteristics to prevent resonance conditions or peak responses, which occur if natural frequencies are at the close proximity of fundamental frequency of an excitation phenomenon, such as earthquakes, waves, and winds. In practice, excitation phenomena are uncontrollable. But, knowing their frequency ranges from recorded or observed data of occurrences is helpful information for designers, such that they can manage a structural design with natural frequencies far away from the excitation frequency region. The natural mode shapes are also important measures of responses since they contribute to response displacements of structural systems under a dynamic loading. The natural frequencies and mode shapes are undamped free vibration characteristics of structures and independent of applied loads. They are calculated from the solution of eigenvalue problems that obtained from free vibrations of structures without damping as explained in this section.

The dynamic equilibrium equation of a structural system is given by Eq. (1.86). For free undamped vibration this equation can be stated:

$$[K]\{D(t)\} + [M]\left\{\ddot{D}(t)\right\} = 0 \tag{1.208}$$

The stiffness matrix $[K]$ and mass matrix $[M]$ of the system are obtained from the assembly process of elements, and the mass matrix also includes concentrated masses applied at some joints of the structure. The solution of Eq. (1.208) suggests a harmonic displacement vector with respect to time t and a vector, say $\{\phi\}$, which is a function of coordinates only. Thus,

$$\{D(t)\} = \{\phi\}\, e^{i\omega t} \quad \rightarrow \quad \omega: \text{ angular frequency (rad/s)} \tag{1.209}$$

Having introduced $\{D(t)\}$ from Eq. (1.209) into Eq. (1.208) it is obtained that

$$\left([K]-\omega^2[M]\right)\{\phi\}=0 \tag{1.210}$$

which is a general eigenvalue problem [58] with symmetrical matrices $[K]$ and $[M]$. Here, ω is a natural frequency and $\{\phi\}$ is the corresponding eigenmode vector. The frequency ω and eigenmode vector $\{\phi\}$ are not independent and they correspond to each other, i.e., for the natural frequency ω_k the eigenmode vector will be $\{\phi\}_k$. and vice versa. The number of eigenmodes of a structural system is equal to the number DOF, each of which satisfies the eigenvalue problem given by Eq. (1.210). There are some properties of eigenmode vectors that used in the response calculation as presented here. For this purpose, let us consider two eigenmodes as k and r so that

$$\left([K]-\omega_k^2[M]\right)\{\phi\}_k=0 \quad \text{and} \quad \left([K]-\omega_r^2[M]\right)\{\phi\}_r=0 \tag{1.211}$$

We multiply the first one by the transpose of eigenmode vector $\{\phi\}_r$, to obtain:

$$\{\phi\}_r^T\left([K]-\omega_k^2[M]\right)\{\phi\}_k=0 \tag{1.212}$$

Having stated the terms of $[K]$ and $[M]$ in terms of other at a time by using Eq. (1.211) and the symmetry property of $[K]$ and $[M]$, and inserted it into Eq. (1.212) the following equations are obtained.

$$\left(\omega_r^2-\omega_k^2\right)\{\phi\}_r^T[M]\{\phi\}_k=0 \quad \text{and} \quad \left(1-\omega_k^2/\omega_r^2\right)\{\phi\}_r^T[K]\{\phi\}_k=0 \tag{1.213a}$$

Since natural frequencies ω_k and ω_r are not zero, for unequal eigenmodes k and r, i.e., $(\omega_k \neq \omega_r)$, the only conditions to satisfy Eq. (1.213a) are:

$$\{\phi\}_r^T[M]\{\phi\}_k=0 \quad \text{and} \quad \{\phi\}_r^T[K]\{\phi\}_k=0 \tag{1.213b}$$

These conditions of eigenmodes are known as the *orthogonality properties* and play very important role in the response calculation. These properties are also valid for orthogonal eigenmodes (free vibrations in perpendicular directions occurring with symmetrical structures) with equal frequencies $(\omega_k=\omega_r)$. If the two eigenmodes are same, i.e., $(k = r)$, and thus $(\{\phi\}_k=\{\phi\}_r)$, Eq. (1.213b) becomes:

$$\{\phi\}_r^T[M]\{\phi\}_r=m_r \quad \text{and} \quad \{\phi\}_r^T[K]\{\phi\}_r=k_r \tag{1.214}$$

Here m_r and k_r are the measures of the mass and stiffness of an equivalent vibration system with a single degree of freedom. These quantities are known as the modal mass and stiffness, or more commonly *the generalized mass and stiffness*, which are also used in this book. Their quantities depend on amplitudes of eigenmodes. In practice, it is customary to specify the largest element of an eigenmode to equal one and, by scaling, adjust the remaining elements accordingly. This is known as *the normalization process* and the resulting vectors are known as the *normal, or normalized, eigenmodes*, which are called just eigenmodes henceforth in this book.

Sometimes, the normalization process is carried out such that ($m_r = 1$) or ($k_r = 1$). In this book, the normalization process with the largest element equal to one is used. In order to find natural frequencies and mode shapes the eigenvalue problem given by Eq. (1.210) must be solved. A simplest one is to use the Rayleigh quotient to calculate approximate natural frequency, which derived from Eq. (1.212). When the eigenmodes k and r are the same, it is expressed that

$$\omega_r^2 = \frac{k_r}{m_r} \rightarrow \omega_r^2 = \frac{\{\phi\}_r^T [K] \{\phi\}_r}{\{\phi\}_r^T [M] \{\phi\}_r} \tag{1.215a}$$

In the light of this expression, for a an arbitrary vector $\{X\}$, the natural frequency is approximated from,

$$\omega^2 = \frac{\{X\}^T [K] \{X\}}{\{X\}^T [M] \{X\}} \tag{1.215b}$$

It can be shown [95] that $\left(\omega_1^2 \leq \omega^2 \leq \omega_q^2\right)$. If we vary the vector $\{X\}$ the minimum of ω^2 will be obtained as ω_1^2 with the eigenvector $\{\phi\}_1$. The following section presents a brief review of solutions and most commonly used methods.

1.6.1 Eigenvalue Solution

Natural frequencies and mode shapes of structural systems are very important characteristics in the dynamic analysis. In practice, the dynamic analysis is usually carried out by using contributions of a few dominant natural modes so that solution of the structural eigenvalue problem is required. The algebraic eigenvalue problems and their solutions were studied for a long time and the solution algorithms were reported in general in many books, see e.g., [92–94]. Some of well-known solution methods, such as QR, Jacobi and Housholder transformation methods [95], deal with calculating all eigenpairs which may be time consuming for large structural systems. From the engineering point of view, only a limited number of dominant eigenpairs are needed in practice so that solution methods which calculate a limited number of dominant eigenpairs are more attractive than other methods. In practice, the inverse iteration (power method), the subspace iteration [95] and the Lanczos algorithm [96] are well-known and largely used to calculate a limited number of eigenpairs. In order to calculate higher eigenpairs by the inverse iteration method a matrix deflation process, or the Gram–Schmidt orthogonalization [95], must be used while the subspace iteration method and the Lanczos algorithm can directly calculate the required number of eigenpairs. The inverse and the subspace iterations are known as numerically stable and the Lanczos algorithm displays a poor stability although its computational performance is rather good [96]. The subspace iteration method can converge in a few steps if the initial start vectors are properly estimated. It is, therefore, said that the method is powerful. If, however, the initial start vectors are poorly conditioned, extra iteration steps need to be carried out. This operation slows down the computation performance considerably, because the Rayleigh–Ritz procedure [58] is applied in each iteration step which may be time consuming. This drawback can be overcome by disregarding some of the Ritz procedure from certain iteration steps and obtaining a higher convergence rate [97]. This accelerated technique is equivalent to combining both the standard subspace and the power iteration methods. Since eigenvalue solution algorithms and related programs are available in many libraries for scientific calculations there are not considered further in this book, except for power and subspace iteration algorithms which are outlined in the next sections for the completeness of the book.

1.6.1.1 Power Iteration Method

Power, or vector, iteration method is an effective and powerful eigenvalue solution method to calculate few lowest natural frequencies and mode shapes that are sufficient to obtain acceptable responses. The method is based on the calculation of an algebraic eigenvalue from the power of a matrix, say $[A]$, using the eigenvalue property:

$$[A]^n\{X\} = \lambda^n\{X\} \tag{1.216}$$

where the power n is a large integer number, $\{X\}$ is an arbitrary vector and λ is the eigenvalue of the matrix $[A]$. The arbitrary vector $\{X\}$ can be stated as a linear combination of the eigenvectors of the matrix $[A]$:

$$\{X\} = \sum_{i=1}^{q} c_i\{\phi\}_i \tag{1.217}$$

in which c_i and $\{\phi\}_i\,(i = 1, 2, \ldots, q)$ are constants and eigenvectors of $[A]$. Having introduced this vector into the left hand side of Eq. (1.216) and rearranging the equation it can be expressed that

$$\sum_{i=1}^{q} c_i\lambda_i^n\{\phi\}_i = \lambda^n\{X\} \;\rightarrow\; \lambda_1^n\left(c_1\{\phi\}_1 + \sum_{i=2}^{q}\left(\frac{\lambda_i}{\lambda_1}\right)^n c_i\lambda_i^n\{\phi\}_i\right) = \lambda^n\{X\} \tag{1.218}$$

We assume that all eigenvalues are ordered from the biggest to the smallest as written $\lambda_1 > \lambda_2 > \cdots > \lambda_q$. It can be seen from Eq. (1.218) that, for a sufficiently large number n, the eigenvalue λ approaches λ_1 and the vector $\{X\}$ approaches the eigenvector $\{\phi\}_1$. This is the basis of the power iteration method with a systematic calculation algorithm. In the structural analysis, the eigenvalue problem is stated from Eq. (1.210) as

$$[K]\{\phi\} = \omega^2[M]\{\phi\} \;\rightarrow\; \frac{1}{\omega^2}\{\phi\} = [K]^{-1}[M]\{\phi\} \tag{1.219}$$

The iterative solution of this equation is known as the inverse power method. The iteration is started with an initial guess vector $\{X\}_0$ of the vibration mode so that

$$[K]\{X\}_{k+1} = \{Y\}_k \;\rightarrow\; \{Y\}_k = [M]\{X\}_k \text{ where } (k = 0, 1, 2, \ldots, n) \tag{1.220}$$

After the vector $\{X\}_{k+1}$ is calculated it is normalized to obtain a unit value for the largest element. Then the natural frequency is estimated from the Rayleigh quotient using Eq. (1.215b). Thus, at the end of iteration $(k + 1)$, it is:

$$\left(\omega^2\right)_{k+1} = \frac{\{X_{k+1}\}^T[K]\{X_{k+1}\}}{\{X_{k+1}\}^T[M]\{X_{k+1}\}} \;\rightarrow\; \left(\omega^2\right)_{k+1} = \frac{\{X_{k+1}\}^T\{Y\}_k}{\{X_{k+1}\}^T\{Y\}_{k+1}} \tag{1.221}$$

The iteration continues until a required precision is obtained. At the end of iteration, $\left(\omega^2\right)_{q+1}$ approaches ω_1^2 and $\{X\}_{k+1}$ approaches the eigenvector $\{\phi\}_1$ of the lowest eigenmode.

For the calculation of higher eigenmodes, a sweeping or mass deflation process is used. It is assumed that the first lowest $(p \quad 1)$ eigenmodes are calculated and we intend to calculate the (p)th eigenmode. For this purpose we use a mass matrix:

$$[M]_p^* = \left(I - \sum_{i=1}^{p-1} \frac{1}{k_i}[K]\{\phi\}_i\{\phi\}_i^T\right)[M] \tag{1.222}$$

in which I denotes a unit matrix and k_i is the generalized stiffness for the mode i. Eq. (1.222) is the definition of the deflated mass matrix from which the first $(p - 1)$ eigenvectors are eliminated. Using this mass matrix the power iteration converges to the (p)th eigenvector. The iteration follows with selecting an initial guess vector $\{X\}_0$ and proceeds similar to Eq. (1.220) as stated:

$$[K]\{X\}_{k+1} = \{Y\}_k^* \;\rightarrow\; \{Y\}_k^* = [M]_p^*\{X\}_k \text{ where } (k = 0, 1, 2, \ldots, n) \tag{1.223a}$$

Having introduced the deflated mass matrix into Eq. (1.223a) the vectors $\{Y\}_k^*$ and $\{X\}_{k+1}$ can be written:

$$\begin{aligned} \{Y\}_k^* &= \{Y\}_k - \sum_{i=1}^{p-1}\left(\frac{\{\phi\}_i^T\{Y\}_k}{k_i}[K]\{\phi\}_i\right) \;\rightarrow\; \{Y\}_k = [M]\{X\}_k \\ \{X\}_{k+1} &= [K]^{-1}\{Y\}_k - \sum_{i=1}^{p-1}\left(\frac{\{\phi\}_i^T\{Y\}_k}{k_i}\{\phi\}_i\right) \end{aligned} \tag{1.223b}$$

For an arbitrary vector $\{X\}$, let us calculate the generalized mass, m_{p-1}^*:

$$m_p^* = \{X\}^T [M]_p^* \{X\} \;\rightarrow\; m_p^* = m_x - \sum_{i=1}^{p-1} \frac{k_{xi}}{k_i} m_{ix} \tag{1.224a}$$

The generalized masses, m_x and m_{ix}, and the generalized stiffness k_{xi} are defined as:

$$m_x = \{X\}^T [M]\{X\}, m_{ix} = \{\phi\}_i^T [M]\{X\} \text{ and } k_{xi} = \{X\}^T [K]\{\phi\}_i \tag{1.224b}$$

After the end of iterations, it is now assume that the vector $\{X\}$ approaches an eigenvector, say $\{\phi\}_r$, i.e., $\{X\} \rightarrow \{\phi\}_r$. In this case, from Eqs. (1.224a) and (1.224b) it is seen that

$$\left.\begin{aligned} m_p^* &= m_r \quad \text{if } (r \neq i) \\ m_p^* &= 0 \quad \text{if } (r = i) \end{aligned}\right\} (i = 1, 2, \ldots p - 1) \tag{1.225}$$

For the eigenmodes $(r < p)$ the generalized mass becomes zero, and thus the power iteration with the inflated mass matrix $[M]_p^*$ produces the next lowest eigenmode with $\left(m_p^* = m_p\right)$. Calculation of the eigenvector $\{\phi\}_p$ follows the iterations given in Eqs. (1.223a, b). The corresponding natural frequency is calculated from the Rayleigh quotient:

$$\left(\omega_p^2\right)_{k+1} = \frac{\{X\}_{k+1}^T [K]\{X\}_{k+1}}{\{X\}_{k+1}^T [M]_p^* \{X\}_{k+1}} \;\rightarrow\; \left(\omega_p^2\right)_{k+1} = \frac{\{X\}_{k+1}^T \{Y\}_k^*}{\{X\}_{k+1}^T \{Y\}_{k+1}^*} \tag{1.226}$$

where the vector $\{Y\}_{k+1}^*$ is calculated from Eq. (1.223b) by using the vector $\{X\}_{k+1}$ instead of $\{X\}_k$. The power iteration algorithm is an efficient tool to calculate few lowest natural frequencies and mode shapes of large structural systems. One other powerful method in the eigenvalue solution is the subspace iteration method to calculate a limited number of lowest eigenvalues. Its calculation algorithm is briefly explained in the next section.

1.6.1.2 Subspace Iteration Method

The subspace iteration method is applied to calculate eigenvalues and eigenvectors of matrices as a generalization of the power iteration method explained above. It iterates simultaneously on a number of initial vectors instead of single vector iteration. During the iteration process the orthogonality condition of vectors is used to avoid a linear dependence. The subspace iteration method is an iterative multistep Rayleigh–Ritz procedure which reduces the original eigenvalue problem to that with a limited number of eigenvalues to be calculated. The Rayleigh–Ritz procedure is an approximate method to calculate a limited number of lowest natural frequencies and mode shapes of a general structural eigenvalue problem [95]. In order to explain the Rayleigh–Ritz procedure, the first (q) eigenvectors are denoted by $\{\phi\}_1$, $\{\phi\}_2$, …, $\{\phi\}_q$ with corresponding natural frequencies ω_1, ω_2, …,

ω_q. In the matrix form, these eigenvectors are shown by $[\Phi]$ with dimensions of $(n \times q)$ where n is the total number of DOF of the structural system and q is the number of natural frequencies to be calculated. This limited number of eigenvectors can be approximated by using the Rayleigh–Ritz procedure [95],

$$[\Phi] = [\Psi]\,[Z] \tag{1.227}$$

where the matrix $[\Psi]$ with $(n \times q)$ dimensions consists of predefined Ritz basis vectors, $\{\Psi\}_1$, $\{\Psi\}_2$, ..., $\{\Psi\}_q$ and $[Z]$ with $(q \times q)$ dimensions consists of Ritz coordinates which will be determined by applying the criterion of using minimum Rayleigh quotients. In the matrix notation, the Rayleigh quotients are written

$$\left([\Phi]^T[M][\Phi]\right)[\Omega] = [\Phi]^T[K][\Phi] \;\rightarrow\; \left([Z]^T[\bar{M}]\,[Z]\right)[\Omega] = [Z]^T[\bar{K}]\,[Z] \tag{1.228a}$$

where $[\Omega]$ is a diagonal matrix which contains ω_i^2 with $(i = 1,2, ..., q)$ and the reduced stiffness and mass matrices with $(q \times q)$ dimensions are defined

$$[\bar{M}] = [\Psi]^T\,[M][\Psi] \quad \text{and} \quad [\bar{K}] = [\Psi]^T[K][\Psi] \tag{1.228b}$$

in which $[K]$ and $[M]$ are the system stiffness and mass matrices, The minimum condition of $[\Omega]$ with respect to the Ritz coordinates is obtained from

$$\delta[\Omega] = 0 \;\rightarrow\; \delta[Z]^T\left([\bar{K}]\,[Z] - [\bar{M}]\,[Z][\Omega]\right) \;\rightarrow\; [\bar{K}]\,[Z] = [\bar{M}]\,[Z][\Omega] \tag{1.229}$$

which is an eigenvalue problem with the stiffness and mass matrices $[\bar{K}]$ and $[\bar{M}]$ with the natural frequencies, $\omega_1, \omega_2, ..., \omega_q$, of the original structural system. As it is seen from Eq. (1.229) the Ritz coordinates are the solution to the eigenvectors stated in Eq. (1.229). The solution of this reduced eigenvalue problem can be easily carried out by using the generalized Jacobi method [95]. Then, the eigenvectors of the original system, $\{\phi\}_1$, $\{\phi\}_2$, ..., $\{\phi\}_q$, will be estimated using Eq. (1.227). The Rayleigh–Ritz procedure is a one-step calculation and its precision depends on the choice of the Ritz basis vectors, $\{\Psi\}_1$, $\{\Psi\}_2$, ..., $\{\Psi\}_q$, so that it may not always produce acceptable results. In order to obtain more accurate results, the subspace iteration is used as explained in the following.

In the subspace iteration method, the Ritz basis vectors are updated after each iteration step and the iteration continues until a required precision is obtained for all eigenvalues that required. In practice, the start vectors can be estimated from the static analysis under Rayleigh loadings which increase the calculation performances considerably. In the power iteration method, the iteration sequence is stated in Eq. (1.220) as written

$$[K][X]_{k+1} = [M][X]_k \quad \text{with} \quad (k = 0, 1, 2, \ldots, n) \tag{1.230}$$

in which $[X]$ is the matrix of iteration vectors, $\{X_1\}$, $\{X_2\}$, ..., $\{X_q\}$ In the subspace iteration method, the algorithm follows the calculation steps:

1. Select a set of initial start vectors, $\{X_1\}_0$, $\{X_2\}_0$, …., $\{X_p\}_0$ included in the matrix $[X]_0$, where (p) is a number of iteration vectors [95] defined from $p = \min. (q + 8, 2q)$.
2. Calculate the Ritz basis vectors, $\{\Psi\}_1$, $\{\Psi\}_2$, …., $\{\Psi\}_p$ in the matrix $[\Psi]$ from the solution of Eq. (1.230) as to be $[\Psi]_{k+1} = [X]_{k+1}$ for the iteration number k.
3. Calculate the reduced stiffness and mass matrices, $[\bar{K}]$ and $[\bar{M}]$, with $(p \times p)$ dimensions using Eq. (1.228b).
4. Solve the eigenvalue problem given in Eq. (1.229) to calculate the Ritz coordinates $[Z]$ with $(p \times p)$ dimensions.
5. Arrange the calculated natural frequencies and eigenvectors, $[Z]$, in ascending order so that $\omega_1 < \omega_2 < \cdots < \omega_p$.
6. Estimate the eigenvectors, $\{\phi\}_1$, $\{\phi\}_2$, …, $\{\phi\}_p$ in the matrix form $[\Phi]_k$ for the iteration k using Eq. (1.227).
7. Check for the convergence of the required (the first q) eigenvalues.
8. If the required eigenvalues converge, then stop the iteration. If not, update the start vectors as $[X]_k = [\Phi]_k$ and go to step (2). Repeat this iteration until a required convergence is obtained.

The performance of the subspace iteration method can be considerably increased using accelerated iterations in which some of the Rayleigh–Ritz procedure is omitted from certain iteration steps and obtaining a higher convergence rate [97, 98]. In the following section, the calculation of natural frequencies and mode shapes of structures with some damaged or deteriorated members is presented.

1.6.2 Eigenvalue Solution of Deteriorated Structures

Deterioration of structural components is a time dependent process that constitutes a failure sequence of structures. The deterioration mechanism reduces kinematic connectivity and cross-sectional properties of members, and thus it leads to member resistance deterioration. Deterioration affects dynamic properties of structures that response results can be changed considerably. Formulation of a calculation model for deteriorated members is presented in Sect. 1.3. In this section, the calculation algorithm of natural frequencies and mode shapes of structures with some deteriorated members is presented. For this purpose, the system stiffness matrix is formulated in a decremental form, and accordingly, the mass is redistributed to member ends in a similar manner. The stiffness and mass matrices of deteriorated members, $[k']$ and $[m']$, are given respectively in Eqs. (1.117b, d). These matrices can be expressed in the form,

$$[k'] = ([k] - [\Delta k]) \quad \text{and} \quad [m'] = ([m] - [\Delta m]) \tag{1.231a}$$

where $[\Delta k]$ is the decremental stiffness matrix and $[\Delta m]$ is a difference mass matrix due to redistribution of the member mass. From Eqs. (1.117b), (d) they are defined

$$[\Delta k] = [k](I - [T]) \quad \text{and} \quad [\Delta m] = \left([m] - [T]^T[m]\,[T]\right) \tag{1.231b}$$

in which $[T]$ is the connectivity matrix of the deteriorated member given in Eq. (1.117a). The stiffness and mass matrices of the structural system can be stated similar to Eq. (1.231a) as written

$$[K'] = ([K] - [\Delta K]\,) \quad \text{and} \quad [M'] = ([M] - [\Delta M]) \tag{1.232}$$

where $[\Delta K]$ and $[\Delta M]$ are obtained from the assembly process of $[\Delta k]$ and $[\Delta m]$ of all deteriorated members, which are symbolically shown for a number of deteriorated members (n_{d}) by

$$[\Delta K] = \sum_{i=1}^{n_{\mathrm{d}}} [\Delta k]_i \quad \text{and} \quad [\Delta M] = \sum_{i=1}^{n_{\mathrm{d}}} [\Delta m]_i \tag{1.233}$$

Having obtained the system stiffness and mass matrices, $[K']$ and $[M']$, the eigenvalue problem can be solved to find natural frequencies and mode shapes of a deteriorated structure by using a solution method explained in previous sections. Since only few members in a structural system may be damaged before a complete failure occurs, the decremental stiffness matrix will be relatively small. Using this property, the calculation algorithm of the eigenvalue problem of a damaged structure can be simplified.

Since the eigenvectors of the damaged structure are close to those of the previous state or undamaged structure, the power iteration method will provide an efficient methodology to calculate natural frequencies and mode shapes of the current state of the damaged structure. The algorithm of the power iteration is given in Eq. (1.223b) in general. By considering the deflated mass matrix this iteration algorithm can be written for the damaged structure:

$$\begin{aligned}
\{Y'\}_k &= ([M]\{X\}_k - [\Delta M]\{X\}_k) \\
\{Y'\}_k^* &= \{Y'\}_k - \sum_{i=1}^{p-1} \left(\frac{\{\phi'\}_i^T \{Y'\}_k}{k_i'} ([K]\{\phi'\}_i - [\Delta K]\,\{\phi'\}_i) \right) \\
\{X\}_{k+1} &= [K']^{-1}\{Y'\}_k^*
\end{aligned} \tag{1.234}$$

where the prime ($'$) denotes the damaged structure. The iteration process requires inversion of the stiffness matrix $[K']$ which will be carried out approximately using Neumann expansion [99]. From Eq. (1.232) it is written

$$[K'] = [K](I - [P]\,) \;\rightarrow\; [P] = [K]^{-1}[\Delta K] \tag{1.235a}$$

and the inversion is

$$[K']^{-1} = (I - [P])^{-1}[K]^{-1} \rightarrow (I - [P])^{-1} \simeq \left(I + \sum_{n=1}^{n_N} [P]^n \right) \quad (1.235b)$$

where n_N is the number of expansion for the approximate inversion. Since the matrix $[P]$ is very small, a few expansion terms are sufficient to produce an acceptable inversion. Having introduced Eq. (1.235b) into Eq. (1.234) it can be stated that

$$\{X\}_{k+1} = \{Z'\}_k + \sum_{n=1}^{n_N} \{Q_n\}_k - \sum_{i=1}^{p-1} \left(\frac{\{\phi'\}_i^T \{Y'\}_k}{k_i'} \{\phi'\}_i \right) \quad (1.236a)$$

where the vectors $\{Z'\}_k$ is calculated from

$$\{Z'\}_k = [K]^{-1} \{Y'\}_k \quad (1.236b)$$

and vector $\{Q_n\}_k$ is defined and calculated from the following recursive relation:

$$\{Q_n\}_k = [P]^n \{Z'\}_k \rightarrow \{Q_n\}_k = [P]\{Q_{n-1}\}_k \text{ with } (\{Q_0\}_k = \{Z'\}_k) \quad (1.236c)$$

or using $[P]$ from Eq. (1.235a) it s stated that

$$\{Q_n\}_k = [K]^{-1} \{\Delta Q_{n-1}\}_k \text{ with } \{\Delta Q_0\}_k = [\Delta K]\{Z'\}_k \quad (1.236d)$$

The iteration is carried out until a required convergence is obtained to calculate natural frequency and mode shape of the damaged structure for the eigenmode p. It is explained step by step below.

The Iteration Steps:

1. Select an initial start vector of the eigenmode (p), which is equal to that of the previous state or undamaged structure, i.e., $\{X\}_0 = \{\phi\}_p$.
2. Calculate the vector $\{Y'\}_k$ from Eq. (1.234).
3. Solve Eq. (1.236b) to obtain the vector $\{Z'\}_k$ and consequently to calculate the vector $\{Q_n\}_k$ from Eq. (1.236d) using the recursive relation.
4. Update the iteration vector to $\{X\}_{k+1}$ using Eq. (1.236a) and normalize it to obtain a unit value at the largest element.
5. Calculate the Rayleigh quotient to estimate the natural frequency from

$$\left(\omega_p^2\right)'_{k+1} = \frac{\{X\}_{k+1}^T [K'] \{X\}_{k+1}}{\{X\}_{k+1}^T [M']_p^* \{X\}_{k+1}} \rightarrow \left(\omega_p^2\right)'_{k+1} = \frac{\{X\}_{k+1}^T \{Y'\}_k^*}{\{X\}_{k+1}^T \{Y'\}_{k+1}^*} \quad (1.237)$$

in which the vector $\{Y'\}_{k+1}^*$ is calculated from Eq. (1.234) by using $\{X\}_{k+1}$ instead of $\{X\}_k$.

6. Check for convergence. If it converges then stop the iteration. If not, go to step 3 for further iteration with $\{Y'\}_{k+1}$ which is calculated at the step 5.

Since the initial start vector is very close to the eigenvector of the damage structure a few iterations (one or two) may be sufficient to obtain correct eigenvectors. The natural frequencies can be roughly calculated from the Rayleigh quotient using eigenvectors of the undamaged structure. For the eigenmode p, it can be calculated approximately from

$$\left(\omega_p^2\right)' = \frac{\{\phi\}_p^T[K']\{\phi\}_p}{\{\phi\}_p^T[M']\{\phi\}_p} \rightarrow \left(\omega_p^2\right)' = \left(\frac{1 - \Delta k_p/k_p}{1 - \Delta m_p/m_p}\right)\omega_p^2 \tag{1.238}$$

$$\Delta k_p = \{\phi\}_p^T[\Delta K]\{\phi\}_p \text{ and } \Delta m_p = \{\phi\}_p^T[\Delta M]\{\phi\}_p$$

where k_p and m_p are the generalized stiffness and mass of the undamaged structure for the eigenmode p while ω_p is the natural frequency. In practice, since the stiffness and mass matrices of the undamaged structure [K] and [M] are available once, only the decremental stiffness and mass matrices [ΔK] and [ΔM] need to be updated for each deterioration state and accordingly the natural frequencies and mode shapes are calculated as explained above. The method presented here is suitable for the reliability analysis and optimization problems and also for the dynamic progressive collapse analysis of structures, since the solution uses the original or previous state stiffness and mass matrices. Having determined the natural frequencies and mode shapes the dynamic response analysis of structures are presented briefly in the following section.

1.7 Dynamic Response Analysis

Structures are frequently subjected to time dependent loading histories known as dynamic loading, which produce responses being also time dependent. Excitations of dynamic loading may be continues, which occur with e.g., sea waves for offshore structures, winds for onshore and offshore structures, traffics for bridges and viaducts, or may be periodic and random occurrences such as earthquakes, impacts, and other natural phenomena that may cause catastrophic results. To prevent such undesirable consequences, structures must be designed to withstand dynamic loadings. The design requirement demands knowledge on dynamic response histories or characteristics that is achieved by the solution of the dynamic equilibrium equation of the structural system, which is given by Eq. (1.86) as rewritten

$$[K]\{D(t)\} + [C]\{\dot{D}(t)\} + [M]\{\ddot{D}(t)\} = \{P(t)\} \tag{1.239}$$

There are different solution methods of this linear differential equation being either in the time or frequency domain. In this section, most commonly used methods will be outlined.

1.7.1 Time-Domain Solution

In the determination of a dynamic response history, time-domain solution algorithms using direct numerical integration are general methods since they can also be applied to the solution of nonlinear problems. Most commonly used direct numerical integration methods in the practice are the central difference, Houbolt, Newmark-β, Wilson-θ methods [95, 100, 101]. However, since only few natural modes are participated in the response calculation of dynamically sensitive structures, the mode superposition method becomes more attractive and efficient in the response calculation. These methods are explained in the following sections.

1.7.1.1 The Central Difference Method

The central difference method is one of the most widely used among numerical integration methods in structural dynamics. The central difference method has the highest accuracy and maximum stability limit [100]. However, its disadvantage is the requirement of small time steps. The central difference method is based on the central difference formulas

$$\begin{aligned} \{\ddot{D}\}_t &= \frac{1}{\Delta t^2}\left(\{D\}_{t+\Delta t} - 2\{D\}_t + \{D\}_{t-\Delta t}\right) \\ \{\dot{D}\}_t &= \frac{1}{2\Delta t}\left(\{D\}_{t+\Delta t} - \{D\}_{t-\Delta t}\right) \end{aligned} \tag{1.240}$$

where $(t - \Delta t)$, t, and $(t + \Delta t)$ are three successive time levels. The displacement solution at time $(t + \Delta t)$ is obtained by considering the discrete equation at time t. Substituting $\{\ddot{D}\}_t$ and $\{\dot{D}\}_t$ from Eq. (1.240) into Eq. (1.239) and rearranging the terms, the displacement vector at the time $(t + \Delta t)$ can be obtained from the solution of the equation

$$[\tilde{K}]\{D\}_{t+\Delta t} = \{\tilde{P}\}_t \tag{1.241a}$$

where $[\tilde{K}]$ and $\{\tilde{P}\}_t$ are effective stiffness matrix and load vector defined as

$$\begin{aligned} [\tilde{K}] &= \left(\frac{1}{\Delta t^2}[M] + \frac{1}{2\Delta t}[C]\right) \\ \{\tilde{P}\}_t &= \{P\}_t - \left([K] - \frac{2}{\Delta t^2}[M]\right)\{D\}_t - \left(\frac{1}{\Delta t^2}[M] - \frac{1}{2\Delta t}[C]\right)\{D\}_{t-\Delta t} \end{aligned} \tag{1.241b}$$

The solution algorithm starts with the initial values of displacements, velocities and accelerations at time ($t = 0$) to calculate displacements at time ($t + \Delta t$) in which displacements at time ($t = -\Delta t$) are also required. Using Eq. (1.240) they are calculated from, for ($t = 0$),

$$\{D\}_{t-\Delta t} = \{D\}_t - \Delta t\{\dot{D}\}_t + \frac{\Delta t^2}{2}\{\ddot{D}\}_t \tag{1.242}$$

The Solution Algorithm

1. Select initial values of the displacements, velocities and accelerations as $\{D\}_0$, $\{\dot{D}\}_0$ and $\{\ddot{D}\}_0$, and the time step Δt such that $\Delta t < \Delta t_{cr}$, where Δt_{cr} is the critical time step [95].
2. Calculate the displacement vector $\{D\}_{-\Delta t}$ from Eq. (1.242) at ($t = 0$).
3. Calculate the effective stiffness matrix and load vector, $[\tilde{K}]$ and $\{\tilde{P}\}_t$, from Eq. (1.241b).
4. Calculate the displacement vector $\{D\}_{t+\Delta t}$ from Eq. (1.241a).
5. If required, calculate acceleration and velocity vectors at time t from Eq. (1.240).
6. Go to step (3) to repeat the process for the next time station until the maximum time station is reached, i.e., ($t \le t_{\max}$).

Since the effective stiffness matrix $[\tilde{K}]$ does not contain the stiffness matrix $[K]$ the central difference method is called as *explicit integration method* [100], and it is very efficient for diagonal mass and damping matrices. Because, in this case, solutions of simultaneous linear equations are not required at time stations. If, however, the damping matrix $[C]$ is proportional to the mass matrix $[M]$ only, then this method remains still efficient since the decomposition of the mass matrix is performed only once to solve the simultaneous linear equations for all time stations.

1.7.1.2 The Houbolt Method

This method uses standard finite difference expressions to approximate velocity and acceleration vectors in terms of the displacement vectors. The Houbolt method is based on fitting a cubic polynomial through values of the current displacement vector, which is to be calculated, and the three previously calculated displacement vectors. Having taken the first and second time derivatives of this polynomial at the current time station, the following finite difference expressions of the velocity and acceleration vectors are obtained.

$$\begin{aligned}\{\dot{D}\}_{t+\Delta t} &= \frac{1}{6\Delta t}\left(11\{D\}_{t+\Delta t}-18\{D\}_{t}+9\{D\}_{t-\Delta t}-2\{D\}_{t-2\Delta t}\right)\\ \{\ddot{D}\}_{t+\Delta t} &= \frac{1}{\Delta t^2}\left(2\{D\}_{t+\Delta t}-5\{D\}_{t}+4\{D\}_{t-\Delta t}-\{D\}_{t-2\Delta t}\right)\end{aligned} \tag{1.243}$$

Now, introducing these statements in the equilibrium equation from Eq. (1.239) at time $(t + \Delta t)$ it is obtained that

$$\left[\tilde{K}\right]\{D\}_{t+\Delta t}= \left\{\tilde{P}\right\}_{t+\Delta t} \tag{1.244a}$$

where the effective stiffness matrix and load vector, $\left[\tilde{K}\right]$ and $\left\{\tilde{P}\right\}_{t+\Delta t}$, are defined

$$\begin{aligned}\left[\tilde{K}\right] &= [K] + \frac{11}{6\Delta t}[C] + \frac{2}{\Delta t^2}[M]\\ \left\{\tilde{P}\right\}_{t+\Delta t} &= \{P\}_{t+\Delta t}+\left(\frac{3}{\Delta t}[C] + \frac{5}{\Delta t^2}[M]\right)\{D\}_{t}-\cdots\\ \cdots &-\left(\frac{3}{2\Delta t}[C] + \frac{4}{\Delta t^2}[M]\right)\{D\}_{t-\Delta t}+\left(\frac{1}{3\Delta t}[C] + \frac{1}{\Delta t^2}[M]\right)\{D\}_{t-2\Delta t}\end{aligned} \tag{1.244b}$$

The solution of this equation at time $(t + \Delta t)$ requires knowledge of $\{D\}_t$, $\{D\}_{t-\Delta t}$ and $\{D\}_{t-2\Delta t}$ so that a special starting procedure is essential. For this purpose, one other numerical integration method, e.g., the central difference method, can be used to calculate displacement vectors at a couple of initial time stations, i.e., at $(t = \Delta t)$ and $(t = 2\Delta t)$, with the initial values of $\{D\}_0$, $\{\dot{D}\}_0$ and $\{\ddot{D}\}_0$ at $(t = 0)$. Then, the Houbolt integration procedure is applied to calculate displacements at time stations $(t = 3\Delta t)$ and forth by using a calculation algorithm similar to that of the central difference method with the effective stiffness matrix and load vector given in Eq. (1.244b). Because of appearance of the stiffness matrix $[K]$ in the effective stiffness matrix $\left[\tilde{K}\right]$, the solution of linear simultaneous equations at each time station requires decomposition process, and therefore the Houbolt method is said to be implicit. It is an unconditionally stable, efficient, and practical method for the numerical integration of the dynamic equilibrium equation.

1.7.1.3 The Newmark-β method

To solve the dynamic equilibrium equation of structures the most widely used implicit method of direct time integration is the Newmark-β method. The Newmark-β method is based on the assumption of linear acceleration within each time step. In this method, the following assumptions are used.

$$\begin{aligned}
\{\dot{D}\}_{t+\Delta t} &= \{\dot{D}\}_t + \Delta t\left[(1-\beta)\{\ddot{D}\}_t + \beta\{\ddot{D}\}_{t+\Delta t}\right] \\
\{D\}_{t+\Delta t} &= \{D\}_t + \Delta t\,\{\dot{D}\}_t + \Delta t^2\left[\left(\frac{1}{2}-\alpha\right)\{\ddot{D}\}_t + \alpha\{\ddot{D}\}_{t+\Delta t}\right]
\end{aligned} \tag{1.245}$$

where β and α are integration parameters, which determine the stability and accuracy of the algorithm. For ($\beta = 1/2$) and ($\alpha = 1/6$), the relations in Eq. (1.245) correspond to the linear acceleration method. Newmark had originally proposed an unconditionally stable algorithm with constant acceleration over the time step Δt [95], which is equal to the average of the accelerations at the ends of the time step. In this case, the corresponding parameters become ($\beta = 1/2$) and ($\alpha = 1/4$). In addition to the relations given in Eq. (1.245), the dynamic equilibrium equation given by Eq. (1.239) must also be satisfied at time station ($t + \Delta t$). The acceleration and displacement vectors at time station ($t + \Delta t$) are calculated from Eq. (1.245) as

$$\begin{aligned}
\{\ddot{D}\}_{t+\Delta t} &= \frac{1}{\Delta t^2\alpha}\{D\}_{t+\Delta t} - \frac{1}{\Delta t^2\alpha}\{D\}_t - \frac{1}{\Delta t\alpha}\{\dot{D}\}_t - \frac{1}{\alpha}\left(\frac{1}{2}-\alpha\right)\{\ddot{D}\}_t \\
\{\dot{D}\}_{t+\Delta t} &= \frac{\beta}{\Delta t\alpha}\{D\}_{t+\Delta t} - \frac{\beta}{\alpha\Delta t}\{D\}_t + \left(1-\frac{\beta}{\alpha}\right)\{\dot{D}\}_t + \Delta t\left(1-\frac{\beta}{2\alpha}\right)\{\ddot{D}\}_t
\end{aligned} \tag{1.246}$$

Having introduced these vectors into the dynamic equilibrium equation in Eq. (1.239) at time station ($t + \Delta t$), the following simultaneous linear equations can be obtained.

$$[\tilde{K}]\{D\}_{t+\Delta t} = \{\tilde{P}\}_{t+\Delta t} \tag{1.247a}$$

where the effective stiffness matrix and load vector, $[\tilde{K}]$ and $\{\tilde{P}\}_{t+\Delta t}$, are defined

$$\begin{aligned}
[\tilde{K}] &= [K] + \frac{\beta}{\alpha\Delta t}[C] + \frac{1}{\alpha\Delta t^2}[M] \\
\{\tilde{P}\}_{t+\Delta t} &= \{P\}_{t+\Delta t} + [C]\left[\frac{\beta}{\alpha\Delta t}\{D\}_t + \left(\frac{\beta}{\alpha}-1\right)\{\dot{D}\}_t + \Delta t\left(\frac{\beta}{2\alpha}-1\right)\{\ddot{D}\}_t\right] + \cdots \\
&\cdots + [M]\left[\frac{1}{\alpha\Delta t^2}\{D\}_t + \frac{1}{\alpha\Delta t}\{\dot{D}\}_t + \left(\frac{1}{2\alpha}-1\right)\{\ddot{D}\}_t\right]
\end{aligned} \tag{1.247b}$$

The step by step solution procedure starts at time ($t = 0$) with initial conditions, $\{D\}_0$, $\{\dot{D}\}_0$ and $\{\ddot{D}\}_0$, and selecting the time step Δt, the parameters β and α such that ($\beta \geq 0.5$) and $\left(\alpha \geq 0.25(0.5+\beta)^2\right)$ for unconditional stability [95, 101]. With these initial set up the effective stiffness matrix and load vector are calculated from Eq. (1.247b), and using Eq. (1.247a) the displacement vector $\{D\}_{t+\Delta t}$ at time station ($t + \Delta t$) is calculated.

1.7.1.4 The Wilson-θ Method

To solve the dynamic equilibrium equation of structures, the Wilson-θ method is also recognized as being widely used implicit method of time integration. The Wilson-θ method is similar to the Newmark-β method in the sense that it is based on a linear acceleration over the time interval ($\theta\Delta t$), where ($\theta \geq 1.0$). If ($\theta = 1$), then the method reduces to the linear acceleration method with conditional stability. In linear problems, the Wilson-θ method becomes unconditionally stable for ($\theta \geq 1.37$) [95, 101] so that ($\theta = 1.40$) is usually applied in practice. For any time τ in the interval ($\theta\Delta t$), by using the assumption of linear acceleration, the acceleration vector $\{\ddot{D}\}_{t+\tau}$ can be obtained as written

$$\{\ddot{D}\}_{t+\tau} = \{\ddot{D}\}_t + \frac{\tau}{\theta\Delta t}\left(\{\ddot{D}\}_{t+\theta\Delta t} - \{\ddot{D}\}_t\right) \tag{1.248a}$$

The velocity and displacement vectors at time τ are obtained from integrations of $\{\ddot{D}\}_{t+\tau}$. Using boundary conditions at ($\tau = 0$) it can be obtained that

$$\begin{aligned}
\{\dot{D}\}_{t+\tau} &= \{\dot{D}\}_t + \tau\{\ddot{D}\}_t + \frac{\tau^2}{2\theta\Delta t}\left(\{\ddot{D}\}_{t+\theta\Delta t} - \{\ddot{D}\}_t\right) \\
\{D\}_{t+\tau} &= \{D\}_t + \tau\{\dot{D}\}_t + \frac{\tau^2}{2}\{\ddot{D}\}_t + \frac{\tau^3}{6\theta\Delta t}\left(\{\ddot{D}\}_{t+\theta\Delta t} - \{\ddot{D}\}_t\right)
\end{aligned} \tag{1.248b}$$

At time ($\tau = \theta\Delta t$) these velocity and acceleration vectors will be

$$\begin{aligned}
\{\dot{D}\}_{t+\theta\Delta t} &= \{\dot{D}\}_t + \frac{\theta\Delta t}{2}\left(\{\ddot{D}\}_t + \{\ddot{D}\}_{t+\theta\Delta t}\right) \\
\{D\}_{t+\theta\Delta t} &= \{D\}_t + \theta\Delta t\{\dot{D}\}_t + \frac{(\theta\Delta t)^2}{6}\left(2\{\ddot{D}\}_t + \{\ddot{D}\}_{t+\theta\Delta t}\right)
\end{aligned} \tag{1.249}$$

from which the velocity and acceleration vectors at time ($t + \theta\Delta t$) are determined in terms of $\{D\}_{t+\theta\Delta t}$ as written

$$\begin{aligned}
\{\dot{D}\}_{t+\theta\Delta t} &= \frac{3}{\theta\Delta t}\left(\{D\}_{t+\theta\Delta t} - \{D\}_t\right) - 2\{\dot{D}\}_t - \frac{\theta\Delta t}{2}\{\ddot{D}\}_t \\
\{\ddot{D}\}_{t+\theta\Delta t} &= \frac{6}{(\theta\Delta t)^2}\left(\{D\}_{t+\theta\Delta t} - \{D\}_t\right) - \frac{6}{\theta\Delta t}\{\dot{D}\}_t - 2\{\ddot{D}\}_t
\end{aligned} \tag{1.250a}$$

In order to apply the dynamic equilibrium equation at time ($t + \theta\Delta t$) to solve the displacement vector $\{D\}_{t+\theta\Delta t}$ the load vector must also be determined at time ($t + \theta\Delta t$). For this purpose, a linear distribution of the load vector in the time interval ($\theta\Delta t$) is assumed and using extrapolation the following expression can be obtained.

$$\{P\}_{t+\theta\Delta t} = \{P\}_t + \theta\left(\{P\}_{t+\Delta t} - \{P\}_t\right) \tag{1.250b}$$

Having introduced Eqs. (1.250a, b) into the dynamic equilibrium equation the displacement vector $\{D\}_{t+\theta\Delta t}$ can be calculated from

$$\left[\tilde{K}\right]\{D\}_{t+\theta\Delta t}=\left\{\tilde{P}\right\}_{t+\theta\Delta t} \tag{1.251a}$$

where the effective stiffness matrix and load vector, $\left[\tilde{K}\right]$ and $\left\{\tilde{P}\right\}_{t+\theta\Delta t}$, are defined as written

$$\begin{aligned}
&\left[\tilde{K}\right]=[K]+\frac{3}{\theta\Delta t}[C]+\frac{6}{(\theta\Delta t)^2}[M]\\
&\left\{\tilde{P}\right\}_{t+\theta\Delta t}=(1-\theta)\{P\}_t+\theta\{P\}_{t+\Delta t}+[C]\left(\frac{3}{\theta\Delta t}\{D\}_t+2\{\dot{D}\}_t+\frac{\theta\Delta t}{2}\{\ddot{D}\}_t\right)+\cdots\\
&\cdots+[M]\left(\frac{6}{(\theta\Delta t)^2}\{D\}_t+\frac{6}{\theta\Delta t}\{\dot{D}\}_t+2\{\ddot{D}\}_t\right)
\end{aligned} \tag{1.251b}$$

The solution algorithm at time $(t+\Delta t)$ is explained step by step in the following.

The solution algorithm:

1. Set initial values of the displacements, velocities and accelerations at $(t=0)$ as $\{D\}_0$, $\{\dot{D}\}_0$ and $\{\ddot{D}\}_0$, the time step Δt, and choose $(\theta = 1.4)$.
2. Calculate the effective stiffness matrix and load vector from Eq. (1.251b).
3. Calculate the displacement vector $\{D\}_{t+\theta\Delta t}$ from Eq. (1.251a).
4. Calculate the velocity and acceleration vectors, $\{\dot{D}\}_{t+\theta\Delta t}$ and $\{\ddot{D}\}_{t+\theta\Delta t}$ from Eq. (1.250a).
5. Calculate the acceleration vector $\{\ddot{D}\}_{t+\Delta t}$ from Eq. (1.248a), the velocity and displacement vectors, $\{\dot{D}\}_{t+\Delta t}$ and $\{\ddot{D}\}_{t+\Delta t}$, from Eq. (1.248b) at time station $(t+\Delta t)$, i.e., for $(\tau=\Delta t)$.

The Wilson-θ method is considered to be the best available unconditionally stable method and is popular in earthquake engineering studies [101]. Its drawback is to be not suitable for problems involving impact or sudden applied loads since it overestimates significantly the exact solution at the first few time stations when large time steps are used.

1.7.1.5 The Mode Superposition Method

When structures are subjected to strong dynamic excitations, corresponding responses become highly dynamic and static response contributions will be of minor importance. Such structural system is defined as dynamic sensitive and its response behavior will be dynamic dominant. For dynamic sensitive structures, the mode superposition method is the most efficient and powerful method applied to determine linear responses of structural systems. The mode superposition method is based on participations of natural mode shapes to the displacements with some scalar time functions to be determined. Thus, the displacement vector $\{D(t)\}$ is expressed as

$$\{D(t)\} = \sum_{j=1}^{q} z_j(t)\,\{\phi\}_j \tag{1.252}$$

where $z_j(t)$ is a time dependent participation factor, which is known as the generalized displacement for the eigenmode j, q is the number eigenmodes to be included and $\{\phi\}_j$ is the corresponding eigenvector. For dynamic sensitive structures, only few eigenmodes are sufficient to obtain correct results. If static responses are also considerable, then more eigenmodes must be included in Eq. (1.252). The exact results are obtained when all eigenmodes are included, even in the quasi-static response calculation. Having substituted Eq. (1.252) into the dynamic equilibrium equation written in Eq. (1.239) it is stated that

$$\sum_{j=1}^{q}\left(z_j(t)\,[K]\{\phi\}_j + \dot{z}_j(t)\,[C]\{\phi\}_j + \ddot{z}_j(t)[M]\,\{\phi\}_j\right) = \{P(t)\} \tag{1.253a}$$

Having multiplied this statement by $\{\phi\}_j^T$ and taking into account the orthogonality conditions, the linear equations can be decoupled as written, for the egenmode j,

$$k_j\,z_j(t) + c_j\,\dot{z}_j(t) + m_j\,\ddot{z}_j(t) = f_j(t) \;\rightarrow\; f_j(t) = \{\phi\}_j^T\{P(t)\} \tag{1.253b}$$

where k_j, c_j, m_j and f_j are respectively the generalized stiffness, damping, mass and force for the mode j. The generalized damping c_j can be stated in terms of the generalized mass and the damping ratio ξ_j to the critical [102, 103], which is $(c_j = 2\xi_j\omega_j m_j)$. By introducing this term into Eq. (1.253b) the differential equation becomes

$$\ddot{z}_j(t) + 2\xi_j\omega_j\dot{z}_j(t) + \omega_j^2\,z_j(t) = \frac{\omega_j^2}{k_j}f_j(t) \tag{1.254}$$

in which ω_j is the natural frequency for the eigenmode j. This differential equation can be solved using one of numerical integration methods explained in previous sections. An analytical solution of this equation is also available using the Duhamel integral [57, 59, 103], from which it is obtained that, for $(0 \le t \le t_{\text{end}})$ shown in Fig. 1.26,

$$\left.\begin{aligned} z_j(t) = & e^{-\beta t}\left[\frac{(\dot{z}_0 + \beta z_0)}{\omega_d}\sin\omega_d t + z_0\cos\omega_d t\right] + \cdots \\ & \cdots + \frac{1}{k_j}\frac{\omega_j^2}{\omega_d}\int_0^t f_j(\tau)e^{-\beta(t-\tau)}\sin\omega_d(t-\tau)\mathrm{d}\tau \end{aligned}\right\} \rightarrow (0 \le t \le t_{\text{end}}) \tag{1.255a}$$

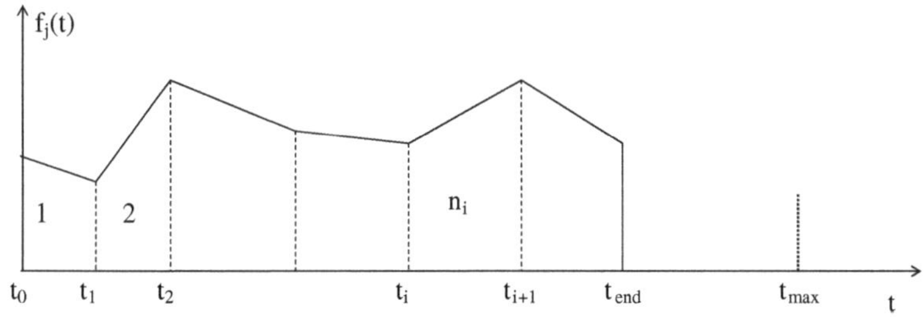

Fig. 1.26 Time discretization of the generalized force $f_j(t)$, and the integration bound t_{max}

and, for $(t_{end} < t \leq t_{max})$ shown in Fig. 1.26, it is

$$z_j(t) = e^{-\beta(t-t_{end})} \left[\begin{array}{l} \frac{(\dot{z}_{end} + \beta z_{end})}{\omega_d} \sin \omega_d (t - t_{end}) + \cdots \\ \cdots + z_{end} \cos \omega_d (t - t_{end}) \end{array} \right] \rightarrow (t_{end} < t \leq t_{max})$$

(1.255b)

In these statements, $\dot{z}_0$ and z_0 are the initial generalized velocity and displacement at $(t = 0)$, $\dot{z}_{end}$ and z_{end} are the generalized velocity and displacement at $(t = t_{end})$, the parameters β and ω_d are respectively a damping term and natural frequency of the damped system which are defined

$$\beta = \xi_j \omega_j \quad \text{and} \quad \omega_d = \omega_j \sqrt{1 - \xi_j^2} \tag{1.255c}$$

For a given, or assumed, function of the generalized force $f_j(t)$, the generalized displacement $z_j(t)$ is calculated from Eqs. (1.255a, b). Having calculated generalized displacements for all eigenmodes included, the system displacement vector is calculated using Eq. (1.252). For linear systems, the mode superposition method is largely used in practice as being the most efficient and powerful calculation tool in dynamic response analysis. For linear structural systems, the dynamic analysis can be better performed in the frequency domain explained in the next section.

1.7.2 Frequency Domain Solution

In the analysis of dynamic responses of linear structures, the frequency domain solution provides a straightforward procedure and enables application of spectral analysis. It is more effective than the time domain solution for linear structures that the principle of superposition holds. The frequency domain solution is based on Fourier transform method, in which a time function $f(t)$ can be stated in its reciprocal frequency function $f(\omega)$. The mutual transformations are written [103] by, for the Fourier transform,

$$f(\omega) = \frac{1}{2\pi} \int_{-\infty}^{\infty} f(t)\, e^{-i\omega t} \mathrm{d}t \tag{1.256a}$$

and for the inverse Fourier transform

$$f(t) = \int_{-\infty}^{\infty} f(\omega)\, e^{i\omega t} \mathrm{d}\omega \tag{1.256b}$$

The displacement and load vectors of a structural system is now expressed using the inverse Fourier transform

$$\{D(t)\} = \int_{-\infty}^{\infty} \{D(\omega)\}\, e^{i\omega t} \mathrm{d}\omega \quad \text{and} \quad \{P(t)\} = \int_{-\infty}^{\infty} \{P(\omega)\}\, e^{i\omega t} \mathrm{d}\omega \tag{1.257}$$

Substituting these statements into the dynamic equilibrium equation, given by Eq. (1.239), the following frequency domain equation can be obtained.

$$\left([K] + i\omega[C] - \omega^2[M]\right) \{D(\omega)\} = \{P(\omega)\} \tag{1.258}$$

from which the displacement vector in the frequency domain is determined as written

$$\{D(\omega)\} = [H(\omega)]_{\mathrm{DP}}\{P(\omega)\} \;\rightarrow\; [H(\omega)]_{\mathrm{DP}} = \left([K] + i\omega[C] - \omega^2[M]\right)^{-1} \tag{1.259}$$

where $[H(\omega)]_{\mathrm{DP}}$ is a matrix of the complex frequency response, which is also known as the structural transfer function matrix between the displacement and load vectors. Eq. (1.259) is the basis of the frequency domain solution. For each variation of the frequency ω, the structural transfer function matrix $[H(\omega)]_{\mathrm{DP}}$ will be calculated, and consequently, the displacement vector $\{D(\omega)\}$ will be calculated. If the time domain solution is required, then it is calculated from the inverse Fourier transform using Eq. (1.257). For each variation of the frequency ω, the calculation of $[H(\omega)]_{\mathrm{DP}}$ from Eq. (1.259) is a time consuming process, and therefore it is not desirable. An alternative and effective solution is to use the mode superposition method in the frequency domain. From Eqs. (1.252) and (1.254) the displacements can be written in the frequency domain as

$$\{D(\omega)\} = \sum_{j=1}^{q} z_j(\omega)\, \{\phi\}_j \quad \text{and} \quad z_j(\omega) = h_j(\omega) f_j(\omega) \tag{1.260a}$$

where $h_j(\omega)$ is the complex frequency response for the eigenmode j, which is obtained to be

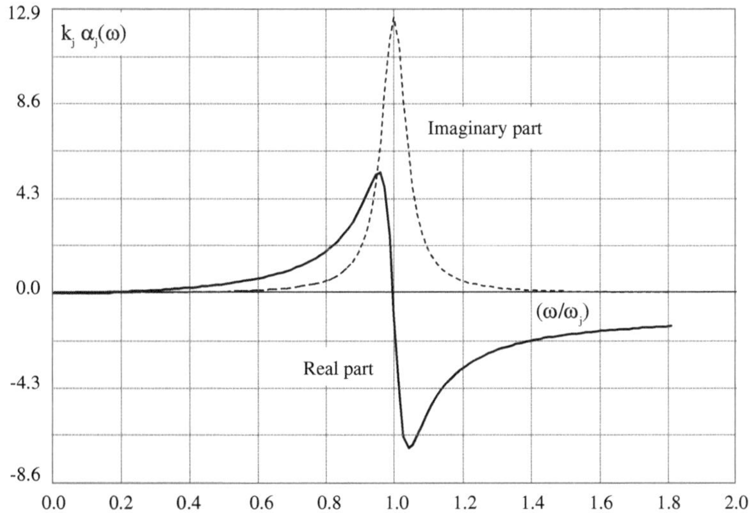

Fig. 1.27 Modal participation factor $\alpha_j(\omega)$ multiplied by k_j for the damping ratio of $(\xi_j = 0.04)$

$$h_j(\omega) = \frac{\omega_j^2}{k_j\left(\omega_j^2 - \omega^2 + 2i\xi_j\omega_j\omega\right)} \tag{1.260b}$$

These formulations correspond to the standard mode superposition method. In order to increase the efficiency of the method, by including only a couple of eigenmodes even in the case of quasi-static response calculation, we modify this standard mode superposition method. To understand the modification, we express Eq. (1.260a) in a different way. By substituting the generalized force $\left(f_j(\omega) = \{\phi\}_j^T\{P(\omega)\}\right)$ into Eq. (1.260a) it can be stated that

$$\{D(\omega)\} = \left(\sum_{j=1}^{q} h_j(\omega)\,\{\phi\}_j\{\phi\}_j^T\right)\{P(\omega)\} \tag{1.261}$$

From the comparison of Eqs. (1.261) and (1.259) it can be written

$$\left([K] + i\omega[C] - \omega^2[M]\right)^{-1} = \left(\sum_{j=1}^{q} h_j(\omega)\,\{\phi\}_j\{\phi\}_j^T\right) \tag{1.262a}$$

which is satisfied if all eigenmodes are included, i.e., $(q = n)$. This statement can be expressed in two terms as being the quasi-static and pure dynamic contributions with the participations of few eigenmodes as

$$\left([K] + i\omega[C] - \omega^2[M]\right)^{-1} = [K]^{-1} + \left(\sum_{j=1}^{q} \alpha_j(\omega)\,\{\phi\}_j\{\phi\}_j^T\right) \tag{1.262b}$$

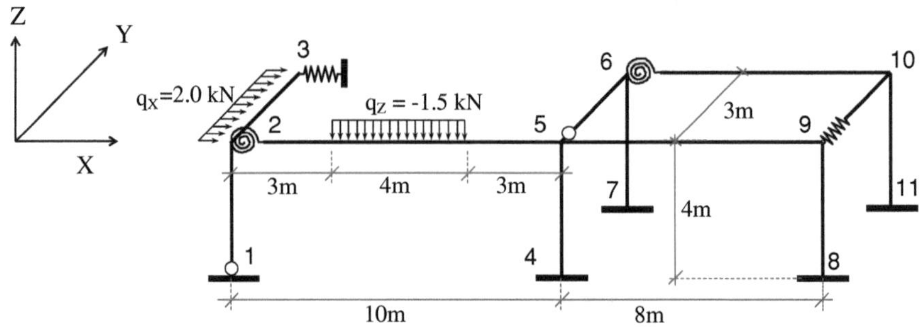

Fig. 1.28 An example portal frame, partial connections, and topological and loading data

Table 1.1 Member dimensions and material properties of the example portal frame, b is the width and h is the height of the cross-section

Members	Cross-sections b/h (cm)	Elasticity modulus E (GPa)	Poisson's ratio ν
Columns	30/30	30.0	0.20
(2–5)	30/40	30.0	0.20
(5–9) and (6–8)	25/40	30.0	0.20
(2–3), (5–6) and (9–10)	20/30	30.0	0.20

in which the dynamic participation factor $\alpha_j(\omega)$ will be determined from the criterion of a matrix inversion, i.e., the product of a matrix by its inverse produces a unit matrix. Thus, from Eq. (1.262b) it is imposed as,

$$\left([K]^{-1} + \left(\sum_{j=1}^{q} \alpha_j(\omega)\,\{\phi\}_j\{\phi\}_j^T\right)\right)\left([K] + i\omega[C] - \omega^2[M]\right) = I \tag{1.263a}$$

Having multiplied this statement by $[K]$ from left and by $\{\phi\}_j$ from right, and then by $\{\phi\}_j^T$ from left it is obtained that

$$\alpha_j(\omega)\,k_j\left(k_j + i\omega c_j - \omega^2 m_j\right) = k_j - \left(k_j + i\omega c_j - \omega^2 m_j\right) \tag{1.263b}$$

from which the eigenmode participation factor $\alpha_j(\omega)$ is determined as written

$$\alpha_j(\omega) = \frac{1}{k_j}\left(\frac{\omega_j^2}{\left(\omega_j^2 - \omega^2 + 2i\xi_j\omega_j\omega\right)} - 1\right) \rightarrow \alpha_j(\omega) = \left(h_j(\omega) - \frac{1}{k_j}\right) \tag{1.264a}$$

This complex scalar function is stated in terms of the real and imaginary parts as

$$\alpha_j(\omega) = \text{Re}\ \alpha_j(\omega) - i\ \text{Im}\ \alpha_j(\omega) \tag{1.264b}$$

where Re $\alpha_j(\omega)$ and Im $\alpha_j(\omega)$ are respectively the real and imaginary parts which are defined below as being functions of the frequency ratio (ω/ω_j).

Table 1.2 Spring coefficients and connectivity ratios of partly connected members

Member	Joint	Released forces	Spring coefficients (r_i)	Connectivity ratios (μ_i)
(1–2)	1	M_x, M_y, M_z	0.0	0.0
(2–3)	3	Q_y	1.45×10^6 N/m	0.3551
(2–5)	2	M_y	6.50×10^6 Nm/rad	0.2535
(5–6)	5	M_x, M_y, M_z	0.0	0.0
(6–10)	6	M_y	9.00×10^6 Nm/rad	0.3115
(9–10)	9	N	3.50×10^8 N/m	0.4929

Table 1.3 Member forces of the example portal frame for rigid connections of members

Member	Joint	N (N)	Q_y (N)	Q_z (N)	M_x (Nm)	M_y (Nm)	M_z (Nm)
1–2	1	2,700.0	−8.4	−1,014.6	115.6	774.4	−11.2
	2	−2,700.0	8.4	1,014.6	−115.6	3,284.0	−22.5
2–3	2	14.9	2,281.6	−13.6	484.9	15.4	357.6
	3	−14.9	3,718.4	13.6	−484.9	25.4	−2,512.7
2–5	2	3,296.2	−23.3	2,713.6	−7.1	−3,769.0	−242.0
	5	−3,296.2	23.3	3,286.4	7.1	6,633.3	9.2
4–5	4	3,468.6	43.6	1,907.3	−50.2	−3,066.8	107.5
	5	−3,468.6	−43.6	−1,907.3	50.2	−4,562.6	66.8
5–6	5	−10.2	−73.4	20.0	−169.3	−49.3	−184.3
	6	10.2	73.4	−20.0	169.3	50.6	−182.6
7–6	7	−37.7	41.6	81.7	−52.7	−145.8	104.8
	6	37.7	−41.6	−81.7	52.7	−181.2	61.5
5–9	5	1,315.5	30.5	162.1	10.4	−1,901.5	125.0
	9	−1,315.5	−30.5	−162.1	−10.4	604.3	119.4
6–10	6	−8.4	31.3	−17.7	10.9	11.9	129.9
	10	8.4	−31.3	17.7	−10.9	129.5	120.8
8–9	8	−169.5	−30.8	1,238.8	−72.2	−2,381.6	−48.2
	9	169.5	30.8	−1,238.8	72.2	−715.4	−28.9
9–10	9	−0.3	−76.7	−7.4	111.0	18.5	−191.6
	10	0.3	76.7	7.4	−111.0	18.4	−192.0
11–10	11	25.1	−31.1	68.3	−71.2	−152.4	−48.4
	10	−25.1	31.1	−68.3	71.2	−18.5	−29.3

$$\left.\begin{aligned} \mathrm{Re}\ \alpha_j(\omega) &= \frac{1}{k_j}\left(\frac{x^2\left(1-x^2-4\xi_j^2\right)}{\left(1-x^2\right)^2+4\xi_j^2x^2}\right) \\ \mathrm{Im}\ \alpha_j(\omega) &= \frac{1}{k_j}\left(\frac{2\xi_j x}{(1-x^2)^2+4\xi_j^2x^2}\right) \end{aligned}\right\} \rightarrow \left(x=\frac{\omega}{\omega_j}\right) \tag{1.264c}$$

Plots of these functions are shown in Fig. 1.27. As it is realized from this figure the dynamic response contribution is effective only in the frequency regions around natural frequencies, i.e., around $(x \rightarrow 1)$ at which $(\omega \rightarrow \omega_j)$. In the regions far from

Table 1.4 Member forces of the example portal frame for partly connected members

Member	Joint	N (N)	Q_y (N)	Q_z (N)	M_x (Nm)	M_y (Nm)	M_z (Nm)
1–2	1	2,402.3	−14.3	−284.7	0.0	0.0	0.0
	2	−2,402.3	14.3	284.7	0.0	1,138.9	−57.1
2–3	2	321.8	4,203.9	11.1	305.6	−33.0	2,695.5
	3	−321.8	1,796.1	−11.1	−305.6	−0.2	916.3
2–5	2	4,488.7	−336.0	2,391.2	−90.0	−1,444.5	−2,695.5
	5	−4,488.7	336.0	3,608.8	90.0	7,532.2	−665.0
4–5	4	3,751.5	187.8	2,400.5	−252.4	−3,962.2	581.7
	5	−3,751.5	−187.8	−2,400.5	252.4	−5,639.7	169.7
5–6	5	−223.9	−27.6	51.3	0.0	0.0	0.0
	6	223.9	27.6	−51.3	0.0	−256.5	−137.8
7–6	7	−81.5	243.9	39.7	−71.5	−100.9	654.2
	6	81.5	−243.9	−39.7	71.5	−58.0	321.3
5–9	5	2,060.6	75.7	91.4	79.6	−1,892.5	412.6
	9	−2,060.6	−75.7	−91.4	−79.6	1,161.4	193.1
6–10	6	−12.2	19.9	−30.2	64.8	58.0	66.3
	10	12.2	−19.9	30.2	−64.8	183.6	93.1
8–9	8	−88.0	−51.9	1,955.9	−83.3	−3,576.6	−59.5
	9	88.0	51.9	−1,955.9	83.3	−1,313.2	−70.4
9–10	9	23.8	−104.7	3.3	151.8	−9.3	−276.5
	10	−23.8	104.7	−3.3	−151.8	−7.5	−246.9
11–10	11	26.9	−43.7	92.5	−153.7	−199.5	−51.9
	10	−26.9	43.7	−92.5	153.7	−31.8	−57.3

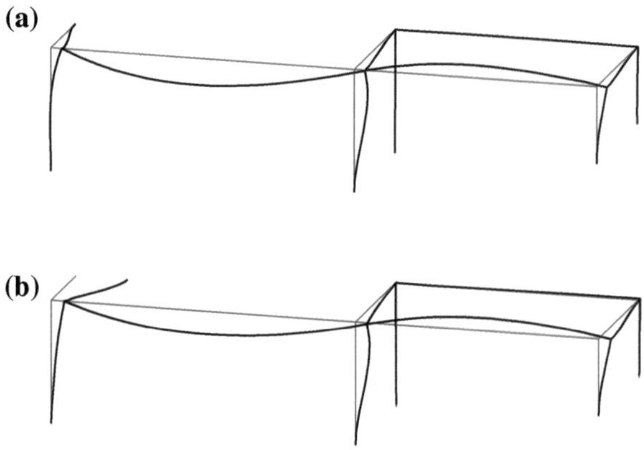

Fig. 1.29 Deformations of the example portal frame. **a** Deformation with rigid member connections. **b** Deformation with partial member connections

natural frequencies, the static response contribution dominates the total response, which is calculated from the equation,

$$\{D(\omega)\} = [K]^{-1}\{P(\omega)\} + \left(\sum_{j=1}^{q} \alpha_j(\omega) f_j(\omega)\{\phi\}_j\right) \tag{1.265}$$

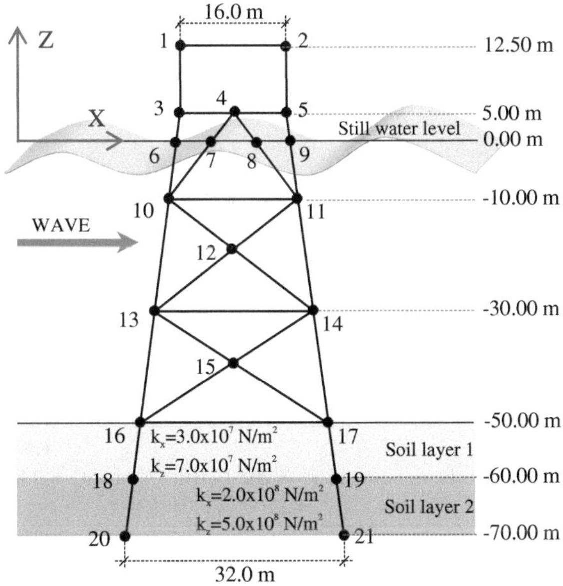

Fig. 1.30 An example 2D jacket structure, joint numbers, and geometrical data

in which the quasi-static contribution is explicitly included. Therefore, it is subtracted from contributions of eigenmodes in the standard mode superposition method as it can be seen from Eq. (1.264a). This modified mode superposition method is more general than the standard mode superposition method since it includes the complete quasi-static contribution and additionally pure dynamic modal contributions. This method will be used further in this book.

1.8 Examples

In this section two examples are presented to demonstrate partly connected members of an example portal space frame, and the static and eigenvalue analyses of an example 2D offshore jacket structure under harmonic wave loading. A simple beam problem is also provided for a hand calculation exercise.

1.8.1 Example of a Portal Frame

In order to demonstrate member releases and partly connected members, which are explained in the Sect. 1.3, an example simple portal frame is analyzed by using the computer program SAPOS [104]. The topology, loadings, and partial connections are shown in Fig. 1.28. The frame is assumed to be reinforced concrete with

Table 1.5 Properties of the soil

Soil layer	Subgrade modulus, k_z (N/m^2)	Axial skin friction, k_x (N/m^2)
1	7.0×10^7	3.0×10^7
2	5.0×10^8	2.0×10^8

Table 1.6 Member dimensions, material properties, and mass of the deck of the 2D jacket

Members	Diameter (mm)	Thickness (mm)
Horizontal bracing at (+12.5 m)	rigid	Rigid
Horizontal bracings at (+5.0 m)	800	8
Horizontal bracings at (−10.0, −30.0, −50.0 m)	1,200	14
Diagonal bracings	1,200	16
Top legs (above +5.0 m)	2,000	50
Inclined legs (below +5.0 m)	1,200	16
Piles	1,400	50
Material properties (steel): $E = 20.5 \times 10^{10}$ N/m^2, $\rho_s = 7{,}800.0$ kg/m^3, $\upsilon_s = 0.25$		
Mass of the deck.......: $M_{\text{deck}} = 2{,}500.0$ ton		

Table 1.7 Wave data for the wave load calculation of 2D jacket

Designation	Value
Wave height	2.5 m
Wave period	6.5 s
Water depth	50.0 m
Drag force coefficient, c_d	1.3
Inertia force coefficient, c_m	2.0
Marine growth thickness	0.25 m
Density of water	1,024.0 kg/m^3

Table 1.8 Wave loads at joints in global directions of the 2D jacket

Joint	Q_X (kN)	Q_Z (kN)	M_Y (kNm)
6	6.08	−0.61	−8.50
7	18.95	−12.00	−38.42
8	16.96	10.74	−37.14
9	20.93	2.09	−31.91
10	8.98	−35.72	124.00
11	44.63	−14.78	−64.57
12	7.65	−8.98	−6.58
13	−0.08	−7.41	28.89
14	9.82	−1.59	6.59
15	0.91	−1.78	−1.75
16	−0.05	−1.02	3.54
17	1.05	0.09	2.19

Table 1.9 Member forces of the left hand-side legs of the 2D jacket under wave loading

Member	Joint	N (kN)	Q_z (kN)	M_y (kNm)
3–1	1	−3.13	0.21	19.78
	3	3.13	−0.21	−18.24
6–3	3	−6.97	7.34	26.63
	6	6.97	−7.34	10.27
10–6	6	−6.97	7.34	−10.27
	10	6.97	1.42	25.52
13–10	10	38.85	2.65	33.28
	13	−38.85	−2.91	16.25
16–13	13	125.35	0.95	10.41
	16	−125.35	−1.73	18.08
18–16	16	171.92	−47.08	−57.80
	18	−129.40	−6.75	−7.81
20–18	18	129.40	6.75	7.81
	20	0.00	0.00	0.00

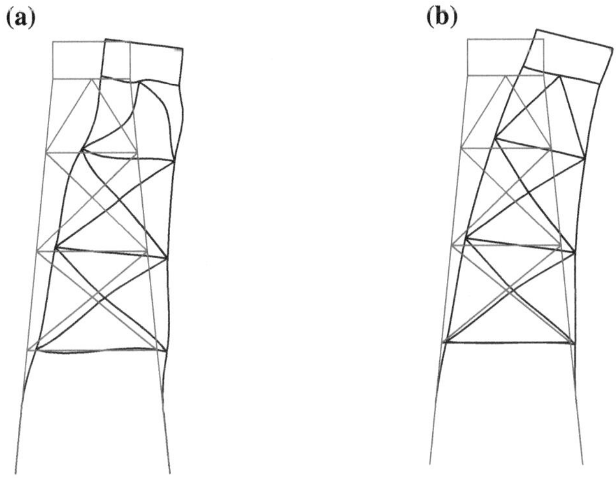

Fig. 1.31 Deformation under wave loading and the first eigenmode shape of the 2D jacket. **a** Deformation under static wave loading **b** First eigenmode shape

rectangle member cross-sections. The member dimensions and material properties are given in Table 1.1. For rectangular cross-sections the effective areas of shear forces, A_y and A_z, and the torsional constant J are calculated from [56, 105],

$$A_y = A_z = \frac{A}{1.2} \quad \text{and} \quad J = \frac{hb^3}{3}\left[1 - 0.63\frac{b}{h}\left(1 - \frac{1}{12}\frac{b^4}{h^4}\right)\right] \quad \text{with} \quad (h \geq b) \tag{1.266}$$

Table 1.10 Lowest three natural frequencies of the 2D jacket

Eigenmode	Natural frequency (rad/s)
1	$\omega_1 = 3.247$
2	$\omega_2 = 12.874$
3	$\omega_3 = 19.332$

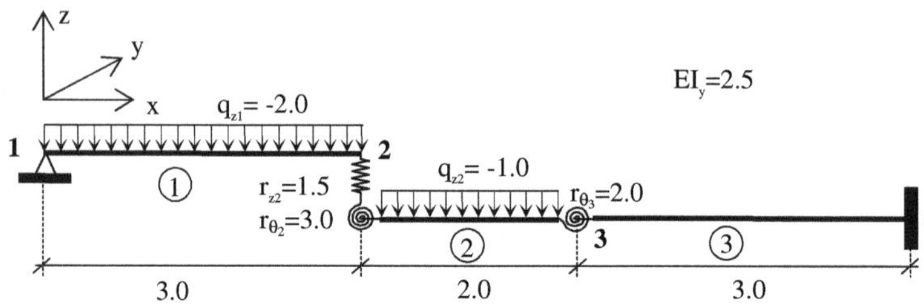

Fig. 1.32 A partly connected simple beam for exercise

The partial connection data of members (spring coefficients and connectivity ratios) are given in Table 1.2. The connectivity ratios are defined as

$$\mu_i = r_i/(r_i + k_{ii}) \quad \text{where} \quad 0.0 \le \mu_i \le 1.0 \tag{1.267}$$

where i a degree of freedom of the member, r_i and k_{ii} are the spring coefficient and diagonal term of the member stiffness matrix for the ith degree of freedom. A zero connectivity ratio means that member is fully released in the related direction and ($\mu_i = 1.0$) indicates a rigid (fixed) connection. The member forces calculated for the rigid and partial connections are presented, respectively in Tables 1.3 and 1.4 whereas corresponding deformations are shown in Fig. 1.29.

1.8.2 Example of 2D Offshore Jacket Structure

The second example is a 2D offshore jacket structure. Static analysis of the structure under harmonic wave loading and eigenvalue solution is carried out using the SAPOS program [104]. The structure is made of tubular steel members and connected on tubular piles in the soil as shown in Fig. 1.30. The underlying soil is modeled as to be of Winkler type consisting of two layers with different characteristics as shown in Fig. 1.30, where joint numbers and geometrical data are also shown. The soil moduli are given in Table 1.5 whereas member dimensions, material properties, and mass of deck are given in Table 1.6. For tubular members, the effective areas of shear forces, A_y and A_z, and the torsional constant J are calculated [105] from,

$$A_y = A_z = \frac{A}{2.0} \quad \text{and} \quad J = \frac{\pi}{32}\left(D_\mathrm{o}^4 - D_\mathrm{i}^4\right) \tag{1.268}$$

where D_o and D_i are respectively outer and inner diameters of the tubular cross-section. The detail of wave load calculation will be presented in Chap. 3. For the wave data given in Table 1.7 the calculated wave loads at joints in the global directions are presented in Table 1.8. The calculated member forces of the legs on the left hand side are written in Table 1.9 and the deformation of the structure is shown in Fig. 1.31a. The lowest three natural frequencies and eigenvectors are calculated and the natural frequencies are given in Table 1.10. The first and third eigenmode shapes are found to be flexural and the second one is in the vertical vibration. The first eigenmode shape is shown in Fig. 1.31b.

1.8.3 A Simple Beam for Exercise

In Fig. 1.32 , a simple beam with partly connected members is given. It is assumed to be a Euler–Bernoulli beam. The joint and member numbers, and all other data necessary for the hand calculation are written in the figure. The beam is free in the (x–z) plane and fixed in the (x–y) plane. It has the same rigidity as EI_y in all parts. Joint displacements and member-end forces are asked to be calculated by using:

1. Differential equations of beams and corresponding boundary conditions.
2. FE calculation with partly connected members, explained in the Sect. 1.3.
3. Standard FE calculation with extra spring members at joints 2 and 3.

Acknowledgments With thanks to ASME for kindly granting permission for the reuse of materials printed in OMAE-2010 [73] as adopted in this chapter in Sect. 1.3.

References

1. Rutenbwg A (1979) Plane frame analysis of laterally loaded asymmetric buildings-an uncoupled solution. Comp Struct 10(3):553–555
2. Wilkinson S, Thambiratnam D (2001) Simplified procedure for seismic analysis of asymmetric buildings. Comp Struct 79(32):2833–2845
3. Kam TY, Lee FS (1986) Nonlinear analysis of steel plane frames with initial imperfections. Comp Struct 23(4):553–557
4. Sophianopoulos DS (2003) The effect of joint flexibility on the free elastic vibration characteristics of steel plane frames. J Constr Steel Res 59(8):995–1008
5. Blandford GE (1988) Static analysis of flexibly connected thin-walled plane frames. Comp Struct 28(1):105–113

6. Dissanayake UI, Burgess IW, Davison JB (2000) Modelling of plane composite frames in unpropped construction. Eng Struct 22(4):287–303
7. Sekulovic M, Salatic R, Nefovska M (2002) Dynamic analysis of steel frames with flexible connections. Comp Struct 80(11):935–955
8. Lui EM, Chen WF (1986) Analysis and behaviour of flexibly-jointed frames. Eng Struct 8(2):107–118
9. Lee HP, Ng TY (1994) In-plane vibration of planar frame structures. J Sound Vib 172(3):420–427
10. Labuschagne A, van Rensburg NFJ, van der Merwe AJ (2009) Comparison of linear beam theories. J Math Comput Modell 49(1–2):20–30
11. The LH (2004) Beam element verification for 3D elastic steel frame analysis. Comput Struct 82(15–16):1167–1179
12. Kim SE, Kim Y, Choi SH (2001) Nonlinear analysis of 3-D steel frames. Thin Walled Struct 39(6):445–461
13. Kumar P, Nukala VV, White DW (2004) A mixed finite element for three-dimensional nonlinear analysis of steel frames. Comput Meth Appl Mech Eng 193(23–26):2507–2545
14. Yang YB (1993) Recent researches on buckling of framed structures and curved beams. J Constr Steel Res 26(2–3):193–210
15. Palaninathan R, Chandrasekharan PS (1985) Curved beam element stiffness matrix formulation. Comput Struct 21(4):663–669
16. Choit J, Lim J (1995) General curved beam elements based on the assumed strain fields. Comput Struct 55(3):379–386
17. Yang SY, Sin HC (1995) Curvature-based beam elements for the analysis of Timoshenko and shear-deformable curved beams. J Sound Vib 187(4):569–584
18. Culver CG, Oestel DJ (1969) Natural frequencies of multispan curved beams. J Sound Vib 10(3):380–389
19. Yoona KY, Parkb NH, Choic YJ, Kangd YJ (2006) Natural frequencies of thin-walled curved beams. Finite Elem Anal Des 42(13):1176–1186
20. Dym CL, Shames IH (1973) Solid mechanics: a variational approach. McGraw·Hill, New York, pp 175–213
21. Boresi AP, Schmidt RJ (2003) Advanced mechanics of materials, 6th edn. Wiley, New York
22. Case J, Chilver L, Ross CTF (1999) Strength of materials and structures, 4th edn. Elsevier, Amsterdam
23. Love AEH (1963) A treatise on the mathematical theory of elasticity, 4th edn. Cambridge University Press, New York
24. Timoshenko SP, Goodier JN (1969) Theory of elasticity, 3rd edn. McGraw-Hill, NewYork
25. Ruge P, Birk C (2007) A comparison of infinite Timoshenko and Euler-Bernoulli beam models on Winkler foundation in the frequency- and time-domain. J Sound Vib 304(3–5):932–947
26. Ganesan N, Engels RC (1992) Hierarchical Bernoulli-Euler beam finite elements. Comput Struct 43(2):297–304
27. Timoshenko SP (1921) On the correction for shear of the differential equation for transverse vibrations of prismatic bars. Philos Mag 41(245):744–746
28. Timoshenko SP (1922) On the transverse vibrations of bars of uniform cross-section. Philos Mag 43(253):125–131
29. Wang CM (1995) Timoshenko beam-bending solutions in terms of Euler–Bernoulli solutions. J Eng Mech 121(6):763–765
30. Davis R, Henshell RD, Warburton GB (1972) A Timoshenko beam element. J Sound Vib 22(4):475–487
31. Bakr EM, Shabana AA (1987) Timoshenko beams and flexible multibody system dynamics. J Sound Vib 116(1):89–107
32. Thomas DL, Wilson JM, Wilson RR (1973) Timoshenko beam finite elements. J Sound Vib 31(3):315–330

33. Horr AM, Schmidt LC (1995) Closed-form solution for the Timoshenko beam theory using a computer-based mathematical package. Comput Struct 55(3):405–412
34. Bazoune A, Khulief YA, Stephen NG (2003) Shape functions of three-dimensional Timoshenko beam element. J Sound Vib 259(2):473–480
35. Lou P, Dai GL, Zeng QY (2006) Finite-element analysis for a Timoshenko beam subjected to a moving mass. Proc Inst Mech Eng Part C J Mech Eng Sci 220(5):669–678
36. Yu W, Hodges DH (2005) Generalized Timoshenko theory of the variational asymptotic beam sectional analysis. J Am Helicopter Soc 50(1):46–55
37. Mikkola A, Dmitrochenko O, Matikainen M (2009) Inclusion of transverse shear deformation in a beam element based on the absolute nodal coordinate formulation. J Comput Nonlinear Dyn 4(1):011004–011012
38. Ghali A, Neville AM, Brown TG (2009) Structural analysis: a unified classical and matrix approach, 6th edn. Spon, London
39. McCormac CJ (2007) Structural analysis: using classical and matrix methods, 4th edn. Wiley, Hoboken
40. Kurrer KE (2008) The history of the theory of structures: from arch analysis to computational mechanics. Ernst & Sohn, Berlin
41. Przemieniecki JS (1968) Theory of matrix structural analysis. McGraw-Hill, New York
42. Zienkiewicz OC, Taylor RL (2005) The finite element method for solid and structural mechanics, 6th edn. Elsevier Butterworth-Heinemann, Oxford
43. Crisfield MA (1986) Finite elements and solution procedures for structural analysis: linear analysis, vol 1. Pineridge Press, Swansea
44. Wunderlich W, Pilkey WD (2003) Mechanics of structures: variational and computational methods, 2nd edn. CRC Press, Boca Raton
45. Slivker VI (2007) Mechanics of structural elements: theory and applications. Springer, Berlin
46. Brenner SC, Scott LR (2008) The mathematical theory of finite element methods, 3rd edn. Springer, New York
47. Chen Z (2005) Finite element methods and their applications. Springer, Berlin
48. Ross CTF (1999) Advanced finite element methods. Horwood, Chichester
49. Bathe KJ, Wilson EL (1976) Numerical methods in finite element analysis. Prentice-Hall, Englewood
50. Omid BN, Rankin CC (1991) Finite rotation analysis and consistent linearization using projectors. Comput Meth Appl Mech Eng 93(3):353–384
51. Ibrahimbegovic A (1997) On the choice of finite rotation parameters. Comput Meth Appl Mech Eng 149:49–71
52. Li M (1998) The finite deformation theory for beam, plate and shell Part III. The three-dimensional beam theory and the FE formulation. Comput Meth Appl Mech Eng 162(1–4):287–300
53. Kim MY, Chang SP, Park HG (2001) Spatial postbuckling analysis of nonsymmetric thin-walled frames. I: theoretical considerations based on semitangential property. J Eng Mech 127(8):769–778
54. Boresi AP, Chong KP (2000) Elasticity in engineering mechanics. Wiley, New York
55. Schramm U, Kitis L, Kang W, Pilkey WD (1994) On the shear deformation coefficient in beam theory. Finite Elem Anal Des 16:141–162
56. Pilkey WD (2002) Analysis and design of elastic beams, computational methods. Wiley, New York
57. Clough RW, Penzien J (1993) Dynamics of structures, 2nd edn. McGraw-Hill, New York
58. Bishop RED, Gladwell GML, Michaelson S (1965) The matrix analysis of vibration. Cambridge University Press, London
59. Meirovitch L (1967) Analytical methods in vibrations. Collier-McMillan Limited, London
60. Woodhouse J (1998) Linear damping models for structural vibration. J Sound Vib 215(3):547–569
61. Liang Z, Lee GC (1991) Representation of damping matrix. J Eng Mech 117(5):1005–1020

62. Hinton E, Rock T, Zienkiewicz OC (1976) A note on mass lumping and related processes. J Earthq Eng Struct Dyn 4:245–249
63. Barltrop NDP, Adams AJ (1991) Dynamics of fixed marine structures, 3rd edn. Butterworth-Heinemann Ltd., Oxford, pp 41–48
64. Chen B, Hu Y, Tan M (1990) Local joint flexibility of tubular joints of offshore structures. Mar Struct 3(3):177–197
65. Bouwkamp JG, Hollings JP, Malson BF, Row DG (1980) Effect of joint flexibility on the response of offshore structures. Proceedings of the offshore technology conference, Texas, USA
66. Fessler H, Mockford PB, Webster JJ (1986) Parametric equations for the flexibility matrices of single brace tubular joint in offshore structures. Proc Inst Civ Eng Part 2 81:659–673
67. Fessler H, Mockford PB, Webster JJ (1986) Parametric equations for the flexibility matrices of multi-brace tubular joint in offshore structures. Proc Inst Civ Eng Part 2 81:675–696
68. Ueda Y, Rashed SMH, Nakacho K (1987) An improved joint model and equations for flexibility of tubular joints. Proceedings of the 6th international symposium offshore Mechanics and Arctic Engineering, Texas, USA
69. Lee DG, Song YH (1993) Efficient seismic analysis of piping systems with joint deformations. Eng Struct 15(1):2–12
70. Hu Y, Chen B, Ma J (1993) An equivalent element representing local flexibility of tubular joints in structural analysis of offshore platforms. Comput Struct 47(6):957–969
71. Chen TY, Zhang HY (1996) Stress analysis of spatial frames with consideration of local flexibility of multiplanar tubular joint. Eng Struct 18(6):465–471
72. Karadeniz H (1994) An algorithm for member releases and partly connected members in offshore structural analysis. Proceedings of the 13th international conference on Offshore Mechanics and Arctic Engineering, OMAE-94, Texas, USA, vol 1, pp 471–476
73. Karadeniz H (2010) A calculation model for deteriorated members of 3D frame structures in the static and dynamic analyses. Proceedings of the 29th international conference on Offshore Mechanics and Arctic Engineering, OMAE-2010, Paper No. OMAE2010-20971, Shanghai, China
74. Baigent AH, Hancock GJ (1982) Structural analysis of assemblages of thin-walled members. Eng Struct 4:207–216
75. Faruque MO (1986) Mechanics of material interfaces. In: Selvadurai APS, Voyiadjis GS (eds) An axisymmetric interface element for soil-structure interaction problems. Elsevier, Amsterdam
76. Boulon M, Garnica P, Vermeer PA (1995) Mechanics of geomaterial interfaces. In: Selvadurai APS, Boulon MJ (eds) Soil-structure interaction: FEM computations. Elsevier, Amsterdam
77. Selvadurai APS (1995) Mechanics of geomaterial interfaces. In: Selvadurai APS, Boulon MJ (eds) Boundary element modeling of geomaterial interfaces. Elsevier, Amsterdam
78. Karadeniz H (2001) Earthquake analysis of buried structures and pipelines based on Rayleigh wave propagation. Int J Offshore Polar Eng 11(2):133–140
79. Miranda C, Nair K (1966) Finite beams on elastic foundation. J Struct Eng 92(ST2):131–141
80. Selvadurai APS (1979) Elastic analysis of soil-foundation interaction. Elsevier, Amsterdam
81. Ting BY (1982) Finite beams on elastic foundation with restrains. J Struct Eng 108(ST3):611–621
82. Ting BY, Mockry EF (1984) Beam on elastic foundation finite element. J Struct Eng 110(10):2324–2339
83. Aydogan M (1995) Stiffness-matrix formulation of beams with shear effect on elastic foundation. J Struct Eng 121(99):1265–1270
84. Yokoyama T (1991) Vibrations of Timoshenko beam-columns on two-parameter elastic foundations. Earthquake Eng Struct Dyn 20:353–370
85. Ishihara K (1996) Soil behaviour in earthquake geotechnics. Oxford University Press, Oxford

86. Toki K, Sato T, Miura F (1981) Separation and sliding between soil and structure during strong ground motion. Earthquake Eng Struct Dyn 9:263–277
87. Lay T, Wallace TC (1995) Modern global seismology. Academic Press, London
88. Takeuchi H, Saito M (1972) Seismic surface waves. In: Bold BA (ed) Methods in computational physics, seismology: surface waves and earth oscillations. Academic Press, London
89. Achenbach JD (1980) Wave propagation in elastic solids. North-Holland, Amsterdam
90. Graff KF (1975) Wave motion in elastic solids. Oxford University Press, London
91. Makris N (1994) Soil-pile interaction during the passage of Rayleigh waves: an analytical solution. Earthq Eng Struct Dyn 23:153–167
92. Wilkinson JH (1965) The algebraic eigenvalue problem. Clarendon Press, Oxford
93. Jennings A (1977) Matrix computation for engineers and scientists. Wiley, London
94. Parlett BN (1980) The symmetric eigenvalue problem. Prentice-Hall, Englewood Cliffs
95. Bathe KJ, Wilson EL (1976) Finite element procedures in engineering analysis. Prentice-Hall, Englewood Cliffs
96. Hudges TJR (1987) The finite element method, linear static and dynamic finite element analysis. Prentice-Hall, Englewood Cliffs
97. Qian Y, Dhatt G (1995) An accelerated subspace method for generalized eigenproblems. Comput Struct 54(6):1127–1134
98. Karadeniz H (1996) An accelerated solution algorithm for the eigenvalue problem in the offshore structural analysis. Proc of ISOPE-96, 26–31 May, Los-Angeles, USA
99. Yamazaki F, Shinozuka M, Dasgupta G (1988) Neumann expansion for stochastic finite element analysis. J Eng Mech 114(8):1335–1354
100. Dokainish MA, Subraraj K (1989) A survey of direct time-integration methods in computational structural dynamics-I: explicit methods. Comput Stuct 32(6):1371–1386
101. Subraraj K, Dokainish MA (1989) A survey of direct time-integration methods in computational structural dynamics-II: implicit methods. Comput Stuct 32(6):1387–1401
102. Biggs JM (1964) Introduction to structural dynamics. McGraw-Hill, New York
103. Humar JL (2002) Dynamics of structures, 2nd edn. Swets & Zeitlinger B.V., Lisse
104. Karadeniz H (2009) SAPOS, spectral analysis program of structures. Report, Structural Mechanics Division, Faculty of Civil Engineering and Geosciences, Delft University of Technology, Delft, The Netherlands
105. Young WC, Budynas RG (2002) Roark's formulas for stress and strain, 7th edn. McGraw-Hill, New York

Chapter 2
Introduction to Random Vibration and Stochastic Analysis

2.1 Introduction

Structures are subjected loading that is mostly time dependent in a weak or strong fashion. Response histories under weakly time-dependent loading may be calculated by using the quasi-static analysis procedure. For moderately or strongly time dependent loading, calculation of response quantities requires a full dynamic analysis procedure as it is presented in the previous chapter assuming that the structure is deterministic and the loading history is fully determined or known, i.e., it obeys a specific rule or a definite function of time such as constant, linear, harmonic, etc. time functions with known properties. Under such a structural and loading case, the corresponding analysis type is called as *the deterministic dynamic analysis* since all necessary parameters of the analysis can be uniquely determined or known. However, the difficulty in the structural dynamic analysis is to determine the loading functions and their properties correctly, such as frequencies, durations, amplitudes, and phases, in practice. Due to lack of sufficient knowledge of dynamic excitations in nature, we possess limited information on loading parameters which is usually obtained from recorded data or observations of occurrences, such as earthquakes and sea waves, which occur in arbitrary fashions. Other examples can be wind loading on high-rise buildings and towers, and traffic loading on bridges and viaducts, which do not follow specific rules. Earthquakes occur periodically in seismic areas with unknown information and sea waves occur continuously with random fluctuation of the sea surface. The only information that we have is based on experiences of past occurrences from which we can predict information of the structural response in a probabilistic manner. When the excitation loading varies arbitrarily in time, the corresponding response will also be arbitrary in time. Such a response process deals with *the random vibration* and its characteristic properties can be determined by using statistical and probabilistic methods. A couple of examples of random loading excitations are shown in Fig. 2.1. For earthquakes, the acceleration of ground motion is recorded

H. Karadeniz, *Stochastic Analysis of Offshore Steel Structures*,
Springer Series in Reliability Engineering, DOI: 10.1007/978-1-84996-190-5_2,

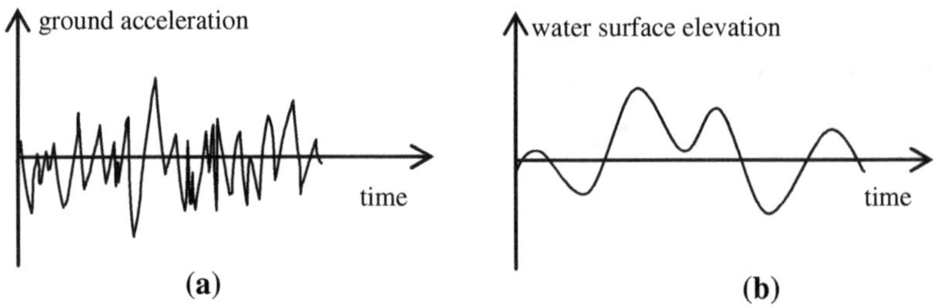

Fig. 2.1 Examples of excitations of random loadings. **a** An example of earthquake record. **b** An example of random sea wave

in different places as illustrated in Fig. 2.1a, and collected for all occurrences to get statistical information. For the sea wave, which always exists, the water surface elevation displays a random fluctuation as shown in Fig. 2.1b that is used as being the input function of wave loads. The problem of random vibrations has been studied for a long time and explained in many text books, see e.g., [1–8], and also reported in numerous papers, see e.g., [9–20]. An overview of historical developments is presented in [21]. In the random vibration theory of linear systems, statistical and probabilistic information of the input function is determined firstly from recorded or observed data that have been collected in the past. Then, by using a stochastic analysis procedure such as a spectral analysis, response statistical characteristics are calculated to be used further in the determination of structural behaviors in probabilistic terms. This subject is closely related to the probability theory and its applications. This chapter is devoted to introduction of the probability theory and stochastic treatment of structural dynamics which is thought to be useful to understand the essence of a probabilistic analysis. In the following sections, basic definitions of random processes and their statistical properties, which are needed in proceeding chapters, are briefly outlined.

2.2 Probability, Random Variables, Stochastic Processes, Probability Distribution and Density Functions

In the theory of random vibrations, some basic principles of the probability theory are applied. They are related to the concept of random phenomena such as random occurrences or outcomes of random experiments. In order to study random vibration, some terminology and definitions from the probability theory are briefly outlined in this section [22, 23].

- *Outcome*. The result of an experiment or occurrence of a natural phenomenon.
- *Random experiment*. An experiment that its outcomes are not predictable in advance.

- *Set.* A collection of individual elements in the domain D. The universal set U is defined if it contains every element in D. The null set $\emptyset$ is defined if it contains no element.
- *Event.* A set of outcomes to which a probability is assigned.
- *Sample space.* A set of all possible outcomes of an experiment, denoted by *S*. Every outcome of the experiment is represented by a point in *S* called as a sample point.
- *Union.* The union of two events *A* and *B*, which is denoted by $(A \cup B)$ or (A or B), is the set of all elements that belong to at least one of the sets *A* and *B*, shown in Fig. 2.2a.
- *Intersection.* The intersection of two events *A* and *B*, which is denoted by $(A \cap B)$ or (*A* and *B*), is the set of elements that belong to both sets *A* and *B*, which is also referred to as a joint event of *A* and *B*, shown in Fig. 2.2b.
- *Complement.* The complement of an event *A*, denoted by $\bar{A}$, is the set containing all points in the sample space *S*, but not in the set *A*, shown in Fig. 2.2c.
- *Mutually exclusive.* Two events *A* and *B* are said to be mutually exclusive if they do not have common elements, i.e., the intersection of *A* and *B* is a null set $(A \cap B = \emptyset)$.
- *Collectively exhaustive.* The events B_1, B_2, …., B_n are said to be collectively exhaustive if their union covers all the events within the entire sample space, i.e., $P(B_1 \cup B_2 \cup \ldots \cup B_n) = S$ where *S* is the sample space.

The union and intersection of the events *A* and *B*, and the compliment of the event A are shown in the Venn diagram in Fig. 2.2.

2.2.1 Probability Measure

The probability is a measure of outcomes of an event *A* among all outcomes of the experiment. It is denoted by *P*(*A*) and defined commonly in two ways as the *relative frequency* and the *classical* definitions [22]. In the frequency definition, a random experiment is repeated n times and if the event *A* occurs n_A times, then the probability of *A* is defined as the fraction of occurrences of *A* in *n* trials. Thus,

$$P(A) = \lim_{n \to \infty} \left(\frac{n_A}{n} \right) \tag{2.1}$$

The precision of *P*(*A*) depends on the number of trials. If the number of trials *n* approaches infinity, then *P*(*A*) tends to a definite limiting value. In the classical definition, all outcomes of the experiment are supposed to be equally likely, i.e., they have the same probability of occurrences. Then, counting the total number *N* of possible outcomes of the experiment, and from which the number N_A of the favorable outcomes to the occurrence of the event *A*, the probability of *A* is defined as,

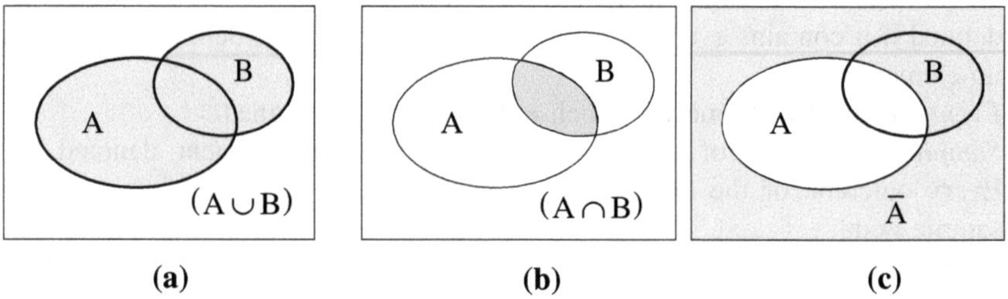

Fig. 2.2 Venn diagrams for two events. **a** Union. **b** Intersection. **c** Complement of A

$$P(A) = \frac{N_A}{N} \tag{2.2}$$

The probability of A, $P(A)$, is a number and satisfies the following three axioms [23].

1. The probability of an event A is a number between zero and one, i.e., $0 \le P(A) \le 1$.
2. For a certain event S, the probability $P(S)$ equals 1, i.e., $P(S) = 1$.
3. The probability of the union of a number of mutually exclusive events, i.e., intersections are null sets, is the sum of probabilities of the events, i.e.,

$$P(\bigcup_{i=1}^{n} A_i) = \sum_{i=1}^{n} P(A_i), \quad \text{where } P(\bigcup_{i=1}^{n} A_i) = P(A_1 \cup A_2 \cup \ldots \cup A_n) \tag{2.3}$$

From these axioms it can be concluded [24] that

$$\begin{aligned} &\text{Probability of null Set:\ldots\ldots} \rightarrow P(\{\varnothing\}) = 0 \\ &\text{Probability of Complement:} \rightarrow P(\bar{A}) = 1 - P(A) \\ &\text{Probability of Union:\ldots\ldots} \rightarrow P(A \cup B) = P(A) + P(B) - P(A \cap B) \end{aligned} \tag{2.4}$$

One other probability measure having practical importance is the *conditional probability*, which is denoted by $P(A|B)$. It is defined as the probability of the event A given that the event B has occurred. In the probability terms, it is

$$\text{Conditional Probability:} \rightarrow \quad P(A|B) = \frac{P(A \cap B)}{P(B)} \tag{2.5}$$

As it can be realized from Eq. (2.5), if the event B is a null (empty) set, i.e., $P(B) = 0$, then a conditional probability is not defined. If the two events, A and B, are not related in any way they are said to be independent events. The only condition of the independence is

$$\text{Independence, if and only if:} \quad \rightarrow \quad P(A \cap B) = P(A)\,P(B) \tag{2.6}$$

Using the conditional probability definition, the *total probability theorem* can be derived. This is expressed as, if B_1, B_2, ..., B_n are collectively exhaustive events of the sample space S and A is an arbitrary event on S, then the total probability of A can be stated as the sum of all intersections, i.e.,

$$P(A) = \sum_{i=1}^{n} P(A \cap B_i) \tag{2.7a}$$

Substituting the intersection from Eqs. (2.5) into (2.7a), the total probability of A is written as,

$$\text{The total probability theorem:} \quad \rightarrow \quad P(A) = \sum_{i=1}^{n} P(A|B_i)P(B_i) \tag{2.7b}$$

which is the statement of the total probability theorem. Using the conditional probability, the total probability theorem and the commutative law of the events A and B, i.e., $P(A \cap B) = P(B \cap A)$, the well-known *Bayes' theorem* can be stated as,

$$\text{The Bayes' theorem:} \quad \rightarrow \quad P(B_k|A) = \frac{P(A|B_k)P(B_k)}{\sum_{i=1}^{n} P(A|B_i)P(B_i)} \tag{2.8}$$

The Bayes' theorem helps for making decision under uncertainties that engineers confront frequently in the practice [24] provided that prior probabilistic models of uncertainties are available or determined previously by experiments.

2.2.2 Random Variables

In practice, outcomes of all experiments, even under equal conditions, are not unique and show discrepancies in values. For example, to find elasticity modulus E of a material, say steel, a number of equal samples have to be tested in laboratory. Each sample produces a specific value of E, which is mostly different than those obtained from other samples, although they may be very close together. As being the design value, we use an average value of E over the values obtained from all experiments made under the same condition. Here, E is a random variable which associates a unique numerical value with every outcome of an experiment. Thus, a *random variable (r.v.)* is a finite single valued function $X(.)$ which associates a real numerical value with every outcome of a random experiment [23]. An r.v. X can be thought as a measurement of outcomes of the random experiment. Its randomness comes from the value that depends on the outcome of the experiment, which cannot be predicted exactly before the experiment is carried out. More generally, an r.v. X is a function that maps the sample space S into a real line with $\pm\infty$ thrown in. There are two types of random variables, discrete and continuous. A *discrete random variable* is defined as, if an r.v. can take only a finite number of distinct values, then it is discrete, i.e., a discrete r.v. takes only a countable number of distinct values. A *continuous random variable* is defined as, if an r.v. can take an infinite number of possible values, then it is continuous, i.e., it is not defined at specific values, instead it is defined over an interval of values. A particular outcome of an r.v. is termed as a *random variate*. In the random vibration theory, an r.v. X is a function of time t, which means that the outcome of an experiment is time dependent. It is denoted by $X(x, t)$ where x is a time-dependent outcome, i.e., $x = x(t)$, which represents an excitation input function or a response function. In the random vibration theory, the probability information of random time functions $x_i(t)$, where $(i = 1,2,3,\ldots n)$, is used to determine statistical characteristics of an event represented by the r.v. $X(x, t)$, which involves in an ensemble process explained in the following section.

2.2.3 Stochastic Processes

As a short definition, a random process is an infinite collection of realizations of an r.v. In a similar way to the definition of an r.v., a random process is a mapping from the sample space into an ensemble of time functions known as sample functions. The r.v. $X(x, t)$ for a fixed random x value, say x_1, is a specific time signal that it is called as *the realization* of the r.v. $X(x, t)$ at $x = x_1$, which is denoted by $x_1(t)$. For a fixed time, say t_1, the r.v. $X(x, t_1)$ is a time-independent r.v. that probability principles are applied. For both fixed values of x and t, say ($x = x_1$ and $t = t_1$), the r.v. $X(x, t)$ will be a mere number with the value of $X(x_1, t_1)$ [23]. The ensemble of all realizations of a time-dependent r.v. represents *the stochastic process* that we use the notation $X(t)$ to indicate it, disregarding its dependence on the outcome x. Such an ensemble, which represents a stochastic process, is shown in Fig. 2.3 with four realizations, or samples, $x_1(t)$, $x_2(t)$, $x_3(t)$ and $x_4(t)$. As indicated above, a stochastic process represents a single number, a time function, an r.v., and a process with time function and r.v. Thus,

1. if x and t are both fixed ($x = x_1$ and $t = t_1$), then $X(x_1, t_1)$ is a single number,
2. if x is fixed ($x = x_1$) and t is a variable, then $X(x_1, t)$ is a time function as $x_1(t)$,
3. if x is a variable and t is fixed ($t = t_1$), then $X(x, t_1)$ is a random variable at $t = t_1$,
4. if x and t are both variables then $X(x, t)$ is a stochastic process.

If we consider an infinite number of samples, at a specific time, say $t = t_1$ as shown in Fig. 2.3, the stochastic process will be a continuous r.v. with the outcomes x, as $(x_1, x_2, \ldots, x_n)$ where $(n \to \infty)$. This r.v. is fully described by its probability characteristics explained in the next section.

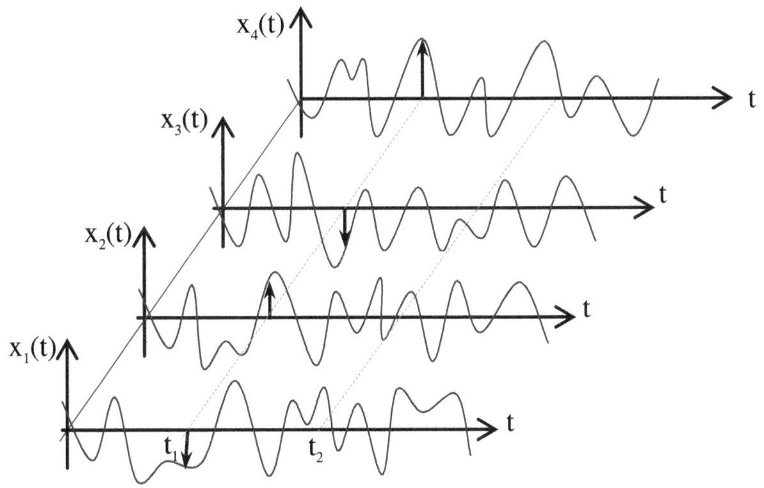

Fig. 2.3 Ensemble of time-dependent random variables with four samples

2.2.4 Probability Distribution and Density Functions

A probability measure of an event has been outlined in the Sect. 2.2.1. In order to determine this measure, probability distributions and related functions are explained in this section. Let us define an event *A* such that all outcomes of the r.v. $X(x, t_1)$, in short *X*, between $(x_1 \leq X \leq x_2)$ fall in, where x_1 and x_2 are the lower and upper bounds of the outcomes. The probability of the event *A* is sated as

$$P(A) = P(x_1 \leq X \leq x_2) \tag{2.9a}$$

It is the probability that the outcomes of *X* fall between the bounds x_1 and x_2. This probability definition of the r.v. *X* can be extended to the probability definition of a stochastic process $X(t)$, of which the event is denoted by $A(t)$, i.e.

$$P(A(t)) = P(x_1(t) \leq X(t) \leq x_2(t)) \tag{2.9b}$$

will be a time function. If the outcomes of the event $A(t)$ fall between $-\infty$ and any realization $x(t)$ in the region, $(-\infty \leq x(t) \leq +\infty)$, of the stochastic process $X(t)$, then by definition the probability is called as the *probability distribution function*, or the *cumulative distribution function* (CDF). It is denoted by $F_X(x, t)$, i.e.,

$$\text{CDF}: \rightarrow \quad F_X(x,t) = P(-\infty \leq X(t) \leq x(t)) \tag{2.10a}$$

For a time independent r.v. *X*, it will be

$$F_X(x) = P(-\infty \leq X \leq x) \tag{2.10b}$$

The probability of the process $X(t)$, which is given by Eq.(2.9b), is stated in terms of the probability distributions at $x_2(t)$ and $x_1(t)$ as

$$P(x_1(t) \leq X(t) \leq x_2(t)) = F_X(x_2, t) - F_X(x_1, t) \tag{2.10c}$$

The probability distribution function $F_X(x, t)$ satisfies the following properties.

$$\left.\begin{array}{l} F_X(-\infty) = 0 \\ F_X(+\infty) = 1 \end{array}\right\} \quad \text{thus} \quad 0 \leq F_X(x,t) \leq 1$$
$$F_X(x_1, t) \leq F_X(x_2, t) \quad \text{for} \quad (x_1(t) \leq x_2(t)) \tag{2.11}$$

One other important definition in the probability theory is the *probability density function* (PDF), which determines probability characteristics of the stochastic process. It is related to the probability that the stochastic process $X(t)$ lies in the interval $x(t)$ to $(x(t) + \mathrm{d}x(t))$. It is stated from Eq. (2.10c)

$$P(x(t) \leq X(t) \leq x(t) + \Delta x(t)) = F_X(x(t) + \Delta x(t)) - F_X(x(t)) \tag{2.12a}$$

and taking the limit of this statement when $(\Delta x(t) \rightarrow 0)$ it will be

$$P(x(t) \leq X(t) \leq x(t) + \mathrm{d}x(t)) = \lim_{\Delta x(t) \rightarrow 0} (F_X(x(t) + \Delta x(t)) - F_X(x(t))) \tag{2.12b}$$

$$\text{or} \qquad P(x(t) \leq X(t) \leq x(t) + dx(t)) = f_X(x, t) dx(t) \tag{2.12c}$$

Thus, the probability that $X(t)$ lies in the interval $x(t)$ to $(x(t) + \mathrm{d}x(t))$ is the area of a function $f_X(x, t)$ in the interval $\mathrm{d}x(t)$. As it may be seen from Eqs. (2.12b) and (2.12c), this function is the derivative of the probability distribution $F_X(x, t)$, which is

$$\text{Probability Density Function (PDF):} \rightarrow f_X(x, t) = \frac{\partial F_X(x, t)}{\partial x} \tag{2.13}$$

The function $f_X(x, t)$ is called as the PDF of the stochastic process $X(t)$. For a *continuous process* $X(t)$, the probability distribution $F_X(x, t)$ can be stated from Eq. (2.13) as,

$$F_X(x, t) = \int_{-\infty}^{x} f_X(\xi, t)\, \mathrm{d}\xi \tag{2.14}$$

With this definition the probability of the process $X(t)$ in the region of $x_1(t) \leq X(t) \leq x_2(t)$ is written

$$P(x_1(t) \leq X(t) \leq x_2(t)) = \int_{x_1}^{x_2} f_X(\xi, t)\, \mathrm{d}\xi \tag{2.15}$$

Since the probability distribution of a continuous stochastic process $X(t)$ is an increasing function and $F_X(+\infty) = 1$ the PDF satisfies the conditions

$$f_X(x,t) \geq 0$$
$$\int_{-\infty}^{\infty} f_X(x,t)\,\mathrm{d}x = 1 \tag{2.16}$$

The probability distribution and density functions of a continuous process are smooth function as shown in Fig. 2.4. For discrete and mixed processes, the probability distribution $F_X(x, t)$ is respectively staircase and discontinuous functions [23]. If the distribution function $F_X(x, t)$ is discontinues but not staircase then the process is called as a mixed process. The probability density function $f_X(x, t)$ of a discrete process is in the form of impulses. Each impulse, which is defined as a probability mass, is equal to the corresponding step size of the distribution function as shown in Fig. 2.5, e.g., $p_X(x_i) = F_X(x_i) - F_X(x_i - \varepsilon)$ at $X = x_i$ where ε is a small positive number. The sum of all probability masses is equal to 1. The probability mass of a continuous process at $X = x_i$ can be stated from the definition as

$$\text{Probability Mass:} \quad \rightarrow \quad p_X(x_i,t) = f_X(x_i,t)\,\mathrm{d}x \tag{2.17}$$

Discrete random processes have the same properties of continuous processes provided that probability masses are used instead of probability density functions and integrations of continuous random processes are replaced with summations for discrete processes. In practical applications, an event such as a structural response contains usually outputs of multiple random variables with joint distributions. This matter and related subjects are outlined briefly in the next section.

2.2.4.1 Distribution and Density Functions of Joint Random Variables

In the previous section, probability distribution and density function of a single random process $X(t)$ is presented. If an event is a collection of outputs of multiple random variables and processes, the probability distribution and density functions are somewhat different than those of a single variable or process. For simplicity, we use only two random variables X and Y. For multiple random variables, the same principles of two random variables can be applied. The joint probability distribution function of multiple random variables are defined as the probability of their intersection for $(-\infty \leq X \leq x)$ and $(-\infty \leq Y \leq y)$, i.e.,

$$\text{Joint CDF:} \quad \rightarrow \quad F_{XY}(x,y) = P(X \leq x \cap Y \leq y) \tag{2.18}$$

In other words, it is the probability that all outcomes of random variables X and Y fall in the region $(X \leq x)$ and $(Y \leq y)$, where x and y may be either time dependent or time invariant. If they are time dependent the random variables $X(t)$ and $Y(t)$ become joint random processes as being similar to a single random

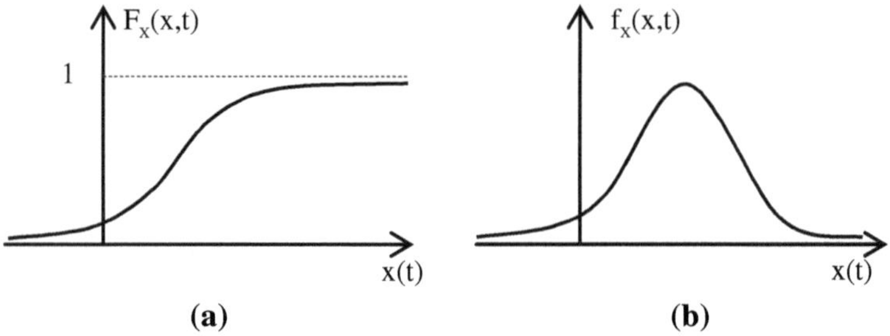

Fig. 2.4 Probability distribution and density functions of a continuous process *X*(*t*). **a** Probability distribution function. **b** Probability density function

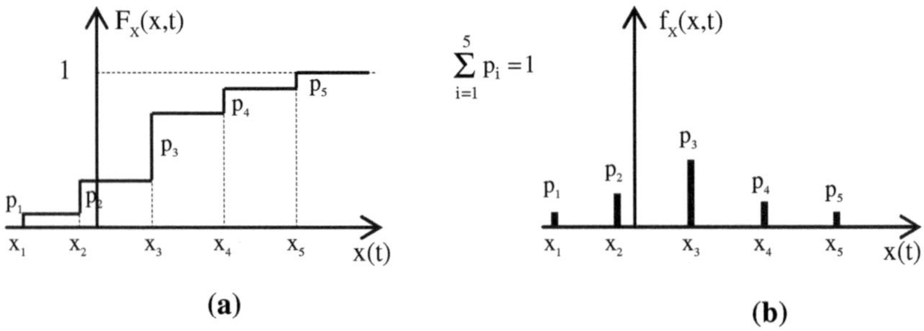

Fig. 2.5 Probability distribution and density functions of a discrete process *X*(*t*). **a** Probability distribution function **b** Probability density function

process explained in the previous section. The joint probability distribution function $F_{XY}(x, y)$ satisfies the following properties.

$$\begin{aligned} &F_{XY}(x,y) \geq 0 \;\; \text{for} \; (-\infty \leq x \leq \infty) \;\; \text{and} \;\; (-\infty \leq y \leq \infty) \\ &F_{XY}(-\infty, y) = F_{XY}(x, -\infty) = 0 \\ &F_{XY}(\infty, \infty) = 1 \\ &F_Y(y) \;= F_{XY}(\infty, y) \quad \text{and} \quad F_X(x) = F_{XY}(x, \infty) \end{aligned} \tag{2.19}$$

For continuous random variables, *X* and *Y*, the joint probability density function $f_{XY}(x, y)$ can be obtained from the derivatives of the joint probability distribution function as written by

$$\text{Joint PDF:} \quad \rightarrow \; f_{XY}(x,y) = \frac{\partial^2 F_{XY}(x,y)}{\partial x \partial y} \tag{2.20}$$

It is related to the probability that *X* and *Y* lie in the intervals x and $(x + \mathrm{d}x)$, and y and $(y + \mathrm{d}y)$ respectively, i.e.,

$$P(\,(x \leq X \leq x + \mathrm{d}x) \cap (y \leq Y \leq y + \mathrm{d}y)) = \; f_{XY}(x,y)\mathrm{d}x\mathrm{d}y \tag{2.21}$$

which defines the probability mass at x and y. For discrete r.v. at $X = x_i$ and $Y = y_j$, it becomes $p_{XY}(x_i, y_j)$ which is

$$\text{Joint Probability Mass:} \quad \rightarrow \quad p_{XY}(x_i, y_i) = f_{XY}(x_i, y_i)\, dx\, dy \tag{2.22}$$

The joint probability and joint probability distribution function are expressed from the joint density function as

$$\begin{aligned} P((x_1 \le X \le x_2) \cap (y_1 \le Y \le y_2)) &= \int_{x_1}^{x_2} \int_{y_1}^{y_2} f_{XY}(\xi, \eta)\, d\eta\, d\xi \\ F_{XY}(x, y) &= \int_{-\infty}^{x} \int_{-\infty}^{y} f_{XY}(\xi, \eta)\, d\eta\, d\xi \end{aligned} \tag{2.23}$$

The joint probability density function satisfies the following requirements

$$f_{XY}(x, y) \ge 0 \quad \text{and} \quad \int_{-\infty}^{\infty} \int_{-\infty}^{\infty} f_{XY}(\xi, \eta)\, d\eta\, d\xi = F_{XY}(-\infty, \infty) = 1 \tag{2.24}$$

Two special cases of joint probability distribution and density functions are presented in the following sections.

2.2.4.2 Marginal Probability Distribution and Density Functions

In the case of multiple random variables, the statistics of an individual variable is called as *marginal*. The related probability distribution and density functions are also called as marginal probability distribution and density functions. From Eq. (2.19), the marginal probability distributions of joint random variables X and Y can be written

$$\text{Marginal CDF of X:} \quad \rightarrow \quad F_X(x) = F_{XY}(x, \infty) = \int_{-\infty}^{x} \int_{-\infty}^{\infty} f_{XY}(\xi, \eta)\, d\eta d\xi \tag{2.25a}$$

$$\text{Marginal CDF of Y:} \quad \rightarrow \quad F_Y(y) = F_{XY}(\infty, y) = \int_{-\infty}^{y} \int_{-\infty}^{\infty} f_{XY}(\xi, \eta)\, d\xi d\eta \tag{2.25b}$$

The marginal probability density functions $f_X(x)$ and $f_Y(y)$ can be obtained from the derivatives of $F_X(x)$ and $F_Y(y)$ or from integrations of the joint density function as stated

$$\text{Marginal PDF of X:} \quad \rightarrow \quad f_X(x) = \frac{\partial F_X(x)}{\partial x} = \int_{-\infty}^{\infty} f_{XY}(x,y)\ \mathrm{dy} \tag{2.26a}$$

$$\text{Marginal PDF of Y:} \quad \rightarrow \quad f_Y(y) = \frac{\partial F_Y(y)}{\partial y} = \int_{-\infty}^{\infty} f_{XY}(x,y)\ \mathrm{dx} \tag{2.26b}$$

Marginal probability density function is used to calculate the probability of particular events required in the *Bayes' theorem* or in the calculation of a conditional probability and in the total probability theorem.

2.2.4.3 Conditional Probability Distribution and Density Functions

For two events A and B, the conditional probability has been defined in Eq. (2.5). Here, the conditional probability distribution and density function of two joint random variables X and Y will be explained. The joint probability distribution (joint cumulative distribution function, CDF) of the random variables X and Y has been defined in Eq. (2.18). It is now assumed that one of them, say Y, takes a specific value, i.e., $Y = y$, and the probability distribution of other one (X) is defined as

$$F_X(x \mid y) = P(X \leq x | Y = y) \tag{2.27}$$

It is the probability that $(-\infty \leq X \leq x)$ on the condition $(Y = y)$, which is called the conditional probability density function. By using the conditional probability given in Eq. (2.5) it can be written

$$\text{Conditional CDF:} \quad \rightarrow \quad \begin{cases} F_X(x \mid y) = \dfrac{F_{XY}(x,y)}{F_Y(y)} \\ F_Y(y \mid x) = \dfrac{F_{XY}(x,y)}{F_X(x)} \end{cases} \tag{2.28}$$

As an example of the conditional CDF function, let us assume that the r.v. Y is defined to be a subset of the r.v. X as $(Y = x \geq a)$ [23, 24]. The intersection of random variables X and Y is defined as

$$(X \leq x) \cap (Y \leq y) \rightarrow (X \leq x) \cap (X \geq a) = \begin{cases} x - a & \text{if } x \geq a \\ 0 & \text{if } x < a \end{cases} \tag{2.29}$$

The marginal distribution of Y is

$$F_Y(y) = P(X \geq a) = F_X(\infty) - F_X(a) = 1 - F_X(a) \tag{2.30a}$$

and the joint distribution of X and Y is obtained by using Eqs. (2.18) and (2.29)

$$F_{XY}(x,y) = \begin{cases} F_X(x) - F_X(a) & \text{if} \quad x \geq a \\ 0 & \text{if} \quad x < a \end{cases} \tag{2.30b}$$

Using Eq. (2.28) the conditional distribution of X, such that $(X|X \geq a)$, is calculated from

$$F_X(x \mid y) = \frac{F_{XY}(x,y)}{F_Y(y)} \rightarrow F_X(x|y) = \begin{cases} \dfrac{F_X(x) - F_X(a)}{1 - F_X(a)} & \textit{if} \ x \geq a \\ 0 & \textit{if} \ x < a \end{cases} \tag{2.30c}$$

The conditional probability density functions of joint r.v. are defined in the same manner as the conditional CDF functions

$$\text{Conditional PDF:} \quad \rightarrow f_X(x \mid y) = \frac{f_{XY}(x,y)}{f_Y(y)} \quad \text{and} \quad f_Y(y \mid x) = \frac{f_{XY}(x,y)}{f_X(x)} \tag{2.31}$$

By using conditional probability density functions the conditional probability distributions are calculated from

$$F_X(x \mid y) = \int_{-\infty}^{x} f_X(\zeta \mid y)\, d\zeta \quad \text{and} \quad F_Y(y|x) = \int_{-\infty}^{y} f_Y(\eta \mid x)\, d\eta \tag{2.32}$$

The conditional density functions satisfy the requirements of an ordinary PDF given in Eq. (2.16), i.e.,

$$\left. \begin{array}{l} f_X(x \mid y) \geq 0 \\ \displaystyle\int_{-\infty}^{\infty} f_X(x \mid y)\, dx = 1 \end{array} \right\} \quad \text{and} \quad \left\{ \begin{array}{l} f_Y(y \mid x) \geq 0 \\ \displaystyle\int_{-\infty}^{\infty} f_Y(y \mid x)\, dy = 1 \end{array} \right. \tag{2.33}$$

If joint random variables are independent, their probability distribution and density functions become products of corresponding marginal functions as similar to the probability of intersection of independent events given in Eq. (2.6). This subject is explained in the following section.

2.2.4.4 Independent Random Variables

Definition of independent events has been given in Eq. (2.6). The only condition is that the probability of their intersection is the product of probabilities of individual events. The same rule also applies for the probability distribution of joint r.v. as to be independent. Thus, for independent random variables X and Y, the joint distribution is stated from Eq. (2.18) as written

$$\text{Joint CDF of independent r.v.:} \quad \rightarrow \quad F_{XY}(x,y) = P(X \leq x)\, P(Y \leq y) \tag{2.34}$$

from which it is seen that the joint probability distribution function is a product of marginal distribution functions, i.e.,

$$F_{XY}(x,y) = F_X(x)\,F_Y(y) \tag{2.35}$$

For independent random variables, conditional probability distributions become marginal as stated from Eq. (2.28),

$$F_X(x \mid y) = F_X(x) \quad \text{and} \quad F_Y(y \mid x) = F_Y(y) \tag{2.36}$$

In a similar way, the joint probability density function of independent continuous random variables X and Y becomes a product of marginal PDF, which is written

$$\text{Joint PDF of independent r.v.:} \;\rightarrow\; f_{XY}(x,y) = f_X(x) f_Y(y) \tag{2.37}$$

These statements can be extended for multiple independent random variables. The calculation of statistical values of events comprising independent r.v. is simply carried out by integrations, or summations for discrete variables, over individual r.v. It does not require a multiple integrations process, which is usually more time consuming and more complicated. In the following section, calculations of the statistical values of r.v. and functions are presented.

2.3 Mean Values, Probability Moments, and Variances of Random Variables and Random Functions

Since outcomes of r.v. are unpredictable and not specific values, the average value, which is also called as the *Mean Value*, of all outcomes of an r.v. is an indicative measure in engineering applications. It is also the *Expected Value* that the most outcomes of the r.v. is likely to occur. It is used as a design value in the deterministic analyses. In the probabilistic analyses, it is one of statistical parameters describing the probability distribution of the r.v. For an r.v. X, the definition of the mean value is as follows,

$$\text{Mean Value of an r.v.:} \;\rightarrow\; m_X = E[X] = \begin{cases} \int_{-\infty}^{\infty} x f_X(x)\,\mathrm{d}x \rightarrow \text{continuous} \\ \sum_{i=1}^{n} x_i p_X(x_i) \rightarrow \text{discrete} \end{cases} \tag{2.38}$$

where m_X is the mean value and $E[.]$ denotes an expected value, $f_X(x)$ is the probability density function of a continuous r.v. X, $p_X(x_i)$ is the probability mass of a discrete r.v. X. In Eq. (2.38), having replaced the marginal PDF $f_X(x)$ by the conditional PDF $f_X(x|y)$, a conditional expected value is defined as

$$\text{Conditional Mean Value:} \quad \rightarrow \quad m_{X|Y} = E[X|Y] = \int_{-\infty}^{\infty} x f_X(x \mid y)\,\mathrm{d}x \qquad (2.39)$$

The mean value of an r.v. determines only the gravity center of its PDF. It does not provide full statistical information of the r.v. which is possible only if its probability moments are known. The *Probability Moment* that corresponds to the nth moment of the area of the PDF of an r.v. with respect to the origin is defined

$$\text{Prob. Moment of an r.v.:} \rightarrow (m_X)_n = E[X^n] = \begin{cases} \displaystyle\int_{-\infty}^{\infty} x^n f_X(x)\,\mathrm{d}x \rightarrow \text{con.} \\ \displaystyle\sum_{i=1}^{n} x_i^n\, p_X(x_i) \rightarrow \text{dis.} \end{cases} \qquad (2.40)$$

where n is an integer number indicating the order (degree) of the probability moment. As it is seen from Eq. (2.38) the expected value $E[X]$ is the first order probability moment. The most statistical informative probability moments are defined with respect to the expected value rather than the origin. They are called as the *Central Probability Moments* which are defined as, for continuous r.v.,

$$\text{Central Prob. Mom.:} \rightarrow (\mu_X)_n = E[(X - m_X)^n] = \int_{-\infty}^{\infty} (x - m_X)^n f_X(x)\,\mathrm{d}x \quad (2.41)$$

As it can be seen from Eq. (2.41) the first order central moment equals zero, i.e., $(\mu_X)_1 = 0$. The second-order central moment is especially important since it defines the variance of the r.v. It provides a measure of the spread or the randomness of an r.v. and it determines the effective width of the PDF. It is defined

$$\text{Variance:} \quad \rightarrow \quad \sigma_X{}^2 = E[(X - m_X)^2] = \int_{-\infty}^{\infty} (x - m_X)^2 f_X(x)\,\mathrm{d}x \qquad (2.42\text{a})$$

The variance can also be stated in terms of the second probability moment and the expected value of the r.v. X. From Eq. (2.42) it is written

$$\sigma_X{}^2 = E[(X - m_X)^2] = E[X^2 - 2m_X X + m_X{}^2] = (m_X)_2 - m_X{}^2 \qquad (2.42\text{b})$$

The square root of the variance is called the *Standard Deviation*, σ_X, and the ratio between σ_X and m_X is called as the *Coefficient of Variation*, which are

$$\begin{aligned} &\text{Standard Deviation (SD)} && \rightarrow \sigma_X = \sqrt{\text{Variance}} \\ &\text{Coefficient Of Variation (COV)} && \rightarrow V_X = \frac{\sigma_X}{m_X} \end{aligned} \qquad (2.43)$$

One other important measure of an r.v. is the *Coefficient of Skewness*, γ_1, which provides information about the shape of the PDF. It is defined as

$$\text{Coefficienet of Skewness:} \quad \rightarrow \quad \gamma_1 = \frac{(\mu_X)_3}{\sigma_X^3} = E\left[(X - m_X)^3 / \sigma_X^3\right] \tag{2.44}$$

If this coefficient is zero, then the PDF is symmetric about its center point, otherwise it is asymmetric as shown in Fig. 2.6. A useful, but less common, measure of an r.v. is the *Coefficient of Kurtosis*, γ_2. It is defined

$$\text{Coefficienet of Kurtosis:} \quad \rightarrow \quad \gamma_2 = \frac{(\mu_X)_4}{{\sigma_X}^4} = E\left[(X - m_X)^4 / {\sigma_X}^4\right] \tag{2.45}$$

Kurtosis determines the flatness of the r.v. It is a measure of whether the PDF is peaked or flat relative to a normal distribution of which ($\gamma_2 = 3$). The PDF of an r.v. with high kurtosis tends to have a distinct peak near the mean declining rather rapidly and have heavy tails. Low kurtosis indicates that the PDF has a flat top near the mean rather than a sharp peak. The CDF and PDF of an r.v. are usually described by its standard deviation and mean value, σ_X and m_X. So far, we have presented properties of r.v., but in most practical applications we encounter functions of r.v. The probabilistic properties of such functions are presented in the next section.

2.3.1 Functions of Random Variables

In engineering applications a response quantity is usually a function of r.v. that needs to be determined in probabilistic terms. The r.v. in its content constitute probabilistic input parameters that fully described by their CDF or PDF. Given the random input parameters or variables, the probabilistic and statistical characteristics of random response functions are required in the calculations of their expected values and probabilities of occurrences of specific events under desired conditions, such as probability of failure due to fatigue damage or ultimate loading conditions. The response functions that contain r.v. are denoted by Y and Z as defined

$$\begin{aligned} &\text{Function of one r.v.X......:} \quad \rightarrow \quad Y = g(X) \\ &\text{Function of multipler.v., i.e. X and Y:} \quad \rightarrow \quad Z = g(X, Y) \end{aligned} \tag{2.46}$$

The statistical values of random functions, $Y = g(X)$ or $Z = g(X, Y)$, are defined and calculated in the same manner of r.v. explained in Sect. 2.3. They are explained below for functions of one r.v. and multiple random variables.

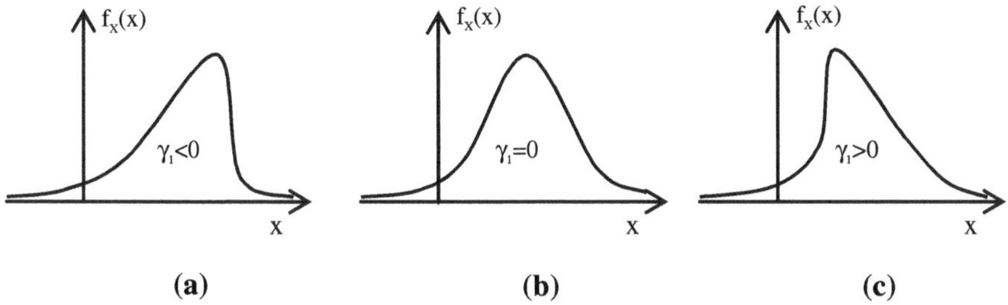

Fig. 2.6 Shapes of PDF with different coefficients of skewness γ_1. **a** Negative skewness. **b** Zero skewness. **c** Positive skewness

2.3.1.1 Function of One Random Variable

The function of one r.v. is defined in Eq. (2.46). Its mean value and variance are defined as

$$\text{Mean of a r.f.:} \quad \rightarrow \quad m_{g(X)} = E[g(X)] = \int_{-\infty}^{\infty} g(X) f_X(x)\,dx \qquad (2.47a)$$

$$\text{Variance of a r.f.:} \quad \rightarrow \quad \sigma^2_{g(X)} = \int_{-\infty}^{\infty} \left(g(X) - m_{g(X)}\right)^2 f_X(x)\,dx \qquad (2.47b)$$

Having used the series expansion of the function *g*(*X*) at the mean value of *X*, it is stated that

$$g(X) = g(m_X) + (X - m_X)\,g'(m_X) + \frac{(X - m_X)^2}{2}\,g''(m_X) + \ldots \qquad (2.48)$$

where $g^{(n)}(m_X)$, with $(n =', '', \ldots)$, is the *n*th derivative of *g*(*X*) evaluated at $X = m_X$. Using Eq. (2.48) the mean and variance of the function *g*(*X*) can be calculated approximately [23] from,

$$\begin{aligned} m_{g(X)} &\simeq g(m_X) + \frac{\sigma_X^2}{2}\,g''(m_X) + \ldots + \frac{(\mu_X)_n}{n!}\,g^{(n)}(m_X) \\ \sigma^2_{g(X)} &\simeq \left(g'(m_X)\right)^2 \sigma_X^2 \end{aligned} \qquad (2.49)$$

The moments of the random function *g*(*X*) are defined as similar to one r.v. *X*,

$$\text{Prob. Moment of a r.f.:} \quad \rightarrow \quad \left(m_{g(X)}\right)_n = \int_{-\infty}^{\infty} g^n(X) f_X(x)\,dx \qquad (2.50a)$$

$$\text{Central Prob. Mom. of a r.f.:} \rightarrow \left(\mu_{g(X)}\right)_n = \int_{-\infty}^{\infty} (g(X) - m_Y)^n f_X(x)\,\mathrm{d}x \tag{2.50b}$$

The PDF of a random function $Y = g(X)$ is calculated in terms of PDF of the r.v. X [23] from,

$$\text{PDF of a r.f. Y = g(X):} \quad \rightarrow f_Y(y) = \sum_{i=1}^{k} \frac{f_X(x_i)}{|g'(x_i)|} \tag{2.51}$$

where x_i ($i = 1, 2, \ldots, k$) are the real roots of the equation $y = g(x)$ in terms of y, $g'(x_i)$ is the derivative of $g(X)$ evaluated at $X = x_i$ and |.| denotes the absolute value. As a demonstration, it is assumed that the function $g(X)$ is given

$$Y = g(X) = ax^{-2} \tag{2.52a}$$

and the PDF of Y will be calculated. The real roots of $g(X)$ are calculated for $(ay > 0)$ as written

$$y = ax^{-2} \quad \rightarrow \quad x_1 = \sqrt{ay^{-1}} \quad \text{and} \quad x_2 = -\sqrt{ay^{-1}} \quad \text{for} \quad (ay > 0) \tag{2.52b}$$

For different sings of y and a, i.e., $(ay < 0)$,the solution is not real. The derivatives evaluated at the roots are

$$g'(x) = -2ax^{-3} \quad \rightarrow \quad g'(x_1) = -2\sqrt{a^{-1}y^3} \quad \text{and} \quad g'(x_1) = 2\sqrt{a^{-1}y^3} \tag{2.52c}$$

Using Eq. (2.51) the PDF of Y is obtained for $(ay > 0)$ as written

$$f_Y(y) = \frac{1}{2}\sqrt{\frac{a}{y^3}}\left[f_X\left(\sqrt{ay^{-1}}\right) + f_X\left(-\sqrt{ay^{-1}}\right)\right] \quad \text{for} \quad (ay > 0) \tag{2.52d}$$

For $(ay < 0)$, the PDF of Y will be zero, i.e., $f_Y(y) = 0$.

2.3.1.2 Functions of Multiple Random Variables

As similar to the case of one continuous r.v. X, a random function Z of two, or more, jointly distributed random variables X and Y is defined to be $Z = g(X, Y)$. Its expected (mean) value is calculated [23, 24] from

$$\text{Mean of a r.f. } Z: \rightarrow E[Z] = E[g(X,Y)] = \int_{-\infty}^{\infty}\int_{-\infty}^{\infty} g(X,Y) f_{XY}(x,y)dx\,dy \tag{2.53}$$

Once the PDF of Z, i.e., $f_Z(z)$, is known, all statistical values and definitions of a r.v. X is also valid for the r.f. Z. However, calculation of the PDF of Z is not as simple as that of a function of one r.v. It can be calculated for each function $Z = g(X, Y)$. A general calculation can also be obtained through the joint PDF of random functions, say Z and W, which are defined

$$Z = g(X, Y) \quad \text{and} \quad W = h(X, Y) \tag{2.54}$$

The joint PDF of the functions Z and W, $f_{ZW}(z, w)$, is calculated [23] from

$$\text{Joint PDF of two r.f } Z \text{ and } W: \quad \rightarrow \quad f_{ZW}(z, w) = \sum_{i=1}^{k} \frac{f_{XY}(x_i, y_i)}{|J(x_i, y_i)|} \tag{2.55}$$

where $(x_i, y_i)(i = 1, 2, \ldots k)$ are all real solutions of the equations

$$g(x_i, y_i) = z \quad \text{and} \quad h(x_i, y_i) = w \tag{2.56}$$

in terms of z and w, and $|J(x_i, y_i)|$ is the absolute value of the determinant of the *Jacobian* for the transformation given in Eq. (2.54) at the solutions (x_i, y_i). It is stated in general as,

$$|J(x, y)| = \begin{vmatrix} \dfrac{\partial g(x, y)}{\partial x} & \dfrac{\partial g(x, y)}{\partial y} \\ \dfrac{\partial h(x, y)}{\partial x} & \dfrac{\partial h(x, y)}{\partial y} \end{vmatrix} \tag{2.57}$$

Having obtained the joint PDF of the random variables Z and W as given in Eq. (2.55) the marginal PDF of Z can be calculated similarly to Eq. (2.26a) from,

$$\text{Marginal PDF of a r.f. } Z: \quad \rightarrow \quad f_Z(z) = \int_{-\infty}^{\infty} f_{ZW}(z, w)\, dw \tag{2.58}$$

A useful application of a random function Z of two random variables X and Y is to calculate joint moments of these random variables. If the function $Z = g(X, Y)$ is defined as

$$Z = g(X, Y) = X^n Y^m \tag{2.59}$$

its expected value will be the joint moments of order $(n + m)$ of the random variables X and Y. From Eq. (2.53) it is written that

$$\text{Joint Mom. of r.v. } X \text{ and } Y: \quad \rightarrow \quad \begin{cases} E[X^n Y^m] = \displaystyle\int_{-\infty}^{\infty} \int_{-\infty}^{\infty} x^n y^m f_{XY}(x, y)\, dx\, dy \\ (m_{XY})_{nm} = E[X^n Y^m] \end{cases} \tag{2.60}$$

As it is seen from Eq. (2.60), if $(n = 1, m = 0)$ the marginal mean $E[X]$ is obtained while for $(n = 0, m = 1)$ gives the marginal mean $E[Y]$. In a similar way to the joint moments, the central joint moments of order $(n + m)$ of the random variables X and Y are defined

$$\begin{aligned} E[(X - m_X)^n (Y - m_Y)^m] &= \int_{-\infty}^{\infty} \int_{-\infty}^{\infty} (x - m_X)^n (y - m_Y)^m f_{XY}(x, y)\, dx\, dy \\ (\mu_{XY})_{nm} &= E[(X - m_X)^n (Y - m_Y)^m] \end{aligned} \tag{2.61}$$

It can be seen from this statement that, if $(n = 2, m = 0)$ the variance σ_X^2 and if $(n = 0, m = 2)$ the variance σ_Y^2 are obtained. The joint central moment for $(n = 1, m = 1)$ is called as the *Covariance* of the joint random variables X and Y, which is defined

$$\begin{aligned} &\text{Covariance of Joint r.v. } X \text{ and } Y: \rightarrow \sigma_{XY} = E[(X - m_X)(Y - m_Y)] \\ &\sigma_{XY} = \int_{-\infty}^{\infty} \int_{-\infty}^{\infty} (x - m_X)(y - m_Y) f_{XY}(x, y)\, dx\, dy = (m_{XY} - m_X m_Y) \end{aligned} \tag{2.62}$$

The ratio between the covariance and the product of marginal standard deviations is called as the *Correlation Coefficient*, which is defined

$$\text{Corralation Coefficint of r.v. } X \text{ and } Y: \quad \rightarrow \quad \rho_{XY} = \frac{\sigma_{XY}}{\sigma_X \sigma_Y} \tag{2.63}$$

The correlation coefficient is the normalized version of the covariance and it satisfies the condition of $(-1 \leq \rho_{XY} \leq 1)$ [24]. It is a measure of dependency between the random variables X and Y. If they are not correlated, then the correlation coefficient becomes zero and, for a full correlation, it becomes $\rho_{XY} = \pm 1$. It can be seen from Eq. (2.62) that, for independent random variables X and Y, the covariance will be zero and accordingly the correlation coefficient will also be zero, therefore, the independent random variables are also uncorrelated. But, uncorrelated random variables are not necessarily independent, i.e.,

$$\begin{aligned} &\text{Uncorrelated r.v. } X \text{ and } Y: \rightarrow \quad \rho_{XY} = 0\ , \ \text{but not necessarily independent} \\ &\text{Independent r.v. X and Y:} \quad \rightarrow \quad \rho_{XY} = 0\ , \ \text{thus uncorrelated} \end{aligned} \tag{2.64}$$

Once the PDF of random variables are determined their statistical values can be calculated as explained above. In the following section, some useful probability distributions that used frequently in practice are presented.

2.3.2 Some Useful Probability Distributions

In this section, some probability distributions of continuous r.v. and their properties, which are used in practical applications mostly, are presented briefly. More information about the distributions and a more complete list of distribution types can be found in text books of statistical distributions, e.g., [25–27].

2.3.2.1 Normal (Gaussian) Distribution

The *Normal* or *Gaussian* univariate probability density function of a r.v. X is the one mostly used in practice. It is defined in general as

$$\text{Normal PDF:} \quad \rightarrow \quad f_X(x) = \frac{1}{\sqrt{2\pi}\,\sigma} \exp\left[-\frac{1}{2} \left(\frac{x-m}{\sigma} \right)^2 \right] \tag{2.65a}$$

in which m and σ are respectively the mean and standard variation of X. The corresponding CDF is calculated from

$$\text{Normal CDF:} \quad \rightarrow F_X(x) = \int_{-\infty}^{x} f_X(\xi)\, d\xi = F_y(x) = \Phi\left(\frac{x-m}{\sigma} \right) \tag{2.65b}$$

where $\Phi(.)$ is called as the *Standard Normal Distribution* function and its PDF is denoted by $\varphi(.)$, which are defined

$$\text{Standard Normal PDF:} \quad \rightarrow \quad \varphi(x) = \frac{1}{\sqrt{2\pi}} e^{-x^2/2} \tag{2.65c}$$

$$\text{Standard Normal CDF:} \quad \rightarrow \quad \Phi(x) = \frac{1}{\sqrt{2\pi}} \int_{-\infty}^{x} e^{-u^2/2}\, du \tag{2.65d}$$

The central moments of X and $|X|$ are calculated [23] from

$$\text{Central Mom. of X:} \rightarrow \begin{cases} \mu_n = 1.3\ldots(n-1)\,\sigma^n & \text{for } n \text{ even} \\ \mu_n = 0 & \text{for } n \text{ odd} \end{cases} \tag{2.65e}$$

$$\text{Central Mom. of } |X|: \rightarrow \begin{cases} (\mu_{|X|})_n = 1.3\ldots(n-1)\,\sigma^n & \text{for } n = 2k \\ (\mu_{|X|})_n = \sqrt{\dfrac{2}{\pi}} 2^k k! \sigma^n & \text{for } n = 2k+1 \end{cases} \tag{2.65f}$$

Using Eq. (2.65e) the coefficients of skewness γ_1 and kurtosis γ_2 of a normal r.v. are calculated from Eqs. (2.44) and (2.45) as written

$$\begin{aligned} &\text{Coefficienet of Skewness of a Normal r.v.:} \quad \rightarrow \quad \gamma_1 = 0 \\ &\text{Coefficienet of Kurtosis of a Normal r.v.:} \quad \rightarrow \quad \gamma_2 = 3 \end{aligned} \tag{2.65h}$$

If multivariate normal variables are involved in a process, then a multivariate normal PDF will be required. In this case, a vector process is used and the multivariate normal PDF is stated as,

$$\text{Multivariate Normal PDF:} \rightarrow f_{\tilde{X}}(\tilde{x}) = \left(\frac{1}{2\pi}\right)^{\frac{p}{2}} \frac{1}{\sqrt{|\rho|}} \exp\left(-\frac{\chi^2}{2}\right) \tag{2.66a}$$

where $\tilde{X}$ is a vector of p-dimensional r.v., $\tilde{x}$ is a vector of their realizations and χ^2 is a scalar calculated from the product

$$\text{Scalar:} \rightarrow \chi^2 = (\tilde{x} - \tilde{m})^T \rho^{-1} (\tilde{x} - \tilde{m}) \tag{2.66b}$$

in which $\tilde{m}$ is a vector of mean values and ρ is the covariance matrix of $\tilde{X}$ and $|\rho|$ in Eq. (2.66a) denotes the determinant of ρ. These definitions are written

$$\begin{aligned} &\text{Vector of Multivariate r.v.:} \quad \rightarrow \quad \tilde{X} = \{X_1, X_2, \ldots, X_p\}^T \\ &\text{Vector of realizations of } \tilde{\text{X}}\text{:} \quad \rightarrow \quad \tilde{x} = \{x_1, x_2, \ldots, x_p\}^T \end{aligned} \tag{2.66c}$$

$$\text{Vector of mean values of } \tilde{\text{X}}\text{:} \quad \rightarrow \quad \tilde{m} = \{m_1, m_2, \ldots, m_p\}^T \tag{2.66d}$$

The covariance matrix ρ is defined as

$$\text{Covariance matrix of } \tilde{\text{X}}\text{:} \quad \rightarrow \quad \rho = \begin{bmatrix} \sigma_1^2 & \sigma_{12} & . & \sigma_{1p} \\ \sigma_{21} & \sigma_2^2 & . & \sigma_{2p} \\ . & . & . & . \\ \sigma_{p1} & \sigma_{p2} & . & \sigma_p^2 \end{bmatrix} \tag{2.66e}$$

As it is seen from Eq. (2.66e), the diagonal terms of this matrix are the variances of the r.v. X_i. For uncorrelated r.v., the off diagonal terms will be zero and the matrix becomes diagonal.

2.3.2.2 Lognormal Distribution

One other commonly used distribution in practice is the *Lognormal Distribution*. If the r.v. X has a Normal distribution with the mean and variance, m_X and σ_X^2, then the r.v. $Y = e^X$ is said to be log normally distributed. It is written as

$$\text{Exponential Function of X:} \quad \rightarrow \quad Y = e^X \quad \text{and} \quad X = \ln Y \tag{2.67a}$$

Using Eq. (2.51), the PDF of the r.v., $Y = e^X$, can be obtained as written

$$\text{Lognormal PDF:} \quad \rightarrow f_Y(y) = \frac{1}{\sigma_X \sqrt{2\pi}\, y} \frac{1}{e}^{-\frac{1}{2}\left(\frac{\ln y - m_X}{\sigma_X}\right)^2} \quad \text{for } (y > 0) \tag{2.67b}$$

In the region of $(y \leq 0)$, PDF of the r.v. Y will be zero, i.e., $f_Y(y) = 0$ for $(y \leq 0)$. The mean and variance of a Lognormal r.v. are calculated from

$$\begin{aligned} &\text{Mean of the r.v. } Y = e^X: && \rightarrow \quad m_Y = e^{m_X} e^{\sigma_x^2/2} \\ &\text{Variance of the r.v. } Y = e^X: && \rightarrow \quad \sigma_Y^2 = m_Y^2 \left(e^{\sigma_X^2} - 1\right) \end{aligned} \tag{2.67c}$$

If m_Y and σ_Y are given, then the variance and mean of X are calculated from the following statements

$$\sigma_x^2 = \ln\left[1 + \left(\sigma_y/\mu_y\right)^2\right] \quad \text{and} \quad \mu_x = \left(\ln \mu_y - \sigma_x^2/2\right) \tag{2.67d}$$

2.3.2.3 Uniform Distribution

The *Uniform Distribution* is also used in practice to describe an r.v. that is equally likely to take any value in the interval a and b, i.e., $(a \leq x \leq b)$. Its density function is constant in the specified region and zero outside the region as defined

$$\text{Uniform PDF:} \quad \rightarrow f_X(x) = \begin{cases} \dfrac{1}{(b-a)} & \text{in the interval } (a \leq x \leq b) \\ 0 & \text{outside the interval } (a \leq x \leq b) \end{cases} \tag{2.68a}$$

The mean and standard deviation of a Uniform r.v. are calculated from

$$\left.\begin{aligned} &\text{Mean:} && \rightarrow \quad m_X = (a+b)/2 \\ &\text{Stand. Dev.:} && \rightarrow \quad \sigma_X = (b-a)/\sqrt{12} \end{aligned}\right\} \rightarrow \left\{\begin{aligned} a &= m_X - \sqrt{3}\,\sigma_X \\ b &= m_X + \sqrt{3}\,\sigma_X \end{aligned}\right. \tag{2.68b}$$

2.3.2.4 Exponential Distribution

The *Exponential Distribution* can be used to describe intermittent variation of an r.v., such as an event occurring continuously and independently at a constant rate. Its distribution and density functions are defined in general as

$$\text{Exponential Distribution, CDF}: \rightarrow F_X(x) = \begin{cases} 1 - e^{-\lambda(x-\alpha)} & \text{for } (x \geq \alpha,\ \lambda > 0) \\ 0 & \text{for } (x < \alpha) \end{cases}$$

$$\text{Exponential Distribution, PDF}: \rightarrow f_X(x) = \begin{cases} \lambda e^{-\lambda(x-\alpha)} & \text{for } (x \geq \alpha,\ \lambda > 0) \\ 0 & \text{for } (x < \alpha) \end{cases} \tag{2.69a}$$

The mean and variance of an Exponential distribution are calculated from

$$\text{Mean} \rightarrow m_X = \alpha + \sigma_X \quad \text{and Variance} \rightarrow \sigma_X^2 = 1/\lambda^2 \tag{2.69b}$$

As an example, if it is assumed that a number of occurrences of an event per year is (n) and the event is described by the r.v. X, then, for independent occurrences, the annual probability distribution of X is stated

$$\text{Annual CDF of X:} \rightarrow F_X^n(x) = \left(1 - e^{-\lambda(x-\alpha)}\right)^n \tag{2.69c}$$

The density function of the annual probability is obtained from the derivative of $F_X^n(x)$ as written

$$\text{Annual PDF of X:} \rightarrow f_{Xn}(x) = \frac{\partial F_X^n(x)}{\partial x} \rightarrow f_{Xn}(x) = n\, F_X^{n-1}(x)\, f_X(x) \tag{2.69d}$$

2.3.2.5 Gamma Distribution

The *Gamma Distribution* represents the sum of r independent exponentially distributed r.v., and r.v. that take always positive values. Its PDF and CDF functions are defined as written

$$\text{Gamma Dist., PDF:} \rightarrow f_X(x) = \begin{cases} \dfrac{\lambda}{\Gamma(r)} (\lambda x)^{r-1} e^{-\lambda x} & \text{if } (x \geq 0, \lambda > 0) \\ 0 & \text{if } x \leq 0 \end{cases} \tag{2.70a}$$

$$\text{Gamma Dist., CDF:} \rightarrow F_X(x) = 1 - \sum_{k=0}^{r-1} \frac{1}{k!} e^{-\lambda x} (\lambda x)^k \quad \text{for } (r = +\text{int.}) \tag{2.70b}$$

in which $\Gamma(.)$ represents a Gamma function [28], which is defined

$$\text{Gamma Function:} \rightarrow \Gamma(x) = \int_0^\infty e^{-u} u^{(x-1)}\, du \tag{2.70c}$$

The mean and variance of the Gamma distribution are calculated to be

$$\text{Mean:} \rightarrow \ m_X = \frac{r}{\lambda} \quad \text{and Variance:} \rightarrow \quad \sigma_X^2 = \frac{r}{\lambda^2} \tag{2.70d}$$

The parameters r and λ are respectively the shape and scale parameters of the distribution. For different values of r and λ, different type of distributions are obtained. When $(r = 1)$, it gives the exponential distribution. If $(r < 1)$, then the distribution is exponentially shaped and asymptotic to both horizontal and vertical axes. If $(r > 1)$, its shape is unimodal and skewed with the mode (location of the peak of the PDF) equals $(x_m = (r-1)/\gamma)$. The skewness reduces with increasing value of r as it is seen from the coefficient of skewness, $(\gamma_1 = 2/\sqrt{r})$. If $(r = s/2)$ and $(\gamma = 1/2)$, then the Gamma distribution becomes the χ^2 (*Chi-square*) distribution with s degrees of freedom. In engineering applications, Gamma distributions occur frequently in models of failure analysis and, in Meteorology, for rainfall studies since the variables are always positive and the results are unbalanced.

2.3.2.6 Rayleigh Distribution

The Rayleigh Distribution is used as a probability model describing the distribution of wind speed over 1-year period. It is often used for the probability model of the absolute value of two components of a random field, e.g., if X and Y are two independent normally distributed random variables, both with zero mean and variance equal to σ^2 and if we define a function $Z = \sqrt{X^2 + Y^2}$ then this function has a Rayleigh distribution with the parameter σ [23]. It also describes the probability distribution of maxima of a narrow band random process with Normal distribution. The PDF and CDF of the Rayleigh distribution are given as

$$\text{Rayleigh PDF:} \rightarrow \ f_X(x) = \begin{cases} \dfrac{x}{\sigma^2} \exp\left(-\dfrac{x^2}{2\sigma^2}\right) & \text{if } x \geq 0 \\ 0 & \text{otherwise} \end{cases} \tag{2.71a}$$

$$\text{Rayleigh CDF:} \rightarrow \ F_X(x) = 1 - \exp\left(-\frac{x^2}{2\sigma^2}\right)$$

in which σ is the only parameter of the distribution, which is equal to the standard deviations of the independent random variables X and Y with Normal distributions and zero means. The mean and variance of the Rayleigh distribution are calculated to be

$$\begin{aligned} &\text{Mean:} && \rightarrow \ m_X = \sigma\sqrt{\frac{\pi}{2}} \\ &\text{Variance:} && \rightarrow \ \sigma_X^2 = 2\sigma^2\left(1 - \frac{\pi}{4}\right) \end{aligned} \tag{2.71b}$$

2.3.2.7 Weibull Distribution

The *Weibull Distribution* with three parameters is the general form of its family, which is also known as the *Extreme Value Type III* distribution. It is originally used in the probability description of strength of materials and fatigue analysis. This distribution is related to extreme value analysis and mostly used in the reliability engineering and failure analysis, survival analysis, industrial engineering, weather forecasting for wind speed variations, extreme value theory. The PDF and CDF of the Weibull distribution are defined as written

$$\text{Weibull PDF:} \rightarrow f_X(x) = \begin{cases} \left(\frac{\alpha}{\beta}\right)\left(\frac{x-\gamma}{\beta}\right)^{(\alpha-1)} \exp\left[-\left(\frac{x-\gamma}{\beta}\right)^{\alpha}\right] & \text{if } (x \geq \gamma) \\ 0 & \text{otherwise} \end{cases} \tag{2.72a}$$

$$\text{Weibull CDF:} \rightarrow F_X(x) = 1 - \exp\left[-\left(\frac{x-\gamma}{\beta}\right)^{\alpha}\right] \quad \text{for } (\alpha > 0, \beta > 0) \tag{2.72b}$$

in which α is the shape parameter, β is the scale parameter and γ is the location parameter. The mean and variance of the Weibull distribution are calculated in terms of its parameters as stated

$$\begin{aligned} &\text{Mean:} \rightarrow m_X = \gamma + \beta\,\Gamma\left(1 + \frac{1}{\alpha}\right) \\ &\text{Variance:} \rightarrow \sigma_X^2 = \beta^2\left[\Gamma\left(1 + \frac{2}{\alpha}\right) - \Gamma^2\left(1 + \frac{1}{\alpha}\right)\right] \end{aligned} \tag{2.72c}$$

If the location parameter is zero, i.e., $(\gamma = 0)$, the distribution is called as two parameters Weibull distribution. If, however, the scale parameter β is taken as a constant $(\beta = C)$ and $(\gamma = 0)$, then the distribution is called as one parameter Weibull distribution. As being special cases, for $(\alpha = 1)$, the Weibull distribution becomes the exponential distribution and, for $(\gamma = 0)$ and $(\alpha = 2)$, it becomes as the Rayleigh distribution.

2.3.2.8 Gumbel Distribution

The *Gumbel Distribution* is usually used to model the distribution of the maximum, or the minimum, of a number of samples or various distributions. It can also be used to find the probability that an extreme event, such as earthquake, flood or other natural disaster, will occur. The Gumbel distribution is also known as the *Extreme Value Type I Distribution*. It has two forms as one is for extreme maximum (*Extreme Value Largest I*) and one is for extreme minimum (*Extreme Value Smallest I*), which are respectively defined below.

$$\text{Gumbel (EV Largest-I):} \rightarrow \begin{cases} f_X(x) = \alpha\, e^{-\alpha(x-\beta)}\, e^{-\exp\,(-\alpha(x-\beta))} \\ F_X(x) = e^{-\exp(-\alpha(x-\beta))} \qquad \text{for } (-\infty < x < \infty) \end{cases}. \tag{2.73a}$$

$$\text{Gumbel (EV Smallest-I):} \rightarrow \begin{cases} f_X(x) = \alpha\, e^{\alpha(x-\beta)}\, e^{-\exp(\alpha(x-\beta))} \\ F_X(x) = 1 - e^{-\exp(\alpha(x-\beta))} \qquad \text{for } (-\infty < x < \infty) \end{cases} \tag{2.73b}$$

in which β is the location parameter and α is the scale parameter, which is defined $(\alpha > 0)$. The Gumbel distribution supports the range of outcomes of the r.v. X between $(-\infty < x < \infty)$. The means and variances of both *Largest-I* and *Smallest-I* distributions are calculated from

$$\text{Mean:} \quad \rightarrow \begin{cases} m_X = \beta + \dfrac{0.57722156649}{\alpha} & \text{(Largest-I)} \\ m_X = \beta - \dfrac{0.57722156649}{\alpha} & \text{(Smallest-I)} \end{cases} \tag{2.73c}$$

$$\text{Variance:} \quad \rightarrow \sigma_X^2 = \frac{\pi^2}{6\alpha^2} \quad \text{(Largest-I and Smallest-I)}$$

The value (0.57722156649) in Eq. (2.73c) is the Euler's constant. More information about the extreme value distributions can be found in text books, e.g., [29–31]. Having presented probability and statistical descriptions of r.v., in the following section, random processes and their probability and statistical descriptions are presented briefly.

2.4 Random Processes, Ensemble Averages, Expected Values, Stationary and Ergodic Processes

Random variables and their probability descriptions are explained in previous sections. In this section, random processes and their properties will be outlined briefly. An r.v. X is a variable which takes values at random and can be thought of as a measurement of outcomes of a random experiment that cannot be predicted beforehand. A random process is a sequence of r.v. that vary in time or 1D space. If the number of random variables are limited then the process is said to be discrete, and if they are infinitely large, a continuous process is obtained. The random process can be thought of as a collection, or ensemble, of functions of time or one dimensional space. Such an ensemble is as shown in Fig. 2.3. If an associated r.v. X is a function of time, i.e., $X(t)$, then the process is called as the stochastic process. Each possible outcome of a stochastic process is called as the realization, or sample, of the process and all realizations constitute the ensemble of

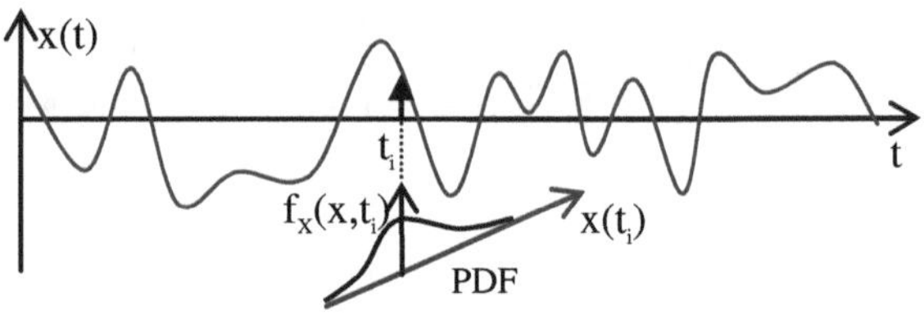

Fig. 2.7 One particular realization of the stochastic process $X(t)$ at $(t = t_i)$ with its PDF

the process. A single realization $x(t_i)$ of the process $X(t)$ at $(t = t_i)$ is shown in Fig. 2.7 with the PDF while the ensemble is as shown in Fig. 2.3. The value of an observed sample of the ensemble at a particular time, say t_i, which is shown in Fig. 2.7, is an r.v. $x(t_i)$ that its probability and statistical characteristics have been explained in previous sections. For a stochastic process, we show explicitly the time dependence of the PDF and CDF in the notation, i.e., $f_X(x, t)$ and $F_X(x, t)$, as presented in Sect. (2.2.4). In this section, ensemble averages and various forms of the stochastic process are explained.

2.4.1 Ensemble Averages and Expected Values

In Sect. 2.3, mean values (or expected values) of r.v. and random functions are explained. In this section, the mean values (or expected values) of stochastic processes will be presented. Since a stochastic process is a collection of time-dependent r.v., $x_i(t)$ where $(i = 1, 2, \ldots, n)$ which are realizations of the process $X(t)$ as shown in Fig. 2.3, its average value, or sample mean, at any time t is called as the *Ensemble Average*. Assuming that the r.v. $x_i(t)$ are independent with identical probability distributions, which are equal to the probability distribution of the r.v. $X(t)$, then the ensemble average is defined as

$$\text{Ensemble Average:} \quad \rightarrow \quad \bar{X}(t_j) = \frac{1}{n}\sum_{i=1}^{n} x_i(t_j) = E[X(t)] \quad (\text{for } n \rightarrow \infty) \qquad (2.74a)$$

which is equal to the expected value of the process $X(t)$ when n approaches infinity according to the law of large numbers [23]. If we have infinite number of realizations, then we can say that the process $X(t)$ is continuous. In this case, Eq. (2.74a) becomes an integral form and the expected value of the process $X(t)$ can be written similarly to the statement of the expected value of an r.v. Thus, for a continuous process $X(t)$, the expected value is

$$\text{Expected (Mean) Value:} \quad \rightarrow \quad E[X(t)] = \int_{-\infty}^{\infty} x(t) f_X(x,t)\, \mathrm{d}x \qquad (2.74b)$$

which is a time function unlike that of an r.v. All other expectations and probability moments presented in Sect. 2.3 are also valid for stochastic processes provided that they are all functions of time t. The probability distribution $F_X(x, t)$ and probability density function $f_X(x, t)$ of a stochastic process $X(t)$ are defined respectively in Eqs. (2.10a) and (2.13) at any time t, which are called as the *First-order* distribution and density functions of the process $X(t)$, which are sufficient to define the range of amplitudes of the process $X(t)$ in a probabilistic sense. In order to observe the variation of the same realization $x(t)$ at different time stations, say t_1 and t_2, the joint distribution of $x(t_1)$ and $x(t_2)$, which is called as the *Second-order* joint distribution of $X(t)$, is defined as

$$\text{Second-order CDF:} \rightarrow F_X(x_1, t_1; x_2, t_2) = P(X(t_1) \leq x_1 \cap X(t_2) \leq x_2) \quad (2.75a)$$

and the corresponding joint PDF is obtained from the derivation

$$\text{Second-order PDF:} \rightarrow f_X(x_1, t_1; x_2, t_2) = \frac{\partial^2 F_X(x_1, t_1; x_2, t_2)}{\partial x_1 \partial x_2} \quad (2.75b)$$

If two different stochastic processes, $X(t)$ and $Y(t)$, are involved, their second-order joint distribution and density functions at different time stations, t_1 and t_2, are defined similarly to those of a single stochastic process as written

$$\text{Second-order joint CDF:} \rightarrow F_{XY}(x, t_1; y, t_2) = P(X(t_1) \leq x \cap Y(t_2) \leq y) \quad (2.75c)$$

$$\text{Second-order joint PDF:} \rightarrow f_{XY}(x, t_1; y, t_2) = \frac{\partial^2 F_{XY}(x, t_1; y, t_2)}{\partial x \partial y} \quad (2.75d)$$

In a similar way, the marginal and conditional distributions of stochastic processes can also be defined as in the case of r.v. These are not presented here and attention is paid further to commonly used definitions that have practical importance. These are first moments (expected values) of the joint distributions defined above. The expected values of the joint r.v. $x(t_1)$ and $x(t_2)$, and $x(t_1)$ and $y(t_2)$ are respectively defined as the *Auto-Correlation* and *Cross-Correlation* functions. Hence, the auto-correlation function is

$$\text{Auto-Correlation:} \rightarrow \begin{cases} R_{XX}(t_1, t_2) = E[X(t_1)\,X(t_2)] \quad \text{or} \\ R_{XX}(t_1, t_2) = \int\limits_{-\infty}^{\infty} \int\limits_{-\infty}^{\infty} x_1\, x_2 f_{XX}(x_1, t_1;\, x_2, t_2)\, \mathrm{d}x_1 \mathrm{d}x_2 \end{cases} \quad (2.76a)$$

and the cross-correlated function is

$$\text{Cross-Correlation:} \rightarrow \begin{cases} R_{XY}(t_1, t_2) = E[X(t_1)\,Y(t_2)] \quad \text{or} \\ R_{XY}(t_1, t_2) = \int\limits_{-\infty}^{\infty} \int\limits_{-\infty}^{\infty} x\, y f_{XY}(x, t_1; y, t_2)\, \mathrm{d}x\, \mathrm{d}y \end{cases} \quad (2.76b)$$

As in the case of r.v. presented in the Sect. 2.3, the first central moments of the joint random variables $x(t_1)$ and $x(t_2)$, and $x(t_1)$ and $y(t_2)$ are respectively defined as the *Auto-Covariance* and *Cross-Covariance* functions, which are

$$\text{Auto-Covariance:} \rightarrow C_{XX}(t_1, t_2) = \begin{cases} E[(X(t_1) - E[X(t_1)])\,(X(t_2) - E[X(t_2)])] \\ R_{XX}(t_1, t_2) - E[X(t_1)]\; E[X(t_2)] \end{cases} \tag{2.77a}$$

$$\text{Cross-Covariance:} \rightarrow C_{XY}(t_1, t_2) = \begin{cases} E[(X(t_1) - E[X(t_1)])\,(Y(t_2) - E[Y(t_2)])] \\ R_{XY}(t_1, t_2) - E[X(t_1)]\; E[Y(t_2)] \end{cases} \tag{2.77b}$$

The correlation and covariance information of stochastic processes is important and frequently used in the spectral analysis of structures. The above mentioned definitions of the correlation and covariance functions are general. For special processes, such as stationary and ergodic processes, they are simplified as presented in the following section.

2.4.2 *Stationary and Ergodic Processes*

Two special forms of a general stochastic process, which are the *Stationary* and *Ergodic* processes, are frequently encountered or assumed in probabilistic analyses of structures. These special forms are outlined in this section.

2.4.2.1 Stationary Process

If the statistical properties of a random process are invariant in time, then the process is said to be *stationary* in the *strict sense* [23]. This implies that the statistics of a stationary process is not affected by a shift in the time origin. Thus, the processes $X(t)$ and $X(t + \tau)$ have the same statistics for any τ value. Similarly, the processes $X(t)$ and $Y(t)$ are *jointly stationary* if their joint statistics are not affected by a shift in the time origin, i.e., if $X(t)$ and $Y(t)$, and $X(t + \tau)$ and $Y(t + \tau)$, have the same joint statistics for any τ value. A random process is said to be *stationary* in the *wide sense* or *weakly* if its expected value is a constant and its auto-correlation is a function of the time difference ($\tau = t_2 – t_1$) which means that it is not affected by a shift in the time origin. From these definitions, it can be stated that a weakly stationary Normal process is also strictly stationary since all statistics of the Normal process are uniquely determined in terms of its mean and auto-correlation. For a stationary process, the statistical characteristics of a general process presented in Sect. 2.4.1 become as written

$$\text{Mean Value:} \quad \rightarrow \quad \mu_X = E[X(t_1)] = E[X(t_2)] = \text{constant} \tag{2.78a}$$

$$\begin{aligned} &\text{Auto-Correlation:} \quad \rightarrow \quad R_{XX}(\tau) = E[X(t)\,X(t+\tau)] \\ &\text{Auto-Covariance:} \quad \rightarrow \quad C_{XX}(\tau) = R_{XX}(\tau) - \mu_X^2 \end{aligned} \tag{2.78b}$$

$$\begin{aligned} &\text{Cross-Correlation:} \rightarrow R_{XY}(\tau) = E[X(t)\,Y(t+\tau)] \\ &\text{Cross-Covariance:} \rightarrow C_{XY}(\tau) = R_{XY}(\tau) - \mu_X\mu_Y \end{aligned} \tag{2.78c}$$

or, in the integral form, the mean value, auto- and cross-correlation function are expressed as

$$\begin{aligned} &\text{Mean Value:} \quad \rightarrow \quad \mu_X = \int_{-\infty}^{\infty} x(t) f_X(x,t)\,\mathrm{d}x = \text{constant} \\ &\text{Auto-Correlation:} \quad \rightarrow \quad R_{XX}(\tau) = \int_{-\infty}^{\infty} x(t)\,x(t+\tau) f_X(x,t)\,\mathrm{d}x \end{aligned} \tag{2.78d}$$

$$\text{Cross-Correlation:} \rightarrow R_{XY}(\tau) = \int_{-\infty}^{\infty}\int_{-\infty}^{\infty} x(t)\,y(t+\tau) f_{XY}(x,y,t)\,\mathrm{d}x\,\mathrm{d}y \tag{2.78e}$$

Since auto- and cross-correlations functions, $R_{XX}(\tau)$ and $R_{XY}(\tau)$, are independent of an absolute time t_i and functions only the time difference τ, it can be verified that these functions become even functions of τ, i.e., they are symmetric with respect to the vertical (function) axis. This property is stated as

$$R_{XX}(\tau) = R_{XX}(-\tau) \quad \text{and} \quad R_{XY}(\tau) = R_{YX}(-\tau) \tag{2.79}$$

An example auto-correlation function $R_{XX}(\tau)$ is shown in Fig. 2.8b. It can be seen from Eq. (2.78) that, when $\tau = 0$, the auto- and cross-covariances of the random processes $X(t)$ and $Y(t)$ become respectively the variance and covariance, i.e.,

$$\begin{aligned} &\text{Variance:} \quad \rightarrow \quad \sigma_X^2 = C_{XX}(0) = R_{XX}(0) - \mu_X^2 \\ &\text{Covariance:} \rightarrow \quad \sigma_{XY} = C_{XY}(0) = R_{XY}(0) - \mu_X\mu_Y \end{aligned} \tag{2.80a}$$

It can also be shown that the auto-correlation function is maximum at $\tau = 0$, i.e.,

$$R_{XX}(0) \geq R_{XX}(\tau) \tag{2.80b}$$

2.4.2.2 Ergodic Process

One important concept in the stochastic analysis is the *ergodic* properties of the process. When the time average of a stationary process is equal to the ensemble average, it is said that this stationary process is ergodic. For an ergodic process,

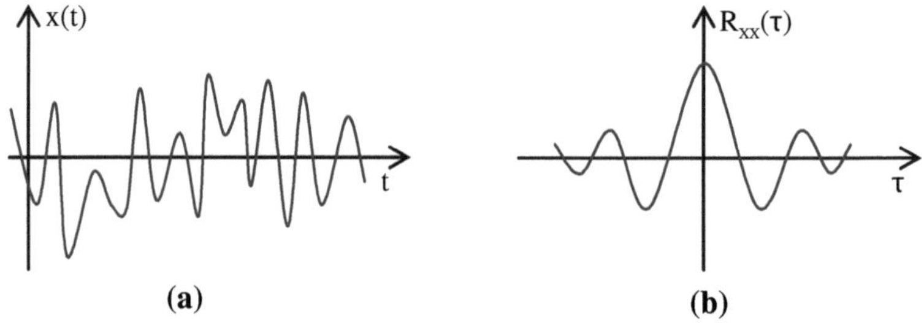

Fig. 2.8 A sample r.v. $x(t)$ and the corresponding autocorrelation function $R_{XX}(\tau)$. **a** A sample r.v. $x(t)$. **b** Autocorrelation function $R_{XX}(\tau)$

any statistic calculated by averaging over all samples (members) of an ensemble at a fixed time is equal to that calculated by averaging over all time on a single sample of the ensemble. Thus, only one sample time function of the ensemble represents the whole ensemble. For continuous ergodic processes, the mean, auto- and cross-correlation functions are calculated from the time averages

$$\text{Mean Value:} \quad \rightarrow \quad \tilde{\mu}_X = E[X(t_1)] = \lim_{T\to\infty} \frac{1}{2T} \int_{-T}^{T} X(t)\,\mathrm{d}t \tag{2.81a}$$

$$\text{Auto-Correlation:} \quad \rightarrow \quad \tilde{R}_{XX}(\tau) = E[X(t)\,X(t+\tau)] = \lim_{T\to\infty} \frac{1}{2T} \int_{-T}^{T} X(t)\,X(t+\tau)\,\mathrm{d}t$$

$$\text{Cross-Correlation:} \quad \rightarrow \quad \tilde{R}_{XY}(\tau) = E[X(t)\,Y(t+\tau)] = \lim_{T\to\infty} \frac{1}{2T} \int_{-T}^{T} X(t)\,Y(t+\tau)\,\mathrm{d}t \tag{2.81b}$$

2.4.2.3 Complex Stationary Processes

In practice, complex random variables and processes can be frequently encountered. A complex random process $Z(t)$ is defined as

$$\text{A complex process:} \rightarrow Z(t) = X(t) + i\,Y(t) \tag{2.82a}$$

where $X(t)$ is the real part and $Y(t)$ is the imaginary part of the process. If both the real and imaginary parts are stationary, then the complex process $Z(t)$ is said to be stationary. The conjugate of the complex process $Z(t)$ is defined

$$\text{Conjugate of } Z(t)\text{:} \rightarrow Z^*(t) = X(t) - i\,Y(t) \tag{2.82b}$$

The mean and auto-correlation function of a complex process are defined as

$$\text{Complex process } Z(t) : \rightarrow \begin{cases} \text{Mean:} & \rightarrow m_Z = E[Z(t)] = m_X + i\, m_Y \\ \text{Auto-correlation:} & \rightarrow R_{ZZ}(\tau) = E[Z(t)\, Z^*(t+\tau)] \end{cases} \tag{2.83a}$$

If two complex stationary processes are defined as $Z(t)$ and $V(t)$, then their cross-correlation function $R_{ZV}(\tau)$ is defined similarly to auto-correlation function as

$$\text{Cross-correlation of Z(t) and V(t):} \rightarrow R_{ZV}(\tau) = E[Z(t)\ V^*(t+\tau)] \tag{2.83b}$$

Using the properties of real processes $X(t)$ and $Y(t)$ given in Eq. (2.79), the following properties of complex processes can be written

$$R_{ZZ}(\tau) = R^*_{ZZ}(-\tau) \quad \text{and} \quad R_{ZV}(\tau) = R^*_{VZ}(-\tau) \tag{2.84}$$

2.4.3 Differentiation of Stochastic Processes

In practice, the derivatives of a stochastic process are also of interest in structural dynamics. They are stochastic processes that properties can be obtained simply using the source process. Stochastic displacements, velocities and accelerations are examples of such processes. Let us assume two general stochastic processes to be $X(t)$ and $Y(t)$. Their time derivatives of the order n and m are called as the *derived processes* which are defined as

$$\text{Derived Processes:} \quad \rightarrow \quad X^{(n)}(t) = \frac{\partial^n X(t)}{\partial t^n} \quad \text{and} \quad Y^{(m)}(t) = \frac{\partial^m Y(t)}{\partial t^m} \tag{2.85}$$

Their mean values and correlation functions are calculated [23] from

$$\text{Mean Value:} \quad \rightarrow \quad E\left[X^{(n)}(t)\right] = \frac{\partial^n E[X(t)]}{\partial t^n} \tag{2.86a}$$

$$\begin{aligned} &\text{Auto-correlation:} \quad \rightarrow \quad R_{X^{(n)}X^{(n)}}(t_1,t_2) = E\left[X^{(n)}(t_1)X^{(n)}(t_2)\right] \\ &\text{Cross-correlation:} \quad \rightarrow \quad R_{X^{(n)}Y^{(m)}}(t_1,t_2) = E\left[X^{(n)}(t_1)Y^{(m)}(t_2)\right] \end{aligned} \tag{2.86b}$$

Having introduced $X^{(n)}(t)$ and $Y^{(m)}(t)$ from Eqs. (2.85) into (2.86b) and taking the average values the auto-correlation function is obtained as written by

$$\text{Auto-correlation:} \quad \rightarrow \quad R_{X^{(n)}X^{(n)}}(t_1,t_2) = \begin{cases} \dfrac{\partial^{2n} E[X(t_1)X(t_2)]}{\partial t_1^n \partial t_2^n} \\ \dfrac{\partial^{2n} R_{XX}(t_1,t_2)}{\partial t_1^n \partial t_2^n} \end{cases} \tag{2.86c}$$

ant the cross-correlation function is obtained as written by

$$\text{Cross-correlation:} \quad \rightarrow \quad R_{X^{(n)}Y^{(m)}}(t_1,t_2) \;=\; \begin{cases} \dfrac{\partial^{(n+m)} E[X(t_1)Y(t_2)]}{\partial t_1^n \partial t_2^m} \\[2ex] \dfrac{\partial^{(n+m)} R_{XY}(t_1,t_2)}{\partial t_1^n \partial t_2^m} \end{cases} \tag{2.86d}$$

where n and m denote the order of derivatives. For jointly stationary processes the cross-correlation function of the derived processes are stated from Eq. (2.86d) as

$$\left.\begin{array}{l}\text{Cross-correlation of} \\ \text{Stationary derived processes}\end{array}\right\}: \rightarrow R_{X^{(n)}Y^{(m)}}(\tau) = (-1)^n \frac{\partial^{n+m} R_{XY}(\tau)}{\partial \tau^{n+m}} \tag{2.87a}$$

In Eq. (2.87a), if the process $Y(t)$ is replaced with the process $X(t)$, then the auto-correlation function of the derived process can be obtained as written

$$\left.\begin{array}{l}\text{Auto-correlation of} \\ \text{Stationary derived processes}\end{array}\right\}: \rightarrow R_{X^{(n)}X^{(m)}}(\tau) = (-1)^n \frac{\partial^{n+m} R_{XX}(\tau)}{\partial \tau^{n+m}} \tag{2.87b}$$

In structural dynamics, the first two time derivatives are mostly used, which are denoted by $\dot{X}(t)$ and $\ddot{X}(t)$. Using Eq. (2.87b) the auto-correlations of $\dot{X}(t)$ and $\ddot{X}(t)$, and the cross-correlations of $X(t)$, $\dot{X}(t)$ and $\dot{X}(t)$, $\ddot{X}(t)$ are stated as

$$\text{Auto-correlations of derived processes:} \rightarrow \begin{cases} R_{\dot{X}\dot{X}}(\tau) = -\dfrac{\partial^2 R_{XX}(\tau)}{\partial \tau^2} \\[2ex] R_{\ddot{X}\ddot{X}}(\tau) = \dfrac{\partial^4 R_{XX}(\tau)}{\partial \tau^4} \end{cases} \tag{2.88a}$$

$$\text{Cross-correlations of derived processes:} \rightarrow \begin{cases} R_{X\dot{X}}(\tau) = \dfrac{\partial R_{XX}(\tau)}{\partial \tau} \\[2ex] R_{\dot{X}\ddot{X}}(\tau) = -\dfrac{\partial^3 R_{XX}(\tau)}{\partial \tau^3} \end{cases} \tag{2.88b}$$

2.5 Spectral Analysis of Stochastic Processes

Correlation functions of stochastic processes are used to get information about their characteristics, but they do not provide direct information about their frequency contents. In structural dynamics, the frequency content of a stochastic process, e.g., a stochastic loading on the structure, plays an important role in the design of the structure such that, for a safe design, fundamental natural frequencies of the structure should be far away from the peak frequencies of the loading process. The frequency content and statistical characteristics, under certain conditions, of a stochastic process can be determined from its *Power Spectrum* or *Spectral Density* function. The power spectrum of a process, which is denoted in general by $S(\omega)$, is simply defined as being the *Fourier transform* of its correlation function $R(\tau)$. Thus, the power spectrum and correlation functions are the Fourier transform pairs

$$S(\omega) = \frac{1}{2\pi}\int_{-\infty}^{\infty} R(\tau)\, e^{-i\omega\tau} d\tau \quad \leftrightarrow \quad R(\tau) = \int_{-\infty}^{\infty} S(\omega) e^{i\omega\tau} d\omega \tag{2.89}$$

Similar to the correlation functions, auto- and cross-power spectral functions are defined using Eq. (2.89). Assuming that $X(t)$ and $Y(t)$ are two stationary stochastic processes. Their correlation and spectral functions are written as, for auto-correlation and auto-spectral functions,

$$R_{XX}(\tau) = \int_{-\infty}^{\infty} S_{XX}(\omega)\, e^{i\omega\tau} d\omega \quad \leftrightarrow \quad S_{XX}(\omega) = \frac{1}{2\pi}\int_{-\infty}^{\infty} R_{XX}(\tau)\, e^{-i\omega\tau} d\tau \tag{2.90a}$$

and for cross-correlation and cross-spectral functions,

$$R_{XY}(\tau) = \int_{-\infty}^{\infty} S_{XY}(\omega) e^{i\omega\tau} d\omega \quad \leftrightarrow \quad S_{XY}(\omega) = \frac{1}{2\pi}\int_{-\infty}^{\infty} R_{XY}(\tau)\, e^{-i\omega\tau} d\tau \tag{2.90b}$$

It can be proved that the power spectrum $S_{XY}(\omega)$ is the expectation of the product $(X^*(\omega)\, Y(\omega))$ where $X^*(\omega)$ is the complex conjugate of $X(\omega)$. For the proof, let us write the Fourier transform pairs of the stationary processes $X(t)$ and $Y(t)$,

$$\left.\begin{aligned} X(t) &= \int_{-\infty}^{\infty} X(\omega)\, e^{i\omega t} d\omega \\ Y(t) &= \int_{-\infty}^{\infty} Y(\omega)\, e^{i\omega t} d\omega \end{aligned}\right\} \leftrightarrow \left\{\begin{aligned} X(\omega) &= \frac{1}{2\pi}\int_{-\infty}^{\infty} X(t) e^{-i\omega t} dt \\ Y(\omega) &= \frac{1}{2\pi}\int_{-\infty}^{\infty} Y(t)\, e^{-i\omega t} dt \end{aligned}\right. \tag{2.91}$$

Using Eq. (2.91) the cross-correlation function $R_{XY}(\tau)$ can be written as

$$R_{XY}(\tau) = E[X(t)Y(t+\tau)] = E\left[\int_{-\infty}^{\infty} X(\omega)\, e^{i\omega t} d\omega \int_{-\infty}^{\infty} Y(\omega)\, e^{i\omega(t+\tau)} d\omega\right] \tag{2.92a}$$

and substituting $X(\omega)$ from Eqs. (2.91) into (2.92a) it can be stated

$$R_{XY}(\tau) = \begin{cases} E\left[\int_{-\infty}^{\infty} \left(\frac{1}{2\pi}\int_{-\infty}^{\infty} X(t)\, e^{-i\omega t} dt\right) e^{i\omega t} d\omega \int_{-\infty}^{\infty} Y(\omega)\, e^{i\omega(t+\tau)} d\omega\right] \\ E\left[\int_{-\infty}^{\infty} \left(\frac{1}{2\pi}\int_{-\infty}^{\infty} X(t)\, e^{i\omega t} dt\right) d\omega \int_{-\infty}^{\infty} Y(\omega)\, e^{i\omega\tau} d\omega\right] \end{cases} \tag{2.92b}$$

Since $\left(\frac{1}{2\pi}\int_{-\infty}^{\infty} X(t)\, e^{i\omega t} dt = X^*(\omega)\right)$ the cross-correlation function can be written as

$$R_{XY}(\tau) = E\left[\int_{-\infty}^{\infty} X^*(\omega)\, Y(\omega) e^{i\omega\tau} d\omega\right] = \int_{-\infty}^{\infty} E[X^*(\omega)\, Y(\omega)] e^{i\omega\tau} d\omega \quad (2.92c)$$

From comparison of Eqs. (2.90b) and (2.92c) it is seen that

$$S_{XY}(\omega) = E[X^*(\omega)\, Y(\omega)] \quad (2.92d)$$

By taking its complex conjugate we can obtain the cross-spectral function $S_{YX}(\omega)$ as stated

$$S^*_{XY}(\omega) = (E[X^*(\omega)\, Y(\omega)])^* = E[Y^*(\omega)\, X(\omega)] = S_{YX}(\omega) \quad (2.92e)$$

The power spectrum of the stationary process $X(t)$ can readily be obtained from Eq. (2.92d) as stated

$$S_{XX}(\omega) = E[X^*(\omega)\, X(\omega)] = E\left[|X(\omega)|^2\right] \quad (2.93)$$

from which it is apparent that the power spectrum $S_{XX}(\omega)$ is a real-valued even function. This can also be verified by using the properties of correlation functions given in Eq. (2.84). The properties of spectral functions are summarized as follows.

$$\begin{aligned} &\text{Auto-spectrum:} \rightarrow S_{XX}(\omega)\text{: even real,} \quad S_{XX}(\omega) = S_{XX}(-\omega) \\ &\text{Cross-spectrum:} \rightarrow S_{YX}(\omega) = S^*_{XY}(\omega) \end{aligned} \quad (2.94)$$

When ($\tau = 0$), i.e., $R_{XX}(0)$ and $R_{XY}(0)$, the variance and covariance of the processes $X(t)$ and $Y(t)$ can be expressed in terms of spectral functions by using Eq. (2.80a) as written

$$\text{Variance:} \quad \rightarrow \quad \sigma_X^2 = \int_{-\infty}^{\infty} S_{XX}(\omega)\, d\omega - \mu_X^2 \quad (2.95a)$$

$$\text{Covariance:} \rightarrow \quad \sigma_{XY} = \int_{-\infty}^{\infty} S_{XY}(\omega)\, d\omega - \mu_X \mu_Y \quad (2.95b)$$

For zero-mean processes, these statistical values become

$$\sigma_{XX}^2 = \int_{-\infty}^{\infty} S_{XX}(\omega)\, d\omega \quad \text{and} \quad \sigma_{XY} = \int_{-\infty}^{\infty} S_{XY}(\omega)\, d\omega \quad (2.95c)$$

A typical correlation function and its counterpart spectral function are shown in Fig. 2.9.

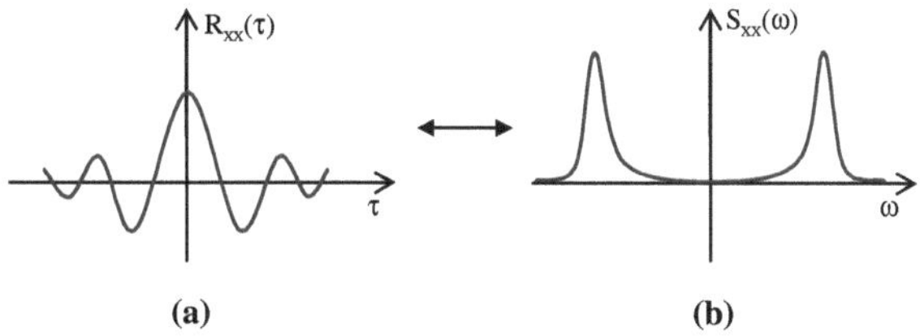

Fig. 2.9 An example correlation function and its corresponding spectral function. **a** A correlation function $R_{XX}(\tau)$. **b** Corresponding spectral function $S_{XX}(\omega)$

2.5.1 Spectral Moments, Variances of Derived Processes, Spectral Bandwidth

In engineering practice, *Spectral Moments* and *Spectral Bandwidth* are used to calculate statistical characteristics of stochastic processes and probability distributions of their extreme values and mean periods.

Spectral moments are defined as calculated from the following integration of the spectrum:

$$\text{Spectral Moments:} \rightarrow m_n = \int_{-\infty}^{\infty} \omega^n S_{XX}(\omega)\, d\omega \tag{2.96}$$

where m_n is called as the nth spectral moment. The first three even moments are used to describe probability distributions of extremes and mean frequencies, or periods, of the process. They are written as

$$m_0 = \int_{-\infty}^{\infty} S_{XX}(\omega)\, d\omega, \quad m_2 = \int_{-\infty}^{\infty} \omega^2\, S_{XX}(\omega)\, d\omega, \quad m_4 = \int_{-\infty}^{\infty} \omega^4\, S_{XX}(\omega)\, d\omega, \tag{2.97}$$

Correlation functions of derived processes are calculated using Eqs. (2.87a) and (2.87b) in general, and Eqs. (2.88a) and (2.88b) in particular for the first and second derivatives. By using the spectral counterparts of these functions from Eqs. (2.90a) and (2.90b) they can be stated for the processes *X*(*t*) and *Y*(*t*) as

$$R_{X^{(n)}Y^{(m)}}(\tau) = (-1)^n\, i^{(n+m)} \int_{-\infty}^{\infty} \omega^{(n+m)} S_{XY}(\omega)\, e^{i\omega\tau} d\omega \tag{2.98}$$

and, for the first and second derivatives, they are, e.g.,

$$R_{X\dot{X}}(\tau) = i \int_{-\infty}^{\infty} \omega S_{XX}(\omega)\, e^{i\omega\tau} d\omega \quad \text{and} \quad R_{\dot{X}\dot{X}}(\tau) = \int_{-\infty}^{\infty} \omega^2 S_{XX}(\omega)\, e^{i\omega\tau} d\omega \tag{2.99a}$$

$$R_{\dot{X}\ddot{X}}(\tau) = i \int_{-\infty}^{\infty} \omega^3 S_{XX}(\omega)\, e^{i\omega\tau} d\omega \quad \text{and} \quad R_{\ddot{X}\ddot{X}}(\tau) = \int_{-\infty}^{\infty} \omega^4 S_{XX}(\omega)\, e^{i\omega\tau} d\omega \tag{2.99b}$$

When ($\tau = 0$), these functions are more meaningful since they are closely related to the variance and covariance of the derived processes of $X(t)$. Thus,

$$\left.\begin{aligned} R_{X\dot{X}}(0) &= i\int_{-\infty}^{\infty} \omega S_{XX}(\omega)\,d\omega = 0 \\ R_{\dot{X}\dot{X}}(0) &= \int_{-\infty}^{\infty} \omega^2 S_{XX}(\omega)\,d\omega = m_2 \end{aligned}\right\} \text{and} \left\{\begin{aligned} R_{\dot{X}\ddot{X}}(0) &= i\int_{-\infty}^{\infty} \omega^3 S_{XX}(\omega)\,d\omega = 0 \\ R_{\ddot{X}\ddot{X}}(0) &= \int_{-\infty}^{\infty} \omega^4 S_{XX}(\omega)\,d\omega = m_4 \end{aligned}\right. \tag{2.100}$$

which are spectral moments of the source process $X(t)$. As it can be seen from these statements that, since $S_{XX}(\omega)$ is an even function, odd spectral moments are all zero. This implies that the related derived processes are uncorrelated, e.g., the processes $X(t)$ and $\dot{X}(t)$ are uncorrelated since $\left(E\left[X(t)\dot{X}(t)\right] = R_{X\dot{X}}(0) = 0\right)$. Thus, the variances and covariances of derived processes can be stated as, for the first two derivatives,

$$\left.\begin{aligned} \sigma_{X\dot{X}} &= -m_X m_{\dot{X}} \\ \sigma_{\dot{X}}^2 &= m_2 - m_{\dot{X}}^2 \end{aligned}\right\} \text{and} \left\{\begin{aligned} \sigma_{\dot{X}\ddot{X}} &= -m_{\dot{X}} m_{\ddot{X}} \\ \sigma_{\ddot{X}}^2 &= m_4 - m_{\ddot{X}}^2 \end{aligned}\right. \tag{2.101}$$

Spectral Bandwidth is a useful informative parameter which defines the relative width of a spectral function. It appears in the formulation of probability distributions of derived processes, and accordingly, in the formulation of extreme value statistics. It is defined in terms of spectral moments, or mean frequencies, or mean periods, of the process as written by

$$\text{Spectral Bandwidth:} \rightarrow \ \varepsilon = \sqrt{1 - \frac{m_2^2}{m_0 m_4}} = \sqrt{1 - \frac{\omega_0^2}{\omega_m^2}} = \sqrt{1 - \frac{T_m^2}{T_0^2}} \tag{2.102}$$

where m_0, m_2, and m_4 are the spectral moments, ω_0 and ω_m are mean frequencies of zero-crossings and maxima of the process, whereas T_0 and T_m are the corresponding mean periods.

2.5.2 Band-Limited, Narrow-Band and Broad-Band Processes

A band-limited process is defined if its spectrum has a uniform magnitude S_0 over a frequency band between ω_1 and ω_2 as shown in Fig. 2.10. This spectrum is also called as the band-limited white noise. Its correlation function is calculated by using Eq. (2.89). It is stated as

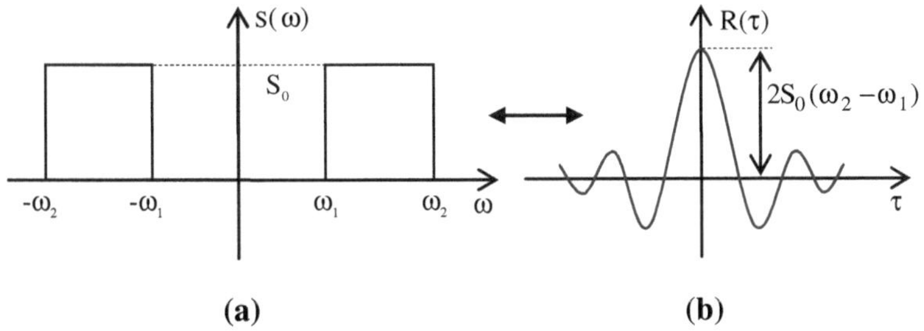

Fig. 2.10 Spectrum of a band-limited process and its corresponding correlation function. **a** A band-limited spectrum. **b** Corresponding correlation function

$$R(\tau)=\begin{cases}\displaystyle\int_{-\infty}^{\infty}S(\omega)\,e^{i\omega\tau}d\omega=S_0\left(\int_{-\omega_2}^{-\omega_1}e^{i\omega\tau}d\omega+\int_{\omega_1}^{\omega_2}e^{i\omega\tau}d\omega\right)\\ \displaystyle\frac{2S_0}{\tau}(\sin\omega_2\tau-\sin\omega_1\tau)=\frac{4S_0}{\tau}\left(\cos\left(\frac{\omega_2+\omega_1}{2}\right)\tau\;\sin\left(\frac{\omega_2-\omega_1}{2}\right)\tau\right)\end{cases}\tag{2.103}$$

which is shown in Fig. 2.10. If this spectrum belongs to the process $X(t)$, its mean square value is calculated from

$$E\left[X^2(t)\right]=R(0)=2S_0(\omega_2-\omega_1)\tag{2.104}$$

The 0th, 2nd, and 4th spectral moments are calculated using Eq. (2.97) and the spectral bandwidth parameter ε is calculated from Eq. (2.102). The results are

$$\left.\begin{aligned}m_0&=2S_0(\omega_2-\omega_1)\\ m_2&=2S_0\left(\omega_2^3-\omega_1^3\right)/3\\ m_4&=2S_0\left(\omega_2^5-\omega_1^5\right)/5\end{aligned}\right\}\;\rightarrow\;\varepsilon=\sqrt{1-\frac{5}{9}\frac{\left(\omega_2^3-\omega_1^3\right)^2}{(\omega_2-\omega_1)\left(\omega_2^5-\omega_1^5\right)}}\tag{2.105}$$

If ($\omega_1=0$), the correlation function, spectral moments and the spectral bandwidth parameter ε are calculated as to be

$$R(\tau)=\frac{2S_0}{\tau}\sin\left(\omega_2\tau\right)\quad\text{and}\quad\left.\begin{aligned}m_0&=2S_0\omega_2\\ m_2&=2S_0\omega_2^3/3\\ m_4&=2S\omega_2^5/5\end{aligned}\right\}\;\rightarrow\;\varepsilon=\frac{2}{3}\tag{2.106}$$

A narrow-band process is defined if its spectrum is a spike with infinite height and zero width so that the area remains finite. Such a spectrum is shown in Fig. 2.11a and represented [2] by the Dirac's delta function which is defined as

$$\text{Delta function:}\rightarrow\delta(\omega-\omega_1)=\begin{cases}1 & \text{at }(\omega=\omega_1)\\ 0 & \text{elsewhere}\end{cases}\tag{2.107a}$$

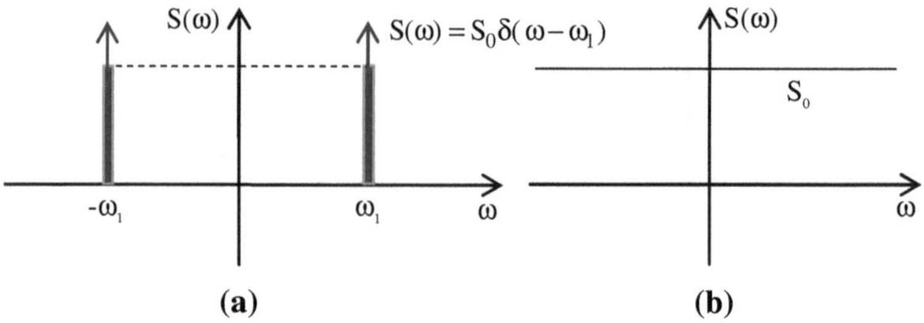

Fig. 2.11 Narrow-band and white noise (broad-band) spectra. **a** A narrow-band spectrum. **b** A white noise spectrum

The delta function has the following property

$$\int_{-\infty}^{\infty} \delta(\omega - \omega_1) f(\omega)\, d\omega = f(\omega_1) \tag{2.107b}$$

where $f(\omega)$ is any function of ω. The narrow-band process can be obtained from the band-limited process when the frequency ω_2 approaches ω_1, i.e., $(\omega_2 \to \omega_1)$, except that the spectral moments are calculated using the property of the delta function given in Eq. (2.107b). The spectral moments and bandwidth parameter ε are calculated to be

$$m_0 = 2S_0, \quad m_2 = 2S_0\omega_1^2, \quad m_4 = 2S_0\omega_1^4 \quad \text{and} \quad \varepsilon = 0 \tag{2.108}$$

As it is seen from Eq. (2.108), the spectral bandwidth parameter ε of a narrow-band process equals zero, which means that the period of zero-crossings is equal to the period of maxima. The correlation function is calculated as stated below.

$$R(\tau) = \int_{-\infty}^{\infty} S(\omega)\, e^{i\omega\tau} d\omega = \int_{-\infty}^{\infty} S_0 \delta(\omega - \omega_1) e^{i\omega\tau} d\omega \tag{2.109a}$$

having carried out the integration it is obtained as written by

$$R(\tau) = S_0\left(e^{i\omega_1\tau} + e^{-i\omega_1\tau}\right) = 2S_0 \cos(\omega_1\tau) \tag{2.109b}$$

which is a cosine function of $(\omega_1\tau)$. If the frequency ω_1 approaches zero, then $R(\tau)$ approaches a constant value, and for $(\omega_1 = 0)$, $R(\tau) = 2S_0$ is obtained, i.e., the correlation function is constant for all frequency ranges with the value of $2S_0$. Here, the coefficient 2 arises from two spikes at the origin, one is from (+) region and other one is from (−) region of the frequency.

A broad-band process is defined if its spectrum is extended over a wide frequency range with smooth variation that does not display sharp peaks, i.e., it produces a spectral bandwidth parameter ε is much greater than zero $(0 < .. < \varepsilon)$. In the case of a large frequency range, the band-limited process is an example of a broad-band process. In the extreme case, when $(\omega_1 = 0)$ and $(\omega_2 \to \infty)$, it covers all frequency ranges between $-\infty$ and ∞, i.e. $(-\infty < \omega < \infty)$. This special case of

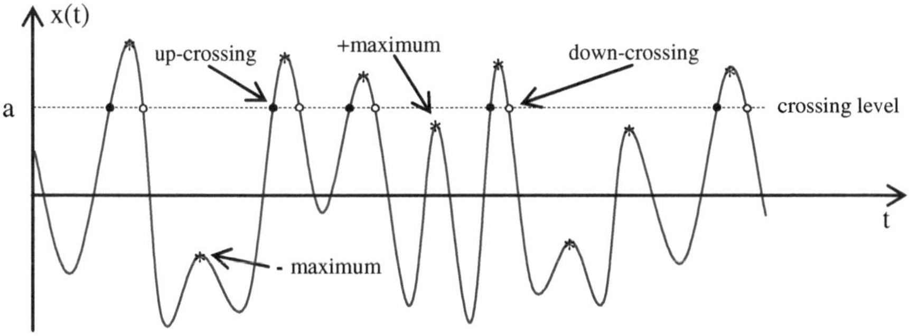

Fig. 2.12 A sample record of time signals, crossing level, maxima, up- and downcrossings

the spectral shape is called as the *white noise spectrum* which is shown in Fig. 2.11b. The correlation function of the band-limited white noise has been given in Eq. (2.106), which displays a peak for large ω_2 at $\tau = 0$ and diminishes rapidly with increasing τ values. As ($\omega_2 \to \infty$), this peak becomes a spike with a constant area of $2\pi S_0$ so that the correlation function of the white noise will be

$$R(\tau) = 2\pi S_0 \delta(\tau) \tag{2.110}$$

which can be proved by back calculation of $S(\omega)$ from the Fourier transform of $R(\tau)$ to obtain S_0 that has been assumed.

2.5.3 Crossing Analysis and Probability Distributions of Maxima

The *crossing analysis* is the central issue to determine average frequencies of a stochastic process and also to determine probability distribution of its peak values as outlined in the following subsections.

2.5.3.1 Average Frequencies of Crossing Levels, Frequency of Maxima

A typical crossing and definition of some terms that used in the crossing analysis are shown in Fig. 2.12. It is required that the average frequency of a stationary process $X(t)$ with amplitudes greater than a crossing level ($X = a$) is to be calculated. An up-crossing, which is shown in Fig. 2.12, occurs whenever the realization $X(t)$ passes through the crossing level ($X = a$) with positive slope. We wish to determine the probability of this up-crossing in the time interval t and $(t + dt)$. The r.v. at t and $(t + dt)$ are $X(t)$ and $X(t + dt)$, respectively. The variable at $(t + dt)$ is calculated using the Taylor expansion [28]. The conditions that an up-crossing occurs in the interval $X(t)$ and $X(t + dt)$ are

$$\begin{aligned} &\text{Crossing condition:} \quad \rightarrow X(t) \leq a \;\; \text{and} \;\; X(t+\mathrm{d}t) \geq a \\ &\text{Up-crossing condition:} \rightarrow \frac{\partial X(t)}{\partial t} \geq 0 \end{aligned} \tag{2.111a}$$

which can also be written as

$$\begin{aligned} &\text{Crossing condition:} \quad \rightarrow \begin{cases} X(t) \leq a \;\; \text{and} \;\; \left(X(t) + \dot{X}(t)\mathrm{d}t\right) \geq a \\ \text{or} \;\; (a - \dot{x}\,\mathrm{d}t) \leq X \leq a \end{cases} \\ &\text{Up-crossing condition:} \rightarrow \dot{X} \geq 0 \end{aligned} \tag{2.111b}$$

Since X and $\dot{X}$ are derived processes with a joint PDF of $f_{X\dot{X}}(x,\dot{x})$, by using Eq. (2.23), the probability that an up-crossing occurs between the time interval t and $(t + \mathrm{d}t)$ is written as

$$dP = P\big(\,(a - \dot{x}\,dt \leq X \leq a) \cap \left(0 \leq \dot{X} \leq \infty\right)\big) = \int_0^{\infty} \left(\int_{(a-\dot{x}d\,t)}^{a} f_{X\dot{X}}(x,\dot{x})\,dx \right) d\dot{x} \tag{2.112}$$

The integration in the brackets (.) with respect to x is the area of $f_{X\dot{X}}(x,\dot{x})$ over the band $(\,a - (a - \dot{x}\,dt) = \dot{x}\,dt)$ and the values of $f_{X\dot{X}}(x,\dot{x})$ at ($X = a$ and $a - \dot{x}\,dt$) are the same as ($\mathrm{d}t \rightarrow 0$). Thus, the area will be $(f_{X\dot{X}}(a,\dot{x})\dot{x}\,dt)$. Having substituted this value into Eq. (2.112), the probability of the up-crossing is obtained as written by,

$$\text{Pr obability of up-crossing at (X = a):} \quad \rightarrow \mathrm{d}P = \left(\int_0^{\infty} \dot{x} f_{X\dot{X}}(a,\dot{x})\,\mathrm{d}\dot{x} \right) \mathrm{d}t \tag{2.113}$$

This probability will be equal to the average number of up-crossings at $X = a$ in time $\mathrm{d}t$ [2]. If we assume that the average frequency of up-crossings in cycle per second at $X = a$ is v_a^+, then the average number of up-crossings in dt will be $(\mathrm{v}_a^+ dt)$ and equalizing this to Eq. (2.113) it can be stated that

$$\text{Average frequency of up-crossings at (X = a):} \; \rightarrow \mathrm{v}_a^+ = \int_0^{\infty} \dot{x} f_{X\dot{X}}(a,\dot{x})\,\mathrm{d}\dot{x} \tag{2.114}$$

Since X and $\dot{X}$ are derived processes, they are uncorrelated. Thus, the joint PDF can be stated as $f_{X\dot{X}}(a,\dot{x}) = f_X(a) f_{\dot{X}}(\dot{x})$. Once these PDF are known, then the average frequency of up-crossings at $X = a$ can be calculated from Eq. (2.114). We now assume that the processes X and $\dot{X}$ have Normal probability distributions with zero means and standard deviations of σ_X and $\sigma_{\dot{X}}$. Using their PDF from Eqs. (2.65a) in (2.114) it can be obtained that

$$\text{Up-crossing frequency:} \rightarrow \begin{cases} \nu_a^+ = \dfrac{1}{2\pi}\dfrac{\sigma_{\dot{X}}}{\sigma_X}\exp\left[-a^2/2\sigma_X^2\right] & \text{at } (X=a) \\ \nu_0^+ = \dfrac{1}{2\pi}\dfrac{\sigma_{\dot{X}}}{\sigma_X} \quad \rightarrow \quad \omega_0 = \dfrac{\sigma_{\dot{X}}}{\sigma_X} & \text{at } (X=0) \end{cases} \tag{2.115}$$

where ν_0^+ is the mean zero-crossings frequency in (cyc/sec) and ω_0 is that in (rad/sec) of a Normal process. For a zero-mean process, the mean zero-crossings frequency can be obtained using Eq. (2.101 for $\sigma_{\dot{X}}$ as

$$\text{Up-crossing frequency at (X = 0):} \rightarrow \omega_0 = \sqrt{m_2/m_0} \ \ (\text{rad/sec}) \tag{2.116}$$

As similar to the average frequency of up-crossings, the average frequency of local maxima of a stationary process can be determined. The maxima of a process is obtained if its first derivative is zero and second derivative (curvature) is negative in all time, i.e., if the conditions, $(\dot{X}=0)$ and $(\ddot{X}<0)$, are satisfied in all time, then maxima of the process X are obtained. The conditions of $(\dot{X}=0)$ and $(\ddot{X}<0)$ implies that the average frequency of maxima of X is equal to the average frequency of down-crossings, see Fig. 2.12, of the derived process $\dot{X}$ at $X = 0$. Thus, using $\dot{X}$ and $\ddot{X}$ instead of X and $\dot{X}$ in Eq. (2.115), the average frequency of maxima of the process X can be obtained as written

$$\text{Average frequency of maxima:} \rightarrow \omega_m = \sigma_{\ddot{X}}/\sigma_{\dot{X}} = \sqrt{m_4/m_2} \tag{2.117}$$

The average frequencies of zero-crossings and maxima are used to determine the spectral bandwidth parameter ε, given in Eq. (2.102), which is a good measure to give information about the spectral shape of a stochastic process.

2.5.3.2 Probability Distribution of Maxima

Maxima of a stochastic process are important values in engineering analyses as they are closely related to amplitudes or highest values of the process. Wave heights and maximum stresses are examples of maxima. In the probabilistic and reliability analyses, their probability distributions are required, which can be determined under certain conditions and probability model of the process.

For a stationary narrow-band process the probability distribution of maxima, or peak values, can be determined easily, while, for a broad-band process, it is relatively complicated. For a narrow-band process, which is shown in Fig. 2.13, the probability that the maxima between $(a \le X(t) \le a + da)$is $f_m(a)da$, where $f_m(a)$ is the PDF of the maxima. The probability of maxima exceeding $(X = a)$ is written by,

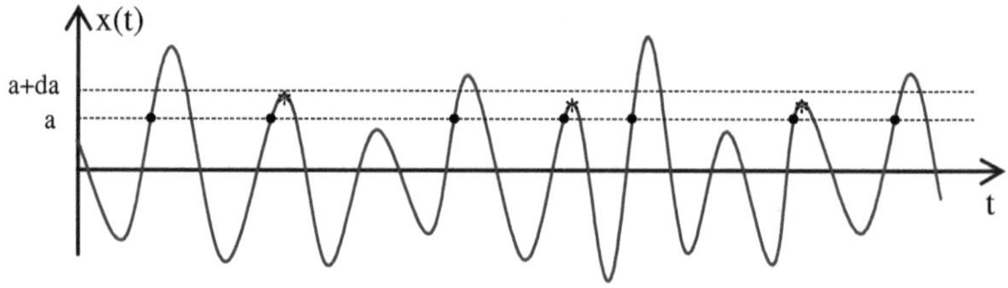

Fig. 2.13 A narrow-band process with maxima between $(a \le X(t) \le a + da)$

$$\text{Probablity of maxima exceeding (X = a):} \rightarrow P_m = \int_a^{\infty} f_m(a)\,\mathrm{d}a \qquad (2.118)$$

This probability is equal to (N_a/N_0) where N_a is the number of up-crossings at the level $X = a$ and N_0 is that at $X = 0$. For a period T, these numbers of up-crossings and the probability P_m are calculated from

$$\left(N_a = \nu_a^+ T \text{ and } N_0 = \nu_0^+ T\right) \rightarrow P_m = \frac{\nu_a^+ T}{\nu_0^+ T} \rightarrow P_m = \frac{\nu_a^+}{\nu_0^+} = 1 - F_m(a) \quad (2.119)$$

in which $F_m(a)$ is the cumulative probability distribution of maxima. Having derived Eq. (2.119) with respect to a, the PDF of maxima of a stationary narrow-band process can be obtained as written by,

$$\text{PDF of maxima of a narrow–band process X(t):} \rightarrow f_m(a) = -\frac{1}{\nu_0^+}\frac{\partial \nu_a^+}{\partial a} \quad (2.120)$$

For a Normal stationary narrow-band process the crossings frequencies have been given in Eq.(2.115), from which and Eq.(2.120) the PDF of the maxima of a narrow-band Normal process $X(t)$ can be obtained as

$$\text{PDF of peaks of a Normal process X(t):} \rightarrow f_m(a) = \frac{a}{\sigma_X^2}\exp\left(-a^2/2\sigma_X^2\right) \qquad (2.121)$$

This is a Rayleigh distribution that explained in Sect. 2.3.2.6.

For a general stationary process, calculation of probability distribution is not so easy. If the process is *stationary* and *Normal*, then the PDF of its maxima can be determined in terms of the spectral bandwidth parameter ε. It is calculated using the joint PDF of $X(t)$, $\dot{X}(t)$ and $\ddot{X}(t)$, i.e., from $f(x, \dot{x}, \ddot{x})$, which is assumed to be Normal. For the maxima above $(X = a)$, the general conditions are

$$\text{Condition of maxima of a process X(t):} \rightarrow \begin{cases} X \ge a \\ \dot{X} = 0 \text{ and } \ddot{X} < 0 \end{cases} \qquad (2.122)$$

The probability that X lies in the band x to $(x + \mathrm{d}x)$ above $(X \geq a)$ is written as

$$\text{Prob. of up-crossings above (X = a)}: \rightarrow P_a = \int_a^{\infty}\left(\int_{-\infty}^{0} f_m(x,\dot{x},\ddot{x})\mathrm{d}\ddot{x}\mathrm{d}\dot{x}\right)\mathrm{d}x \tag{2.123a}$$

Since $\dot{X} = 0$ and $d\dot{x} \simeq |\ddot{x}|dt$, the probability of a maximum above $(X = a)$ can be written from Eq. (2.123a) as

$$\text{Prob. of maxima above (X = a)}: \rightarrow P_m = \int_a^{\infty}\left(\int_{-\infty}^{0} |\ddot{x}|f_m(x,0,\ddot{x})\mathrm{d}\ddot{x}\right)\mathrm{d}x\,\mathrm{d}t \tag{2.123b}$$

which is equal to the number of cycles of maxima in dt above $(X = a)$, i.e., $(n_a = \nu_m(a)\,\mathrm{d}t)$ so that the frequency of maxima above $(X = a)$ can be stated as

$$\text{Freq. of maxima above (X = a)}: \rightarrow \nu_m(a) = \int_a^{\infty}\left(\int_{-\infty}^{0} |\ddot{x}|f_m(x,0,\ddot{x})\mathrm{d}\ddot{x}\right)\mathrm{d}x \tag{2.124a}$$

In a similar way, the frequency of maxima above a given crossing, say $(X = b)$ is written as

$$\text{Frequency of maxima above (X = b)}: \rightarrow \nu_m(b) = \int_b^{\infty}\left(\int_{-\infty}^{0} |\ddot{x}|f_m(x,0,\ddot{x})d\ddot{x}\right)dx \tag{2.124b}$$

In a period of time T, the probability of maxima for $(X > a)$ can be stated in terms of numbers of maxima, which leads to

$$\int_a^{\infty} f_m(a)da = 1 - F_m(a) = \frac{\nu_m(a)}{\nu_m(b)} \rightarrow f_m(a) = -\frac{1}{\nu_m(b)}\frac{\partial\nu_m(a)}{\partial a} \tag{2.125}$$

Having substituted $\nu_m(a)$ and $\nu_m(b)$ from Eqs. (2.124a) and (2.124b) into (2.125), the PDF of maxima above a crossing level $(X = b)$ can be obtained as written by

$$\text{PDF of maxima above (X = b)}: \rightarrow f_m(a) = \frac{\int_{-\infty}^{0} |\ddot{x}|f_m(a,0,\ddot{x})\mathrm{d}\ddot{x}}{\int_b^{\infty}\left(\int_{-\infty}^{0} |\ddot{x}|f_m(x,0,\ddot{x})\mathrm{d}\ddot{x}\right)\mathrm{d}x} \tag{2.126}$$

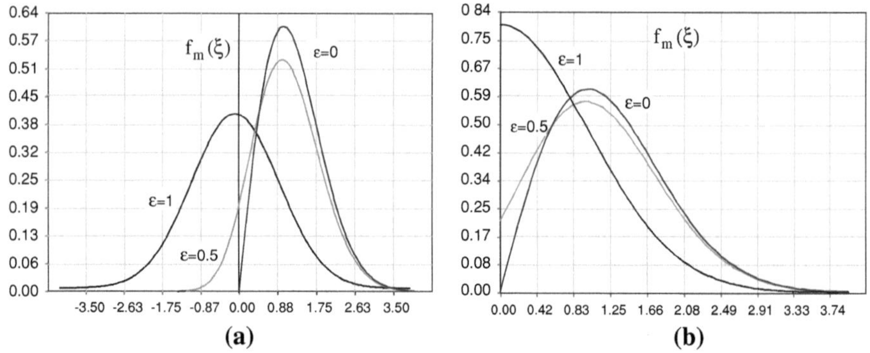

Fig. 2.14 Probability density functions of maxima of a general process, where $\xi = a/\sqrt{m_0}$. **a** PDF of all maxima between ($-\infty$ and ∞). **b** PDF of +maxima between (0 and ∞)

In Eq. (2.126), if ($b = 0$), then all positive maxima, i.e. maxima above ($X = 0$), are considered, and if ($b = -\infty$), then all maxima (positive and negative) of the process $X(t)$ are considered. The PDF of these two cases, which are of practical interest, are given [30] in the following.

$$\text{PDF of all maxima, i.e. for } (-\infty \leq \xi \leq \infty):$$

$$f_m(\xi) = \frac{\varepsilon}{\sqrt{2\pi}} e^{-\xi^2/2\varepsilon^2} + \sqrt{1-\varepsilon^2}\,\xi e^{-\xi^2/2} \Phi\left(\frac{\sqrt{1-\varepsilon^2}}{\varepsilon}\xi\right) \tag{2.127a}$$

$$\text{PDF of positive maxima, i.e. for } (0 \leq \xi \leq \infty):$$

$$f_m(\xi) = \frac{2}{1+\sqrt{1-\varepsilon^2}} \left[\frac{\varepsilon}{\sqrt{2\pi}} e^{-\xi^2/2\varepsilon^2} + \sqrt{1-\varepsilon^2}\,\xi e^{-\xi^2/2} \Phi\left(\frac{\sqrt{1-\varepsilon^2}}{\varepsilon}\xi\right)\right] \tag{2.127b}$$

in which $\left(\xi = a/\sqrt{m_0}\right)$, where a is the *amplitude* and m_0 is the zeroth spectral moment, $\Phi(.)$ is the *Standard Normal Distribution* function given in Eq. (2.65c). The plots of these two PDF are shown in Figs. 2.14a and 2.14b, respectively. For the limiting case for $\varepsilon = 0$ (narrow-band process) and $\varepsilon = 1$ (extreme broad-band process), these distributions become respectively the Rayleigh and Normal distributions. As it is seen from Fig. 2.14b, for $\varepsilon > 0$, the distributions obtained from Eq. (2.127b) are truncated at the zero value of the variable ($\xi = 0$). For the spectral bandwidth parameter ε between 0 and 1, ($0 < \varepsilon < 1$), the distributions have mixed forms of the Normal and Rayleigh distributions. As it is expected, for an infinitely narrow-band process ($\xi = 0$), the PDF of the +maxima and all maxima will be the same since, for a narrow-band process, all maxima are located in the positive side of the process as shown in Fig. 2.13.

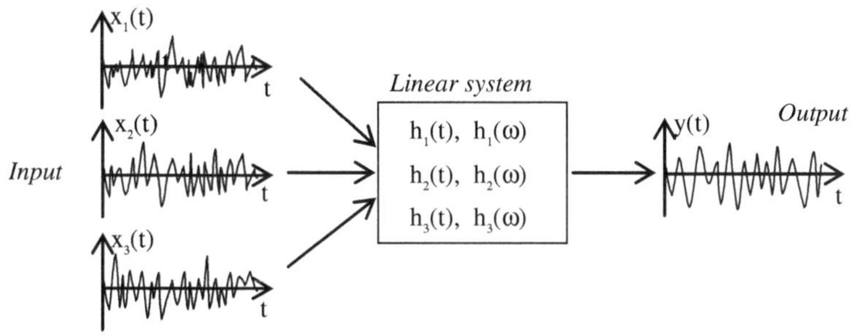

Fig. 2.15 A linear system with three input excitations and one output response

2.6 Input–Output Relations of Stochastic Processes, Transfer Functions

In the stochastic analysis of structures, the loadings exerted by natural phenomena are mostly stochastic processes such as wave, earthquake, and wind loadings. The stochastic loadings produce structural responses that will be also stochastic. We assume that the loading process is fully determined stochastically and probabilistically. Our purpose is to determine a response process under this loading condition which is only known or available information that we can use. In the stochastic structural analysis, the loading is the *input* and a response to be calculated is the *output*. The inputs are also called *excitations* of a response process. Multiple input (loadings) processes can be applied to produce one response (output) process or samples of responses. The main issue in this context is that the input and output relation is assumed to be linear, and they form a *linear system* together as shown schematically in Fig. 2.15 where $h_i(t)$ and $h_i(\omega)$ with $(i = 1, 2, 3)$ are respectively impulse response functions and frequency response functions [2]. A linear system in structural dynamics is defined if the response function is related to the input by a linear differential equation [2] in the form of

$$\text{A linear system:} \rightarrow \sum_{k=0}^{n} a_k \frac{\partial^k y}{\partial t^k} = \sum_{k=0}^{n1} b_{1k} \frac{\partial^k x_1}{\partial t^k} + \sum_{k=0}^{n2} b_{2k} \frac{\partial^k x_2}{\partial t^k} + \ldots + \sum_{k-0}^{nr} b_{rk} \frac{\partial^k x_r}{\partial t^k} \tag{2.128}$$

which is composed of one output and superposition of (nr) number of inputs. The dynamic equilibrium equation, given by Eq. (1.86) in Chap. 1, is a simple example of a linear system in the time domain, and that given by Eq. (1.258) in the frequency domain. The output process of a linear system follows the type of the input process [2], which means that, if the input process is stationary, then the output process becomes also stationary, and if it is ergodic, then the corresponding

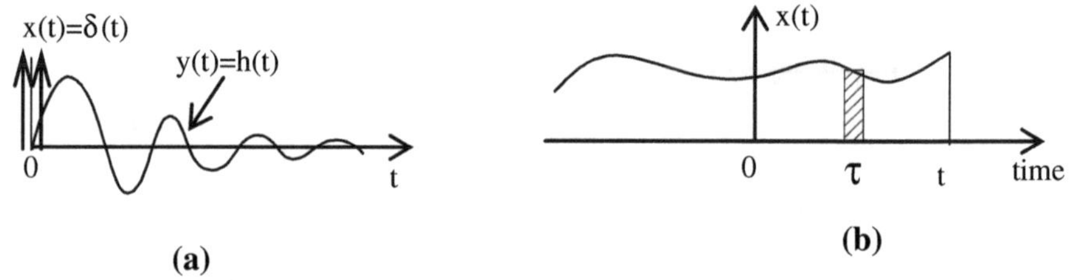

Fig. 2.16 An impulse response function and an arbitrary input consisted of infinite impulses. **a** An impulse response function. **b** an arbitrary input function

ouptput process is also ergodic. The *impulse response* function $h(t)$ is the response of the system at time t due to a unit impulse applied at time $(t = 0)$ [2, 32] as shown in Fig. 2.16a. For an arbitrary input $x(t)$ shown in Fig. 2.16b, the response at time t can be considered as infinite collections of impulse response functions due to impulses applied at $(t = \tau)$, where $(-\infty \leq \tau \leq t)$. Thus, the response function $y(t)$ at time t can be written as

$$\text{Response to an arbitrary input } x(t): \rightarrow y(t) = \begin{cases} \int_{-\infty}^{t} h(t-\tau)x(\tau)\,d\tau \\ \int_{-\infty}^{\infty} h(t-\tau)x(\tau)\,d\tau \end{cases} \tag{2.129a}$$

The integration is extended from t to $+\infty$, since $h(t-\tau) = 0$ for $(\tau > t)$ and $y(t)$ remains unchanged. In the frequency domain, the response to an arbitrary input $x(\omega)$ can be stated [2] as

$$\text{Response to an arbitrary input } x(\omega):, \rightarrow \quad y(\omega) = h(\omega)x(\omega) \tag{2.129b}$$

where $h(\omega)$ is called as the *frequency response* function. It can be shown that, see e.g., [2], the impulse and frequency response functions, $h(t)$ and $h(\omega)$, are the Fourier transform pairs. Our intention is to determine the response spectral function $S_Y(\omega)$ in terms of the input spectrum $S_X(\omega)$. Since, in the structural analysis, we deal mostly with multi degrees-of-freedom systems, henceforth we use vector processes instead of scalar to determine spectral outputs of structural responses. The output of a stationary input vector process in the time domain is written from Eq. (2.129a) as,

$$\text{Response to input } \{x(t)\}: \quad \rightarrow \{y(t)\} = \begin{cases} \int_{-\infty}^{\infty} [h(\theta)]\{x(t-\theta)\}\,d\theta \\ \text{where } \ (\theta = t - \tau) \end{cases} \tag{2.130a}$$

and, in the frequency domain from Eq. (2.129b), it is

$$\text{Response to input}\{x(\omega)\}: \ \rightarrow \ \{y(\omega)\} = [h(\omega)]\{x(\omega)\} \tag{2.130b}$$

In general, let us determine the cross-correlation matrix $[R_{YZ}(\tau)]$ of two output stationary vector processes, $\{Y(t)\}$ and $\{Z(t)\}$, from which the auto-correlation matrix can be obtained by equalizing the two processes. Using the relation given in Eq. (2.78c) it is written that

$$\text{Cross-correlation of } Y(t) \text{ and } Z(t): \rightarrow [R_{YZ}(\tau)] = E\left[\{Y(t)\}\{Z(t+\tau)\}^T\right] \tag{2.131a}$$

or using the relation of input–output written in Eq. (2.130a), it is stated as

$$[R_{YZ}(\tau)] = E\left[\int_{-\infty}^{\infty}\int_{-\infty}^{\infty} [H_Y(\theta_Y)]\{X_Y(t-\theta_Y)\}\{X_Z(t-\theta_Z+\tau)\}^T[H_Z(\theta_Z)]^T d\theta_Z d\theta_Z\right] \tag{2.132a}$$

By using a variable transformation of $(t_1 = t - \theta_Y)$ and taking the ensemble average, Eq. (2.132a) can be written as

$$[R_{YZ}(\tau)] = \int_{-\infty}^{\infty}\int_{-\infty}^{\infty} [H_Y(\theta_Y)][R_{X_YX_Z}(\theta_Y - \theta_Z + \tau)][H_Z(\theta_Z)]^T d\theta_Y d\theta_Z \tag{2.132b}$$

in which $[R_{X_YX_Z}(\theta_Y - \theta_Z + \tau)]$ is the cross-correlation matrix of the input processes $\{X_y(t)\}$ and $\{X_z(t)$. In order to calculate the cross-spectral matrix of the output processes $\{Y(t)\}$ and $\{Z(t\}$, the Fourier transform of Eq. (2.132b) will be taken. By using the relation written in Eq. (2.90b), the cross-spectral matrix of the outputs $\{Y(t)\}$ and $\{Z(t\}$ can be obtained as written by

$$\text{Cross-spectral matrix of outputs: } \rightarrow [S_{YZ}(\omega)] = [H_Y^*(\omega)][S_{X_YX_Z}(\omega)][H_Z(\omega)]^T \tag{2.133}$$

in which (*) denotes a complex conjugate and $[S_{X_YX_Z}(\omega)]$ is the cross spectral matrix of the inputs $\{X_Y(t)\}$ and $\{X_Z(t)\}$. The output processes in the frequency domain, $\{Y(\omega)\}$ and $\{Z(\omega)\}$, are linked to input processes $\{X_Y(\omega)\}$ and $\{X_Z(\omega)\}$ by the following relations

$$\text{Input-Output relations:} \quad \rightarrow \begin{cases} \{Y(\omega)\} = [H_Y(\omega)]\{X_Y(\omega)\} \\ \{Z(\omega)\} = [H_Z(\omega)]\{X_Z(\omega)\} \end{cases} \tag{2.134}$$

which means that the inputs $\{X_Y(\omega)\}$ and $\{X_Z(\omega)\}$ are transferred to the outputs $\{Y(\omega)\}$ and $\{Z(\omega)\}$ by means of frequency dependent matrices $[H_Y(\omega)]$ and $[H_Z(\omega)]$. These matrices are called as the *transfer function matrices* of linear stochastic processes. Eq. (1.259) in Chap. 1, is a good example of the input–output relation in structural random vibrations, which is written below for convenience.

$$\{D(\omega)\} = [H(\omega)]_{DP}\{P(\omega)\} \quad \rightarrow \quad [H(\omega)]_{DP} = \left([K] + i\omega[C] - \omega^2[M]\right)^{-1} \tag{2.135}$$

where $\{D(\omega)\}$ is the structural displacements vector, $\{P(\omega)\}$ is the applied load vector and $[H(\omega)]_{DP}$ is the transfer function matrix between the load and displacement vectors. The spectral matrix of displacements can be readily determined using the spectral relation given in Eq. (2.133) as

$$\text{Spectral matrix of displacements:} \rightarrow [S_D(\omega)] = [H(\omega)]^*_{DP}[S_P(\omega)][H(\omega)]^T_{DP} \tag{2.136}$$

in which $[S_P(\omega)]$ is the spectral matrix of applied loads. As it can be realized from Eq. (2.136), $[S_D]$ is the auto-spectral matrix of the displacements.

2.7 Examples

Three stationary processes are defined as written

$$X(t) = \sin(\omega t + x), \quad Y(t) = \cos^2(\omega t + y) \quad \text{and} \quad Z(t) = X(t) + Y(t) \tag{2.137}$$

in which x and y are random phase angles with a uniform joint distribution between $(-\pi \leq x \leq \pi)$ and $(-\pi \leq y \leq \pi)$. The following items are asked to be calculated:

1. Expected values of $X(t)$, $Y(t)$ and $Z(t)$
2. Auto- and cross-correlation functions of $X(t)$, $Y(t)$ and $Z(t)$
3. Show that these processes are also ergodic.

Solution

1. In order to use the ensemble averages, the PDF function of the random variables x and y need to be determined. Since a uniform distribution for both x and y is used, the joint PDF is defined as

$$f_{XY}(x,y) = \begin{cases} \dfrac{1}{4\pi^2} & \rightarrow \quad (-\pi \leq x \leq \pi) \text{ and } (-\pi \leq y \leq \pi) \\ 0 & \text{elsewhere} \end{cases} \tag{2.138a}$$

The marginal PDF of x and y are calculated from

$$f_X(x) = \int_{-\pi}^{\pi} f_{XY}(x, y)\, dy = \frac{1}{2\pi} \quad \text{and} \quad f_Y(y) = \int_{-\pi}^{\pi} f_{XY}(y)\, dx = \frac{1}{2\pi} \tag{2.138b}$$

The expected values of $X(t)$, $Y(t)$ are calculated from

$$\begin{aligned} \mu_X &= E[X(t)] = \int_{-\pi}^{\pi} x f_X(x)\, dx \quad \rightarrow \quad \mu_X = \frac{1}{2\pi} \int_{-\pi}^{\pi} \sin(\omega t + x)\, dx = 0 \\ \mu_Y &= E[Y(t)] = \int_{-\pi}^{\pi} y f_Y(y)\, dy \quad \rightarrow \quad \mu_Y = \frac{1}{2\pi} \int_{-\pi}^{\pi} \cos^2(\omega t + y)\, dy = \frac{1}{2} \end{aligned} \tag{2.139a}$$

and the expected value of $Z(t)$ is calculated from the sum of two processes

$$\mu_Z = E[Z(t)] = E[X(t) + Y(t)] = E[X(t)] + E[Y(t)] = \frac{1}{2} \tag{2.139b}$$

2. The auto- and cross-correlation functions of the processes $X(t)$ and $Y(t)$ are calculated from, for auto-correlation of $X(t)$,

$$R_{XX}(\tau) = \begin{cases} E[X(t)\, X(t+\tau)] = \displaystyle\int_{-\pi}^{\pi} x(t)\, x(t+\tau) f_X(x)\, dx \\ \displaystyle\frac{1}{4\pi} \left| x \cos(\omega\tau) - \frac{1}{2} \sin(2\omega t + 2x + \omega\tau) \right|_{-\pi}^{\pi} = \frac{1}{2} \cos(\omega\tau) \end{cases} \tag{2.140a}$$

and for auto-correlation of Y(t),

$$R_{YY}(\tau) = \begin{cases} E[X(t)\, Y(t+\tau)] = \displaystyle\int_{-\pi}^{\pi} y(t)\, y(t+\tau) f_Y(y)\, dy \\ \displaystyle\frac{1}{16\pi} \left| \begin{aligned} & y \cos(2\omega\tau) + \frac{1}{4} \sin(4\omega t + 4y + 2\omega\tau) + \ldots \\ & \ldots + \sin(2\omega t + 2y) + \sin(2\omega t + 2\,\omega\tau + 2y) + 2y \end{aligned} \right|_{-\pi}^{\pi} \\ \displaystyle\frac{1}{4} \left(\cos^2(\omega\tau) + \frac{1}{2} \right) \end{cases} \tag{2.140b}$$

The cross-correlations of $X(t)$ and $Y(t)$ are calculated from

$$R_{XY}(\tau)=\begin{cases} E[X(t)\,Y(t+\tau)]=\int\limits_{-\pi}^{\pi}\int\limits_{-\pi}^{\pi} x(t)\,y(t+\tau)f_{XY}(x,y)\,dx\,dy \\ \dfrac{1}{4\pi^2}\left|\cos(\omega t+x)\left(\begin{array}{l}\frac{1}{2}\cos(\omega t+\omega\tau+y)\sin(\omega t+\omega\tau+y)+\ldots \\ \ldots+\frac{1}{2}(\omega t+\omega\tau+y)\end{array}\right)\right|_{-\pi}^{\pi}=0 \end{cases} \quad (2.140c)$$

$$R_{YX}(\tau)=\begin{cases} E[Y(t)\,X(t+\tau)]=\int\limits_{-\pi}^{\pi}\int\limits_{-\pi}^{\pi} y(t)\,x(t+\tau)f_{XY}(x,y)\,dx\,dy \\ -\dfrac{1}{8\pi^2}|\cos(\omega t+\omega\tau+x)(\cos(\omega t+y)\sin(\omega t+y)+\omega t+y)|_{-\pi}^{\pi}=0 \end{cases} \quad (2.140d)$$

The auto-correlation and cross-correlation functions, $R_{ZZ}(\tau)$, $R_{XZ}(\tau)$ and $R_{YZ}(\tau)$, are calculated using the sum of two processes

$$R_{ZZ}(\tau)=\begin{cases} E[(X(t)+Y(t))(X(t+\tau)+Y(t+\tau))] \\ R_{XX}(\tau)+R_{YY}(\tau)+R_{XY}(\tau)+R_{YX}(\tau)=\dfrac{1}{4}\left(2\cos(\omega\tau)+\cos^2(\omega\tau)+\dfrac{1}{2}\right) \end{cases} \quad (2.141a)$$

the cross-correlation function of $X(t)$ and $Z(t)$ is calculated from

$$R_{XZ}(\tau)=\begin{cases} E[X(t)(X(t+\tau)+Y(t+\tau))]=R_{XX}(\tau)+R_{XY}(\tau) \\ \dfrac{1}{2}\cos(\omega\tau) \end{cases} \quad (2.141b)$$

and the cross-correlation function of $Y(t)$ and $(Z(t)$ is calculated from

$$R_{YZ}(\tau)=\begin{cases} E[Y(t)(X(t+\tau)+Y(t+\tau))]=R_{YX}(\tau)+R_{YY}(\tau) \\ \dfrac{1}{4}\left(\cos^2(\omega\tau)+\dfrac{1}{2}\right) \end{cases} \quad (2.141c)$$

3. In order to show that the processes $X(t)$, $Y(t)$ and $Z(t)$ are also ergodic, we need to calculate time averages. If the statistics are equal to the ensemble averages calculated above, these processes are said to be ergodic. The statistics of $X(t)$ and $Y(t)$ are sufficient to prove ergodicity of the processes $X(t)$, $Y(t)$ and $Z(t)$. Using Eq. (2.81a) the mean values are calculated from

$$\tilde{\mu}_X = \begin{cases} \lim\limits_{T\to\infty} \dfrac{1}{2T} \displaystyle\int_{-T}^{T} X(t)\, dt = \lim\limits_{T\to\infty} \dfrac{1}{2T} \displaystyle\int_{-T}^{T} \sin\,(\omega t + x)\, dt \\ \lim\limits_{T\to\infty} \left| -\dfrac{1}{2\omega T} (\cos\,(\omega t + x) \right|_{t=-T}^{T} = 0 \end{cases} \tag{2.142a}$$

$$\tilde{\mu}_Y = \begin{cases} \lim\limits_{T\to\infty} \dfrac{1}{2T} \displaystyle\int_{-T}^{T} Y(t)\, dt = \lim\limits_{T\to\infty} \dfrac{1}{2T} \displaystyle\int_{-T}^{T} \cos^2(\omega t + y)\, dt \\ \lim\limits_{T\to\infty} \left| \dfrac{1}{4\omega T} (\cos\,(\omega t + y) \sin\,(\omega t + y) + \omega t + y \right|_{t=-T}^{T} = \dfrac{1}{2} \end{cases}$$

and the correlation functions are calculated from

$$\tilde{R}_{XX}(\tau) = \begin{cases} \lim\limits_{T\to\infty} \dfrac{1}{2T} \displaystyle\int_{-T}^{T} \sin(\omega t + x)\, \sin(\omega t + \omega\tau + x) dt \\ \lim\limits_{T\to\infty} \left| -\dfrac{1}{8\omega T} (-2t\omega \cos(\omega\tau) + \sin(2\omega t + 2x + \omega\tau)) \right|_{t=-T}^{T} = \dfrac{1}{2} \cos(\omega\tau) \end{cases} \tag{2.142b}$$

$$\tilde{R}_{YY}(\tau) = \begin{cases} \lim\limits_{T\to\infty} \dfrac{1}{2T} \displaystyle\int_{-T}^{T} \cos^2(\omega t + y)\cos^2(\omega t + \omega\tau + y)\, dt \\ \lim\limits_{T\to\infty} \left| \dfrac{1}{64\omega T} \begin{pmatrix} 4t\omega\cos(2\omega\tau) + \sin(4\omega t + 4y + 2\omega\tau) + \ldots \\ \ldots + 4\sin(2\omega t + 2y) + 4\sin(2\omega t + 2\omega\tau + 2y) + 8\omega t \end{pmatrix} \right|_{t=-T}^{T} \\ \dfrac{1}{4}\left(\cos^2(\omega\tau) + \dfrac{1}{2}\right) \end{cases} \tag{2.142c}$$

$$\tilde{R}_{XY}(\tau) = \begin{cases} \lim\limits_{T\to\infty} \dfrac{1}{2T} \displaystyle\int_{-T}^{T} \sin(\omega t + x)\cos^2(\omega t + \omega\tau + y)\, dt \\ \lim\limits_{T\to\infty} \left| -\dfrac{1}{24\omega T} \begin{pmatrix} \cos(3\omega t + x + 2\omega\tau + 2y) - \ldots \\ \ldots - 3\cos(\omega t - x + 2\omega\tau + 2y) + 6\cos(\omega t + x) \end{pmatrix} \right|_{t=-T}^{T} = 0 \end{cases} \tag{2.142d}$$

$$\tilde{R}_{YX}(\tau) = \begin{cases} \lim\limits_{T\to\infty} \dfrac{1}{2T} \displaystyle\int_{-T}^{T} \cos^2(\omega t + y)\sin(\omega t + \omega\tau + x)\, dt \\ \lim\limits_{T\to\infty} \left| -\dfrac{1}{24\omega T} \begin{pmatrix} \cos(3\omega t + \omega\tau + x + 2y) - \ldots \\ \ldots - 3\cos(\omega t - \omega\tau - x + 2y) + 6\cos(\omega t + \omega\tau + x) \end{pmatrix} \right|_{t=-T}^{T} = 0 \end{cases} \tag{2.142e}$$

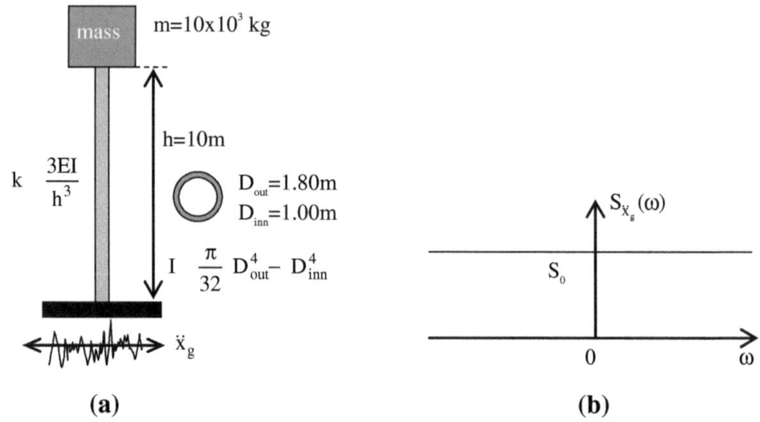

Fig. 2.17 A tower subjected to random earthquake motion, and the white noise spectrum $\ddot{x}_g$. **a** A tower with input information. **b** White noise spectrum of $\ddot{x}_g$

It seen from these statements that the statistics calculated from the time averages are same as those calculated from the ensemble averages, which proves that the processes are ergodic.

Exercise 1

The following processes are defined

$$X(t) = \sin(\omega t + x/2)\,, \quad Y(t) = \cos(\omega t + y/2) \tag{2.143}$$

in which x and y are independent random phase angles with a uniform distributions defined in $(-\pi \leq x \leq \pi)$and $(-2\pi \leq y \leq 2\pi)$.

1. Show that the process $X(t)$ is not stationary.
2. Show that the process $Y(t)$ is stationary in the wide sense and also ergodic.

Exercise 2

A water tower is given as shown in Fig. 2.17a. It is made of reinforced concrete with hollow circular cross-section of D_{out} (outer diameter) and D_{inn} (inner diameter), and the height h with numeric values given in Fig. 2.17a. The mass at the top is denoted by m and the stiffness k of the tower is calculated as written in Fig. 2.17a. It is assumed that the mass of the tower is neglected and only the mass at the top is considered with the value written in Fig. 2.17a. The tower is subjected to a random earthquake motion with an acceleration of $\ddot{x}_g$, which is Normally distributed with a zero mean. It is assumed that the earthquake acceleration has a white noise spectrum with a magnitude of S_0 as shown in Fig. 2.17b. The elasticity modulus E of the reinforced concrete is assumed to be random with a Lognormal probability distribution presented in Sect. 2.3.2.2. It is further assumed that the r.v. E (elasticity modulus) and $\ddot{x}_g$ (acceleration of the ground) are independent. The mean and coefficient of variation of the elasticity modulus E are assumed to be,

$$\begin{aligned}&\text{Mean of Elasticity modulus}, E \quad &&: \rightarrow \mu_E = 30.0\ GPa \quad (30.0\ \times\ 10^9\, N/m^2)\\&\text{COV of Elasticity modulus}\, (\sigma_E/\mu_E)\text{:} &&\rightarrow V_E = 0.05\end{aligned} \tag{2.144}$$

The damping ratio of the material is assumed to be $(\xi = 0.04)$. Under these conditions, the following items are required to be calculated and determined.

1. Natural frequency of the tower as being function of the elasticity modulus E of the concrete, which is a random function
2. Mean value and variance of the natural frequency
3. Correlation function and spectrum of the horizontal displacement d_{top} at the top as depending on the spectrum of ground acceleration
4. Variance of the ground acceleration in terms of the magnitude S_0 of its spectrum, which is shown in Fig. 2.17b
5. Spectral bandwidth parameter ε of the horizontal response displacement at the top, and its mean frequencies of zero-crossings and maxima.
6. Probability density function of the amplitude (maxima) of the horizontal displacement at the top
7. Mean value and variance of the amplitude of the displacement d_{top}.

References

1. Crandall SH, Mark WD (1963) Random vibration in mechanical systems. Academic Press, New York
2. Newland DE (1984) An introduction to random vibrations and spectral analysis, 2nd edn. Longman, London
3. Yang CY (1986) Random vibration of structures. Wiley, New York
4. Roberts JB, Spanos PD (1990) Random vibration and statistical linearization. Wiley, Chichester
5. Soong TT, Grigoriu M (1993) Random vibration of mechanical and structural systems. PTR Prentice-Hall, Englewood Cliffs
6. Preumont A (1994) Random vibration and spectral analysis. Kluwer Academic Publishers, Dordrecht
7. Paz M, Leigh W (2004) Structural dynamics: theory and computation, 5th edn. Kluwer Academic Publishers, Boston
8. Wijker J (2009) Random vibrations in spacecraft structures design. Theory and Applications. Springer, Dordrecht
9. Shinozuka M, Yang JN (1969) Random vibration of linear structures. Int J Solids Struct 5:1005–1036
10. Iyengar RN, Dash PK (1976) Random vibration analysis of stochastic time-varying systems. J Sound Vib 45(1):69–89
11. Robson JD (1980) The response relationships of random vibration analysis. J Sound Vib 73(2):312–315
12. Lin YK, Kozin F, Wen YK et al (1986) Methods of stochastic structural dynamics. Struct Saf 3:167–194

13. Ziegler F (1987) Random vibrations: a spectral method for linear and nonlinear structures. Probab Eng Mech 2(2):92–99
14. Shihab S, Preumont A (1989) Non-stationary random vibrations of linear multi-degree-of-freedom systems. J Sound Vib 132(3):457–471
15. Chen SH, Liu ZS, Zhang NF (1992) Random vibration analysis for large-scale structures with random parameters. Comput Struct 43(4):681685
16. Elishakoff I, Zhu L (1993) Random vibration of structures by the finite element method. Comput Meth Appl Mech Eng 105:359–373
17. Chen MT, Ali A (1998) An efficient and robust integration technique for applied random vibration analysis. Comput Struct 66(6):785–798
18. Alkhaleefi AM, Ali A (2002) An efficient multi-point support-motion random vibration analysis technique. Comput Struct 80(22):1689–1697
19. Li QS, Zhang YH, Wu JR et al (2004) Seismic random vibration analysis of tall buildings. Eng Struct 26:1767–1778
20. de la Fuente E (2008) An efficient procedure to obtain exact solutions in random vibration analysis of linear structures. Eng Struct 30:2981–2990
21. Paez TL (2006) The history of random vibrations through 1958. Mech Syst Sig Process 20:1783–1818
22. Hines WW, Montgomery DC (1980) Probability and statistics in engineering and management science, 2nd edn. Wiley, New York
23. Papoulis A, Pillai SU (2002) Probability, random variables and stochastic processes, 4th edn. McGraw Hill, New York
24. Benjamin JR, Cornell CA (1970) Probability, statistics, and decision for civil engineers. McGraw Hill, New York
25. Krishnamoorthy K (2006) Handbook of statistical distributions with applications. Chapman and Hall/CRC, Boca Raton
26. Pal N, Jin C, Lim WK (2006) Handbook of exponential and related distributions for engineers and scientists. Chapman and Hall/CRC, Boca Raton
27. Evans M, Hastings N, Peacock B (1993) Statistical distributions, 2nd edn. Wiley, New York
28. Abramowitz M, Stegun IA (eds) (1972) Handbook of mathematical functions with formulas. Graphs and Mathematical Tables. Dover, New York
29. Kotz S, Nadarajah S (2002) Extreme value distributions: theory and applications. Imperial College Press, London
30. Ochi MK (1990) Applied probability and stochastic processes in engineering and physical sciences. Wiley, New York
31. Coles S (2001) An introduction to statistical modeling of extreme values. Springer, London
32. Meirovitch L (1967) Analytical methods in vibrations. Collier-McMillan, London

Chapter 3
Water Wave Theories and Wave Loads

3.1 Introduction

Oceans and seas are significant parts of the Earth that have been used by human beings for different purposes in different periods. They occupy about 70 % of the Earth [1] and contain rich ingredients and unexploited energy resources that gradually getting scarce on lands. Since early periods, oceans and seas have been used for transportations, food resources, military purposes, etc. [2]. Next to other activities in oceans, in the last era, much attention has been focused on discovery and utilization of undersea deposits of hydrocarbon [3], and recently to produce renewable energy [4]. Ocean and sea environments are very complex and chaotic, sometimes calm, mostly disturbed by atmospheric phenomenon, and occasionally excited by seismic motion of the Earth in tectonic areas leading to catastrophic consequences [1] in terms of life and economy. The disturbances on ocean surfaces are irregular water waves that generated mostly by winds, and rarely by other sources such as storms, earthquakes, gravitational attraction of the Moon and the Sun [2]. Physical ocean waves can be classified in different types as depending on the generating force, motion of the wave form, water depth, wavelength, wave height and period, etc. The gravitational attraction of the Moon and the Sun generates tidal waves. The wind generated waves on the water surface, from the smallest to the biggest in order, are the capillary waves or ripples, ultra-gravity, gravity and infragravity waves. Storms, landslides, underwater earthquakes and explosions, and other phenomenal instances generate *tsunamis* that have much longer wavelengths and periods, as well as more catastrophic, than wind generated waves. Besides these waves, sometimes very big waves occur in sea states. These are called as *rogue* or *freak* waves [5]. Although their causes are not yet clearly known, the physical mechanisms that suggested as possible explanations of the freak wave phenomenon may be wave–current interaction, geometrical and diffractive focusing, focusing due to dispersion and modulational instability, soliton collision, and atmospheric action [5]. Most commonly occurring waves in ocean

H. Karadeniz, *Stochastic Analysis of Offshore Steel Structures*,
Springer Series in Reliability Engineering, DOI: 10.1007/978-1-84996-190-5_3,

environments are due to winds which develop waves in different stages as being capillary, sea, and swell waves. The size of a wind generated wave depends on the wind speed, duration of the wind and the *fetch* (the distance of water over which the wind blows). On a perfectly calm sea, the wind has practically no grip and as it slides over the water surface the water moves developing eddies and small ripples to form *capillary waves*. Due to ripples the wind grips the water surface better and it becomes rougher. Gradually, it develops short-crested irregular waves traveling in different directions. This wave condition is called as the *sea waves*. Depending on duration and fetch of the wind, the wave develops into a *fully developed sea*, in which it reaches maximum size, speed and period beyond the fetch, i.e., when the wind has imparted its maximum energy to the waves, the sea is said to be fully developed. When the sea wave moves out of the wind generating area it is called as the *swell wave*. It stops to grow in size and to gain the energy usually having smooth and well-rounded profile. Since it looses energy, during the propagation, the swell wave starts decaying. Sea waves have irregular surface fluctuations that can be composed of many regular harmonic waves with different amplitudes, periods, lengths, and directions, which form altogether a random surface elevation. The condition of these wave forms defines a random sea [6, 7] that a stochastic analysis approach is essentially applied to determine. Sea waves can also be classified into deepwater, shallow water, and very shallow water waves regarding to the relative water depth, d/L, where d is the still water depth and L is the wavelength [8]. From the theoretical point of view the regular waves can be determined from the wave equation which satisfies a number of boundary conditions on the water surface and at the bottom of the sea. According to the simplification procedure of boundary conditions the wave theories are classified in two groups as linear and nonlinear (finite amplitude) wave theories, which have been explained in many text books, see e.g., [8–21]. Since sea waves are random, a stochastic description is essentially used in practice [6, 7], which constitutes basic inputs to the stochastic analysis of offshore structures [22, 23]. Among others sea waves are the most important phenomena exerting loads on offshore structures in a random fashion that structures must be designed to withstand for a long period of time. Since sea waves produce predominant dynamic loads on structures in the short term as well as in the long term which cause fatigue damages, in this chapter, an introduction to water waves and their stochastic representation will be presented. Then, the calculation procedure of wave loads for the stochastic analysis and water–structure interaction, which is based on wave load formulation, is highlighted with emphasis on added mass and hydrodynamic damping.

3.2 Introduction to Wave Theories

Water surface waves are obtained from inviscid, incompressible and irrotational flow under certain boundary conditions by using the principles of hydrodynamics [24]. To obtain wave equations *the continuity* of water is used. The continuity in fluid dynamics states that the fluid mass is conserved, i.e., in a steady state process, the rates of mass entering and leaving a system are equal which leads to the *continuity* equation [10], in general,

$$\frac{\partial \rho}{\partial t} + \nabla \cdot (\rho V) = 0 \tag{3.1a}$$

where ρ is the mass density and V is the velocity vector of the water, ∇ is the *divergence* operator defined as

$$\nabla = \frac{\partial}{\partial x} i + \frac{\partial}{\partial y} j + \frac{\partial}{\partial z} k \tag{3.1b}$$

For incompressible fluid, i.e., ρ is constant, the mass conservation, or the continuity equation, becomes equivalent to the conservation of volume which is, in terms of water velocities, stated as

$$\text{Continuity equation} :\to \nabla \cdot V = \frac{\partial u}{\partial x} + \frac{\partial v}{\partial y} + \frac{\partial w}{\partial z} = 0 \tag{3.1c}$$

where u, v, and w are the components of the water velocity vector V in x, y, and z directions respectively. In addition to the translational motion, one other important definition in fluid motion is the rotation of fluid particles. The rotation vector is defined as

$$\text{Rotation vector:} \to \Omega = \frac{1}{2} \nabla x V \tag{3.2a}$$

where (x) denotes a vector product. The components in x, y, and z directions are

$$\Omega_x = \frac{1}{2}\left(\frac{\partial w}{\partial y} - \frac{\partial v}{\partial z}\right), \quad \Omega_y = \frac{1}{2}\left(\frac{\partial u}{\partial z} - \frac{\partial w}{\partial x}\right) \quad \text{and} \quad \Omega_z = \frac{1}{2}\left(\frac{\partial v}{\partial x} - \frac{\partial u}{\partial y}\right) \tag{3.2b}$$

The irrotationality of the flow states that the rotation components are all zero, i.e., $\Omega_x = 0, \Omega_y = 0$ and $\Omega_z = 0$, and the flow undergoes only the translational motion. The continuity and irrotationality conditions of the flow, given by Eqs. (3.1c) and (3.2b), are the basic equations of water waves that satisfy a number of boundary conditions at the bottom and on the free surface of the sea:

- At the bottom, the velocity components normal to the bottom surface will be zero and, for a body in the wave, the normal velocity of the water is equal to the normal velocity of the body. If the body is fixed, then the normal water velocity will be zero [14]. These are stated as, if the z axis is in vertical direction,

$$\begin{aligned} &\text{At the bottom} \quad \to \quad w = 0 \\ &\text{At a body} \quad \to \quad v_n = v_{nbody} \end{aligned} \tag{3.3a}$$

where v_n denotes the normal velocity component on the surface of the body.

- For inviscid and irrotational unsteady flows, the Bernoulli equation can be obtained from the spatial integration of the Navier–Stokes equation [10] as written in vector notations by

$$\int \frac{\partial V}{\partial t} \mathrm{d}S + \frac{p}{\rho} + gz + \frac{V^2}{2} = f(t) \tag{3.3b}$$

in which the first term is the contribution of the local fluid acceleration (local inertia term), p is the pressure equal to the atmospheric pressure at the free surface which is assumed to be zero, g is the gravitational acceleration, V is the velocity vector, and $f(t)$ is a function of time only. Eq. (3.3b) provides the dynamic boundary condition at the free surface, i.e.,

$$\text{Dynamic boundary condition at free surface:} \rightarrow \int \frac{\partial V}{\partial t} \mathrm{d}S + g\eta + \frac{V^2}{2} = f(t) \tag{3.3c}$$

in which η is the water surface elevation above the still water level and S denotes a spatial variation. One other free surface boundary condition is the kinematic condition which states that the free surface moves with the fluid. The equation of the free surface at any spatial point and time is defined as,

$$z = \eta(x, y, t) \tag{3.3d}$$

from which the vertical velocity of the fluid at the free surface is obtained from the total derivative,

$$\text{Kinematic boundary condition at free surface:} \rightarrow \frac{\mathrm{d}z}{\mathrm{d}t} = w = \frac{\partial \eta}{\partial t} + \frac{\partial \eta}{\partial x} u + \frac{\partial \eta}{\partial y} v \tag{3.3e}$$

The wave equation defined by the continuity equation, Eq. (3.1c), or the condition of zero rotations, Eq. (3.2b), are solved under the boundary conditions given by Eqs. (3.3a), (3.3c) and (3.3e). There are two approaches to solve the wave equation, one is to use a scalar velocity potential function, Φ, and the other one is to use a scalar stream function, Ψ, from which the components of the fluid velocity are derived as written by,

$$\text{Potential function :}\rightarrow \begin{cases} u = \dfrac{\partial \Phi}{\partial x} \\ v = \dfrac{\partial \Phi}{\partial y} \\ w = \dfrac{\partial \Phi}{\partial z} \end{cases}, \quad \text{Stream function :}\rightarrow \begin{cases} u = \dfrac{\partial \Psi}{\partial y} \\ v = -\dfrac{\partial \Psi}{\partial x} \end{cases} \tag{3.3f}$$

The stream function approach is suitable for 2D flow and satisfies the continuity equation automatically, and from the condition of zero rotation, $\Omega_z = 0$, it is obtained that,

$$\Omega_z = -\frac{1}{2}\left(\frac{\partial^2 \Psi}{\partial x^2} + \frac{\partial^2 \Psi}{\partial y^2}\right) = 0 \quad \rightarrow \quad \nabla^2 \Psi = 0 \tag{3.4}$$

The potential function approach is more general and also suitable for 3D flow. It satisfies the condition of zero rotations automatically. Using the continuity condition, the Laplace equation is obtained as written,

$$\frac{\partial u}{\partial x} + \frac{\partial v}{\partial y} + \frac{\partial w}{\partial z} = 0 \quad \rightarrow \quad \frac{\partial^2 \Phi}{\partial x^2} + \frac{\partial^2 \Phi}{\partial y^2} + \frac{\partial^2 \Phi}{\partial z^2} = 0 \quad \rightarrow \quad \nabla^2 \Phi = 0 \tag{3.5a}$$

In 2D flow, the potential and stream functions are orthogonal functions. Since the potential function approach is more general, it is used further in this book. Having used velocity components defined in Eq. (3.3d) in the boundary conditions, the following boundary equations are obtained [8, 14].

$$\begin{aligned}
&\text{Bottom boundary condition...} \quad : \rightarrow \frac{\partial \Phi}{\partial z} = 0 \\
&\text{Free surface dynamic condition.} : \rightarrow \frac{\partial \Phi}{\partial t} + g\eta + \frac{1}{2}\left[\left(\frac{\partial \Phi}{\partial x}\right)^2 + \left(\frac{\partial \Phi}{\partial y}\right)^2 + \left(\frac{\partial \Phi}{\partial z}\right)^2\right] = 0 \\
&\text{Free surface kinematic condition} : \rightarrow \frac{\partial \Phi}{\partial z} = \frac{\partial \eta}{\partial t} + \frac{\partial \eta}{\partial x}\frac{\partial \Phi}{\partial x} + \frac{\partial \eta}{\partial y}\frac{\partial \Phi}{\partial y}
\end{aligned} \tag{3.5b}$$

To obtain a water wave, the Laplace equation given by Eq. (3.5a) will be solved under the boundary conditions given by Eq. (3.5b). As it is seen from Eq. (3.5b), the dynamic and kinematic free surface boundary conditions are nonlinear and the solution of the Laplace equation under these conditions, $\nabla^2 \Phi = 0$, is a difficult task and numerical solution is usually applied [8, 16, 19, 25–31]. A typical regular periodic wave profile is shown in Fig. 3.1 in which the illustrated and some other definitions of wave properties are written in Table 3.1. Some of these parameters are not independent and they are related to each other. The dependent parameters are the angular wave frequency, ω, wave number, m, wave celerity, C, and wave steepness, α. Their interrelations are stated as,

$$\begin{aligned}
&\text{Wave frequency} : \rightarrow \quad \omega = \frac{2\pi}{T}, \qquad \text{wave number} : \rightarrow \quad m = \frac{2\pi}{L} \\
&\text{Wave celerity} : \quad \rightarrow \quad C = \frac{L}{T} = \frac{\omega}{m}, \qquad \text{wave steepness} : \rightarrow \quad \alpha = \frac{H}{L}
\end{aligned} \tag{3.6a}$$

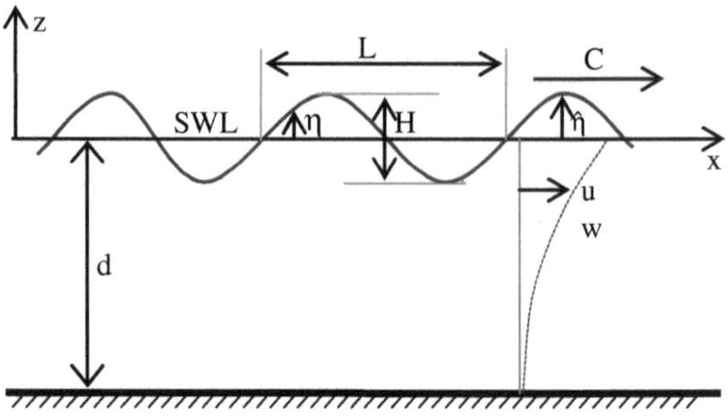

Fig. 3.1 A 2D wave profile and definition of some symbols

Table 3.1 Definition of symbols used in the wave profile and properties

Symbol	Designation
SWL	Still water level
η	Free surface elevation
H	Wave height
L	wavelength
$\hat{\eta}$	Wave amplitude
d	Water depth
C	Wave celerity
C_g	Wave group velocity
u, w	Water particle velocities in horizontal and vertical directions
T	Wave period
ω	Wave frequency (rad/s)
m	Wave number
α	Wave steepness

For regular waves, the amplitude $\hat{\eta}$ is equal to the halve of the wave height, i.e.,

$$\text{Wave amplitude} : \rightarrow \; \hat{\eta} = H/2 \tag{3.6b}$$

as shown in Fig. 3.1. The wave group velocity is defined as,

$$\text{Wave group velocity} : \rightarrow \quad C_g = \frac{\partial \omega}{\partial m} \tag{3.6c}$$

Being depending on the ratio of the water depth and wavelength, *d*/*L*, waves are classified into groups of deep, intermediate, and shallow water waves [8, 14] as defined in Table 3.2. For deep and shallow water waves, asymptotic values of functions are used whereas, for intermediate water depth, functions are not simplified. For the solution of the Laplace equation, Eq. (3.5a), under the boundary conditions given in Eq. (3.5b), some simplifying assumptions are made. The simplest one is the linearization of the nonlinear boundary conditions that associates with the linear Airy wave theory, which will be used in this book. Since it will be explained in Sect. 3.3 in detail, some important nonlinear wave theories are outlined further in this section.

Table 3.2 Wave classification with respect to the ratio of water depth and wavelength (*d*/*L*)

Definition region	Water wave type	Simplification of the function, tan*h*(x)
$1/2 < d/L$	Deep water	1
$1/25 < d/L < 1/2$	Intermediate depth	tan*h*(*x*)
$d/L < 1/25$	Shallow water	*x*

3.2.1 Stokes Wave Theory

Due to nonlinearities in the free surface boundary conditions, Eq. (3.5b), the analytical solution of the wave equation becomes a difficult task. Therefore, simplifying assumptions are made for the solution. The Stokes wave theory assumes that the velocity potential Φ and water elevation η are stated as power series in terms of a nondimensional small perturbation parameter ε which is defined as the product of wave number and wave amplitude. Thus,

$$\Phi = \sum_{n=1}^{\infty} \Phi_n \varepsilon^n \quad \text{and} \quad \eta = \sum_{n=1}^{\infty} \eta_n \varepsilon^n \quad \text{with} \quad (\varepsilon = w\hat{\eta} \quad \text{or} \quad \varepsilon = \pi\alpha) \tag{3.7a}$$

in which the terms, Φ_n and η_n, are calculated from solutions of differential equations. This wave theory is valid for small wave steepness, $(\alpha \ll 1)$, and also for deep and intermediate water depth, i.e., for $(d/L > 1/25)$. Because, the wave steepness α is an increasing function of the inverse of (d/L), i.e., the wave steepness increases with decreasing value of (d/L). One other approximation is the Taylor expansion of the velocity potential function Φ with respect to z at the still water level $(z = 0)$ so that the free surface conditions are simplified. Thus,

$$\Phi(x, z, t) = \Phi + z\frac{\partial \Phi}{\partial z} + \frac{1}{2}z^2\frac{\partial^2 \Phi}{\partial z^2} + \frac{1}{6}z^3\frac{\partial^3 \Phi}{\partial z^3} + \ldots \tag{3.7b}$$

in which Φ denotes the velocity potential function at the still water level, $z = 0$. Having introduced Φ from Eq. (3.7a) into Eq. (3.5a) and bottom boundary condition $(\partial\Phi/\partial z = 0)$, it is obtained that,

$$\nabla^2 \Phi_n = 0, \quad \text{at the bottom:} \ \frac{\partial \Phi_n}{\partial z}\Big|_{z=-d} = 0 \quad \rightarrow \quad (n = 1, 2, \ldots, \infty) \tag{3.8}$$

The free nonlinear boundary conditions, which are given in Eq. (3.5b), are reformulated by using Eqs. (3.7a) and (3.7b), as depending on perturbation terms less than the current one, i.e., in terms of (Φ_k and η_k with $k < n$). By equalizing the terms of powers of the perturbation parameter ε in both sides of the boundary conditions new surface boundary conditions can be obtained as stated by, for kinematic boundary condition of 2D wave in $(x\text{–}z)$ plane:

$$\text{Kinematic condition: } \rightarrow \frac{\partial \Phi}{\partial z} = \frac{\partial \eta}{\partial t} + \frac{\partial \eta}{\partial x}\frac{\partial \Phi}{\partial x}$$
$$\frac{\partial \Phi_n}{\partial z} - \frac{\partial \eta_n}{\partial t} = G(\Phi_k, \eta_k) \quad \rightarrow \quad \text{with } (k<n,\ n = 1, 2, \ldots, \infty) \tag{3.9a}$$

and for dynamic boundary condition of 2D wave in (*x*–*z*) plane:

$$\text{dynamic condition. : } \rightarrow \frac{\partial \Phi}{\partial t} + g\eta + \frac{1}{2}\left[\left(\frac{\partial \Phi}{\partial x}\right)^2 + \left(\frac{\partial \Phi}{\partial z}\right)^2\right] = 0$$
$$\frac{\partial^2 \Phi_n}{\partial t^2} + g\frac{\partial \Phi_n}{\partial z} = F(\Phi_k, \eta_k) \quad \rightarrow \quad \text{with } (k<n, \quad n = 1, 2, \ldots, \infty) \tag{3.9b}$$

Since the functions $F(\Phi_k, \eta_k)$and $G(\Phi_k, \eta_k)$ are known from previous solutions of perturbation terms, the Laplace equation given in Eq. (3.8) can be solved with corresponding boundary conditions so that the complete solution requires a successive solution algorithm. The number of perturbation terms, which are included in the power series, defines the degree of the Stokes wave theory. For the first- and second-order Stokes wave theories, $(n = 1)$ and $(n = 2)$, the functions, $F(\Phi_k, \eta_k)$, $G(\Phi_k, \eta_k)$ and the water elevation η_k are presented in Table 3.3. Using the perturbation method the velocity potential function Φ and water elevation η for the *Stokes Third-Order Wave Theory* can be stated as written by

$$\left.\begin{aligned} \Phi &= \Phi_{z1}\sin\theta + \Phi_{z2}\sin 2\theta + \Phi_{z3}\sin 3\theta \\ \eta &= \hat{\eta}\cos\theta + \eta_{c2}\cos 2\theta + \eta_{c3}\cos 3\theta \end{aligned}\right\} \quad \rightarrow \quad \theta = (mx - \omega t) \tag{3.10a}$$

in which the functions Φ_{z1}, Φ_{z2}, Φ_{z3} are obtained [32–34] to be:

$$\begin{aligned} \Phi_{z1} &= \hat{\eta}\frac{\omega}{m}\frac{\cosh m(z+d)}{\sinh md} \\ \Phi_{z2} &= \frac{3}{8}\hat{\eta}^2\omega\frac{\cosh^2 m(z+d)}{\sinh^4 md} \\ \Phi_{z3} &= \frac{1}{64}\hat{\eta}^3 m\omega\frac{\left(11 - 2\cosh^2 md\right)}{\sinh^7 md}\cosh^3 m(z+d) \end{aligned} \tag{3.10b}$$

and the constants η_{c2} and η_{c3} are:

$$\begin{aligned} \eta_{c2} &= \frac{1}{4}\hat{\eta}^2 m\frac{\left(2 + \cosh^2 md\right)\ \cos h\, md}{\sinh^3 md} \\ \eta_{c3} &= \frac{3}{64}\hat{\eta}^3 m^2\frac{\left(1 + 8\cosh^6 md\right)}{\sinh^6 md} \end{aligned} \tag{3.10c}$$

The dispersion relationship for the first- and second-order Stokes wave theories are the same, which means that, under the same condition, wave periods remain unchanged while wave amplitudes are changed. For the third-order Stokes wave theory, the dispersion relationship is also different from those for the first- and second-order wave theories so that the wave period and amplitude are both

Table 3.3 Water elevation and free surface boundary conditions of Stokes wave theories

Function	First order (n = 1)	Second order (n = 2)
η_n	$-\frac{1}{g}\frac{\partial \Phi_1}{\partial t}$	$-\frac{1}{g}\left\{\frac{\partial \Phi_2}{\partial t}+\eta_1\frac{\partial^2\Phi_1}{\partial t\partial z}+\frac{1}{2}\left[\left(\frac{\partial \Phi_1}{\partial x}\right)^2+\left(\frac{\partial \Phi_1}{\partial z}\right)^2\right]\right\}$
Kinematic boundary condition		
$G(\Phi_k,\eta_k)=\frac{\partial \Phi_n}{\partial z}-\frac{\partial \eta_n}{\partial t}$	0	$\frac{\partial \eta_1}{\partial x}\frac{\partial \Phi_1}{\partial x}-\eta_1\frac{\partial^2\Phi_1}{\partial z^2}$
Dynamic boundary condition		
$F(\Phi_k,\eta_k)=\frac{\partial^2\Phi_n}{\partial t^2}+g\frac{\partial \Phi_n}{\partial z}$	0	$-\frac{1}{g}\frac{\partial}{\partial t}\left\{\eta_1\frac{\partial^2\Phi_1}{\partial t\partial z}+\frac{1}{2}\left[\left(\frac{\partial \Phi_1}{\partial x}\right)^2+\left(\frac{\partial \Phi_1}{\partial z}\right)^2\right]\right\}$

changed. Thus, they are not dispersive. The dispersion relationship for the Stokes third-order wave theory becomes,

$$\text{Dispersion relationship:} \rightarrow \omega^2 = mg\tanh\, md\left(1+\hat{\eta}^2 m^2\frac{\cosh^4 md+8}{8\sinh^4 md}\right) \tag{3.10d}$$

Using the velocity potential function Φ in Eq. (3.10a), the water particle velocities u and w are calculated from Eq. (3.3d), and then, the corresponding water particle accelerations are calculated from the derivatives:

$$\text{Water particle accelerations:} \rightarrow \begin{cases} \dot{u} = \dfrac{\partial u}{\partial t}+\dfrac{1}{2}\dfrac{\partial}{\partial x}(u^2+w^2) \\ \dot{w} = \dfrac{\partial w}{\partial t}+\dfrac{1}{2}\dfrac{\partial}{\partial z}(u^2+w^2) \end{cases} \tag{3.11}$$

Having determined water particle velocities and accelerations wave forces on structural members can be calculated using Morison's equation [35], which will be explained in Sect. 3.7.

3.2.2 *Other Wave Theories*

As presented in Sect. 3.2.1 the Stokes wave theories are most suitable for deep and intermediate water depth. Even higher order terms in the Stokes theory for steeper waves produce unrealistic results. For shallower water, a finite-amplitude wave theory is required. *Cnoidal* wave theory [9, 36–38] and, in very shallow water, *solitary* wave theory [9, 39, 40], are the analytical wave theories most commonly used for shallow water. Solutions in the cnoidal wave theory are obtained in terms of elliptical integrals of the first kind. The solitary wave theory is a special case of the cnoidal wave theory at one limit, and at the other limit it is identical with the linear wave theory. As the relative depth decreases the cnoidal wave becomes the

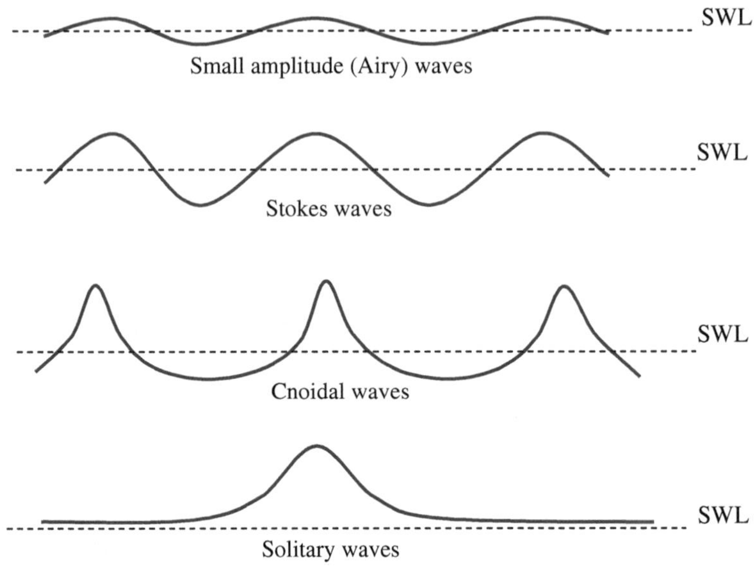

Fig. 3.2 Different wave forms (forms of the surface elevation)

solitary wave, which has a crest that is completely above the still water level and has no trough. The cnoidal wave theory covers a large class of long waves with finite amplitudes. It is presented in terms of two parameters, k^2 and U_r, where k^2 depends on the water depth, the wave length and height, which is one of the independent variables in the elliptical function. The parameter U_r is the Ursell parameter defined as $U_r = L^2H/d^3$ which depends on the wave steepness and the relative water depth. This parameter defines the range of application of the wave theories. In general, the cnoidal wave theory is applicable for $U_r > 25$, the Stokes theory is applicable for $U_r < 10$ and both theories are applicable for $U_r = 10 - 25$. The limiting case of ($k^2 = 0$) results in the small amplitude wave theory while ($k^2 = 1$) results in the solitary wave theory. The forms (water elevation) of different waves are shown in Fig. 3.2. Since the nonlinear wave theories are not the main topic covered in this book they are not further presented herein. The interested readers should consult related references [8, 9, 14, 16, 27, 32–41]. As the linear (Airy) wave theory is used for the stochastic analysis of offshore structures further in this book, it is presented below in detail

3.3 Linear (Airy) Wave Theory

The linear wave theory (also known as the Airy wave theory) is the simplest and most useful theory among other wave theories. It assumes small wave steepness and small relative water depth (*H*/*d*), which allows the free surface boundary

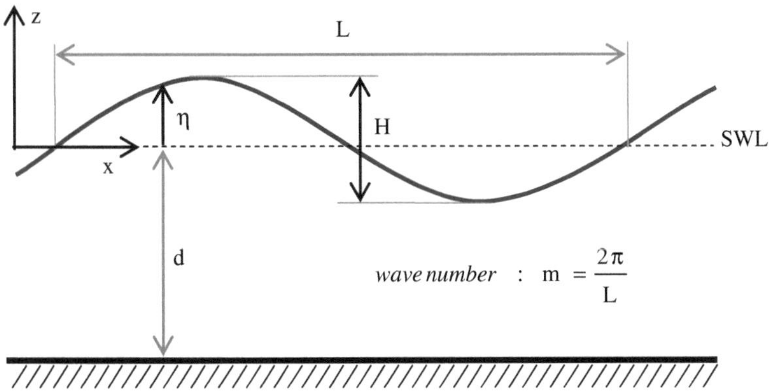

Fig. 3.3 2D small amplitude wave (Airy wave)

conditions to be linearized and satisfied at the mean water level (still water level, SWL). It is equivalent to the first-order Stokes wave theory. The linearized differential equations are stated from Eqs. (3.5a) and (3.5b) as written by,

$$\nabla^2\Phi = 0 \rightarrow \begin{cases} \text{Bottom boundary condition.... :} \rightarrow & \dfrac{\partial\Phi}{\partial z}\Big|_{z=-d} = 0 \\ \text{Free surface boundary condition :} \rightarrow & \left(\dfrac{\partial^2\Phi}{\partial t^2} + g\dfrac{\partial\Phi}{\partial z}\right)_{z=0} = 0 \end{cases} \tag{3.12}$$

The solution of the differential equation given by Eq. (3.12) is carried out by using separation technique of the velocity potential function Φ as stated by,

$$\Phi = Q(z)U(x)F(t) \quad \rightarrow \quad \nabla^2\Phi = \frac{\partial^2 Q}{\partial z^2}U + \frac{\partial^2 U}{\partial x^2}Q = 0 \tag{3.13a}$$

from which and using the free surface boundary condition, it can be written that

$$\frac{1}{Q}\frac{\partial^2 Q}{\partial z^2} = -\frac{1}{U}\frac{\partial^2 U}{\partial x^2} = m^2 \quad \text{and} \quad \left(\frac{\partial^2 F}{\partial t^2}Q + g\frac{\partial Q}{\partial z}F\right)_{z=0} = 0 \tag{3.13b}$$

In this equation, m is a constant (wave number). A typical Airy wave with the water depth d, water elevation η, wave length L, and height H is shown in Fig. 3.3. From the solution of differential equations given in Eq. (3.13b) and using the bottom boundary condition given in Eq. (3.12), the velocity potential function Φ for a progressive wave traveling in the positive x direction is obtained as written by

$$\text{Velocity potential function:} \rightarrow \Phi = D \cosh m(z+d)\, e^{i(\omega t - mx)} \tag{3.14a}$$

where ω is the wave frequency which satisfies the dispersion relationship,

$$\text{Dispersion relationship} :\rightarrow \omega^2 = mg \tanh md \tag{3.14b}$$

The water elevation η is calculated from,

$$\text{Water elevation} \rightarrow \eta = -\frac{1}{g}\frac{\partial \Phi}{\partial t}|_{z=0} = -i\,\frac{\omega D}{g}\cosh\; md\; e^{i(\omega t - mx)} \tag{3.14c}$$

The coefficient D is obtained in terms of the wave amplitude $\hat{\eta}$ as written by,

$$\hat{\eta} = \frac{H}{2} = \max \eta = |\eta| \quad \rightarrow \quad D = \hat{\eta}\frac{g}{\omega\ \cosh\ md} \tag{3.14d}$$

Having introduced Eq. (4.14d) in Eqs. (3.14a) and (3.14c) the velocity potential function and water elevation are obtained as written by,

$$\Phi = \hat{\eta}\frac{g}{\omega}\frac{\cosh m(z+d)}{\cosh md} e^{i(\omega t - mx)} \text{ and } \eta = -i\,\hat{\eta}\ e^{i(\omega t - mx)} \tag{3.15}$$

The horizontal and vertical water particle velocities are calculated from,

$$\text{Water particle velocities} :\rightarrow \begin{cases} u = \dfrac{\partial \Phi}{\partial x} = -i\ \hat{\eta}\omega \dfrac{\cosh\ m(z+d)}{\sinh\ md} e^{i(\omega t - mx)} \\ w = \dfrac{\partial \Phi}{\partial z} = \hat{\eta}\omega \dfrac{\sinh\ m(z+d)}{\sinh\ md} e^{i(\omega t - mx)} \end{cases} \tag{3.16a}$$

The corresponding accelerations for small amplitude waves are calculated from the time derivatives of these velocities as written

$$\text{Water particle accelerations:} \rightarrow \begin{cases} \dot{u} = \dfrac{\partial u}{\partial t} = \hat{\eta}\omega^2 \dfrac{\cosh\ m(z+d)}{\sinh\ md} e^{i(\omega t - mx)} \\ \dot{w} = \dfrac{\partial w}{\partial t} = i\,\hat{\eta}\omega^2 \dfrac{\sinh\ m(z+d)}{\sinh\ md} e^{i(\omega t - mx)} \end{cases} \tag{3.16b}$$

The water particle displacements are calculated from the time integration of velocities as written by,

$$\text{Water particle displacements:} \rightarrow \begin{cases} \xi_x = \displaystyle\int u\, dt = -\,\hat{\eta} \frac{\cosh\ m(z+d)}{\sinh\ md} e^{i(\omega t - mx)} \\ \xi_z = \displaystyle\int w\, dt = -i\,\hat{\eta} \frac{\sinh\ m(z+d)}{\sinh\ md} e^{i(\omega t - mx)} \end{cases} \tag{3.16c}$$

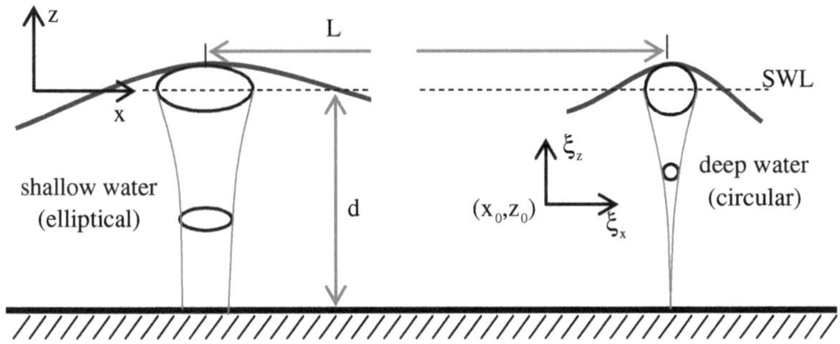

Fig. 3.4 Particle orbits for shallow and deep water waves

The real parts of these displacements form an elliptical orbit around a fixed point (x_0, z_0) as shown in Fig. 3.4. The equation of the orbit is

$$\text{Water particle orbit:} \rightarrow \left(\frac{\left(\xi_{x_0}-x_0\right)^2}{\cosh^2 m(z_0+d)}+\frac{\left(\xi_{z_0}-z_0\right)^2}{\sinh^2 m(z_0+d)}\right)=\frac{\hat{\eta}^2}{\sinh^2 md} \tag{3.16d}$$

With these velocity and acceleration information the wave forces on structural members can be calculated using the Morison's equation [35] as it will be explained later in this chapter. However, since the deep water condition is used further in this book, the velocity potential function, velocity and acceleration components, and the orbit equation for deep water condition are presented below.

3.3.1 Formulation for Deep Water Condition

In ocean environments far away the shoreline, the deep water wave condition is applied adequately. As given in Table 3.2, the deep water condition is defined when $(d/L > 1/2)$. This condition can also be stated alternatively as $(md > \pi)$, in which case, the following approximations can be used:

$$\text{Deep water approximations:} \rightarrow \begin{cases} \tanh md \approx 1 \\ \sinh m(z+d) \approx e^{mz} \sinh md \\ \cosh m(z+d) \approx e^{mz} \sinh md \end{cases} \tag{3.17}$$

Having used these approximations in the formulations presented above it can be obtained that

$$\text{Velocity potential function}: \rightarrow \quad \Phi = \hat{\eta}\frac{g}{\omega}e^{mz}e^{i(\omega t - mx)} \tag{3.18a}$$

$$\begin{aligned} &\text{Water elevation} &&:\rightarrow\ \eta = -i\,\hat{\eta}\ e^{i(\omega t - mx)} \\ &\text{Dispersion relationship} &&:\rightarrow\ \omega^2 = mg \end{aligned} \tag{3.18b}$$

$$\begin{aligned} &\text{Water particle velocities...} &&:\rightarrow \begin{cases} u = \hat{\eta} - i\,\omega\, e^{mz}\, e^{i(\omega t - mx)} \\ w = \hat{\eta}\,\omega\, e^{mz}\, e^{i(\omega t - mx)} \end{cases} \\ &\text{Water particle accelerations} &&:\rightarrow \begin{cases} \dot{u} = \hat{\eta}\,\omega^2\, e^{mz}\, e^{i(\omega t - mx)} \\ \dot{w} = i\,\hat{\eta}\,\omega^2\, e^{mz}\, e^{i(\omega t - mx)} \end{cases} \end{aligned} \tag{3.18c}$$

$$\text{Water particle displacements} : \rightarrow \begin{cases} \xi_x = -\hat{\eta}\, e^{mz}\, e^{i(\omega t - mx)} \\ \xi_z = -i\,\hat{\eta}\, e^{mz}\, e^{i(\omega t - mx)} \end{cases} \tag{3.18d}$$

$$\text{Water particle orbit} : \rightarrow \left(\xi_{x_0} - x_0\right)^2 + \left(\xi_{z_0} - z_0\right)^2 = \hat{\eta}^2 e^{2mz_0} \tag{3.18e}$$

For deep water condition, the water particle orbit becomes circular as shown in Fig. 3.4. For the stochastic analysis of offshore structures in deep water environments, which is used in this book, Eqs. (3.18a) trough (3.18c) will be used. In order to carry out a stochastic structural analysis in random ocean environments, the stochastic description of sea waves need to be defined first. The description of stochastic sea waves is presented shortly in the following section.

3.4 Stochastic Description of Ocean Waves and Short-Term Sea States

Sea waves have irregular profiles changing randomly in time and space. They cannot be determined in a deterministic way like those presented in above sections. Sea waves are random in nature and mostly short crested. During a relatively short period, they are composed of infinite number of independent regular waves with different amplitudes, frequencies, traveling directions, and phases that are all random. Since we assume that these individual regular waves are determined from the linear wave theory, the superposition rule is applied to form a random wave group traveling along an arbitrary direction with individual waves scattered around it. Since there is infinite number of waves in a random wave group with different amplitudes, in the light of central limit theorem, the random surface displacement (surface elevation) becomes a Gaussian random process [7, 42–46] with zero mean value. A random wave profile obtained from the superposition of a number of individual independent regular waves is shown in Fig. 3.5. Over a time interval of about 3 h at some locations on the sea surface, the surface elevations η are measured relative to the still water level and, in this way, a number of records are obtained. Under the same sea condition, over another time intervals with about 3 h

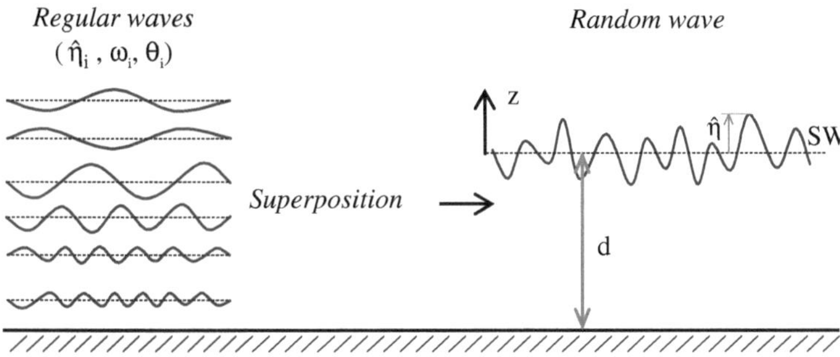

Fig. 3.5 A number of regular wave profiles and representation of a random wave pattern

long at different locations, the measurements of the surface elevations are repeated and different records are obtained. From the comparison of all records of the surface elevation it can be seen that the surface elevation η would have a similar appearance and its general properties would not change. Thus, a single record of the surface elevation represents its random characteristics. This observation reveals that, in a short term, the water surface elevation η is a stationary ergodic process. Similarly, the water particle velocities and accelerations become also stationary ergodic processes. In a short term, random waves are usually described in terms of sea states. A unidirectional sea state is a stationary ergodic process described by the parameters, the significant wave height H_s and mean zero-crossings period T_z, and also by a spectral function $S_{\eta\eta}(\omega)$ of the surface elevation η, which is defined between $(0 \leq \omega \leq \infty)$ as known to be the *sea spectrum*, or wave *energy spectral density function* [14, 22]. The total energy E_{tot} of random waves per unit surface area can be stated as,

$$\text{Total wave energy} : \rightarrow \quad E_{\text{tot}} = \frac{1}{2} g \rho_w \eta^2 \tag{3.19a}$$

which is also a stochastic function. Since the mean value of the surface elevation η is zero, the mean value of the total wave energy per unit surface area can be calculated from,

$$\text{Mean total wave energy} : \rightarrow \quad E[E_{\text{tot}}] = \frac{1}{2} g \rho_w E[\eta^2] = \frac{1}{2} g \rho_w \sigma_\eta^2 \tag{3.19b}$$

where ρ_w denotes the water density, g is the gravitational acceleration and σ_η^2 is the variance of the surface elevation η. As it is seen from Eq. (3.19b), the mean value of the total wave energy per unit surface area is proportional to the variance of the surface elevation. In terms of the correlation and spectral functions of η the mean total wave energy can be stated as,

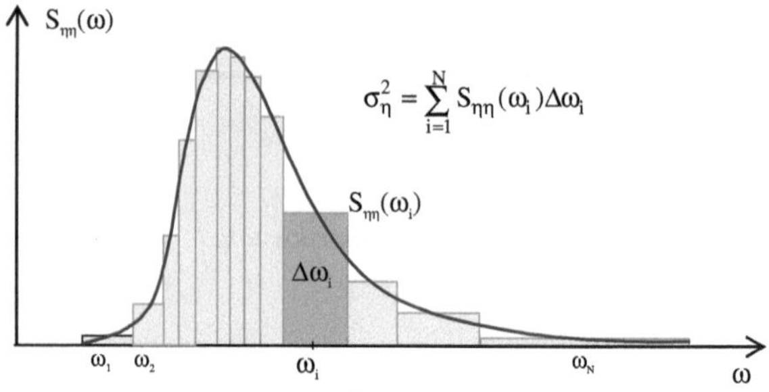

Fig. 3.6 Calculation of the variance of the water elevation η from the sea spectrum

$$E[E_{tot}] = \frac{1}{2} g\rho_w R_{\eta\eta}(0) \quad \rightarrow \quad E[E_{tot}] = \frac{1}{2} g\rho_w \int_0^\infty S_{\eta\eta}(\omega)\,\mathrm{d}\omega \tag{3.19c}$$

and, therefore, the spectral function $S_{\eta\eta}(\omega)$ can also be called as the *variance density spectrum*. But, in this book, it is preferably called as the *sea spectrum* since it characterizes the stochastic description of the water surface elevation η (in short, water elevation). The sea spectrum $S_{\eta\eta}(\omega)$ describes a short-term random sea in the frequency domain. In the time domain, the random sea is represented by superposition of many individual independent linear waves with random amplitudes, frequencies and phase angles. For a 2D random sea, (assuming that all waves are traveling in the x-direction), the real part of the water elevation η can be stated from Eq. (3.18a) as,

$$\text{Random water elevation} : \rightarrow \ \eta = \sum_{i=1}^{\infty} \hat{\eta}_i \sin(\omega_i t - m_i x + \varphi_i) \tag{3.20}$$

where $\hat{\eta}_i$, ω_i, m_i, and $\varphi_{\rm i}$ are respectively the amplitude, frequency, wave number, and phase angle of the $i^{\rm th}$ wave component. The phase angle $\varphi_{\rm i}$ is an independent random variable which is uniformly distributed between 0 and 2π. For the deep water condition, the wave number m_i is given in terms of the frequency as, from Eq. (3.18a), $m_i = \omega_i^2/g$. Since the random sea is an ergodic process, the correlation function of the water elevation η is calculated as a time average from,

$$\text{Correletaion function of } \eta : \rightarrow \ R_{\eta\eta}(\tau) = \lim_{T\to\infty} \frac{1}{2T} \int_{-T}^{T} \eta(t)\,\eta(t+\tau)\,\mathrm{d}t \tag{3.21a}$$

Having introduced η from Eq. (3.20) into Eq. (3.201a) and carried out the integration and with the limit $(T \rightarrow \infty)$ it can be obtained that

$$R_{\eta\eta}(\tau) = \sum_{i=1}^{\infty} \frac{1}{2} \hat{\eta}_i^2 \cos(\omega_i \tau) \quad \rightarrow \quad \sigma_\eta^2 = R_{\eta\eta}(0) = \sum_{i=1}^{\infty} \frac{1}{2} \hat{\eta}_i^2 \tag{3.21b}$$

This variance can also be calculated from the discretized spectrum of the water elevation shown in Fig. 3.6 in which each frequency band corresponds to a harmonic component of the random water elevation. The variance is stated as,

$$\sigma_\eta^2 = \sum_{i=1}^{N} S_{\eta\eta}(\omega_i)\Delta\omega_i \tag{3.21c}$$

with $(N \to \infty)$. By equalizing Eqs. (3.21b) and (3.21c), the amplitude of the ith wave component can be obtained as written by,

$$\hat{\eta}_i = \sqrt{2\, S_{\eta\eta}(\omega_i)\Delta\omega_i} \tag{3.21d}$$

Once the sea spectrum $S_{\eta\eta}(\omega)$ is known a short-term random sea can be fully determined. Before it is explained briefly, some transfer functions related to a random wave in deep water environments are presented in the following section.

3.4.1 Transfer Functions of a Random Wave in Deep Water

Formulation of deep water regular (linear) waves is presented in Sect. 3.3.1. In this section, relations between the random water elevation η and random water particle velocities and accelerations are presented. Since the water elevation η is an ergodic process, its randomness is not location dependable, i.e., a time record taken at any location on the sea surface represents the randomness of η. If we choose a measurement station of η at the origin of x, i.e., $(x = 0)$, and denoting its random fluctuations by η_0 in time, for a linear wave form, it is written from Eq. (3.18a) as,

$$\eta_0 = -i\,\hat{\eta}\; \mathrm{e}^{i\omega t} \tag{3.22}$$

The randomness in the sea wave is represented by this reference water elevation η_0. Using the linear wave theory, other properties of the deep water wave can be related linearly to this random water elevation η_0 and some other frequency dependent functions as stated, for η and Φ,

$$\begin{aligned} &\text{Water elevation at any location } x: \rightarrow \quad \eta = \mathrm{e}^{-imx}\eta_0(\omega) \\ &\text{Velocity potential function}\ldots \quad : \rightarrow \quad \Phi = H_{\Phi\eta}(\omega)\,\eta_0(\omega) \end{aligned} \tag{3.23a}$$

where $H_{\Phi\eta}(\omega)$ is the transfer function between the velocity potential function and the water elevation defined as,

$$\text{Transfer function of } \Phi :\rightarrow \quad H_{\Phi\eta}(\omega) = i\,\frac{g}{\omega}\,\mathrm{e}^{mz}\mathrm{e}^{-imx} \tag{3.23b}$$

From Eq. (3.23a) it can be shown that the spectrum of η is equal to the spectrum of η_0, i.e., $\left(S_{\eta\eta}(\omega) = S_{\eta_0\eta_0}(\omega)\right)$, and therefore η_0 is replaced by η henceforth in this book. Thus, the velocities are,

$$\text{Water particle velocities:} \rightarrow \begin{cases} u = H_{u\eta}(\omega)\,\eta \\ w = H_{w\eta}(\omega)\,\eta \end{cases} \tag{3.24a}$$

with the transfer functions of $H_{u\eta}(\omega)$ and $H_{w\eta}(\omega)$ defined as,

$$\text{Transfer functions of velocities:} \rightarrow \begin{cases} H_{u\eta}(\omega) = \omega\,\mathrm{e}^{mz}\,\mathrm{e}^{-i\,mx} \\ H_{w\eta}(\omega) = i\;\omega\,e^{mz}\,\mathrm{e}^{-i\,mx} \end{cases} \tag{3.24b}$$

The accelerations are,

$$\text{Water particle accelerations :} \rightarrow \begin{cases} \dot{u} = H_{\dot{u}\eta}(\omega)\,\eta \\ \dot{w} = H_{\dot{w}\eta}(\omega)\,\eta \end{cases} \tag{3.25a}$$

with the transfer functions of $H_{\dot{u}\eta}(\omega)$ and $H_{\dot{w}\eta}(\omega)$ defined as,

$$\text{Transfer functions of accelerations :} \rightarrow \begin{cases} H_{\dot{u}\eta}(\omega) = i\;\omega^2\,\mathrm{e}^{mz}\,\mathrm{e}^{-i\,mx} \\ H_{\dot{w}\eta}(\omega) = -\omega^2\,\mathrm{e}^{mz}\,\mathrm{e}^{-i\,mx} \end{cases} \tag{3.25b}$$

With these transfer functions water particle velocities and accelerations, which are used in the wave force calculation, are related to the water elevation of which spectral description is available. In the following section, commonly used spectral functions of the water elevation η, which describe a short-term sea state, are presented.

3.4.2 Statistics and Spectral Functions of the Water Elevation in the Short Term

Short-term stationary sea states are described by a spectral function of the water elevation with an analytic formula having parametric constants determined from measured spectra as depending on the geographic area with local measurements and the severity of the sea state. A sea spectrum is generally described in terms of two parameters, the significant wave height H_s and the mean zero-crossings period T_z. The significant wave height H_s is the measure commonly used for representing the severity of the sea state. It is defined as the average value of the one-third (1/3) highest wave heights. Since a short-term sea state is assumed to be a narrow-banded Gaussian process the probability distribution of the wave height follows the Rayleigh distribution defined as,

$$\text{PDF of wave height } H: \rightarrow \; f_H(x) = \frac{x}{4\sigma_\eta^2}\,\exp\left(-\frac{x^2}{8\sigma_\eta^2}\right) \quad (0 \leq x < \infty) \tag{3.26a}$$

The lower limit, x_*, of the highest one-third of the PDF is calculated from

$$P(H \geq x_*) = \int_{x_*}^{\infty} f_H(x)\,\mathrm{d}x = \frac{1}{3} \quad \rightarrow \quad x_* = \sqrt{8\sigma_\eta^2 \ln 3} \tag{3.26b}$$

and the significant wave height is calculated by definition from

$$\frac{1}{3}H_s = \int_{x_*}^{\infty} x f_H(x)\,\mathrm{d}x \quad \rightarrow \quad H_s = \left[3\sqrt{2\pi}\left(1 - \mathrm{erf}(\sqrt{\ln 3})\right) + \sqrt{8\ln 3}\right]\sigma_\eta \tag{3.26c}$$

from which H_s is obtained to be

$$\text{Significant wave height}: \quad \rightarrow \quad H_s = 4.0043\sigma_\eta \simeq 4\sigma_\eta \tag{3.26d}$$

where σ_η is the standard deviation of the water elevation calculated from the integration of the sea spectrum. This calculation is based on the narrow-band assumption of the random sea process. If, however, the non narrow-banded process is considered, the same procedure is applied to calculate H_s using the probability density function of the positive maxima given by Eq. (2.127b). In this case, the value obtained must be doubled since the related PDF is stated for the positive amplitude only. Having determined the significant wave height H_s, some useful spectral measures for the water elevation, i.e., the spectral moments of the water elevation η, are calculated in general from the spectrum integration,

$$\text{Spectral moments of water elevation } \eta :\rightarrow m_n = \int_0^{\infty} \omega^n S_{\eta\eta}(\omega)\,\mathrm{d}\omega \tag{3.27}$$

Statistics of the water elevation can be estimated in terms of the spectral moments. The standard deviation of the water elevation is calculated from,

$$\text{Standard deviation of } \eta : \quad \rightarrow \quad \sigma_\eta = \sqrt{m_0} \tag{3.28}$$

The mean frequencies of zero-crossings and maxima of the water elevation are calculated from,

$$\begin{aligned} &\text{Mean zero-crossings wave frequency} : \rightarrow \omega_0 = \sqrt{m_2/m_0} \\ &\text{Mean frequency of wave maxima}\ldots :\rightarrow \omega_m = \sqrt{m_4/m_2} \end{aligned} \tag{3.29a}$$

The mean periods of zero-crossings and maxima of the wave are calculated from the corresponding frequencies as written,

$$\begin{aligned} &\text{Mean zero-crossings wave period}: \rightarrow T_Z = 2\pi\sqrt{m_0/m_2} \\ &\text{Mean period of wave maxima}\ldots :\rightarrow T_m = 2\pi\sqrt{m_2/m_4} \end{aligned} \tag{3.29b}$$

The spectral band-width of the wave is calculated from,

$$\text{Spectral band-width of wave:} \rightarrow \quad \varepsilon_\eta = \sqrt{1 - \frac{m_2^2}{m_0 m_4}} = \sqrt{1 - \frac{T_m^2}{T_Z^2}} \tag{3.30}$$

There are many spectral functions available in the literature [7, 14, 16, 23, 47–56] among which two peaks spectra describe a composed sea state for both wind sea and swell. However, the wind sea is the most dominant and commonly used part, and therefore most spectral shapes have only one peak describing it. In this section, two spectral functions, which are mostly used in the design of offshore structures, are explained.

3.4.2.1 The Pierson–Moskowitz Sea Spectrum

One of the most commonly used spectra for wind seas is the Pierson–Moskowitz (PM) sea spectrum [47]. It is developed to describe short-term fully developed sea conditions. Its general form is stated as

$$\text{PM sea spectrum:} \rightarrow \quad S_{\eta\eta}(\omega) = \frac{A}{\omega^5} \exp\left(-\frac{B}{\omega^4}\right) \tag{3.31a}$$

in which A and B are constant parameters of the sea condition, which are defined in general as

$$A = \alpha_\eta g^2 \quad \text{and} \quad B = 5\omega_p^4/4 \tag{3.31b}$$

where α_η is the shape parameter obtained from data, with $\alpha_\eta = 0.0081$ for the North Sea condition, g is the gravitational acceleration and ω_p is the peak frequency of the wave spectrum. If the severity of the sea state is represented by the significant wave height H_s only, the peak wave frequency can be calculated in terms of α_η and H_s as written,

$$\text{Peak wave frequency } (H_s \text{ represents severity}): \rightarrow \quad \omega_p = \left(\frac{16}{5}\frac{\alpha_\eta g^2}{H_s^2}\right)^{1/4} \tag{3.31c}$$

Thus, in this case, the shape parameter α_η and the significant wave height H_s determine the sea spectrum fully. If, however, the severity of the sea state is represented by both the significant wave height H_s and the mean zero-crossings period T_z the constant parameter A and the peak wave frequency ω_p are calculated by using Eqs. (3.26d) and (3.29b). It is obtained that,

Fig. 3.7 A typical PM sea spectrum for $H_s = 3.0$ m

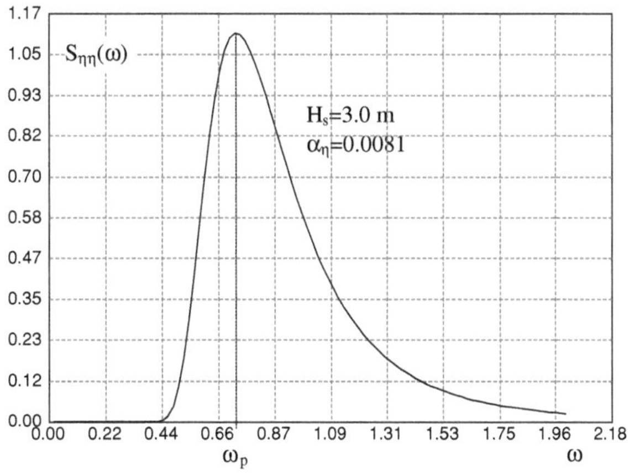

$$H_s \;\&\; T_z \text{ represent severity: } \begin{cases} A = \alpha_\eta g^2 = \dfrac{4\pi^3 H_s^2}{T_z^4} \\ \omega_p = \dfrac{2\pi}{T_z}\left(\dfrac{4}{5\pi}\right)^{1/4} \rightarrow \text{peak wave frequency} \end{cases} \tag{3.31d}$$

A typical PM sea spectrum is shown in Fig. 3.7 for the significant wave height $H_s = 3.0$ m, which corresponds to $\omega_p = 0.725$ rad/sec and $T_z = 6.156$ sec. In the case of sea state representation by H_s only, as it can be seen from Eq. (3.31c), for higher significant wave heights the peaks of the waves move toward the zero frequency with higher peak values and the spectral shapes become narrower. This indicates that for higher waves the wave process becomes more narrow-banded. In the case of the representation of the sea state severity by both H_s and T_z, as it can be seen from Eq. (3.31d), the peak moves toward the zero frequency with increasing T_z values only, and the peak values become higher by increasing H_s values.

3.4.2.2 The JONSWAP Sea Spectrum

The JONSWAP sea spectrum is developed during a joint North Sea wave project for fetch limited seas [48]. It has a similar form to the PM sea spectrum with an enhancement factor around the peak value. It describes developing wind sea in most severe states. The general form of the JONSWAP sea spectrum can be written similarly to the PM spectrum as,

$$\text{JONSWAP sea spectrum}: \rightarrow \quad S_{\eta\eta}(\omega) = \frac{A}{\omega^5}\exp\left(-\frac{B}{\omega^4}\right)\gamma^{g(\omega)} \tag{3.32a}$$

in which A and B are constant parameters of the sea condition as defined in Eq. (3.31b), γ is called as the peak shape parameter [7], $g(\omega)$ is a frequency function defined as.

$$g(\omega) = \exp\left[-\frac{1}{2}\left(\frac{\omega - \omega_p}{\omega_p \sigma}\right)^2\right] \tag{3.32b}$$

where ω_p is the peak frequency of the spectrum, σ is a constant defined as [7, 14, 45, 51],

$$\sigma = \begin{cases} 0.07 & \text{if } \omega \leq \omega_p \\ 0.09 & \text{if } \omega > \omega_p \end{cases} \tag{3.32c}$$

The function $\gamma^{g(\omega)}$ is called as the peak enhancement factor. The peak shape parameter γ is random varying between $(1 \leq \gamma \leq 7)$ with an average value of $(\gamma = 3.3)$ [14, 45]. For the limiting case of $(\gamma = 1)$ the JONSWAP sea spectrum becomes identical to the PM sea spectrum. If the severity of the sea state is represented by the significant wave height H_s only, the peak frequency ω_p can be calculated in terms of α_η, H_s and the peak shape parameter γ. Based on the analysis of numerically generated data the peak frequency ω_p can be approximated from,

$$\text{Peak wave frequency } (H_s \text{ represents severity}): \rightarrow \quad \omega_p = \left(\frac{f_p(\gamma)\,\alpha_\eta g^2}{H_s^2}\right)^{1/4} \tag{3.33a}$$

where $f_p(\gamma)$ is a function of γ obtained as written by,

$$f_p(\gamma) = \frac{3.19714}{(1 - 0.286\ \ln \gamma)} \tag{3.33b}$$

For a given value of γ, the shape parameter α_η and the significant wave height H_s determine the sea spectrum. If the severity of the sea state is represented by both the significant wave height H_s and the mean zero-crossings period T_z, the constant parameter A and the peak frequency ω_p are calculated approximately using the following statements obtained from the analysis of numerically generated data.

$$H_s \ \& \ T_z \text{ represent severity:} \begin{cases} A = \alpha_\eta g^2 = \dfrac{4\pi^3 H_s^2}{T_z^4}(1 - 0.286 \ln \gamma)\, f^2(\gamma) \\ \omega_p = \dfrac{2\pi}{T_z}\sqrt{\dfrac{f(\gamma)}{1.98255}} \end{cases} \tag{3.33c}$$

in which $f(\gamma)$ is a function of the peak shape parameter γ obtained to be,

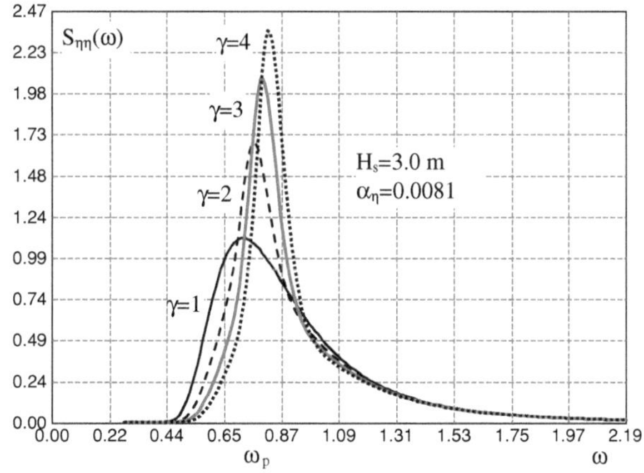

Fig. 3.8 Typical JONSWAP sea spectra for $H_s = 3.0$ m and different γ values

$$f(\gamma) = \frac{1}{(1 - 0.13763587 \ln \gamma)} \tag{3.33d}$$

As it can be verified through Eqs. (3.33a) and (3.33d), for the value of ($\gamma = 1$), the JONSWAP sea spectrum becomes the PM sea spectrum. The peak enhancement factor $\gamma^{g(\omega)}$ controls the spectral shape only around the peak frequency region, and the tail is not affected much. Thus, for sea states with higher frequencies, the PM and JONSWAP sea spectra produce identical results. The JONSWAP sea spectrum and the effect of the peak enhancement factor on the spectral shape are illustrated in Fig. 3.8. As it is seen from Fig. 3.8 and Eq. (3.33a), the peak frequency depends on the peak shape parameter γ. For higher γ values, the peak frequency becomes higher with increasing peak values although the tails remain unchanged. The spectral functions of the random sea presented above are associated with sea waves that all wave energy is concentrated in one wave direction and, therefore, these spectral functions are considered as unidirectional sea spectra [7]. In the reality, there are also individual waves traveling in different directions around a major wave propagation direction, which form altogether a multidirectional wave group with a directional spreading. Its spectral description is presented briefly in the following section.

3.4.2.3 Directional Wave Spectrum

The wind wave as a group, which consists of many individual waves, propagates in the same direction of the wind. But there are also some individual waves developed around a major (main) wave traveling in any direction between ($+\pi/2$) and ($-\pi/2$) as shown in Fig. 3.9 in a horizontal (x–y) plane. This property of wave groups can be taken into account in the sea spectrum formulation by expressing it in terms of both the frequency ω and the direction angle θ [7, 44, 51, 57–62]. The sea spectrum describing multidirectional waves can be expressed as,

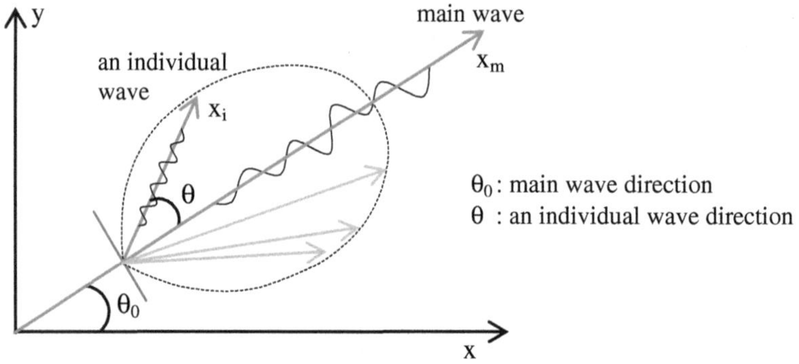

Fig. 3.9 Main and individual waves traveling in the horizontal plane

$$S_{\eta\eta}(\omega,\theta) = D(\omega,\theta)\, S_{\eta\eta}(\omega) \tag{3.34a}$$

in which $D(\omega,\theta)$ is a directionality function dependent on both ω and θ, $S_{\eta\eta}(\omega)$ is the unidirectional sea spectrum explained in previous sections. The directionality function $D(\omega,\theta)$ may be considered as the distribution of wave energy for a given frequency ω over the direction θ, and therefore it is also called as the directional spreading of the sea spectrum. It satisfies the condition,

$$\int_{\theta_{\min}}^{\theta_{\max}} D(\omega,\theta)\, d\theta = 1 \tag{3.34b}$$

Since $D(\omega,\theta)$ is a periodic function of θ, it can be conveniently represented by Fourier series in general. For practical use, different models are available to represent the wave directional spreading [7]. Due to simplicity and efficiency in representing the wave directional spreading, the cosine power model is often used in practice [51] with the approximation of $D(\omega,\theta) \approx D(\theta)$. This is given as,

$$\text{Directional spreading function:} \;\rightarrow D(\theta) = \frac{\Gamma\left(1+\frac{n}{2}\right)}{\sqrt{\pi}\,\Gamma\left(\frac{1+n}{2}\right)} \cos^n\theta \tag{3.34c}$$

where Γ is the Gamma function and θ is the angle of an individual wave (wave spreading angle) measured from the main wave direction as shown in Fig. 3.9. This wave spreading angle is defined between $(-\pi/2 \le \theta \le \pi/2)$. In practical applications, the most commonly used form of $D(\theta)$ is the cosine square function which is written as

$$D(\theta) = \frac{2}{\pi}\cos^2\theta \quad \text{where} \quad \left(-\frac{\pi}{2} \le \theta \le \frac{\pi}{2}\right) \tag{3.34d}$$

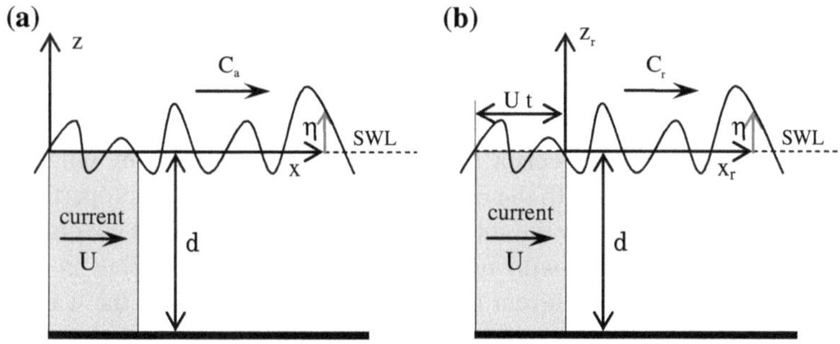

Fig. 3.10 Random wave on a uniform current with reference frames

The main wave direction, θ_0, is the same with the wind direction. This direction can also vary arbitrarily, and therefore it is conveniently assumed as random variable with uniform distribution between $(0 \leq \theta_0 \leq 2\pi)$. If all wave energy is concentrated in one direction (unidirectional sea state) then the structural response is overestimated and the design becomes more conservative.

3.5 Wave–Current Interaction

In ocean and sea environments, currents are usually exist due to tides, ocean circulations, winds and water density differences. In general, currents are complex phenomena as being turbulent and mostly vary in water depth with different directions. For the simplicity in the analysis, they are approximated to a corresponding mean water flow [63] and often considered to be uniform along the water depth. Currents influence waves significantly near continental margins and in shallow seas [64]. The strong ocean currents can strongly affect ocean waves, especially waves propagating and against such currents, which become shorter and steeper by the adverse current. Due to increasing number of offshore structures located in areas of currents, the wave–current interaction will be of great importance in the structural design. In the calculation of response statistical characteristics of structures, the effect of current is reflected on response results in two ways as (a) existence of currents alter the sea spectrum and (b) wave loadings and damping ratios will be dependent on the current velocity which introduces a nonzero mean value response. Wave–current interaction has a profound effect on the modeling of random waves. For deterministic analysis, this interaction has extensively been studied in the literature, see e.g., [64–71]. For stochastic analysis, the problem is rather complicated and, only under certain conditions, it can be solved [71–78]. It is assumed that the random wave is first generated on a quiescent (zero-current) condition and then it crosses from the quiescent area into the current region. Then the spectral density of water elevation for the current region, $S_{\eta\eta}(\omega_a, U)$, is determined in terms of the spectral density of water elevation in the quiescent area, $S_{\eta\eta}(\omega_a)$, by using the *wave action* conservation [79]. It is further assumed that (a) the current is uniform and does not change direction along the water depth and horizontal, (b) it is in-line with the wave propagation, and (c) waves are not refracted by the current. Definition of the wave–current field and reference frames are shown in Fig. 3.10. The fixed reference frame is a stationary frame in which wave properties are denoted by subscript a meaning absolute, and the moving frame is defined as a frame moving with the current velocity U. The wave properties in this reference frame are denoted by the subscript r meaning relative. The governing wave equations will be developed with reference to the current traveling in the same direction of the wave. If the current is in the opposite direction of the wave propagation then the current velocity will be negative. In a stationary reference frame, the waves have an apparent angular frequency, ω_a, and in the frame moving with the current velocity U, the frequency becomes the relative or intrinsic frequency denoted by ω_r. In this moving frame, the velocity potential function, Φ_r and the water elevation η obtained from the linear wave theory can be written as,

$$\Phi_r = \hat{\eta}\frac{g}{\omega_r}\frac{\cosh m(z+d)}{\cosh md}\,\mathrm{e}^{i(\omega_r t - m x_r)} \quad \text{and} \quad \eta = -i\hat{\eta}\,\mathrm{e}^{i(\omega_r t - m x_r)} \tag{3.35a}$$

From Fig. 3.10, the horizontal coordinate x_r in the moving reference frame can be written as,

$$x_r = x - Ut \tag{3.35b}$$

Having introduced Eq. (3.35b) into Eq. (3.35a) it can be obtained that

$$\Phi_r = \hat{\eta}\frac{g}{\omega_r}\frac{\cosh m(z+d)}{\cosh md}\,\mathrm{e}^{i(\omega_a t - m x)} \quad \text{and} \quad \eta = -i\hat{\eta}\,\mathrm{e}^{i(\omega_a t - m x)} \tag{3.35c}$$

in which the absolute frequency ω_a is defined to be,

$$\omega_a = \omega_r + mU \tag{3.36a}$$

The dispersion relation given by Eq. (3.14b) becomes as

$$\omega_r^2 = mg\tanh md \tag{3.36b}$$

The absolute frequency ω_a does not change when the wave travels from one current region to another. On the contrary, the relative frequency ω_r and the wave number m change. They can be solved from Eqs. (3.36a) and (3.36b) as written by, for the deep water condition,

$$\left.\begin{aligned} \omega_r &= \omega_a\frac{2}{(1+\zeta)} \\ m &= \frac{\omega_a^2}{g}\frac{4}{(1+\zeta)^2} \end{aligned}\right\} \quad \text{where} \quad \zeta = \sqrt{1 + \frac{4\omega_a U}{g}} \tag{3.36c}$$

The absolute frequency ω_a can be stated in terms of the relative frequency ω_r as written by, for the deep water condition,

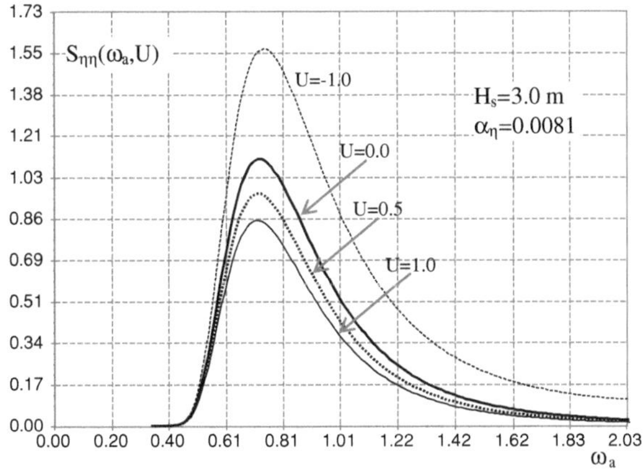

Fig. 3.11 PM sea spectrum in a current area for the current velocities U = (−1.0, 0.0, 0.5, 1.0) m/s

$$\omega_a = \omega_r\left(1 + \frac{\omega_r U}{g}\right) \tag{3.36d}$$

The change in the sea spectrum can be obtained by using the wave action conservation [79] which is defined as,

$$\text{Wave action conservation:} \rightarrow \left(E_a C_{ga}/\omega_a\right) = \left[E\left(U + C_{gr}\right)/\omega_r\right] \tag{3.37a}$$

in which E_a and E are respectively wave energy densities in the quiescent (zero-current) and current regions, and C_{ga} and C_{gr} are the corresponding group velocities which are calculated by using Eq. (3.6c) as written by, for the deep water condition,

$$\text{Group vel.of deep water waves:} \rightarrow C_{ga} = g/(2\omega_a) \text{ and } C_{gr} = g/(2\omega_r) \tag{3.37a}$$

The wave energy densities in the quiescent and current regions, E_a and E, are

$$\text{Wave energy densities:} \rightarrow \begin{cases} E_a = S_{\eta\eta}(\omega_a)\, d\omega_a \\ E = S_{\eta\eta}(\omega_a, U)\, d\omega_a \end{cases} \tag{3.37c}$$

in which $S_{\eta\eta}(\omega_a)$ and $S_{\eta\eta}(\omega_a, U)$ are respectively the sea spectra in the quiescent and current water regions. From Eqs. (3.37a)–(3.37c), the sea spectrum in the current water region, $S_{\eta\eta}(\omega_a, U)$ can be obtained as stated by,

$$\text{Sea spectrum in current area:} \rightarrow S_{\eta\eta}(\omega_a, U) = \frac{4}{\zeta(1+\zeta)^2} S_{\eta\eta}(\omega_a) \tag{3.38a}$$

where ζ is as defined in Eq. (3.36c). Assuming a PM sea spectrum for $S_{\eta\eta}(\omega_a)$ with $H_s = 3.0$ m, the spectral shapes of $S_{\eta\eta}(\omega_a, U)$ are illustrated in Fig. 3.11 for the current velocities of $U = -1.0, 0.0, 0.5$, and 1.0 m/s. As it is seen from this figure, for opposite (negative) current, the peak of the sea spectrum becomes bigger and waves become shorter with increasing heights. As it is also seen from Eq. (3.38a), for a zero value of ζ, the spectrum will be infinite corresponding to a relative frequency of ($\omega_r = 2\omega_a$) and the wave number becomes four times larger than that in the quiescent (zero-current) area. For this special case, the absolute and relative frequencies, ω_a and ω_r, can be determined from Eq. (3.36c) as stated by

$$\zeta = 0 \quad \rightarrow \quad \omega_a = -g/(4U) \quad \text{and} \quad \omega_r = -g/(2U) \tag{3.38b}$$

At this frequency point and beyond, the energy of a particular wave component cannot propagate onto the current and the wave breaks at the current boundary [71, 75, 76]. In the wave breaking region, the sea spectrum is given for the deep water condition [71] by,

$$\text{Equlibrium range spectrum:} \rightarrow S_{\eta\eta\text{ER}}(\omega_a, U) = \frac{\beta g^2}{\omega_r^5}\left(1 + \frac{2U\omega_r}{g}\right)^{-1} \tag{3.38c}$$

in which β is a numerical constant in the range 0.008–0.015 with a mean value of 0.0123 [76]. When $\left(S_{\eta\eta}(\omega_a, U) > S_{\eta\eta\text{ER}}(\omega_a, U)\right)$ then the sea spectrum in the current region is taken to be $\left(S_{\eta\eta}(\omega_a, U) = S_{\eta\eta\text{ER}}(\omega_a, U)\right)$. Having determined the sea spectrum in the current region, transfer functions of the water particle velocities and accelerations will also be determined to calculate wave forces. The velocities in the stationary reference frame can be stated as written by,

$$\left.\begin{aligned} u_a &= U + \left(\omega_r \frac{\cosh m(z+d)}{\sinh md} \mathrm{e}^{-imx}\right)\eta_0 \\ w_a &= i\,\omega_r \frac{\sinh m(z+d)}{\sinh md} \mathrm{e}^{-imx}\eta_0 \end{aligned}\right\} \rightarrow \eta_0 = -i\,\hat{\eta}\;\mathrm{e}^{i\omega_a t} \tag{3.39a}$$

and the accelerations in the stationary reference frame will be calculated from the total time derivative of the corresponding velocities. They are obtained as,

$$\begin{aligned} \dot{u} &= i\omega_r^2 \frac{\cosh m(z+d)}{\sinh md} \mathrm{e}^{-imx}\eta_0 \\ \dot{w} &= -\omega_r^2 \frac{\sinh m(z+d)}{\sinh md} \mathrm{e}^{-imx}\eta_0 \end{aligned} \tag{3.39b}$$

From Eqs. (3.39a) and (3.39b) the water particle velocities and accelerations can be stated in terms of the transfer functions as written by,

$$\begin{aligned} u_a &= U + H_{u\eta}(\omega)\eta_0 \\ w_a &= H_{w\eta}(\omega)\eta_0 \end{aligned} \quad \text{and} \quad \begin{aligned} \dot{u}_a &= H_{\dot{u}\eta}(\omega)\eta_0 \\ \dot{w}_a &= H_{\dot{w}\eta}(\omega)\eta_0 \end{aligned} \tag{3.39c}$$

where the transfer functions for the deep water condition are obtained to be

$$\begin{aligned} H_{u\eta}(\omega) &= \omega_r\,\mathrm{e}^{mz}\,\mathrm{e}^{-i\,mx} \\ H_{w\eta}(\omega) &= i\,\omega_r\,\mathrm{e}^{mz}\,\mathrm{e}^{-i\,mx} \end{aligned} \quad \text{and} \quad \begin{aligned} H_{\dot{u}\eta}(\omega) &= i\,\omega_r^2\,\mathrm{e}^{mz}\,\mathrm{e}^{-i\,mx} \\ H_{\dot{w}\eta}(\omega) &= -\omega_r^2\,\mathrm{e}^{mz}\,\mathrm{e}^{-i\,mx} \end{aligned} \tag{3.39d}$$

As it is seen from Eqs. (3.39c) and (3.39d) the presence of the current modifies the water particle velocities and accelerations. The zero-mean property of the

Table 3.4 Table of an example scatter diagram (number of occurrences of sea states)

H_s (m)	T_z (second)								Sum
	1–3	3–6	6–9	9–12	12–15	15–18	18–21	21–24	
0–1	12	358	578	162	53	13	2	0	1178
1–2	0	756	2485	673	277	98	14	1	4304
2–3	0	83	1887	893	233	90	12	0	3198
3–4	0	3	646	711	170	34	0	0	1564
4–5	0	0	117	418	146	16	1	0	698
5–6	0	0	12	196	89	7	1	0	305
6–7	0	0	0	59	56	6	0	0	121
7–8	0	0	0	9	25	5	0	0	39
8–9	0	0	0	2	16	4	0	0	22
9–10	0	0	0	0	10	2	0	0	12
10–11	0	0	0	0	0	1	0	0	1
11–12	0	0	0	0	0	1	0	0	1
Sum	12	1,200	5,725	3,123	1,075	277	30	1	11,443

horizontal velocity in the zero-current area is destroyed and the presence of the current makes the velocity as a nonzero mean value process. Consequently, wave forces on offshore structures are also affected and their mean values will not be zero any more due to the velocity dependent drag terms.

3.6 Probabilistic Description of Sea States in the Long Term

Requirements of performances of offshore structures in the long term, such as fatigue damage accumulation and the structural behavior under an extreme sea condition, necessitate information about long-term wave statistics. The long-term description of a wave environment is defined by the statistical properties of the severity of sea states, which is characterized by the significant wave height H_s or both H_s and the mean zero-crossings period T_z. These parameters are assumed to be constants in the short-term sea states as explained in Sect. 3.4, and they are probabilistic in the long term. Their statistical information is obtained from measured, or observed, wave data which are accumulated over a long period [7, 44, 80, 81]. The number of wave observations defines the reliability of the data, and the larger the data accumulations the more reliable the data [82]. The data are tabulated as the number of occurrences of specific short-term sea states in the axes of the significant wave height H_s and mean zero-crossings period T_z from which a wave scatter diagram is constructed [51, 83, 84]. The wave scatter diagram is a 2D histogram of the significant wave height and mean zero-crossings wave period for all sea states. Different wave scatter diagrams can be obtained to be used in different geographic locations [51, 85–87] as depending on the availability of wave

Table 3.5 Parameters of the fitted Weibull distributions of H_s in Eq. (3.41)

Station	A (m)	B (m)	C	Location
India	0.80	2.70	1.22	Atlantic Ocean
Juliett	0.90	2.70	1.24	
Sevenstones	0.60	1.67	1.21	
Morecambe Bay	0.00	0.78	1.05	Irish Sea
Mersey Bar	0.00	0.69	1.01	
Varne	0.00	1.05	1.30	North Sea
Smith's Knoll	0.08	0.89	1.28	

data. The table of an example wave scatter diagram is as shown in Table 3.4 where the marginal occurrences of sea states for H_s and T_z are given respectively in the column and row indicated by 'Sum'. The probability distributions of H_s and T_z can be easily calculated from the scatter diagram to be used in the long-term behavior of structural responses.

If the severity of the sea state is represented by H_s only, then the probability mass of H_s can be calculated from the numbers of marginal occurrences of the sea states as written by,

$$p_{H_s}(h_i) = N_i/N \tag{3.40}$$

where N_i is the number of marginal occurrences of sea states in the interval Δh_i of H_s and N is the total number of occurrences. For example, in Table 3.4, for the interval of $H_s = 4$–5, $N_i = 698$, $N = 11443$, and $p_{H_s}(h_i) = 0.061$. For a continuous probability distribution, various probability functions have been proposed [7]. The general trend is to use the log-normal probability distribution to represent the greater part of significant wave height data. For large significant wave heights, the data diverge from the log-normal distribution. Instead, the Weibull probability distribution is well fitted to represent large significant wave heights [51, 86, 88]. For the extremes of the significant wave height, or maximum individual wave height, a Gumbel distribution can be assumed [51]. Based on wave data for a 1-year period, a three-parameter Weibull distribution function is suggested [86] to represent the long-term marginal probability distribution of the significant wave height H_s. It is written as,

$$F_{H_s}(h) = 1 - \exp\left[-\left(\frac{h-A}{B}\right)^C\right] \tag{3.41}$$

The parameters A, B, and C in Eq. (3.41) are to be calculated from wave data as depending on locations of data collection. Their values reported in [86] are presented in Table 3.5.

Joint probability models may often be used to represent long-term variations of wave characteristics. These models can be discrete like those in the form of wave scatter diagrams, or continues obtained by fitting distributions to available wave data from measurements or observations. In both cases, the severity of sea states is represented by H_s and T_z together. In the case of discrete joint probability models, the probability mass of the joint distribution of H_s and T_z is calculated from the numbers of occurrences of sea states as written by

$$p_{H_sT_z}(h_i, t_j) = N_{ij}/N \quad (3.42)$$

where N_{ij} is the number of occurrences of sea states in the intervals Δh_i for H_s and Δt_j for T_z, and N is the total number of occurrences. For example, in Table 3.4, for the intervals of $H_s = 4$–5 and $T_z = 9$–12, $N_{ij} = 418$, $N = 11443$, the probability mass is calculated to be $p_{H_sT_z}(h_i, t_i) = 0.0365$. For a continuous joint probability distribution of H_s and T_z, the joint density function is defined in terms of a marginal distribution and a series of conditional density functions that obtained by fitting parameters to data. For a specific location, such a joint distribution is usually approximated by a marginal distribution for the significant wave height H_s, which is typically a three parameters Weibull distribution, and conditional log-normal distributions for the mean zero-crossings period T_z given the significant wave height H_s [84, 89]. Thus, the joint probability density function of H_s and T_z can be stated as,

$$f_{H_sT_z}(h, t) = f_{H_s}(h) f_{T_z}(t|h - H_s) \quad (3.43a)$$

where $f_{H_s}(h)$ is the marginal probability density function of H_s for which the distribution is a three parameters Weibull function given in Eq. (3.41), and $f_{T_z}(t|h = H_s)$ is a conditional probability density function of T_z which fits a log-normal probability density function written as,

$$f_{T_z}(t|h = H_s) = \frac{1}{\sigma t\sqrt{2\pi}} \exp\left[-\frac{1}{2}\left(\frac{\ln t - \mu}{\sigma}\right)^2\right] \quad (3.43b)$$

where σ and μ are respectively the standard deviation and mean value of $(\ln T_z)$ which are estimated from actual data as depending on H_s [84, 89]. Having determined the long-term distribution of sea states, structural response behavior in the long-term like probabilistic fatigue damage accumulation can be estimated.

3.7 Morison's Equation and Wave Forces on Structural Members

In general, wave forces are calculated from the water pressure under the wave propagation which can be calculated from the Bernoulli equation given by Eq. (3.3b). The function $f(t)$ in Eq. (3.3b) is given by Eq. (3.3c) as the dynamic boundary condition at the free surface where the pressure is zero. Since velocities and accelerations are both functions of time and space, and $f(t)$ is a function of time alone it must be equal zero [90]. Thus, the total water pressure at any point can be stated from Eq. (3.3b) as,

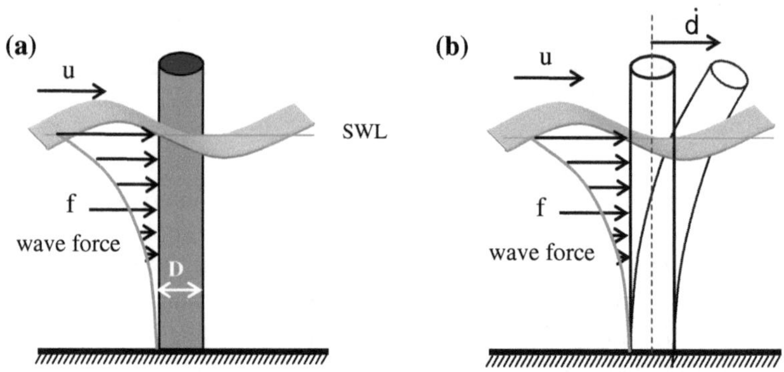

Fig. 3.12 Rigid and flexible cylinders fixed at the bottom under wave actions (**a**) Rigid cylinder (**b**) Flexible cylinder

$$\text{Total pressure:} \rightarrow p_{\text{tot}} = -\rho_w \left(gz + \int \frac{\partial V}{\partial t} dS + \frac{V^2}{2} \right) \tag{3.44a}$$

where the first term (gz) in the bracket is the hydrostatic pressure and the others are the dynamic pressure produced by waves. The dynamic pressure p at any point is a function of water particle acceleration and velocity square. It can be written in terms of the velocity potential function Φ as,

$$\text{Dynamic pressure :} \rightarrow p = -\left(\rho_w \frac{\partial \Phi}{\partial t} + \frac{\rho_w}{2} \left[\left(\frac{\partial \Phi}{\partial x} \right)^2 + \left(\frac{\partial \Phi}{\partial z} \right)^2 \right] \right) \tag{3.44b}$$

In the presence of an object in water, if the velocity potential function Φ is known then the dynamic pressure p on the surface of the object can be calculated using Eq. (3.44b). But, due to complexity of wave interactions with the object in water, calculation of the velocity potential function Φ is a difficult task analytically. In the analysis of slender offshore structures, the wave loads are calculated using the Morison's equation [35] given for a rigid cylindrical pile fixed at the bottom as shown in Fig. 3.12a. The Morison's equation is defined as a distributed wave force per unit length of the pile and is normal (perpendicular) to the pile. It is written as,

$$\text{Morison's equation :} \rightarrow f = C_D |u| u + C_M \dot{u} \tag{3.45a}$$

in which the first term is the drag force contribution and the second one is the inertia force contribution, u and $\dot{u}$ are respectively the velocity and acceleration components of a water particle which are normal to the pile, $|.|$ denotes an absolute value, C_D and C_M are respectively drag and inertia force constants defined as,

$$\text{Drag and inertia force constants:} \rightarrow \begin{cases} C_D = D\rho_w c_d/2 \\ C_M = \pi D^2 \rho_w c_m/4 \end{cases} \tag{3.45b}$$

in which D is the diameter of the pile, ρ_w is the density of water, c_d and c_m are respectively the drag and inertia force coefficients of the Morison's equation. These coefficients are functions of the Reynolds number (R_e), Keulegan–Carpenter number (K_c) and the roughness of the cylinder [14, 22, 23, 51, 91, 92]. API recommends the following drag and inertia values for circular cylinders [93, 94] for large Keulegan–Carpenter number ($K_c > 30$):

- for smooth cylinders: $c_d = 0.65$ and $c_m = 1.6$
- for rough cylinders: $c_d = 1.05$ and $c_m = 1.2$.

For ($K_c < 30$) these values are modified by a wake amplification factor. The DNV rules [51] accept the same values for the smooth and rough cylinders, except the inertia force coefficient c_m. The DNV rules define higher c_m values than the API recommended practices. For ($K_c < 3$), c_m can be assumed to be independent of K_c and equal to the theoretical value of ($c_m = 2.0$) for both smooth and rough cylinders. In the case of ($K_c > 3$), it is calculated from [51] as,

$$c_m = \max\begin{cases} 1.6, \; [2 - 0.044(K_c - 3)] \rightarrow \text{smooth cylinder} \\ 1.2, \; [2 - 0.044(K_c - 3)] \rightarrow \text{rough cylinder} \end{cases} \tag{3.45c}$$

For low Keulegan–Carpenter numbers, a detailed discussion on the drag and inertia force coefficients is presented in [95]. It is reported [96] that, for ($K_c < 10$), the inertia force becomes dominant, for ($10 < K_c < 20$), both inertia and drag force components are significant and for ($K_c > 20$), the drag force becomes dominant. The diameter of the cylinder, D, is also an influential factor in the wave force regime. The Morison's equation is appropriate for slender members. However, for large members, it can also be applied with modification of the inertia force coefficient c_m. This is done by using wave diffraction theory to calculate dynamic pressure, which is explained in the following section.

3.7.1 Wave Forces on Large Structural Members

The Morison's equation is valid for slender cylindrical members. For members with large diameters, or short period waves, the Morison's equation needs to be modified. For large members, the inertia force becomes dominant. It is calculated by using wave diffraction theory. Based on the classical work of linear diffraction theory for water waves [97] linear wave forces on a large cylindrical pile immersed in the ocean are calculated [98]. Nonlinear wave forces are also calculated using the wave diffraction theory [99–101]. Since linear forces are used in the spectral analysis, the calculation of linear wave forces on a large cylindrical

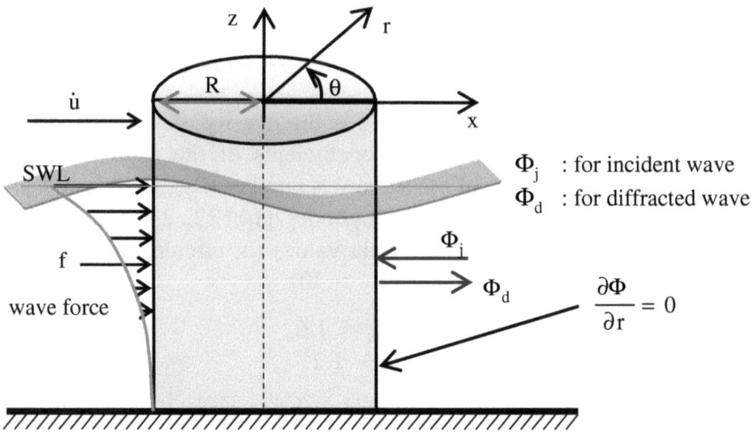

Fig. 3.13 Wave force on a large cylindrical object and velocity potential functions

object in the ocean is outlined briefly in this section by using the wave diffraction theory.

In the wave diffraction theory, an incident wave traveling in a radial direction is diffracted at the surface of the cylindrical body traveling in the opposite direction of the incident wave. The velocity potential functions of the incident and diffracted waves are respectively denoted by Φ_j and Φ_d. The total velocity potential function is obtained from the superposition of the potential functions Φ_j and Φ_d as,

$$\text{Total velocity potential function:} \rightarrow \Phi = \Phi_j + \Phi_d \tag{3.46a}$$

From the boundary condition on the cylindrical surface, $(\partial\Phi/\partial r = 0)$ where r is the radial coordinate as shown in Fig. 3.13, the following condition can be written.

$$\text{Boundary condition at the surface:} \rightarrow \frac{\partial\Phi_j}{\partial r} + \frac{\partial\Phi_d}{\partial r} = 0 \quad \text{at} \quad (r = R) \tag{3.46b}$$

The velocity potential function of an incident wave in the horizontal x direction is given in Eq. (3.15). By using the coordinate transformation written as, see Fig. 3.13,

$$x = r\cos\theta \tag{3.47}$$

the velocity potential function of an incident wave can be stated in terms of the radial coordinate as written by, from Eq. (3.15),

$$\Phi_j = A\mathrm{e}^{i\omega t}\mathrm{e}^{-imr\cos\theta} \quad \text{where} \quad A = \hat{\eta}\frac{g}{\omega}\frac{\cosh m(z+d)}{\cosh md} \tag{3.48a}$$

Using the expansion of the function $(\exp(-imr\cos\theta))$ in terms of Bessel functions [102], the velocity potential function Φ_j can be written as,

$$\text{Incident wave:} \rightarrow \Phi_j = Ae^{i\omega t} \sum_{n=-\infty}^{\infty} (-i)^n J_n(mr)\, e^{-in\theta} \tag{3.48b}$$

in which $J_n(.)$ is the Bessel function of the first kind. In a similar way, the velocity potential function of the diffracted wave Φ_d can be stated in terms of the Bessel functions of the third kind [22] as written by,

$$\text{Diffracted wave:} \rightarrow \Phi_d = Ae^{i\omega t} \sum_{n=-\infty}^{\infty} B_n H_n^{(2)}(mr)\, e^{-in\theta} \tag{3.48c}$$

in which $H_n^{(2)}(.)$ is the Bessel function of the third kind (Hankel function) and B_n is a constant which is determined from the boundary condition on the surface of the cylinder. Having used the boundary condition given by Eq. (3.46b), the total velocity potential function can be obtained from Eq. (3.46a) as written by,

$$\Phi = Ae^{i\omega t} \sum_{n=-\infty}^{\infty} (-i)^n \left(J_n(mr) - \frac{J_n'(mR)}{H_n^{(2)'}(mR)} H_n^{(2)}(mr) \right) e^{-in\theta} \tag{3.48d}$$

where R is the radius of the cylinder, $J_n'(.)$ and $H_n^{(2)'}(.)$ are the derivatives of the Bessel functions of the first and third kind. The dynamic pressure on the cylinder is calculated from Eq. (3.44b) in which higher order (nonlinear) terms are neglected, i.e., $(p = -\rho_w\, \partial\Phi/\partial t)$. Having taken into account the following relation of Bessel functions [14, 99, 102],

$$\begin{aligned} &J_n(x)\, H_n^{(2)'}(x) - J_n'(x)\, H_n^{(2)}(x) = -i\frac{2}{\pi x} \\ &H_{-n}^{(2)}(x) = (-1)^n H_n^{(2)}(x) \end{aligned} \tag{3.49}$$

the dynamic pressure on the surface of the cylinder can be obtained as written by,

$$p = -\frac{2\rho_w \omega A}{\pi m R} \left[\frac{1}{H_0^{(2)'}(mR)} + \sum_{n=1}^{\infty} (-i)^n \frac{2}{H_n^{(2)'}(mR)} \cos n\theta \right] e^{i\omega t} \tag{3.50a}$$

The force per unit length acting on the cylinder in the wave traveling direction, i.e., horizontal x direction, is calculated from,

$$f = -\int_0^{2\pi} p \cos\theta\, R d\theta \tag{3.50b}$$

Having introduced the dynamic pressure from Eq. (3.50a) into Eq. (3.50b) and carried out the integration the force, f, per unit length is obtained as written by,

$$f = -i\frac{4\rho_w \omega A}{m} \frac{1}{H_1^{(2)'}(mR)} e^{i\omega t} \tag{3.51a}$$

Using Eq. (3.16a) the horizontal acceleration of water particles at the center of the cylinder ($x = 0$) can be stated as,

$$\dot{u} = A\,\omega\, m\, e^{i\omega t} \tag{3.51b}$$

in which A is as defined in Eq. (3.48a). Having written the force given by Eq. (3.51b) in terms of the acceleration, and comparing it with the inertia force term of the Morison's equation given by Eq. (3.45a), the inertia force coefficient c_m can be stated as,

$$\text{Inertia force coefficient:} \quad \rightarrow \quad c_m = 2 f_m(x) \tag{3.52a}$$

where $f_m(x)$ is a complex reduction function defined as,

$$f_m(x) = \frac{2}{\pi}\frac{1}{x^2}\frac{e^{-i\alpha}}{\left|H_1^{(2)'}(x)\right|} \quad \text{with} \quad (x = mR) \tag{3.52b}$$

in which the definitions are,

$$\left|H_1^{(2)'}(x)\right| = \sqrt{J_1'^2(x) + Y_1'^2(x)} \quad \text{and} \quad \tan\alpha = \frac{J_1'(x)}{Y_1'(x)} \tag{3.52c}$$

The derivatives of the Bessel functions can be calculated from the following polynomial approximations [102].

$$\text{for } (x \le 3) :\rightarrow \begin{cases} J_1' = 0.5 + \sum_{j=1}^{6} a_j \left(\frac{x}{3}\right)^{2j} \\ Y_1' = \frac{2}{\pi}\ln\left(\frac{x}{2}\right) J_1' + \frac{b_7}{x^2} + \sum_{j=0}^{6} b_j \left(\frac{x}{3}\right)^{2j} \end{cases} \tag{3.53a}$$

where the constants are given in Table 3.6. For large values of x, they are calculated from the following statements [102].

$$\text{for } (x \ge 3) :\rightarrow \begin{cases} J_1' = \frac{1}{\sqrt{x}}\left(f_0(x)\cos\theta_0(x) - \frac{f_1(x)}{x}\cos\theta_1(x)\right) \\ Y_1' = \frac{1}{\sqrt{x}}\left(f_0(x)\sin\theta_0(x) - \frac{f_1(x)}{x}\sin\theta_1(x)\right) \end{cases} \tag{3.53b}$$

where the functions $f_0(x)$, $\theta_0(x)$, $f_1(x)$, and $\theta_1(x)$ are calculated from the following polynomial approximations [102].

$$f_0(x) = \sum_{j=0}^{6} a_j \left(\frac{3}{x}\right)^j \quad \text{and} \quad \theta_0(x) = x + \sum_{j=0}^{6} b_j \left(\frac{3}{x}\right)^j \tag{3.53c}$$

Table 3.6 Constants of derivatives of Bessel functions for $(x \leq 3)$, Eq. (3.53a)

Constants (i)	J_1' (a_i)	Y_1' (b_i)
0		0.3428881211
1	−1.68749985	0.3646746711
2	1.05468507	−0.5972279844
3	−0.27684371	0.2182906033
4	0.04001471	−0.0381568511
5	−0.00362679	0.0039694600
6	0.00019891	−0.0002484600
7		0.6366198000

Table 3.7 Constants of the functions $f_0(x)$, $\theta_0(x)$, $f_1(x)$ and $\theta_1(x)$ for $(x \geq 3)$, Eq. (3.53b–d)

Constants (i)	$f_0(x)$ (a_i)	$\theta_0(x)$ (b_i)	$f_1(x)$ (c_i)	$\theta_1(x)$ (d_i)
0	0.79788456	−0.78539816	0.79788456	−2.35619449
1	−0.00000077	−0.04166397	0.00000156	0.12499612
2	−0.00552740	−0.00003954	0.01659667	0.00005650
3	0.00009512	0.00262573	0.00017105	−0.00637879
4	0.00137237	-0.00054125	−0.00249511	0.00074348
5	−0.00072805	−0.00029333	0.00113653	0.00079824
6	0.00014476	0.00013558	−0.00020033	−0.00029166

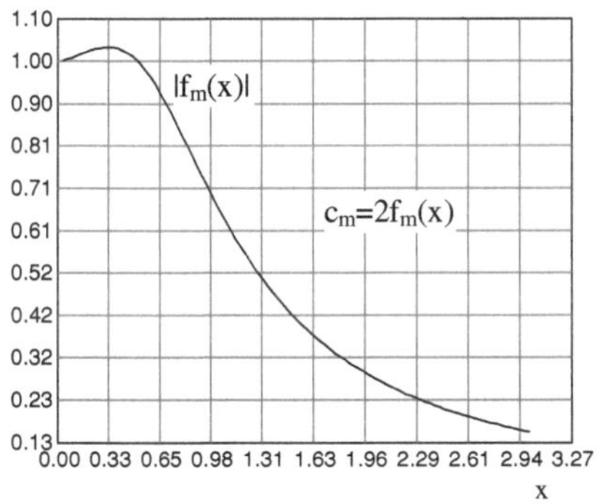

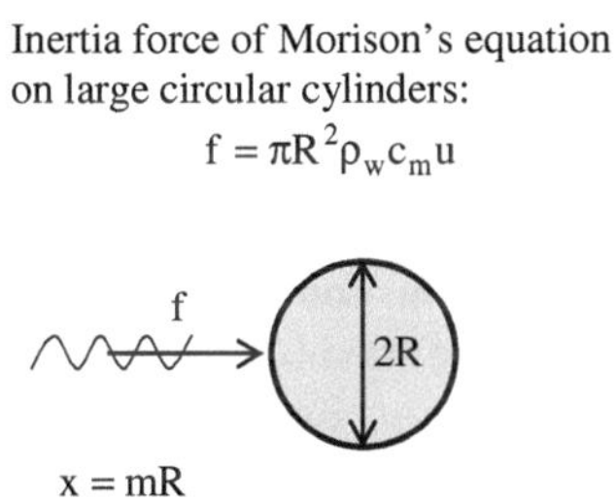

Fig. 3.14 Modulus of the reduction function of the inertia force coefficient cm, $|f_m(x)|$

$$f_1(x) = \sum_{j=0}^{6} c_j \left(\frac{3}{x}\right)^j \quad \text{and} \quad \theta_1(x) = x + \sum_{j=0}^{6} d_j \left(\frac{3}{x}\right)^j \tag{3.53d}$$

in which the constants are given in Table 3.7. The absolute value (modulus) of the reduction function $f_m(x)$ is shown in Fig. 3.14.

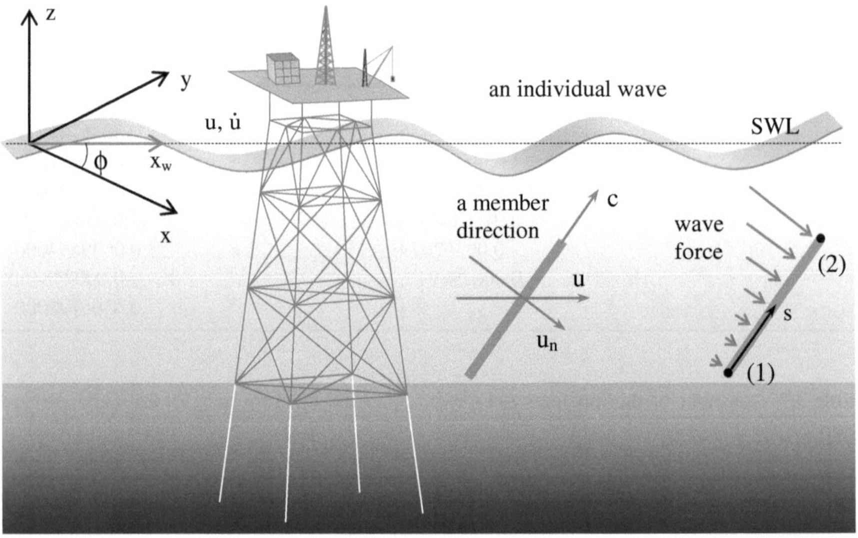

Fig. 3.15 A schematic offshore jacket structure, normal water velocity, and wave force on an inclined member

3.7.2 Wave Forces on Inclined Structural Members

The wave force presented in aforementioned sections is acting on a vertical cylindrical member in the sea and it is in the perpendicular direction to the member axis. The offshore steel structures are complex having many members in any direction (not necessarily vertical) as shown schematically in Fig. 3.15 so that the wave forces acting on these structural members need to be defined properly. The Morison's equation can be still used for such members by taking the velocity and acceleration components of water particles in the perpendicular (normal) direction to the member axes. These water velocity and acceleration components are denoted by u_n and $\dot{u}_n$. The wave forces on inclined members are also perpendicular to the member axes in the same direction of normal water velocity and acceleration components. In order to calculate member consistent wave forces, these distributed wave forces will be transformed to the member local coordinates. For an inclined member with the cosine direction vector $\vec{c}$ in the global coordinates (x, y, z) shown as in Fig. 3.15, the normal water velocity vector $\vec{u}_n$ which is defined in the ($\vec{u}$ and $\vec{c}$) plane can be written from Fig. 3.15 as,

$$\vec{u}_n = \vec{c} \times \vec{u} \times \vec{c} \tag{3.54}$$

in which $\vec{u}$ is the water velocity vector in the global (x, y, z) coordinates. This vector product can be stated in the vectorial form as,

$$\{u_n\} = [T_n]\,\{u\} \tag{3.55a}$$

in which the transformation matrix $[T_n]$ can be obtained as written by,

$$[T_n] = \begin{bmatrix} (1-c_x^2) & -c_x c_y & -c_x c_z \\ -c_x c_y & (1-c_y^2) & -c_y c_z \\ -c_x c_z & -c_y c_z & (1-c_z^2) \end{bmatrix} \tag{3.55b}$$

in which c_x, c_y, and c_z are the cosine directions of the member axis X_L (components of the vector $\vec{c}$ in the global coordinates) as defined in Eq. (1.93). The components of the water velocity vector $\vec{u}$ in the global coordinates are written from Fig. 3.15 as,

$$\left.\begin{aligned} u_x &= u\ \cos\phi \\ u_y &= u\ \sin\phi \\ u_z &= w \end{aligned}\right\} \rightarrow \{u\} = \begin{Bmatrix} u\ \cos\phi \\ u\ \sin\phi \\ w \end{Bmatrix} \tag{3.55c}$$

in which u is the horizontal water velocity in the wave propagation direction, ϕ is the angle of the wave propagation direction measured from the global x axis as shown in Fig. 3.15, and w is the water velocity in the vertical direction, which are given in Eq. (3.16a). The water acceleration vectors can be obtained from the time derivatives of Eqs. (3.55a) and (3.55c) as written by,

$$\{\dot{u}_n\} = [T_n]\,\{\dot{u}\} \quad \text{and} \quad \{\dot{u}\} = \begin{Bmatrix} \dot{u}\ \cos\phi \\ \dot{u}\ \sin\phi \\ \dot{w} \end{Bmatrix} \tag{3.55d}$$

Having determined normal water velocity and acceleration vectors, the distributed wave force vector on an inclined member can be calculated from the Morison's equation in the global coordinates as written by,

$$\{f_n\}_G = [T_n]\ (C_D|u_n|\{u\} + C_M\{\dot{u}\}) \tag{3.56a}$$

in which the drag and inertia force constants C_D and C_M are defined in Eq. (3.45a), and $|u_n|$ is the modulus (absolute value) of the normal water velocity u_n which is calculated from,

$$|u_n| = \sqrt{\{u_n^*\}^T\{u_n\}} \tag{3.56b}$$

where $\{u_n^*\}$ denotes the complex conjugate of the velocity vector $\{u_n\}$ and $\{.\}^T$ denotes the transpose of a vector. By using the following property of the transformation matrix $[T_n]$,

$$[T_n] = [T_n]^T[T_n] \tag{3.56c}$$

the modulus $|u_n|$ can be calculated in terms of the water velocity vector $\{u\}$ as written by,

$$|u_n| = \sqrt{\{u^*\}^T[T_n]\{u\}} \tag{3.56d}$$

As it is seen from Eq. (3.56a) the drag force term of the Morison's equation is nonlinear. For the spectral analysis, the Morison's equation will be linearized from which $|u_n|$ will be replaced by a linearization parameter denoted by A. This parameter is dependent on the variance of the water velocity which is a function of the vertical coordinate Z. Since the linearization will be explained in Sect. 3.8, it is considered here that the linearization parameter A is a function of Z. By using this assumption the calculation of consistent member wave forces is explained in the following section.

3.7.3 Calculation of Consistent Wave Forces of Members

Formulation of the consistent load vector of a structural member has been explained in Chap. 1 as presented by Eq. (1.74). In this section, the calculation of the consistent member wave-forces of offshore jacket structures is explained for the deep water condition. The calculation is carried out in the frequency domain. By using Eqs. (3.24a) and (3.24b) for the deep water condition, the water velocity and acceleration vectors $\{u\}$ and $\{\dot{u}\}$ can be stated as,

$$\left.\begin{aligned}\{u\} &= \omega\, \mathrm{e}^{m(z-i x_w)}\{\phi\}\,\eta \\ \{\dot{u}\} &= \omega\, \mathrm{e}^{m(z-i x_w)}\{\phi\}\,\dot{\eta}\end{aligned}\right\} \quad \rightarrow \quad \{\phi\} = \left\{\begin{array}{c}\cos\phi \\ \sin\phi \\ i\end{array}\right\} \tag{3.57a}$$

where x_w is the horizontal coordinate in the wave propagation direction. By using η from Eq. (3.18b) into Eq. (3.57a) the water acceleration vector $\{\dot{u}\}$ can be expressed as,

$$\{\dot{u}\} = i\,\omega\,\{u\} \tag{3.57b}$$

Having used $\{\dot{u}\}$ from Eq. (3.57b) and replaced $|u_n|$ by the linearization parameter A in Eq. (3.56a), the Morison's equation can be written in the global coordinates as,

$$\{f_n\}_G = R\,[T_n]\,\{u\} \quad \rightarrow \quad R = (C_D A + i\,\omega\,C_M) \tag{3.58a}$$

in which R is a scalar complex function of the vertical coordinate Z and the frequency ω. By using the velocity vector $\{u\}$ from Eq. (3.57a) into Eq. (3.58a) the Morison's equation can be stated in terms of the water elevation η as written by,

$$\{f_n\}_G = \omega\, \mathrm{e}^{m(z-i x_w)}\,R\,[T_n]\,\{\phi\}\,\eta \quad \rightarrow \quad R = (C_D A + i\,\omega\,C_M) \tag{3.58b}$$

which is defined in the global coordinates. In order to calculate the member consistent forces using Eq. (1.74), this wave force vector must be transformed to the member local coordinates. It is written as,

$$\{f_n\}_L = [t]\{f_n\}_G \quad \rightarrow \quad \{f_n\}_L = \omega\, \mathrm{e}^{m(z-i x_w)}\,R\,[t][T_n]\,\{\phi\}\,\eta \tag{3.58c}$$

in which the transformation matrix $[t]$ has been given in Eq. (1.91). As a special case, for tubular members which are used in offshore steel structures, this transformation matrix will be equivalent to the transformation matrix $[t_2]$ given by Eq. (1.96a). Thus, it is written as,

$$[t] = \begin{bmatrix} c_x & c_y & c_z \\ -\dfrac{c_x c_y}{\sqrt{1-c_y^2}} & \sqrt{1-c_y^2} & -\dfrac{c_y c_z}{\sqrt{1-c_y^2}} \\ -\dfrac{c_z}{\sqrt{1-c_y^2}} & 0 & \dfrac{c_x}{\sqrt{1-c_y^2}} \end{bmatrix} \tag{3.58d}$$

By using Eq. (1.74) the consistent wave force vector of a member in the member local coordinates is calculated from the integration,

$$\{p_w\}_L = \int_0^{\ell} [N_u]^T \{f_n\}_L \, ds \tag{3.59a}$$

in which $[N_u]$ is the shape function vector of the member, ℓ is the member length and s denotes the variable of the member axial coordinate as shown in Fig. (3.15). It can be easily verified from the matrix multiplication in Eq. (3.58c) that the axial force component is zero, and therefore the corresponding terms in the shape function matrix are taken to be zero. For the calculation of consistent forces, the Euler–Bernoulli beam theory is used and the shape function matrix can be written from Eq. (1.50a) as,

$$[N_u] = \begin{bmatrix} 0 & 0 & 0 & 0 & 0 & 0 & 0 & 0 & 0 & 0 & 0 & 0 \\ 0 & N_1 & 0 & 0 & 0 & N_3 & 0 & N_2 & 0 & 0 & 0 & N_4 \\ 0 & 0 & N_1 & 0 & -N_3 & 0 & 0 & 0 & N_2 & 0 & -N_4 & 0 \end{bmatrix} \tag{3.59b}$$

in which N_1, N_2, N_3, and N_4 are the shape functions defined as,

$$\left.\begin{aligned} N_1 &= 1 - 3\xi^2 + 2\xi^3 \\ N_2 &= 3\xi^2 - 2\xi^3 \end{aligned}\right\} \text{and} \left\{\begin{aligned} N_3 &= \ell(\xi - 2\xi^2 + \xi^3) \\ N_4 &= \ell(-\xi^2 + \xi^3) \end{aligned}\right. \quad \text{where} \left(\xi = \frac{s}{\ell}\right) \tag{3.59c}$$

The transpose of the shape function matrix $[N_u]$ can be stated in terms of sub-matrices as written by,

$$[N_u]^T = \begin{bmatrix} N_1[I_1] \\ N_3[I_2] \\ N_2[I_1] \\ N_4[I_2] \end{bmatrix} \rightarrow [I_1] = \begin{bmatrix} 0 & 0 & 0 \\ 0 & 1 & 0 \\ 0 & 0 & 1 \end{bmatrix} \text{ and } [I_2] = \begin{bmatrix} 0 & 0 & 0 \\ 0 & 0 & -1 \\ 0 & 1 & 0 \end{bmatrix} \tag{3.59d}$$

Having substituted $[N_u]^T$ from Eq. (3.59d) and $\{f_n\}_L$ from Eq. (3.58c) into Eq. (3.59a) the consistent wave force vector $\{p_w\}_L$ in the member local coordinates can be written as,

$$\{p_w\}_L = \int_0^{\ell} \left(\begin{Bmatrix} N_1[I_1][t] \\ N_3[I_2][t] \\ N_2[I_1][t] \\ N_4[I_2][t] \end{Bmatrix} \{f_n\}_G \right) ds \tag{3.59e}$$

This member consistent force vector must be transformed to the global coordinates to calculate the system load vector. This transformation is given in Eq. (1.102) as written by,

$$\{p_w\}_G = [T]^T \{p_w\}_L \quad \rightarrow \quad [T] = \begin{bmatrix} [t] & 0 & 0 & 0 \\ 0 & [t] & 0 & 0 \\ 0 & 0 & [t] & 0 \\ 0 & 0 & 0 & [t] \end{bmatrix} \tag{3.60a}$$

in which $[t]$ is given by Eq. (3.58d). Having carried out this transformation the member consistent force vector in the global coordinates can be stated as,

$$\{p_w\}_G = \int_0^{\ell} \left(\begin{Bmatrix} N_1[t]^T[I_1][t] \\ N_3[t]^T[I_2][t] \\ N_2[t]^T[I_1][t] \\ N_4[t]^T[I_2][t] \end{Bmatrix} \{f_n\}_G \right) ds \tag{3.60b}$$

By using the following properties of the matrix multiplications,

$$\left. \begin{aligned} [T_n] &= [t]^T[I_1][t] \\ [T_\theta] &= [t]^T[I_2][t] \end{aligned} \right\} \quad \rightarrow \quad [T_\theta] = \begin{bmatrix} 0 & -c_z & c_y \\ c_z & 0 & -c_x \\ -c_y & c_x & 0 \end{bmatrix} \tag{3.60c}$$

Eq. (3.60b) can be simplified as written by,

$$\{p_w\}_G = \int_0^{\ell} \left(\begin{Bmatrix} N_1[T_n] \\ N_3[T_\theta] \\ N_2[T_n] \\ N_4[T_\theta] \end{Bmatrix} \{f_n\}_G \right) ds \tag{3.60d}$$

This equation will be used to calculate consistent wave forces of members in the global coordinates with wave–current and wave–structure interactions. In this section, the consistent wave forces without interactions (i.e., only the water velocities and accelerations are used) will be calculated. Having introduced $\{f_n\}_G$ from Eq. (3.58b) into Eq. (3.60d) the consistent wave forces in the global coordinates can be stated as,

Table 3.8 Integration functions of consistent wave forces in the global coordinates, q_1, q_2, q_3, q_4

Function	For $(0 \le \lvert\alpha\rvert \le 0.5)$	For $(\lvert\alpha\rvert \ge 0.5)$
q_1	$\beta_1\left[y_1\left(\frac{7}{20}+\frac{13\alpha}{60}\right)+y_2\left(\frac{3}{20}+\frac{\alpha}{30}\right)\right]$	$\beta_2\left[y_1\left(\frac{e^{\alpha}}{2}-f_1\right)+y_2\left(f_1-\frac{1}{2}\right)\right]$
q_2	$\beta_1\left[y_1\left(\frac{3}{20}+\frac{7\alpha}{60}\right)+y_2\left(\frac{7}{20}+\frac{4\alpha}{30}\right)\right]$	$\beta_2\left[y_1\left(\frac{e^{\alpha}}{2}-f_2\right)+y_2\left(f_2-\frac{1}{2}\right)\right]$
q_3	$\beta_1\,\ell\left[y_1\left(\frac{1}{20}+\frac{\alpha}{30}\right)+y_2\left(\frac{1}{30}+\frac{\alpha}{120}\right)\right]$	$\beta_2\,\ell\left[y_1\left(\frac{e^{\alpha}}{12}-f_3\right)+y_2\left(f_3-\frac{1}{12}\right)\right]$
q_4	$-\beta_1\,\ell\left[y_1\left(\frac{1}{30}+\frac{\alpha}{40}\right)+y_2\left(\frac{1}{20}+\frac{\alpha}{60}\right)\right]$	$\beta_2\,\ell\left[-y_1\left(\frac{e^{\alpha}}{12}+f_4\right)+y_2\left(f_4+\frac{1}{12}\right)\right]$
definitions:	$\beta_1=\frac{\ell}{(1+\alpha/2)}$	$\beta_2=\frac{\ell}{(e^{\alpha}-1)}$

Table 3.9 Functions f_1, f_2, f_3, f_4 in q_1, q_2, q_3, q_4 for $(\lvert\alpha\rvert \ge 0.5)$ presented in Table 3.8

Function	For $(\lvert\alpha\rvert \ge 0.5)$	Definition
f_1	$\beta_f[-\alpha^3+6\alpha+12+e^{\alpha}(6\alpha-12)]$	
f_2	$\beta_f[-(6\alpha+12)+e^{\alpha}(\alpha^3-6\alpha+12)]$	$\left(\beta_f=\frac{1}{\alpha^4}\right)$
f_3	$\beta_f[\alpha^2+4\alpha+6+e^{\alpha}(2\alpha-6)]$	
f_4	$\beta_f[2\alpha+6+e^{\alpha}(-\alpha^2+4\alpha-6)]$	

$$\left.\begin{array}{l}\text{Consistent wave forces}\\ \text{in global coordinates}\end{array}\right\} \rightarrow \{p_w\}_G = \omega\begin{Bmatrix} q_1[T_n] \\ q_3[T_\theta] \\ q_2[T_n] \\ q_4[T_\theta] \end{Bmatrix}\{\phi\}\,\eta \tag{3.60e}$$

for which the following properties of matrix multiplications are used.

$$[T_n] = [T_n][T_n] \quad \text{and} \quad [T_\theta] = [T_\theta][T_n] \tag{3.60f}$$

In the consistent force vector given by Eq. (3.60e), q_1, q_2, q_3 and q_4 are complex functions of the frequency ω which are calculated from the following integrations,

$$q_j = \int_0^{\ell} e^{m(z-i x_w)} R\, N_j\, ds \quad \text{where} \quad (j = 1,\ 2,\ 3,\ 4) \tag{3.61a}$$

For a member, the coordinates z and x_w are functions of the member axial coordinate s between the member ends (1) and (2) shown in Fig. (3.15). For the integration, the following exponential function is used between the member ends (1) and (2).

$$y(s) = e^{m(z-i x_w)} R \quad \rightarrow \quad y(s) = \frac{1}{(e^{\alpha}-1)}\left[y_1\left(e^{\alpha}-e^{\alpha\xi}\right)+y_2\left(e^{\alpha\xi}-1\right)\right] \tag{3.61b}$$

where ξ is a dimensionless variable defined in Eq. (3.59c), y_1 and y_2 are the values of the function $y(s)$ at the member ends (1) and (2) which are defined as,

$$y_1 = \mathrm{e}^{m(z_1 - i x_{w1})} R_1 \quad \text{and} \quad y_2 = \mathrm{e}^{m(z_2 - i x_{w2})} R_2 \tag{3.61c}$$

and α is a complex parameter defined as written by,

$$\alpha = m(\Delta z - i\,\Delta x_w) \rightarrow \begin{cases} \Delta z = (z_2 - z_1) \\ \Delta x_w = (x_{w2} - x_{w1}) \end{cases} \tag{3.61d}$$

The horizontal coordinates, x_{w1} and x_{w2}, in the wave propagation direction at the member ends (1) and (2) are calculated from Fig. (3.15) as written by,

$$x_{w1} = (x_1 \cos\phi + y_1 \sin\phi) \quad \text{and} \quad x_{w2} = (x_2 \cos\phi + y_2 \sin\phi) \tag{3.61e}$$

where x_1, y_1 and x_2, y_2 are the horizontal coordinates of the member ends (1) and (2) respectively. With the function $y(s)$ defined by Eq. (3.61b), the wave force functions q_j (j = 1,2,3,4) defined by Eq. (3.61a) is stated as,

$$q_j = \int_0^{\ell} y(s)\, N_j\, \mathrm{d}s \quad \text{where} \quad (j = 1,\ 2,\ 3,\ 4) \tag{3.61f}$$

which can be integrated analytically. The results are presented in Table 3.8. The functions f_j (j = 1,2,3,4) for ($|\alpha| \geq 0.5$) in Table 3.8 are given in Table 3.9. As it is seen from Eq. (3.61d) α is a complex parameter which depends on the wave number m so that functions q_j (j = 1,2,3,4) will be complex and functions of the frequency ω due to the wave number m. For each variation of the frequency ω, the q_j functions can be calculated easily from the statements given in Table 3.8, and then by substituting them in Eq. (3.60e), the member consistent wave forces can be calculated in the global coordinates.

3.8 Linearization of the Morison's Equation

In the spectral analysis of structures, a linear relation is used between the inputs and outputs. If the relation is not linear, then it must be linearized by using a linearization criterion. Since stochastic inputs and outputs are considered in this book a stochastic linearization method is employed. It is well explained generally in literature, see e.g., [103, 104], by using the criterion that the error, which is denoted by ε, remains at the minimum level. In the stochastic variables space, this is achieved if the mean square value of the error is a minimum, i.e.,

$$\min.\left(E[\varepsilon^2]\right) \rightarrow \quad (\varepsilon = g(x) - y(x) \quad \text{where} \quad y = a + bx) \tag{3.62a}$$

in which $E[.]$ denotes a mean value, $g(x)$ is a nonlinear function of x, a and b are linearization constants calculated from the minimization of the mean square error with respect to the constants a and b, i.e., $\partial E[\varepsilon^2]/\partial a = 0$ and $\partial E[\varepsilon^2]/\partial b = 0$. From these conditions it is obtained that,

$$a = \eta_g - \frac{\mu_x}{\sigma_x^2}\left(E[x\,g] - \mu_x \eta_g\right) \quad \text{and} \quad b = \frac{1}{\sigma_x^2}\left(E[x\,g] - \mu_x \eta_g\right) \tag{3.62b}$$

in which μ_g and μ_x are the mean values defined as $\mu_g = E[g]$ and $\mu_x = E[x]$, and σ_x is the standard deviation of the stochastic variable x.

The wave force of the Morison's equation for a vertical cylinder has been given by Eq. (3.45a) and, for an inclined member, it is given by Eq. (3.56a) The drag force terms of these equations are nonlinear in terms of the water velocities and must be linearized in order to be able to apply the spectral analysis. Different criteria may be used for this purpose [105–113]. The nonlinear function of the drag force term of an inclined member is replaced by a linear function as stated by,

$$|u_n| u_n \rightarrow A\, u_n \tag{3.63a}$$

in which u_n is the normal water velocity and A is a linearization coefficient to be determined. It is assumed that the water velocity u_n is an ergodic Gaussian process with zero mean from which it can be obtained that the mean values are zero, i.e.,

$$E[|u_n| u_n] = A\, E[u_n] = 0 \tag{3.63b}$$

By using the criterion of the minimum mean square error, the linearization coefficient A can be readily obtained from Eq. (3.62b) as written by,

$$A = \frac{E[\,|u_n| u_n^2]}{\sigma_{u_n}^2} = \frac{E[\,|u_n|^3]}{\sigma_{u_n}^2} \quad \rightarrow \quad A = \sqrt{\frac{8}{\pi}}\,\sigma_{u_n} \tag{3.63c}$$

in which σ_{u_n} is the standard deviation of the normal water velocity u_n. This is calculated from,

$$\sigma_{u_n} = \sqrt{E[\,|u_n|^2]} \tag{3.64a}$$

Having introduced $\{u\}$ from Eq. (3.57a) into Eq. (3.56d) and carried out the vector multiplications it can be obtained that,

$$E\left[\,|u_n|^2\right] = \left[1 + \left(c_x \sin\phi - c_y \cos\phi\right)^2\right] E\left[\,|u|^2\right] \tag{3.64b}$$

from which σ_{u_n} is stated as,

$$\sigma_{u_n} = \sqrt{1 + \left(c_x \sin\phi - c_y \cos\phi\right)^2}\;\sigma_u \tag{3.64c}$$

In this equation, σ_u is the standard deviation of the horizontal water velocity component calculated from the zero spectral moment of u as written by,

$$\sigma_u^2 = \int_0^{\infty} \omega^2\, e^{2mz}\, S_{\eta\eta}(\omega)\mathrm{d}\omega \tag{3.64d}$$

It can be found from Eq. (3.64c) that, as depending on the wave traveling direction ϕ and the member orientation, σ_{u_n} varies between,

$$\sigma_u \leq \sigma_{u_n} \leq \sqrt{2}\,\sigma_u \tag{3.64e}$$

The linearization coefficient A given by Eq. (3.63c) can also be obtained from the Borgman's series solution of the covariance function of $(W = |u_n|u_n)$ which can be written [105] as,

$$R_W(\tau) = \sigma_{u_n}^4 G(r) \;\rightarrow\; \begin{cases} G(r) = \dfrac{1}{\pi}\left(8r + \dfrac{4r^3}{3} + \dfrac{r^5}{15} + \ldots\right) \\ r = R_{u_n}(\tau)/\sigma_{u_n}^2 \end{cases} \tag{3.65a}$$

By taking only the first term of $G(r)$ it is obtained that,

$$R_W(\tau) = \frac{8}{\pi}\sigma_{u_n}^2 R_{u_n}(\tau) \quad\rightarrow\quad W = |u_n|u_n = \sqrt{\frac{8}{\pi}}\sigma_{u_n} u_n \tag{3.65b}$$

which results in the same linearization coefficient A as given by Eq. (3.63c).

In the stochastic analysis of structures, statistical quantities of response outputs are required usually under given statistical information of inputs. These quantities are mostly the mean values and variances of response outputs since probability distributions are determined in terms of these quantities. Since, under a stochastic loading case such as wave loads, fatigue damages are formulated as depending on stress variances, this subject becomes especially important in the estimate of mean value fatigue damages. Other issue in this context is that if the peak distribution of wave forces follows a Rayleigh distribution, then the expected value of the largest force in a number of wave cycles is a linear function of the standard deviation of wave forces [113]. Starting from these arguments one other criterion of the equivalent linearization procedure may be to use equivalent variances of the nonlinear and the linearized drag force terms of the Morison's equation [110, 114]. This criterion produces somewhat more conservative results than those obtained from the minimization of the mean square error. Since the mean values of the nonlinear and linearized drag forces are zero their variances can be calculated from Eq. (3.63a) and equalized as written by,

$$E\left[\left(|u_n|u_n\right)^2\right] = A^2\,E[u_n^2] \tag{3.66a}$$

from which the linearization coefficient A is calculated as,

$$A^2 = \frac{E\left[\left(|u_n|u_n\right)^2\right]}{E[u_n^2]} \quad\rightarrow\quad A = \sqrt{\frac{E[|u_n|^4]}{E[|u_n|^2]}} = \sqrt{3}\,\sigma_{u_n} \tag{3.66b}$$

where σ_{u_n} is calculated from Eq. (3.64c).

The fatigue damage of offshore structures is calculated by using the fracture mechanics approach or an experimentally determined S–N line approach. In both approaches, the damage is formulated as depending on the stress range which is assumed to be double of the stress amplitude. This is correct if the stress process is narrow banded, and approximate if the process is not narrow banded in which case some correction factors are used. Since stress amplitudes in linear systems are related to the applied forces by means of transfer functions it can be argued [113] that, for the fatigue analysis, the linearization process of the Morison's equation can be based on mean values of amplitudes of the wave forces. In order to apply this concept the maximum value of the wave forces needs to be determined first. For this purpose, the real parts of the water elevation η, wave velocities and accelerations are used. From Eqs. (3.15), (3.16a) and (3.16b) it can be written that,

$$\left.\begin{aligned} \eta &= \hat{\eta}\ \sin(\omega t - m x_w) \\ u &= \hat{\eta}\ h(\omega)\ \sin(\omega t - m x_w) \\ \dot{u} &= \hat{\eta}\,\omega\ h(\omega)\ \cos(\omega t - m x_w) \end{aligned}\right\} \rightarrow h(\omega) = \omega\frac{\cosh m(z+d)}{\sinh md} \tag{3.67a}$$

in which $\hat{\eta}$ is the amplitude of the water elevation, x_w is the horizontal coordinate in the wave propagation direction, u and $\dot{u}$ are the water velocity and acceleration in the horizontal x_w direction. The water velocity and acceleration, which are normal to the member, can be stated by using Eq. (3.64b) as written by,

$$\begin{aligned} u_n &= \hat{\eta}\ H_n(\omega)\ \sin(\omega t - m x_w) \\ \dot{u}_n &= \hat{\eta}\,\omega\, H_n(\omega)\ \cos(\omega t - m x_w) \end{aligned} \tag{3.67b}$$

where the function $H_n(\omega)$ is defined as,

$$H_n(\omega) = \sqrt{1 + \left(c_x \sin\phi - c_y \cos\phi\right)^2}\ h(\omega) \tag{3.67c}$$

Having introduced u_n and $\dot{u}_n$ from Eq. (3.67b) into the Morison's equation given by Eq. (3.45a) it can be written that,

$$\begin{aligned} f_n &= \hat{u}_n(C_D\hat{u}_n|\ \sin\vartheta|\ \sin\vartheta + \omega C_M\ \cos\vartheta) \\ \vartheta &= (\omega t - m x_w) \end{aligned} \tag{3.68a}$$

in which $\hat{u}_n$ is the amplitude of the normal velocity u_n. For the maximum value, the condition is $(\partial f_n/\partial t = 0)$ must be satisfied from which it can be obtained that,

$$(f_n)_{\max} = \hat{f}_n = \begin{cases} \omega\, C_M \hat{u}_n & \rightarrow \ \text{for}\ (\hat{u}_n \leq \lambda(\omega)) \\ C_D(\hat{u}_n^2 + \lambda^2(\omega)) & \rightarrow \ \text{for}\ (\lambda(\omega) \leq \hat{u}_n) \\ \lambda(\omega) = (\omega C_M)/(2C_D) & \end{cases} \tag{3.68b}$$

This maximum force is also the amplitude of the nonlinear wave force of the Morison's equation. The linearized form of the Morison's equation is stated as,

Table 3.10 Mean values of powers of nonlinear wave force amplitude, $E\left[\hat{f}_n^k\right]$

k	$E\left[\hat{f}_n^k\right]$
1	$\frac{\sigma_{u_n}}{\sqrt{2}} C_M\,\omega\left[\sqrt{\pi}\,\mathrm{erf}(\mu) - 2\mu e^{-\mu^2}\right] + 2\sigma_{u_n}^2 C_D(2\mu^2+1)e^{-\mu^2}$
2	$2\sigma_{u_n}^2 C_M^2\,\omega^2\left[1-(\mu^2+1)e^{-\mu^2}\right] + 8\sigma_{u_n}^4 C_D^2(2\mu^4+2\mu^2+1)e^{-\mu^2}$
3	$\frac{\sigma_{u_n}^3}{\sqrt{2}} C_M^3\,\omega^3\left[3\sqrt{\pi}\,\mathrm{erf}(\mu) - \mu(4\mu^2+6)e^{-\mu^2}\right] + 16\sigma_{u_n}^6 C_D^3(4\mu^6+6\mu^4+6\mu^2+3)e^{-\mu^2}$
4	$4\sigma_{u_n}^4 C_M^4\,\omega^4\left[2-(\mu^4+2\mu^2+2)e^{-\mu^2}\right] + 128\sigma_{u_n}^8 C_D^4(2\mu^8+4\mu^6+6\mu^4+6\mu^2+3)e^{-\mu^2}$
5	$\frac{\sigma_{u_n}^5}{\sqrt{2}} C_M^5\,\omega^5\left[15\sqrt{\pi}\,\mathrm{erf}(\mu) - (8\mu^5+20\mu^3+30\mu)e^{-\mu^2}\right] + \ldots$ $\ldots + 256\,\sigma_{u_n}^{10} C_D^5(4\mu^{10}+10\mu^8+20\mu^6+30\mu^4+30\mu^2+15)e^{-\mu^2}$

where $\mu = \lambda/(\sqrt{2}\,\sigma_{u_n})$, λ is given in Eq. (3.68b), and $\left(\mathrm{erf}\,(x) = \frac{2}{\sqrt{\pi}}\int_0^x e^{-t^2}\,dt\right)$

$$f_L = \hat{u}_n(C_D A\ \sin\vartheta + \omega\,C_M\ \cos\vartheta) \tag{3.69a}$$

where ϑ is defined in Eq. (3.68a). The maximum value (the amplitude) of this linearized force is obtained as written by,

$$(f_L)_{\max} = \hat{f}_L = \sqrt{(\omega\,C_M)^2 + (C_D A)^2}\ \hat{u}_n \ \rightarrow \text{ for } (0 \le \hat{u}_n \le \infty) \tag{3.69b}$$

Having determined the amplitudes of the nonlinear and the linearized wave forces the stochastic linearization process can now be applied. As it is seen from Eqs. (3.68b) and (3.69b), the force amplitudes are functions of the amplitude $\hat{u}_n$ of the normal velocity u_n, which has a Rayleigh distribution for a stationary zero-mean Gaussian process for u_n. From Eq. (2.121) it is stated as,

$$f_{\hat{u}_n}(\hat{u}_n) = \frac{\hat{u}_n}{\sigma_{u_n}^2}\ \exp\left(-\frac{\hat{u}_n^2}{2\sigma_{u_n}^2}\right) \ \rightarrow \text{ for } \ (0 \le \hat{u}_n \le \infty) \tag{3.70}$$

In the fatigue damage calculation, the mean values of powers of the wave force amplitudes, namely $E\left[\hat{f}_n^k\right]$ and $E\left[\hat{f}_L^k\right]$ can be equalized to find the linearization coefficient A. Here, k denotes the slope of the S–N line which can be assumed to be an integer number in the linearization process. In fact, it is a real number between 3 and 4. But, for simplicity of the calculations, it is assumed as an integer. Then, if it is required, an extrapolation may be used for real value of k. The mean values of powers are calculated from,

Table 3.11 Linearization coefficient (A/σ_u) of the drag force term of a vertical cylinder at the SWL shown in Fig. (3.16) for different frequencies

k	$\omega = 0$	Peak frequency of $S_{\eta\eta}(\omega)$ $\omega = 0.562$ rad/s	An assumed natural frequency $\omega = 3.0$ rad/s
1	1.5957	1.5218	0.4259
2	2.0000	1.9184	0.6105
3	2.3372	2.2600	0.8175
4	2.6322	2.5606	1.0430
5	2.8976	2.8309	1.2822

$$E\left[\hat{f}_n^k\right] = \int_0^\infty \hat{f}_n^k f_{\hat{u}_n}(\hat{u}_n)\, d\hat{u}_n \quad \text{or}$$
$$E\left[\hat{f}_n^k\right] = \omega^k C_M^k \int_0^\lambda \hat{u}_n^k f_{\hat{u}_n}(\hat{u}_n)\, d\hat{u}_n + C_D^k \int_\lambda^\infty \left(\hat{u}_n^2 + \lambda^2\right)^k f_{\hat{u}_n}(\hat{u}_n)\, d\hat{u}_n \tag{3.71a}$$

in which λ denotes $\lambda(\omega)$ defined in Eq. (3.68b). For the linearized force it is stated as,

$$E\left[\hat{f}_L^k\right] = \int_0^\infty \hat{f}_L^k f_{\hat{u}_n}(\hat{u}_n)\, d\hat{u}_n = \left(\omega^2 C_M^2 + A^2 C_D^2\right)^{\frac{k}{2}} \int_0^\infty u_n^k f_{\hat{u}_n}(\hat{u}_n)\, d\hat{u}_n \tag{3.71b}$$

or having carried out the integration it is obtained as written by,

$$E\left[\hat{f}_L^k\right] = \left(\omega^2 C_M^2 + A^2 C_D^2\right)^{\frac{k}{2}} \left(\sqrt{2}\sigma_{u_n}\right)^k \Gamma(1 + k/2) \tag{3.71c}$$

in which $\Gamma(.)$ denotes the Gamma function. The general calculation of $E\left[\hat{f}_n^k\right]$ from Eq. (3.71a) is not as simple as that of $E\left[\hat{f}_L^k\right]$. It is calculated for the values of ($k = 1$, 2, 3, 4, 5) and the results are presented in Table 3.10. Having determined mean values of powers of the force amplitudes, the linearization coefficient A of the drag force term can be calculated by equalizing these mean values as written by,

$$E\left[\hat{f}_n^k\right] = E\left[\hat{f}_L^k\right] \;\rightarrow\; E\left[\hat{f}_n^k\right] = \left(\omega^2 C_M^2 + A^2 C_D^2\right)^{\frac{k}{2}} \left(\sqrt{2}\sigma_{u_n}\right)^k \Gamma(1 + k/2) \tag{3.72a}$$

from which A can be obtained as depending on the frequency ω, which is written by,

$$A = \frac{1}{C_D} \sqrt{\frac{1}{2\sigma_{u_n}^2} \left(\frac{E\left[\hat{f}_n^k\right]}{\Gamma(1 + k/2)}\right)^{\frac{2}{k}} - \omega^2 C_M^2} \tag{3.72b}$$

For the evaluation of A, the frequency ω can be selected to be the peak frequency

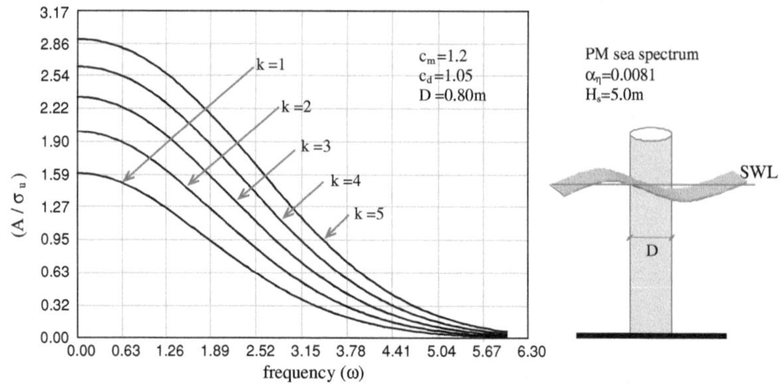

Fig. 3.16 Frequency variation of the linearization coefficient A of the drag force term of a vertical cylinder, which is given by (3.72b), at the SWL

of the sea spectrum $S_{\eta\eta}(\omega)$, or the lowest natural frequency, as depending on the dynamic sensitivity of the structural response behavior. Thus,

$$\omega = \begin{cases} \omega_p \text{ (peak frequency of } S_{\eta\eta}(\omega)) \rightarrow \text{ for quasi-static analysis} \\ \omega_r \text{ (lowest natural frequency)} \rightarrow \text{ for dynamic analysis} \end{cases} \tag{3.72c}$$

In fact, the linearization coefficient A is a frequency function as formulated above and can be taken into account in the transfer function of the wave force vector given by Eq. (3.58b) through the scalar function R which becomes to be,

$$R = (C_D A(\omega) + i\,\omega\, C_M) \tag{3.72d}$$

Since the parameter μ in $E\left[\hat{f}_n^k\right]$ given in Table 3.10 is a linear function of ω, it can be seen from the inspection of Eq. (3.72b) and Table 3.10 that, in the small frequency region, the linearization coefficient A will be small and, in the high frequency region, it will be large. This means that the small frequency region is dominated by the drag force term and the high frequency region is dominated by the inertia force term. For the limit case, when the frequency is zero ($\omega = 0$) the wave force becomes purely the drag force in which case the linearization coefficient A reaches at its highest value. For a vertical rough cylinder ($c_m = 1.2$ and $c_d = 1.05$) with the diameter of ($D = 0.80$ m), the linearization coefficient A is demonstrated in Table 3.11 for different frequencies at the still water level ($z = 0$). For this purpose, the Pierson Markowitz sea spectrum is used with $\alpha_\eta = 0.0081$ and $H_s = 5.0$ m. The frequency variation of the linearization coefficient (A/σ_{u}) is also illustrated in Fig. (3.16). As it is seen from Table 3.11 and Fig. (3.16), the mean value approach of the drag-force-amplitude ($k = 1$ and $\omega = 0$) produces the same result as that ($A/\sigma_{\text{u}} = 1.5957$) obtained from the minimization of the mean square error, which constitutes the lower bound of A. In the fatigue damage calculation, this lower value underestimates fatigue damages, and therefore Eq. (3.72b) may be used more appropriately as being a frequency function with the correct value of k (the slope of the S–N line).

For wave–current and wave–structure interaction problems, the linearization of the drag force term is more complicated than that presented above. This subject is presented in the next section.

3.9 Linearization of the Morison's Equation Under Wave–Current and Wave–Structure Interactions

In offshore environments, besides random waves there are also currents and the structures to be built in these environments are usually flexible, as shown in Fig. 3.12b. The current and structural velocities and accelerations should also be taken into account in the calculation of the forces using the Morison's equation. Although currents in ocean environments are not uniform and unidirectional generally, it is assumed here that they are unidirectional with horizontal constant velocities denoted by U. But, their directions which are denoted by ϕ_c may be different from the wave traveling direction ϕ. As it is shown in Fig. 3.12b, the structural velocity is denoted by $\dot{d}$ and the acceleration is denoted by $\ddot{d}$, which are zero-mean processes under a constant deterministic current condition that is assumed in this book. As similar to the water velocity vector given in Eq. (3.55c), the current and structural velocity vectors in the structural global coordinates can be stated as,

$$\{U_c\} = \begin{Bmatrix} U \cos\phi_c \\ U \sin\phi_c \\ 0 \end{Bmatrix} \quad \text{and} \quad \{\dot{d}\} = \begin{Bmatrix} \dot{\mathrm{d}}_x \\ \dot{\mathrm{d}}_y \\ \dot{\mathrm{d}}_z \end{Bmatrix} \tag{3.73a}$$

A relative water velocity vector $\{u_r\}$ with a zero mean is introduced as it is defined in the structural global coordinates by,

$$\{u_r\} = \{u\} - \{\dot{d}\} \tag{3.73b}$$

The total water velocity vector in the global coordinates is expressed in terms of $\{u_r\}$ and $\{U_c\}$ as written by,

$$\{w\} = \{u_r\} + \{U_c\} \tag{3.73c}$$

The normal water velocity and acceleration vectors, which are used to calculate member forces using the Morison's equation, can be stated similarly to Eq. (3.55a) as written by,

$$\{w_n\} = [T_n]\{w\} \quad \text{and} \quad \{\dot{w}_n\} = [T_n]\{\dot{u}_r\} \tag{3.73d}$$

Table 3.12 Mean values of powers of the absolute value of a general Gaussian variable w_n, $E[|w_n|^k]$ defined by Eq. (3.75d) with $(E[w_n] = U_{cn})$

k	$E\left[\lvert w_n\rvert^k\right]$
1	$\sigma_{u_{rn}}\left(\sqrt{\frac{2}{\pi}}e^{-\mu^2} + \sqrt{2}\mu\,\mathrm{erf}(\mu)\right)$
2	$\sigma_{u_{rn}}^2(1+2\mu^2)$
3	$\sigma_{u_{rn}}^3\left[\sqrt{\frac{8}{\pi}}(1+\mu^2)e^{-\mu^2} + \sqrt{2}\,\mu\,(3+2\mu^2)\,\mathrm{erf}(\mu)\right]$
4	$\sigma_{u_{rn}}^4[3+2\mu^2(6+2\mu^2)]$
5	$\sigma_{u_{rn}}^5\left[\sqrt{\frac{2}{\pi}}(8+18\,\mu^2+4\mu^4)e^{-\mu^2} + \sqrt{2}\,\mu\,(15+20\mu^2+4\mu^4)\,\mathrm{erf}(\mu)\right]$
where $\mu = \frac{U_{cn}}{\sqrt{2}\sigma_{u_{rn}}}$ and $\left(\mathrm{erf}(x) = \frac{2}{\sqrt{\pi}}\int_0^x e^{-t^2}dt\right)$	

in which $[T_n]$ is defined in Eq. (3.55b), or in terms of random and deterministic components, it is stated as,

$$\{w_n\} = \{u_{rn}\} + \{U_{cn}\} \qquad \text{and} \qquad \{\dot{w}_n\} = \{\dot{u}_{rn}\} \tag{3.73e}$$

where $\{u_{rn}\}$, $\{U_{cn}\}$ and $\{\dot{u}_{rn}\}$ are the vectors of the velocity and acceleration components which are normal to the member associated. With these velocity and acceleration vectors, the force vector of the Morison's equation which is given by Eq. (3.56a) can be stated as,

$$\{f_n\}_G = [T_n]\,(C_D|w_n|(\{u_r\} + \{U_c\}) + C_M\{\dot{u}_r\}) \tag{3.74a}$$

or in scalar form it is written as,

$$f_n = C_D|w_n|w_n + C_M\dot{w}_n \tag{3.74b}$$

in which $|w_n|$ is calculated from,

$$|w_n| = \sqrt{\{w_n^*\}^T\{w_n\}} = \sqrt{(\{u_r^*\}^T + \{U_c\}^T)[T_n](\{u_r\} + \{U_c\})} \tag{3.74c}$$

and w_n can be written as,

$$w_n = u_{rn} + U_{cn} \tag{3.74d}$$

where u_{rn} is the relative normal velocity with zero-mean Gaussian process and U_{cn} is a constant normal current velocity. The linearization of the nonlinear force given by Eq. (3.74b) can now be carried out as explained in Sect. 3.8. The only difference is that the normal velocity w_n is not a zero-mean value process unlike the normal water velocity u_n in Eq. (3.63a).

By applying the minimization of the mean square error, the linearization coefficient A can be obtained as written by,

$$A = \frac{E[|w_n|w_n^2]}{E[w_n^2]} = \frac{E[|w_n|^3]}{E[|w_n|^2]} \tag{3.75a}$$

The mean values can be calculated by using the function $|w_n|$ defined as,

$$|w_n| = \begin{cases} -(u_{rn} + U_{cn}) & \text{for } (u_{rn} \leq -U_{cn}) \\ +(u_{rn} + U_{cn}) & \text{for } (u_{rn} \geq -U_{cn}) \end{cases} \tag{3.75b}$$

The probability distribution of u_{rn} is stated from Eq. (2.65a) as written by,

$$f_{u_{rn}}(x) = \frac{1}{\sqrt{2\pi}\,\sigma_{u_{rn}}} \exp\left[-\frac{x^2}{2\sigma_{u_{rn}}^2}\right] \tag{3.75c}$$

and the mean values of powers of the absolute value $|w_n|$ is calculated from the following integration,

$$E\left[|w_n|^k\right] = (-1)^k \int_{-\infty}^{-U_{cn}} (x + U_{cn})^k f_{u_{rn}}(x)\,dx + \int_{-U_{cn}}^{\infty} (x + U_{cn})^k f_{u_{rn}}(x)\,dx \tag{3.75d}$$

The results for (k = 1, 2, 3, 4, 5) are presented in Table 3.12. Having introduced $E[|w_n|^3]$ and $E[|w_n|^2]$ from Table 3.12 into Eq. (3.75a) the linearization coefficient A can be obtained as written by,

$$A = \frac{1}{(1 + 2\mu^2)} \left[\sqrt{\frac{8}{\pi}}(1 + \mu^2)e^{-\mu^2} + \sqrt{2}\,\mu\,(3 + 2\mu^2)\ \mathrm{erf}(\mu)\right] \sigma_{u_{rn}} \tag{3.75e}$$

where the parameter μ is defined as $\left(\mu = U_{cn}/\left(\sqrt{2}\sigma_{u_{rn}}\right)\right)$. For a zero-current velocity ($U_{cn} = 0$) Eq. (3.75e) produces the same result of the non-current case given by Eq. (3.63c).

By using the equivalent second moment criterion for the nonlinear and the linearized forces, i.e., $E[f_n^2] = E[f_L^2]$, the linearization coefficient A can be obtained as written by,

$$A = \sqrt{\frac{E[|w_n|^4]}{E[|w_n|^2]}} \quad \rightarrow \quad A = \sqrt{\frac{1}{(1 + 2\mu^2)}[3 + 2\mu^2(6 + 2\mu^2)]}\ \sigma_{u_{rn}} \tag{3.76}$$

which produces the same result of the non-current case given by Eq. (3.66a) for a zero-current velocity. One other variation of the equivalent second moment criterion is to use the equivalent central second moments. This is the criterion of using the equivalent variances of the nonlinear and linearized forces. The variances of the nonlinear and linearized forces, f_n and f_L, are calculated from,

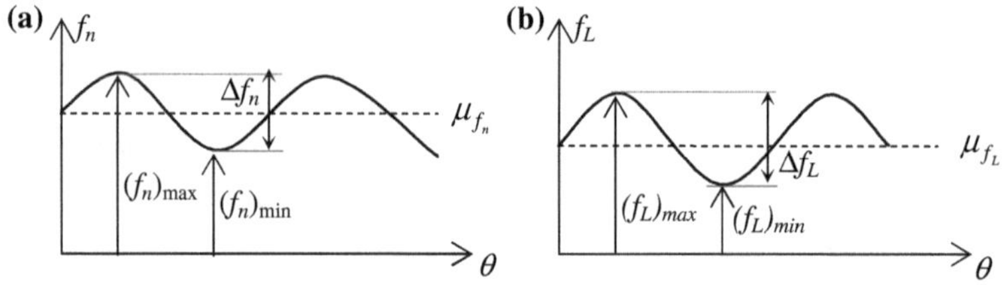

Fig. 3.17 Nonlinear and linearized wave forces under wave–current combination (a) Nonlinear wave force, f_n (b) Linearized wave force, f_L

$$\begin{aligned} \sigma_{f_n}^2 &= E\left[\left(f_n - \mu_{f_n}\right)^2\right] = E\left[f_n^2\right] - \mu_{f_n}^2 \\ \sigma_{f_L}^2 &= E\left[\left(f_L - \mu_{f_L}\right)^2\right] = E\left[f_L^2\right] - \mu_{f_L}^2 \end{aligned} \tag{3.77a}$$

Having carried out the related mean values they can be obtained as written by,

$$\begin{aligned} \sigma_{f_n}^2 &= C_D^2\, E\left[|w_n|^4\right] + C_M^2\, E\left[\dot{w}_n^2\right] - \mu_{f_n}^2 \\ \sigma_{f_L}^2 &= A^2 C_D^2\, E\left[|w_n|^2\right] + C_M^2\, E\left[\dot{w}_n^2\right] \end{aligned} \tag{3.77b}$$

where the mean values of $|w_n|^4$ and $|w_n|^2$ are taken from Table 3.12 and the mean value μ_{f_n} is calculated from,

$$\begin{aligned} \mu_{f_n} &= E[f_n] = C_D E[|w|w] + C_M E[\dot{w}] = C_D E[|u_{rn} + U_{cn}|(u_{rn} + U_{cn})] \\ \mu_{f_n} &= C_D \left[\int_{-\infty}^{-U_{cn}} -(x + U_{cn})^2 f_{u_{rn}}(x)\mathrm{d}x + \int_{-U_{cn}}^{\infty} (x + U_{cn})^2 f_{u_{rn}}(x)\,\mathrm{d}x \right] \end{aligned} \tag{3.77c}$$

in which the probability density function $f_{u_{rn}}(x)$ is given by Eq. (2.75c). Having carried out the integration in Eq. (3.77c) the mean value can be obtained as written by,

$$\mu_{f_n} = C_D \sigma_{u_{rn}}^2 \left[\sqrt{\frac{4}{\pi}}\,\mu\; \mathrm{e}^{-\mu^2} + \left(1 + 2\mu^2\right)\mathrm{erf}(\mu) \right] \tag{3.77d}$$

where μ and erf(.) are defined in Table 3.12. It can be realized from Eq. (3.77d) that, for a zero-current case, this mean value will be zero. By equalizing the variances the linearization coefficient A can be obtained as stated by,

$$A = \sqrt{\frac{C_D^2\, E\left[|w_n|^4\right] - \mu_{f_n}^2}{C_D^2\, E\left[|w_n|^2\right]}} \tag{3.77e}$$

Having introduced $|w_n|^4$, $|w_n|^2$ and μ_{f_n} into Eq. (3.77e) it can be obtained that,

$$A = \sigma_{u_{rn}} \sqrt{\frac{[3+2\mu^2(6+2\mu^2)] - \left[\sqrt{\frac{4}{\pi}}\,\mu\ \mathrm{e}^{-\mu^2} + (1+2\mu^2)\mathrm{erf}(\mu)\right]^2}{(1+2\mu^2)}} \tag{3.77f}$$

which produces the same result of the non-current case given by Eq. (3.66a) for a zero-current velocity.

Since the fatigue damage is formulated as a function of the stress range it can be argued that the linearization process of the nonlinear Morison's equation may be based on the force range criterion. The force ranges, Δf_n and Δf_L, are defined as being the differences between the maxima and the minima of the nonlinear and the linearized forces. They are written as,

$$\Delta f_n = (f_n)_{\max} - (f_n)_{\min} \quad \text{and} \quad \Delta f_L = (f_L)_{\max} - (f_L)_{\min} \tag{3.78}$$

and also shown in Fig. 3.17. To calculate the maxima and the minima of the forces, a single harmonic function is assumed for the relative normal water velocity. With this assumption the total normal velocity can be stated from Eq. (3.74d) as,

$$w_n = \hat{u}_{rn} \sin \vartheta + U_{cn} \tag{3.79}$$

where ϑ is defined in Eq. (3.68a). Then, the nonlinear and the linearized Morison's equations can be written as,

$$\begin{aligned} f_n &= C_D|\hat{u}_{rn} \sin \vartheta + U_{cn}|(\hat{u}_{rn} \sin \vartheta + U_{cn}) + C_M\omega\,\hat{u}_{rn} \cos \vartheta \\ f_L &= C_D A(\hat{u}_{rn} \sin \vartheta + U_{cn}) + C_M\omega\,\hat{u}_{rn} \cos \vartheta \end{aligned} \tag{3.80a}$$

The extremum values of these functions are obtained from their derivatives with respect to ϑ as written by,

$$\left.\begin{aligned} \frac{\partial f_n}{\partial \vartheta} &= 2C_D\hat{u}_{rn}(\hat{u}_{rn}|\sin \vartheta + \beta| \cos \vartheta - \lambda \sin \vartheta) = 0 \\ \frac{\partial f_L}{\partial \vartheta} &= C_D\hat{u}_{rn}(A \cos \vartheta - 2\lambda \sin \vartheta) = 0 \end{aligned}\right\} \rightarrow \left\{\begin{aligned} \lambda &= \frac{\omega C_M}{2C_D} \\ \beta &= \frac{U_{cn}}{\hat{u}_{rn}} \end{aligned}\right. \tag{3.80b}$$

For the nonlinear force f_n, the solution of ϑ can be obtained from the following equations,

$$\left\{\left(\cos \vartheta = \frac{\lambda}{\hat{u}_{rn}} \frac{\sin \vartheta}{|\sin \vartheta + \beta|}\right) \text{or} \left(\tan \vartheta = \frac{\hat{u}_{rn}}{\lambda}|\sin \vartheta + \beta|\right)\right\} \rightarrow \left\{\begin{aligned} 0 < \vartheta < \pi/2 \\ \pi \le \vartheta \le 3\pi/2 \end{aligned}\right. \tag{3.80c}$$

By using these relations of ϑ into Eq. (3.80a), the maximum nonlinear wave force $(f_n)_{\max}$ can be obtained as written by,

$$(f_n)_{\max} = C_D \frac{\hat{u}_{rn}}{\lambda} \cos\vartheta \left[u_{rn}^2 \frac{(\sin\vartheta + \beta)^3}{\sin\vartheta} + 2\lambda^2 \right] \quad \text{where} \quad (0 \leq \vartheta \leq \pi/2) \tag{3.80d}$$

Having used an auxiliary variable x for sin ϑ, an equation of x can be obtained to solve it between $(0 \leq x \leq 1)$. It is written as,

$$\left(\cos\vartheta = \sqrt{1 - x^2} = \frac{\lambda}{\hat{u}_{rn}} \frac{x}{|x + \beta|} \right) \rightarrow \begin{cases} x = \sin\vartheta \\ (0 \leq x \leq 1) \end{cases} \tag{3.81a}$$

Then, the maximum value of the nonlinear force $(f_n)_{\max}$ can be obtained from Eq. (3.80d) as a function of x. It is written as,

$$(f_n)_{\max} = \frac{C_D}{|x + \beta|} \left[\hat{u}_{rn}^2 (x + \beta)^3 + 2\lambda^2 x \right] \quad \text{where} \quad (0 \leq x \leq 1) \tag{3.81b}$$

Since $(-1 \leq \cos\vartheta \leq 1)$, there are boundary conditions of the validity of Eq. (3.81b). These conditions can be stated as, from Eq. (3.81a),

$$\left(-1 \leq \frac{\lambda}{\hat{u}_{rn}} \frac{x}{|x + \beta|} \leq 1 \right) \rightarrow \left(0 \leq \frac{x}{|x + \beta|} \lambda \leq \hat{u}_{rn} \right) \tag{3.81c}$$

Since β is a function of $\hat{u}_{rn}$, having introduced β from Eq. (3.80b) into Eq. (3.81c), the boundary condition will be further arranged as written by,

$$\left(0 \leq \frac{x}{|x + \beta|} \lambda \leq \hat{u}_{rn} \right) \rightarrow \left(0 \leq \lambda \leq \left| \hat{u}_{rn} + \frac{U_{cn}}{x} \right| \right) \rightarrow \left(\lambda - \frac{U_{cn}}{x} \right) \leq \hat{u}_{rn} \tag{3.81d}$$

Outside this boundary region, $(f_n)_{\max}$ will be entirely due to the inertia force term. Thus, it is defined in the whole region of $\hat{u}_{rn}$ $(0 \leq \hat{u}_{rn} \leq \infty)$ as written by,

$$(f_n)_{\max} = \begin{cases} \omega C_M \hat{u}_{rn} & \text{if } \hat{u}_{rn} \leq (\lambda - U_{cn}/x) \\ \dfrac{C_D}{|x + \beta|} \left[\hat{u}_{rn}^2 (x + \beta)^3 + 2\lambda^2 x \right] & \text{if } (\lambda - U_{cn}/x) \leq \hat{u}_{rn} \end{cases} \tag{3.81e}$$

The amplitude of the nonlinear force $\hat{f}_n$ is simply calculated as being the difference between the maximum and mean values. Thus, it is written as,

$$\hat{f}_n = \begin{cases} \omega C_M \hat{u}_{rn} & \text{if } \hat{u}_{rn} \leq (\lambda - U_{cn}/x) \\ \left(\dfrac{C_D}{|x + \beta|} \left[\hat{u}_{rn}^2 (x + \beta)^3 + 2\lambda^2 x \right] - \mu_{f_n} \right) & \text{if } (\lambda - U_{cn}/x) \leq \hat{u}_{rn} \end{cases} \tag{3.82a}$$

in which the mean value μ_{f_n} is given in Eq. (3.77d). Then, the mean values of powers of the amplitude $\hat{f}_n$ will be calculated from the following integration,

$$E\left[\hat{f}_n^k\right] = \omega^k C_M^k \int_0^{(\lambda - U_{cn}/x)} \hat{u}_{rn}^k f_{\hat{u}_{rn}}(\hat{u}_{rn})\, d\hat{u}_{rn} + \ldots$$
$$\ldots + \int_{(\lambda - U_{cn}/x)}^{\infty} \left(\frac{C_D}{|x+\beta|} \left[\hat{u}_{rn}^2 (x+\beta)^3 + 2\lambda^2 x \right] - \mu_{f_n} \right)^k f_{\hat{u}_{rn}}(\hat{u}_{rn})\, d\hat{u}_{rn} \tag{3.82b}$$

in which $f_{\hat{u}_{rn}}(\hat{u}_{rn})$ is the probability density function of the relative normal water velocity amplitude, which is a Rayleigh density function. Calculation of this integration is analytically cumbersome and, therefore, it will be performed numerically.

The maximum value and the amplitude of the linearized Morison's equation can be calculated by using the same procedure explained above for the nonlinear Morison's equation. The $(\tan \vartheta)$ value for maximum force $(f_L)_{\max}$ is calculated from Eq. (3.80b) as written by,

$$\tan \vartheta = A/(2\lambda) \quad \rightarrow \quad (0 \le \vartheta \le \pi/2) \tag{3.83a}$$

Having used this relation of ϑ in Eq. (3.80a) for the linearized force f_L, its maximum value and amplitude can be calculated as stated by,

$$(f_L)_{\max} = C_D \left(\hat{u}_{rn} \sqrt{4\lambda^2 + A^2} + A U_{cn} \right) \tag{3.83b}$$

$$\hat{f}_L = \left[C_D \left(\hat{u}_{rn} \sqrt{4\lambda^2 + A^2} + A U_{cn} \right) - \mu_{f_L} \right] \rightarrow \hat{f}_L = \hat{u}_{rn} \sqrt{\omega^2 C_M^2 + A^2 C_D^2} \tag{3.83c}$$

The mean values of powers can be written readily from Eq. (3.71c) as,

$$E\left[\hat{f}_L^k\right] = \left(\omega^2 C_M^2 + A^2 C_D^2\right)^{\frac{k}{2}} \left(\sqrt{2}\sigma_{u_{rn}}\right)^k \Gamma(1 + k/2) \tag{3.83d}$$

Then, the linearization coefficient A can be obtained by equalizing the means of powers of the force amplitudes, i.e., from $E\left[\hat{f}_n^k\right] = E\left[\hat{f}_L^k\right]$, as written by,

$$A = \frac{1}{C_D} \sqrt{\frac{1}{2\sigma_{u_{rn}}^2} \left(\frac{E\left[\hat{f}_n^k\right]}{\Gamma(1 + k/2)} \right)^{\frac{2}{k}} - \omega^2 C_M^2} \tag{3.83e}$$

As it may be seen through Eqs. (3.75c)–(3.83e) the linearization coefficient A is a function of the standard deviation of the relative normal water velocity u_{rn}. The calculation of this standard deviation $(\sigma_{u_{rn}})$ is presented in the next section.

3.9.1 Calculation of Standard Deviation of the Relative Normal Water Velocity

The relative water velocity vector $\{u_r\}$ at a location in the sea is given by Eq. (3.73b). By using Eq. (3.55a) the relative normal water velocity vector to a member is calculated from,

$$\{u_{rn}\} = [T_n](\{u\} - \{\dot{d}\}) \rightarrow \begin{cases} \{u_{rn}\} = \{u_n\} - \{\dot{d}_n\} \\ \{u_n\} = [T_n]\{u\} \quad \text{and} \quad \{\dot{d}_n\} = [T_n]\{\dot{d}\} \end{cases} \tag{3.84a}$$

in which $[T_n]$ is given by Eq. (3.55b) and $\{\dot{d}\}$ is the velocity vector of the structural member in the global coordinates. Since both the water and structural velocity vectors, $\{u\}$ and $\{\dot{d}\}$, are with zero means, the variance of the relative normal velocity is calculated from,

$$\sigma^2_{u_{rn}} = E\left[|u_{rn}|^2\right] = E\left[\{u^*_{rn}\}^T\{u_{rn}\}\right] \tag{3.84b}$$

Having introduces $\{u_{rn}\}$ from Eq. (3.84a) into Eq. (3.84b) it can be written that,

$$\sigma^2_{u_{rn}} = E\left[|u_n|^2\right] + E\left[|\dot{d}_n|^2\right] - E\left[\{u^*_n\}^T\{\dot{d}_n\}\right] - E\left[\{\dot{d}^*_n\}^T\{u_n\}\right] \tag{3.84c}$$

and having used the following property of the covariance $\sigma_{\dot{d}_n u_n}$, the variance and the standard deviation of the relative normal water velocity can be obtained as written by,

$$\left(\sigma_{\dot{d}_n u_n} = \sigma^*_{u_n \dot{d}_n}\right) \rightarrow \begin{cases} \sigma^2_{u_{rn}} = \sigma^2_{u_n} + \sigma^2_{\dot{d}_n} - 2\mathrm{Re}\left(\sigma_{u_n \dot{d}_n}\right) \\ \sigma_{u_{rn}} = \sqrt{\sigma^2_{u_n} + \sigma^2_{\dot{d}_n} - 2\mathrm{Re}\left(\sigma_{u_n \dot{d}_n}\right)} \end{cases} \tag{3.84d}$$

in which Re(.) denotes the real part of the covariance, the variance $\sigma^2_{u_n}$ will be calculated from Eq. (3.64c), the calculations of $\sigma^2_{\dot{d}_n}$ and $\sigma_{u_n \dot{d}_n}$ are explained here. In order to calculate these variances member velocities in translational directions $(\dot{d}_x, \dot{d}_y, \dot{d}_z)$ need to be calculated first. They are derived from the corresponding member displacements in the frequency domain as written by,

$$\{\dot{d}\} = i\,\omega\,\{d\} \quad \text{where} \quad \{d\}^T = \{d_x,\ d_y,\ d_z\} \tag{3.85a}$$

The member displacement vector $\{d\}$ will be extracted from the displacement vector $\{D\}$ of the structural system in the frequency domain, which is given generally by Eq. (1.259) in Chap. 1. For convenience, it is rewritten here for the wave load application in the frequency domain as stated by,

$$\{D\} = [H]_{DP}\{P\} \quad \rightarrow \quad \begin{cases} \{D\} = [H]_{DP}\{H\}_{P\eta}\eta \ \text{ or} \\ \{D\} = \{H\}_{D\eta}\eta \end{cases} \tag{3.85b}$$

in which $[H]_{\mathrm{DP}}$ is a matrix of the complex frequency response (see Sect. 1.7.2), $\{H\}_{P\eta}$ is the transfer function vector between global wave forces and the water elevation η (explained in Chap. 4) and $\{H\}_{D\eta}$ is the transfer function vector between global displacements $\{D\}$ and the water elevation η. With this information, the member displacement vector $\{d\}$ can be as,

$$\{d\} = \{h\}_{d\eta}\eta \quad \text{where} \quad \{h\}_{d\eta} \quad \rightarrow \quad \text{extracted from} \quad \{H\}_{D\eta} \tag{3.85c}$$

Having introduced $\{d\}$ from Eq (3.85c) into Eq. (3.85a) and transformed it to the member normal direction, the member normal velocity vector $\{\dot{d}_n\}$ can be obtained for a zero-current condition in the frequency domain as written by,

$$\{\dot{d}_n\} = [T_n]\{\dot{d}\} \ \rightarrow \ \{\dot{d}_n\} = \{h\}_{\dot{d}_n\eta}\eta \ \text{ where } \ \{h\}_{\dot{d}_n\eta} = i\,\omega[T_n]\{h\}_{d\eta} \tag{3.85d}$$

in which $\{h\}_{\dot{d}_n\eta}$ is the transfer function vector between $\{\dot{d}_n\}$ and η. On the other hand, by using Eqs. (3.55a) and (3.57a) the normal water velocity vector $\{u_n\}$ can be stated for a zero-current condition as,

$$\{u_n\} = \{h\}_{u_n\eta}\,\eta \quad \rightarrow \quad \{h\}_{u_n\eta} = \omega\ \mathrm{e}^{m(z-i\,x_w)}[T_n]\{\phi\} \tag{3.85e}$$

in which the vector $\{\phi\}$ of the wave propagation direction is given in Eq. (3.57a). Then, by using Eqs. (3.85d) and (3.85e), the auto- and cross-spectral functions of $\dot{d}_n$ and u_n can be stated in terms of the sea spectrum for a zero-current condition as written by,

$$\begin{aligned} S_{\dot{d}_n\dot{d}_n}(\omega) &= \{h^*\}^T_{\dot{d}_n\eta}\{h\}_{\dot{d}_n\eta}\ S_{\eta\eta}(\omega) = \omega^2\{h^*\}^T_{d\eta}[T_n]\,\{h\}_{d\eta}\ S_{\eta\eta}(\omega) \\ S_{u_n\dot{d}_n}(\omega) &= \{h^*\}^T_{u_n\eta}\{h\}_{\dot{d}_n\eta}\ S_{\eta\eta}(\omega) = i\,\omega^2\ \mathrm{e}^{m(z+i\,x_w)}\{\phi^*\}^T[T_n]\{h\}_{d\eta}\,S_{\eta\eta}(\omega) \end{aligned} \tag{3.86a}$$

in which $\{h\}_{d\eta}$ is given in Eq. (3.85c), $[T_n]$ is given by Eq. (3.55b) and $\{\phi\}$ is given in Eq. (3.57a). Then, the auto- and covariances, $\sigma^2_{\dot{d}_n}$ and $\sigma_{u_n\dot{d}_n}$, will be calculated from the following integrations of these spectral functions,

$$\sigma^2_{\dot{d}_n} = \int_0^\infty S_{\dot{d}_n\dot{d}_n}(\omega)\,\mathrm{d}\omega \quad \text{and} \quad \sigma_{u_n\dot{d}_n} = \int_0^\infty S_{u_n\dot{d}_n}(\omega)\,\mathrm{d}\omega \tag{3.86b}$$

In the case of wave–current interaction, the spectral functions, $S_{\dot{d}_n\dot{d}_n}(\omega)$ and $S_{u_n\dot{d}_n}(\omega)$, will be modified to include current effects on the spectral shapes. In this case, the sea spectrum becomes $S_{\eta\eta}(\omega_a, U_c)$ as given by Eq. (3.38a) and the transfer functions $\{h\}_{\dot{d}_n\eta}$ and $\{h\}_{u_n\eta}$ can be obtained in the relative and absolute frequency domains (ω_r) and (ω_a), as written below, see Eq. (3.39d).

$$\text{With wave-current interaction} \rightarrow \begin{cases} \{h\}_{u_n\eta} = \omega_r \, e^{m(z-ix_w)}[T_n]\{\phi\} \\ \{h\}_{\dot{d}_n\eta} = i\,\omega_a[T_n]\{h(\omega_a)\}_{d\eta} \end{cases} \quad (3.87a)$$

provided that the structural response is formulated in the absolute frequency (ω_a) domain. In Eq. (3.87a), the wave number m is related to the relative frequency (ω_r) through the dispersion relation given by Eq. (3.36b) and ω_r is stated in terms of ω_a in Eq. (3.36c). Then, the spectral functions, $S_{\dot{d}_n\dot{d}_n}(\omega_a)$ and $S_{u_n\dot{d}_n}(\omega_a)$, under wave–current interaction, can be obtained as given by,

$$\begin{aligned} S_{\dot{d}_n\dot{d}_n}(\omega_a) &= \omega_a^2\{h^*\}_{d\eta}^T[T_n]\,\{h\}_{d\eta}\, S_{\eta\eta}(\omega_a, U_c) \\ S_{u_n\dot{d}_n}(\omega_a) &= i\,\omega_a\omega_r\, e^{m(z+ix_w)}\{\phi^*\}^T[T_n]\{h\}_{d\eta}\, S_{\eta\eta}(\omega_a, U_c) \end{aligned} \quad (3.87b)$$

The variance and covariance, $\sigma^2_{\dot{d}_n}$ and $\sigma_{u_n\dot{d}_n}$, are dependent on structural response velocities that makes the linearization coefficient A also dependent on the structural response. Therefore, an iterative solution is required for a correct result. At each iteration step, the linearization coefficient A is updated. The first iteration starts without wave–structure interaction. From numerical experiments it seems that a few iteration cycles are sufficient for correct results. However, it depends on the loading regime. For a drag force sensitive response, higher iteration cycles may be required.

3.10 Calculation of Consistent Current Forces of Members, Hydrodynamic Damping Ratio and Added Mass Matrices of Members

Before we present the calculation of hydrodynamic damping ratio and the added mass matrix, we make a little modification of the force acting on a flexible structural member, i.e., member is moving, shown in Fig. 3.12b. The total force acting on a flexible member is consisted of two parts as being the wave force due to Morison's equation and the inertia force of water due to structural acceleration. Thus, the force of Morison's equation given by Eq. (3.45a) is modified as written by,

$$\text{Force on a flexible member:} \rightarrow f = (C_D|u|u + C_M\dot{u}) + (\pi D^2\rho_w/4)\ddot{d} \quad (3.88a)$$

in which $\ddot{d}$ is the acceleration of the member and u is the water velocity at the location where the force is calculated. The term in the first bracket is the contribution of the Morison's equation and the term in second bracket is the contribution of the water inertia force per unit member length. By using the total relative water velocity, $(u - \dot{d} + U_c)$, instead of u in Eq. (3.88a) it can be stated as,

Table 3.13 Integration functions of the consistent current forces, q_{c1}, q_{c2}, q_{c3}, q_{c4}

Function	For $(0 \leq \lvert\alpha\rvert \leq 0.5)$	For $(\lvert\alpha\rvert \geq 0.5)$
q_{c1}	$\ell(0.5 + 3\alpha/20 + \alpha^2/30 + \alpha^3/168)$	$\beta_c[-\alpha^3 + 6\alpha + 12 + e^\alpha(6\alpha - 12)]$
q_{c2}	$\ell(0.5 + 7\alpha/20 + 2\alpha^2/15 + \alpha^3/28)$	$\beta_c[-6\alpha - 12 + e^\alpha(\alpha^3 - 6\alpha + 12)]$
q_{c3}	$\ell^2(1/12 + \alpha/30 + \alpha^2/120 + \alpha^3/630)$	$\beta_c[\alpha^2 + 4\alpha + 6 + e^\alpha(2\alpha - 6)]$
q_{c4}	$-\ell^2(1/12 + \alpha/20 + \alpha^2/60 + \alpha^3/252)$	$\beta_c[2\alpha + 6 + e^\alpha(-\alpha^2 + 4\alpha - 6)]$
Definition:	$\beta_c = (\ell/\alpha^4)$	

$$f = C_D|u - \dot{d} + U_c|(u - \dot{d} + U_c) + C_M\dot{u} - C_A\ddot{d} \tag{3.88b}$$

in which C_A is the added mass constant which is defined as,

$$\text{Added mass constant:} \rightarrow C_A = \frac{\pi D^2}{4}\rho_w c_a \rightarrow c_a = (c_m - 1) \tag{3.88c}$$

where c_a is the added mass coefficient. In the vectorial form, similar to Eq. (3.74a), the linearized normal force of the member can be written in three parts in the global coordinates as,

$$\{f_n\}_G = \{f_n\}_{G,\text{wave}} + \{f_n\}_{G,\text{current}} - \{f_{dn}\}_{G,\text{struc.}} \tag{3.88d}$$

The first term is due to waves, the second term is due to current and the third one is due to structural deformations, which are defined in the global coordinates as,

$$\begin{aligned} \{f_n\}_{G,\text{wave}} &= [T_n](C_D A\{u\} + C_M\{\dot{u}\}) \\ \{f_n\}_{G,\text{current}} &= C_D A[T_n]\{U_c\} \\ \{f_{dn}\}_{G,\text{struc.}} &= [T_n](C_D A\{\dot{d}\} + C_A\{\ddot{d}\}) \end{aligned} \tag{3.88e}$$

The member consistent force vector due to wave forces $\{f_n\}_{G,\text{wave}}$ is given by Eq. (3.60e). Calculation of the consistent current forces of members is presented in the following section.

3.10.1 Calculation of Consistent Current Forces of Members

Having introduced the current force vector $\{f_n\}_{G,\text{current}}$ from Eq. (3.88e) into Eq. (3.60d), the consistent current force vector in the global coordinates can be obtained as written by,

$$\{p_c\}_G = C_D\left(\int_0^\ell A\begin{Bmatrix} N_1[T_n] \\ N_3[T_\theta] \\ N_2[T_n] \\ N_4[T_\theta] \end{Bmatrix} ds\right)\{U_c\} \tag{3.89a}$$

in which N_i (i = 1, 2, 3, 4) are defined in Eq. (3.59c) and the vector $\{U_c\}$ is given in Eq. (3.73a). The integrations are carried out by assuming the following exponential distribution of A.

$$A = A_1 e^{\alpha\xi} \quad \rightarrow \quad \alpha = \ln(A_2/A_1) \tag{3.89b}$$

in which A_1 and A_2 are the values of A at the member ends (1) and (2), and ξ is defined in Eq. (3.59c). The result can be written as,

$$\left.\begin{array}{l}\text{Consistent current forces}\\ \text{in global coordinates}\end{array}\right\} \rightarrow \{p_c\}_G = C_D A_1 \begin{Bmatrix} q_{c1}[T_n] \\ q_{c3}[T_\theta] \\ q_{c2}[T_n] \\ q_{c4}[T_\theta] \end{Bmatrix} \{U_c\} \tag{3.89c}$$

where the integration functions q_{ci} (i = 1, 2, 3, 4) are presented in Table 3.13.

3.10.2 *Calculation of Hydrodynamic Damping Ratio*

The hydrodynamic damping ratio is calculated by using hydrodynamic damping matrices of members. The hydrodynamic damping matrix of a member in the global coordinates is calculated by introducing the velocity related part of the force vector $\{f_{dn}\}_{G,\text{struc.}}$ due to structural deformations given in Eq. (3.88e), into the member consistent force vector given by Eq. (3.60d). It can be obtained as written by,

$$\{p_{\dot{d}}\}_G = C_D \int_0^{\ell} \left(A \begin{Bmatrix} N_1[T_n] \\ N_3[T_\theta] \\ N_2[T_n] \\ N_4[T_\theta] \end{Bmatrix} \{\dot{d}\} \right) ds \tag{3.90a}$$

in which $\{\dot{d}\}$ is the translational velocity vector at a point of the member. It is stated in terms of the velocity vectors $\{\dot{d}_L\}$ and $\{\dot{d}_G\}$ at the member ends in the local and global coordinates as written by,

$$\{\dot{d}\} = [t]^T[N_u]\{\dot{d}_L\} \quad \rightarrow \quad \{\dot{d}\} = [t]^T[N_u][T]\{\dot{d}_G\} \tag{3.90b}$$

where the transformation matrix $[t]$ is given by Eq. (3.58d), the shape function matrix $[N_u]$ is given in Eqs. (3.59b) and (3.59d), the transformation matrix $[T]$ is given in Eq. (3.60a). Having introduced $\{\dot{d}\}$ from Eq. (3.90b) into Eq. (3.90a) and carried out some related matrix and vector multiplications, the member consistent force vector due to structural velocities can be obtained [115, 116] as written by,

$$\{p_{\dot{d}}\}_G = [c_h]\{\dot{d}_G\} \tag{3.90c}$$

where the matrix $[c_h]$ is a symmetric hydrodynamic damping matrix for a member defined as,

Table 3.14 Coefficients of the hydrodynamic damping matrix in Eq. (3.90d) for ($|\alpha| \geq 0.50$)

Function	For ($|\alpha| \geq 0.50$)
c_{11}	$\beta_h\left[e^{\alpha}(216\alpha^2 - 1440\alpha + 2880) - \alpha^6 + 12\alpha^4 + 24\alpha^3 - 216\alpha^2 - 1440\alpha - 2880\right]$
c_{12}	$6\,\beta_h[e^{\alpha}(\alpha^4 - 2\alpha^3 - 36\alpha^2 + 240\alpha - 480) - \alpha^4 - 2\alpha^3 + 36\alpha^2 + 240\alpha + 480]$
c_{13}	$\ell\,\beta_h\left[e^{\alpha}(72\alpha^2 - 600\alpha + 1440) + \alpha^5 + 4\alpha^4 - 12\alpha^3 - 192\alpha^2 - 840\alpha - 1140\right]$
c_{14}	$2\,\ell\beta_h[e^{\alpha}(-9\alpha^3 + 96\alpha^2 - 420\alpha + 720) + \alpha^4 + 3\alpha^3 - 36\alpha^2 - 300\alpha - 720]$
c_{22}	$\beta_h\left[e^{\alpha}\left(\alpha^6 - 12\alpha^4 + 24\alpha^3 + 216\alpha^2 - 1440\alpha + 2880\right) - 216\alpha^2 - 1440\alpha - 2880\right]$
c_{23}	$2\,\ell\beta_h[e^{\alpha}(\alpha^4 - 3\alpha^3 - 36\alpha^2 + 300\alpha - 720) + 9\alpha^3 + 96\alpha^2 + 420\alpha + 720]$
c_{24}	$\ell\,\beta_h\left[e^{\alpha}\left(-\alpha^5 + 4\alpha^4 + 12\alpha^3 - 192\alpha^2 + 840\alpha - 1440\right) + 72\alpha^2 + 600\alpha + 1440\right]$
c_{33}	$2\ell^2\beta_h[e^{\alpha}(12\alpha^2 - 120\alpha + 360) - \alpha^4 - 12\alpha^3 - 72\alpha^2 - 240\alpha - 360]$
c_{34}	$6\,\ell^2\beta_h[e^{\alpha}(-\alpha^3 + 12\alpha^2 - 60\alpha + 120) - \alpha^3 - 12\alpha^2 - 60\alpha - 120]$
c_{44}	$2\,\ell^2\beta_h[e^{\alpha}(\alpha^4 - 12\alpha^3 + 72\alpha^2 - 240\alpha + 360) - 12\alpha^2 - 120\alpha - 360]$
Definition:	$\beta_h = A_1\ell/\alpha^7$

Table 3.15 Coefficients of the hydrodynamic damping matrix in Eq. (3.90d) for ($0 \leq |\alpha| \leq 0.50$)

Function	for ($0 \leq |\alpha| \leq 0.50$)
c_{11}	$A_1\ell(13 + 3\alpha + 19\alpha^2/36 + 11\alpha^3/144)/35$
c_{12}	$A_1\ell(9 + 9\alpha/2 + 23\alpha^2/18 + 19\alpha^2/72)/70$
c_{13}	$A_1\ell^2(11/3.5 + \alpha + 17\alpha^2/84 + \alpha^3/31.5)/60$
c_{14}	$-A_1\ell^2(13/6 + \alpha + 19\alpha^2/72 + 11\alpha^3/216)/70$
c_{22}	$A_1\ell[(13/5 + 2\alpha + 29\alpha^2/36)/7 + 23\alpha^3/720]$
c_{23}	$A_1\ell^2(13/7 + \alpha + 5\alpha^2/16.8 + \alpha^3/15.75)/60$
c_{24}	$-A_1\ell^2[(11/7.5 + \alpha + 13\alpha^2/36)/28 + 7\alpha^3/1440]$
c_{33}	$A_1\ell^3(1/1.5 + \alpha/4 + \alpha^2/18 + \alpha^3/108)/70$
c_{34}	$-A_1\ell^3(1 + \alpha/2 + \alpha^2/7.2 + \alpha^3/36)/140$
c_{44}	$A_1\ell^3[(1 + \alpha/1.6 + \alpha^2/4.8)/105 + \alpha^3/2160]$

$$[c_h] = C_D \begin{bmatrix} c_{11}[T_n] & -c_{13}[T_\theta] & c_{12}[T_n] & -c_{14}[T_\theta] \\ c_{13}[T_\theta] & c_{33}[T_n] & c_{23}[T_\theta] & c_{34}[T_n] \\ c_{12}[T_n] & -c_{23}[T_\theta] & c_{22}[T_n] & -c_{24}[T_\theta] \\ c_{14}[T_\theta] & c_{34}[T_n] & c_{24}[T_\theta] & c_{44}[T_n] \end{bmatrix} \tag{3.90d}$$

in which the matrices $[T_n]$ and $[T_\theta]$ are given by Eqs. (3.55b) and (3.60c) respectively. The coefficients c_{ij} ($i = 1$ to 4 and $j = i$ to 4) are calculated from the following integrations,

$$c_{ij} = \int_0^{\ell} A\, N_i N_j ds \quad \text{where} \quad (i = 1 \text{ to } 4 \;\text{ and } j = i \text{ to } 4) \tag{3.90e}$$

By assuming an exponential distribution of the linearization coefficient A given in Eq. (3.89b), the coefficients c_{ij} are calculated and presented in Tables 3.14 and 3.15. In the dynamic analysis of structures, if the damping matrix is used, then the hydrodynamic damping matrix given by Eq. (3.90d) for an element will be added to the member structural damping matrix. Since, in the mode superposition method of the dynamic analysis (see Sect. 1.7.1.5), the damping ratios to the critical are used, the hydrodynamic damping ratios to the critical will be calculated first to add them to the structural damping ratios. For this purpose, generalized hydrodynamic dampings will be calculated for the eigenmodes as stated by, for an eigenmode r,

$$(c_h)_r = \{\phi\}_r^T [C_h] \{\phi\}_r \quad \rightarrow \quad (c_h)_r = \sum_{i=1}^{NLM} \{\phi_i\}_r^T [c_h]_i \{\phi_i\}_r \tag{3.91a}$$

in which $(c_h)_r$ is the generalized hydrodynamic damping for the eigenmode r, $[C_h]$ is the hydrodynamic damping matrix of the structural system, $\{\phi\}_r$ is the eigenmode vector, NLM denotes the number of loaded members, $[c_h]_i$ is given by Eq. (3.90d) for the member i. Having calculated the generalized hydrodynamic damping for an eigenmode, the corresponding damping ratio is calculated from,

$$\text{Hydrodynamic damping ratio:} \rightarrow (\xi_h)_r = \frac{(c_h)_r}{2\omega_r m_r} \tag{3.91b}$$

in which ω_r is the natural frequency and m_r is the generalized mass for the eigenmode r. Then, the total vibration damping ratio for the eigenmode r can be obtained by adding the structural and hydrodynamic ratios as written by,

$$\text{Total damping ratio:} \rightarrow \quad \xi_r = (\xi_s)_r + (\xi_h)_r \tag{3.91c}$$

where $(\xi_s)_r$ denotes the structural damping ratio for the eigenmode r. Having obtained the vibration damping ratio ξ_r, the mode superposition method can be used to find a dynamic response as it is explained in Sect. 1.7.1.5.

3.10.3 Calculation of Added Mass Matrices of Members

The added mass matrix of a member in the global coordinates is calculated by introducing the acceleration related part of the force vector $\{f_{dn}\}_{G,\text{struc.}}$, which is given in Eq. (3.88e), into the member consistent force vector given by Eq. (3.60d). The consistent force vector due to structural accelerations can be obtained as,

$$\{p_{\ddot{d}}\}_G = C_A \int_0^{\ell} \left(\begin{Bmatrix} N_1[T_n] \\ N_3[T_\theta] \\ N_2[T_n] \\ N_4[T_\theta] \end{Bmatrix} \{\ddot{d}\} \right) ds \tag{3.92a}$$

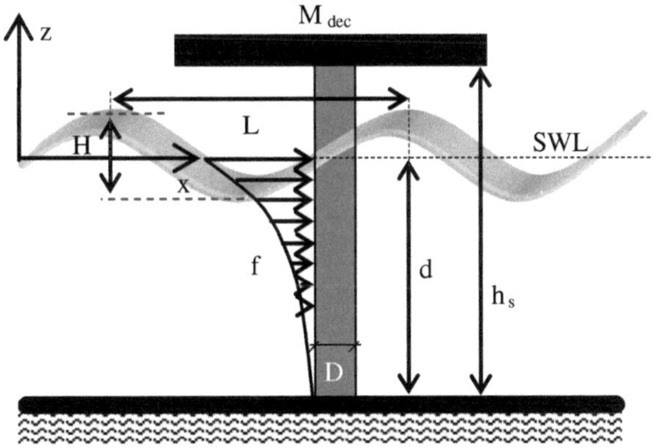

Fig. 3.18 A monopod tower under a linear wave and the wave loading

in which $\{\ddot{d}\}$ is the translational acceleration vector at a point of the member. Having introduced $\{\ddot{d}\}$ into Eq. (3.92a) the consistent force vector $\{p_{\ddot{d}}\}_G$ can be obtained as written by, similar to Eq. (3.90c),

$$\{p_{\ddot{d}}\}_G = [m_a]\{\ddot{d}_G\} \tag{3.92b}$$

where the matrix $[m_a]$ is a symmetric added mass matrix of the member, which is produced by the co-vibrating water around the member. As similar to the hydrodynamic damping matrix of a loaded member, the added mass matrix can be obtained in the global coordinate system as stated by,

$$[m_a] = \frac{\ell C_A}{35}\begin{bmatrix} 13\,[T_n] & -(11\ell/6)\,[T_\theta] & (9/2)\,[T_n] & (13\ell/12)\,[T_\theta] \\ (11\ell/6)[T_\theta] & (\ell^2/3)\,[T_n] & (13\ell/12)\,[T_\theta] & -(\ell^2/4)\,[T_n] \\ (9/2)\,[T_n] & -(13\ell/12)[T_\theta] & 13\,[T_n] & (11\ell/6)[T_\theta] \\ -(13\ell/12)[T_\theta] & -(\ell^2/4)\,[T_n] & -(11\ell/6)[T_\theta] & (\ell^2/3)[T_n] \end{bmatrix} \tag{3.92c}$$

Having calculated the added mass matrix, the total mass matrix of a loaded member will be calculated simply by superimposing the added and structural mass matrices. Thus, in the global coordinates, it is written as,

$$[m] = [m_s] + [m_a] \tag{3.92d}$$

in which $[m_s]$ is the structural mass matrix of the member in the global coordinates. In the following section, application of wave forces to a monopod offshore tower is presented.

3.11 Examples

The static analysis of a 2D jacket structure under wave loads has been presented in Sect. 1.8.2 by using the SAPOS program [117]. In this section, static analysis of a monopod tower will be presented by using the Airy (linear) wave theory. The monopod tower is shown in Fig. 3.18. It is subjected to a linear wave with wavelength L, wave height H and wave number m. The water depth is denoted by d, the height and diameter of the tower are denoted by h_s and D respectively, the wave load is denoted by f. SWL and M_{dec} indicate respectively the still water level and the mass of the deck. The maximum bending moment and shear force of the tower at the bottom are required under:

- Deterministic Wave Loading
- Random Wave Loading

The calculations of maximum values of the bending moment and shear force of the tower at the bottom are presented for these two cases in the following sections.

3.11.1 Example of a Monopod Tower Under Deterministic Wave Loading

The wave loading acting on a monopod tower is calculated by using the Morison's equation given by Eq. (3.45a) assuming that the wave profile is sinusoidal as given in Eq. (3.67a) and the structural response is static, i.e., dynamic response contribution is not considered. Under these conditions, the velocity and acceleration are stated from Eq. (3.67a) as,

$$\left.\begin{aligned} u &= \hat{\eta}\ h(z)\ \sin\theta \\ \dot{u} &= \hat{\eta}\ \omega\ h(z)\ \cos\theta \end{aligned}\right\} \rightarrow\ h(z) = \omega\frac{\cosh m(z+d)}{\sinh md}\quad \text{and}\ (\theta = \omega t - mx) \tag{3.93}$$

where $\hat{\eta}$ is the wave amplitude. Having used these quantities in Eq. (3.45a), the distributed wave force *f*(*z*) can be written as,

$$f(z) = \hat{\eta}\ h(z)[C_D\hat{\eta}\ h(z)\,|\,\sin\theta|\ \sin\theta + C_M\,\omega\ \cos\theta] \tag{3.94}$$

in which the drag and inertia force constants, C_D and C_M, are defined in Eq. (3.45b). Since maximum bending moment and shear force of the tower will be calculated here, the maximum value of the Morison's wave loading is not calculated at this time. The bending moment of the tower at the bottom is calculated from the following integration,

$$M = \int_{z=-d}^{0} (d+z) f(z)\, \mathrm{d}z \tag{3.95a}$$

Having introduced $f(z)$ from Eq. (3.94) into Eq. (3.95a), the bending moment is written as,

$$\begin{aligned} M = & \left(\hat{\eta}^2 C_D \int_{z=-d}^{0} (d+z)\ h^2(z)\ \mathrm{d}z \right) |\sin\theta|\ \sin\theta + \ldots \\ & \ldots + \left(\hat{\eta}\omega C_M \int_{z=-d}^{0} (d+z)\ h(z)\ \mathrm{d}z \right) \cos\theta \end{aligned} \tag{3.95b}$$

from which it can be calculated as expressed by,

$$M = \hat{\eta}\frac{g}{m}(\hat{\eta}\, B_d|\ \sin\theta|\ \sin\theta + B_m cos\,\theta) \tag{3.95c}$$

where the constants B_d and B_m of the drag and inertia terms are obtained as written below.

$$\begin{aligned} B_d &= \frac{C_D}{4}\left[2md\left(1 + \frac{md}{\sinh(2md)}\right) - \tanh md \right] \\ B_m &= C_M\left(md\tanh(md) + \frac{1}{\cosh(md)} - 1 \right) \end{aligned} \tag{3.95d}$$

The maximum value of M occurs at the time station obtained from the derivative of M written as,

$$\frac{\partial M}{\partial t} = 0 \quad \rightarrow \quad (2\hat{\eta}B_d|\ \sin\theta| cos\,\theta - B_m \sin\theta = 0) \tag{3.96a}$$

from which the θ value is obtained from the solution of,

$$\cos\theta = \frac{B_m}{2\hat{\eta}B_d}\frac{\sin\theta}{|\sin\theta|} \quad \rightarrow \quad \cos\theta = \begin{cases} B_m/(2\hat{\eta}B_d) & \text{if } (\sin\theta > 0) \\ -B_m/(2\hat{\eta}B_d) & \text{if } (\sin\theta < 0) \end{cases} \tag{3.96b}$$

The $(-)$ value of $\cos\theta$ corresponds to the minimum value of M, and therefore we take only the (+) value which corresponds to the maximum value of M. Thus,

$$\cos\theta = \frac{B_m}{2\hat{\eta}B_d} \quad \rightarrow \quad (0 \le \theta \le \pi/2) \quad \rightarrow \left(0 \le \frac{B_m}{2\hat{\eta}B_d} \le 1 \right) \tag{3.96c}$$

The value of θ is calculated from Eq. (3.96c) provided that the indicated boundary condition is satisfied. Otherwise, the maximum moment is produced purely due to the inertia force term of the Morison's equation. Having used this θ value in Eq.

(3.95c) and by taking into account the validity condition indicated, the maximum bending moment can be stated as,

$$M_{\max} = \begin{cases} \dfrac{g}{m}\left(\hat{\eta}^2 B_d + \dfrac{B_m^2}{4B_d}\right) & \text{if} \quad \left(\dfrac{B_m}{2B_d} \leq \hat{\eta}\right) \\ \hat{\eta}\dfrac{g}{m}B_m & \text{if} \quad \left(\dfrac{B_m}{2B_d} > \hat{\eta}\right) \end{cases} \tag{3.96d}$$

The total shear force at the bottom of the tower is calculated from the following integration,

$$V = \int\limits_{z=-d}^{0} f(z)\,\mathrm{d}z \tag{3.97a}$$

or, having introduced $f(z)$ from Eq. (3.94) into Eq. (3.97a) and carried out the integration, it can be obtained that,

$$V = \hat{\eta}g(\hat{\eta}V_d|\sin\theta|\sin\theta + V_m\cos\theta) \rightarrow \begin{cases} V_d = \dfrac{C_D}{2}\left(1 + \dfrac{2md}{\sinh(2md)}\right) \\ V_m = C_M\tanh(md) \end{cases} \tag{3.97b}$$

which is similar to the statement of the bending moment given by Eq. (3.95c). Using this similarity, the maximum shear force at the bottom of the tower can be expressed as written by,

$$V_{\max} = \begin{cases} g\left(\hat{\eta}^2 V_d + \dfrac{V_m^2}{4V_d}\right) & \text{if} \quad \left(\dfrac{V_m}{2V_d} \leq \hat{\eta}\right) \\ g\hat{\eta}V_m & \text{if} \quad \left(\dfrac{V_m}{2V_d} > \hat{\eta}\right) \end{cases} \tag{3.97c}$$

For a given wave (with wave height H and wavelength L), maximum values of the static bending moment and total shear force at the bottom of the tower can be calculated from Eqs. (3.96d) and (3.97c). For the deep water condition, the constants of the drag and inertia terms of the maximum bending moment and shear force can be simplified as written below.

$$\text{For deep water condition:} \rightarrow \begin{Bmatrix} V_d = \dfrac{C_D}{2} \\ V_m = C_M \end{Bmatrix} \text{ and } \begin{Bmatrix} B_d = \dfrac{C_D}{4}(2md - 1) \\ B_m = C_M(md - 1) \end{Bmatrix} \tag{3.98}$$

As it can be seen from these statements and Eqs. (3.97c) and (3.96d), for the deep water condition, the maximum of the static shear force depends only on the wave amplitude while the maximum static bending moment depends on both wave amplitude and wave number.

3.11.2 Example of a Monopod Tower Under Random Wave Loading

In this section, the bending moment and shear force of the tower at the bottom is calculated under random wave loading. It can be considered in two cases as:

- random maximum values of the bending moment and shear force at a given fixed frequency, i.e., a given wave profile with random amplitudes
- random bending moment and shear force in the whole frequency region, i.e., random wave profiles and amplitudes as given by Eq. (3.20), at the bottom of the tower

In the first case, the randomness in the maximum values is introduced by the wave amplitude $\hat{\eta}$ only as it can be seen from Eqs. (3.96a) and (3.97c). Since the water elevation η is assumed to be a stationary zero-mean narrow-banded Gaussian process, the probability distribution of its amplitude will be a Rayleigh distribution. Thus, the probability density function of $\hat{\eta}$ can be stated as similar to Eq. (3.70) by,

$$f_{\hat{\eta}}(\hat{\eta}) = \frac{\hat{\eta}}{\sigma_\eta^2} \exp\left(-\frac{\hat{\eta}^2}{2\sigma_\eta^2}\right) \quad \rightarrow \text{for } (0 \le \hat{\eta} \le \infty) \tag{3.99}$$

in which σ_η is the standard deviation of the water elevation η. The mean values of the maximum bending moment and shear force are calculated by using Eqs. (3.96d) and (3.97c) from the following integrations.

$$\begin{aligned} E[M_{\max}] &= \frac{g}{m}\left(B_m \int_0^{\alpha_m} \hat{\eta}\, f_{\hat{\eta}}(\hat{\eta})\, d\hat{\eta} + \int_{\alpha_m}^{\infty} \left(B_d \hat{\eta}^2 + \alpha_m^2\right) f_{\hat{\eta}}(\hat{\eta})\, d\hat{\eta}\right) \\ E[V_{\max}] &= g\left(V_m \int_0^{\alpha_v} \hat{\eta}\, f_{\hat{\eta}}(\hat{\eta})\, d\hat{\eta} + \int_{\alpha_v}^{\infty} \left(V_d \hat{\eta}^2 + \alpha_v^2\right) f_{\hat{\eta}}(\hat{\eta})\, d\hat{\eta}\right) \end{aligned} \tag{3.100a}$$

in which the parameters α_m and α_v are defined as,

$$\left(\alpha_m = \frac{B_m}{2B_d}\right) \quad \text{and} \quad \left(\alpha_v = \frac{V_m}{2V_d}\right) \tag{3.100b}$$

Having carried out the integrations in Eq. (3.100a), the required means of maximum values can be obtained as written below.

$$E[M_{\max}] = \frac{g}{m}\left[\begin{array}{l}\sqrt{\frac{\pi}{2}}B_m\sigma_\eta\left(\operatorname{erf}(\mu_m) - \frac{2\mu_m}{\sqrt{\pi}}e^{-\mu_m^2}\right) - \ldots \\ \ldots - \left[\alpha_m^2 + 2B_d\sigma_\eta^2(\mu_m^2+1)\right]e^{-\mu_m^2} + 2\sigma_\eta^2(B_d + \mu_m^2)\end{array}\right] \tag{3.101a}$$

$$E[V_{\max}] = g\left[\begin{array}{l}\sqrt{\frac{\pi}{2}}V_m\sigma_\eta\left(\operatorname{erf}(\mu_v) - \frac{2\mu_v}{\sqrt{\pi}}e^{-\mu_v^2}\right) - \ldots \\ \ldots - \left[\alpha_v^2 + 2V_d\sigma_\eta^2(\mu_v^2+1)\right]e^{-\mu_v^2} + 2\sigma_\eta^2(V_d + \mu_v^2)\end{array}\right] \tag{3.101b}$$

in which the parameters μ_m and μ_v are defined by,

$$\left(\mu_m^2 = \alpha_m^2/\left(2\sigma_\eta^2\right)\right) \quad \text{and} \quad \left(\mu_v^2 = \alpha_v^2/\left(2\sigma_\eta^2\right)\right) \tag{3.101c}$$

For a given wave period T (wave number is calculated from the dispersion relation given by Eq. (3.14b)), the mean values of the maximum bending moment and shear force are calculated from Eqs. (3.101a) and (3.101b).

In the second case, when the wave profile and amplitude are both random, the calculation of the static bending moment and shear force will be somewhat different than those presented above in the sense that they will be expressed in terms of the random water elevation η. For this purpose, we use the real and imaginary parts of the water elevation η as stated from Eq. (3.15) by,

$$(\eta = \hat{\eta}\sin\theta) \quad \text{and} \quad (i\eta = \hat{\eta}\cos\theta) \tag{3.102}$$

Having introduced these relations into Eqs. (3.95c) and (3.97b), the static bending moment and shear force can be written as,

$$M = \frac{g}{m}(B_d|\eta|\,\eta + iB_m\eta) \quad \text{and} \quad V = g(V_d\,|\eta|\,\eta + i\,V_m\eta) \tag{3.103}$$

which are nonlinear functions of the random water elevation η. In order to apply the spectral method one of the linearization technique presented in Sect. 3.8 is used, e.g., linearization by using the criterion of the minimum mean square error. Then, the linearization constant A will be, from Eq. (3.63c),

$$A = \sqrt{\frac{8}{\pi}}\,\sigma_\eta \tag{3.104}$$

With this value of A, Eq. (3.103) can be stated in the linearized form as written by,

$$M = \frac{g}{m}(B_dA + iB_m)\eta \quad \text{and} \quad V = g(V_dA + i\,V_m)\eta \tag{3.104b}$$

which are stationary zero-mean Gaussian processes as they follow the same process of η. The spectra of M and V are calculated from,

$$S_M(\omega) = \left(\frac{g}{m}\right)^2 \left(B_d^2 A^2 + B_m^2\right) S_{\eta\eta}(\omega)$$
$$S_V(\omega) = g^2 \left(V_d^2 A^2 + V_m^2\right) S_{\eta\eta}(\omega) \tag{3.105}$$

Their variances are calculated from the following integrations of their spectra.

$$\sigma_M^2 = \int_0^\infty S_M(\omega)\,\mathrm{d}\omega \quad \text{and} \quad \sigma_V^2 = \int_0^\infty S_V(\omega)\,\mathrm{d}\omega \tag{3.106}$$

As similar to the amplitude of η, the maxima of M and V have Rayleigh probability distributions from which the means and variances can be readily calculated from Eq. (2.71b) as written by,

$$\left\{\begin{aligned} \mu_{M_{\max}} &= \sqrt{\pi/2}\,\sigma_M \\ \sigma_{M_{\max}}^2 &= (2-\pi/2)\sigma_M^2 \end{aligned}\right\} \quad \text{and} \quad \left\{\begin{aligned} \mu_{V_{\max}} &= \sqrt{\pi/2}\,\sigma_V \\ \sigma_{V_{\max}}^2 &= (2-\pi/2)\sigma_V^2 \end{aligned}\right\} \tag{3.107}$$

in which $\mu_{M_{\max}}$ and $\mu_{V_{\max}}$ denote mean values of maxima of M and V respectively which are different than those presented by Eqs. (3.101a) and (3.101b).

Exercise 1

For given waves, evaluate the mean values of $M_{\max}$ and $V_{\max}$ presented in Eqs. (3.107) (3.101a) and (3.101b). Explain the differences in these mean values.

Exercise 2

The spectra of M and V presented in Eq. (3.105) are obtained assuming that the linearization process is carried out on the bending moment and shear force given in Eq. (3.103). Their spectra can also be calculated by using a linearization process of the Morison's equation. The followings are required.

- Find out the differences of the variances of M and V calculated from the different spectral approaches, i.e., from Eq. (3.105) and those based on a linearization process of the Morison's equation.
- Explain these differences.

Exercise 3

The aforementioned examples are carried out by using the Airy (linear) wave theory. In this exercise, the Stokes Second-order Wave Theory will be used. The following horizontal water particle velocity and acceleration are given.

$$\left.\begin{aligned} u &= \hat{\eta}\,\omega\left[\frac{\cosh y}{\sinh(md)}\sin\theta + \frac{3}{4}\hat{\eta}\,m\frac{\cosh 2y}{\sinh^4(md)}\sin 2\theta\right] \\ \dot{u} &= \hat{\eta}\,\omega^2\left[\frac{\cosh y}{\sinh(md)}\cos\theta + \frac{3}{2}\hat{\eta}m\frac{\cosh 2y}{\sinh^4(md)}\cos 2\theta\right] \end{aligned}\right\} \rightarrow y = m(Z+d) \tag{3.108}$$

It is required that,

- Repeat the example under deterministic wave loading presented in Sect. 3.11.1 by using the Stokes Second-order Wave Theory.
- Linearize the nonlinear velocity and acceleration given in Eq. (3.108) in terms of the first-order water elevation η, which is assumed as $(\eta = \hat{\eta}\,\sin\theta)$.
- Linearize Morison's equation in terms of η assuming that it is stated by using the velocity acceleration given in Eq. (3.108).
- Repeat the example under random wave loading presented in Sect. 3.11.2 by using the linearizations made above.

References

1. Smith CB (2006) Extreme waves. Joseph Henry Press, Washington
2. Reddy MPM (2001) Descriptive physical oceanography. A.A Balkema, Leiden
3. Brandt W, Crowley D, Hodder M, Juiniti R, Ohara S, Rushton S (1998) Deepening the search for offshore hydrocarbons. Schlumberger Oilfield Rev 10(1):2–21
4. REN21 (2009) Renewables global status report: update, Paris
5. Kharif C, Pelinovsky E, Slunyaev A (2009) Rogue waves in the ocean. Springer, Berlin
6. Goda Y (1985) Random seas and design of maritime structures, 2nd edn. World Scientific, Singapore
7. Ochi MK (1998) Ocean waves—the stochastic approach. Cambridge University Press, Cambridge
8. Sawaragi T (1995) Coastal engineering—waves, beaches, wave-structure interactions. Elsevier, Amsterdam
9. Wiegel RL (1964) Oceanographical engineering. Prentice-Hall, London
10. LeMehaute B (1976) An Introduction to hydrodynamics and water waves. Springer, Heidelberg
11. Newman JN (1977) Marine hydrodynamics. MIT Press, Cambridge
12. LeBlond PH, Mysak LA (1978) Waves in the ocean. Elsevier, Amsterdam
13. Mei CC (1983) The applied dynamics of ocean surface waves. Wiley, New York
14. Chakrabarti SK (1987) Hydrodynamics of offshore structure. Springer, Heidelberg
15. Dean RG, Dalrymple RA (1991) Water wave mechanics for engineers and scientists. World Scientific, Singapore
16. Sorensen RM (1993) Basic wave mechanics: for coastal and ocean engineers. Wiley, New York
17. Tucker MJ, Pitt EG (2001) Waves in ocean engineering. Elsevier, Amsterdam
18. Falnes J (2002) Ocean waves and oscillating systems: linear interactions including wave-energy. Cambridge University Press, Cambridge
19. Mader CL (2004) Numerical modeling of water waves, 2nd edn. CRC Press, Boca Raton
20. Mei CC, Stiassnie M, Yue DK-P (2005) Theory and applications of offshore surface waves-part 1: linear aspects. World Scientific, Singapore
21. Mei CC, Stiassnie M, Yue DK-P (2005) Theory and applications of offshore surface waves-part 2: nonlinear aspects. World Scientific, Singapore
22. Brebbia CA, Walker S (1979) Dynamic analysis of offshore structures. Butterworths, London

23. Barltrop NDP, Adams AJ (1991) Dynamics of fixed marine structures. Butterworth-Heinemann, Oxford
24. Lamb H (1993) Hydrodynamics. Cambridge University Press, Cambridge
25. Fenton JD (1988) The numerical solution of steady water wave problems. Comput Geosci 14(3):357–368
26. Newman JN (1990) Numerical solutions of the water-wave dispersion relation. Appl Ocean Res 12(1):14–18
27. Fenton JD (1999) Numerical methods for nonlinear waves. In: Liu PL-F (ed) Advances in coastal and ocean engineering, vol 5. World Scientific, Singapore, pp 241–324
28. Bai KJ, Choo SM, Chung SK, Kim DY (2005) Numerical solutions for nonlinear free surface flows by finite element methods. Appl Math Comput 163:941–959
29. Toffolia A, Onorato M, Bitner-Gregersena E, Osborne AR, Babanin AV (2008) Surface gravity waves from direct numerical simulations of the Euler Equations: a comparison with second-order theory. Ocean Eng 35:367–379
30. Xi-zeng Z, Zhao-chen S, Shu-xiu L, Chang-hong H (2009) A numerical method for nonlinear water waves. J. Hydrodynamics 21(3):401–407
31. Xu L, Guyenne P (2009) Numerical simulation of three-dimensional nonlinear water waves. J Comput Phys 228:8446–8466
32. Skjelbreia L (1959) Gravity waves, stokes' third order approximation; tables of functions. Engineering Foundation Council of Wave Research, Berkeley
33. Laitone EV (1962) Limiting conditions for cnoidal and Stokes waves. J Geophys Res 67(4):1555–1564
34. Hsu JRC, Tsuchiya Y, Silvester (1979) Third-order approximation to short-crested waves. J Fluid Mech 90(1):179–196
35. Morison JR, O'Brien MP, Johnson JW, Schaaf SA (1950) The force exerted by surface wave on piles. Pet Trans, AIME 189:149–154
36. Korteweg DJ, de Vries G (1895) On the change of form of long waves advancing in a rectangular canal, and on a new type of long stationary waves. Phil Mag Ser 5(39):422–443
37. Fenton JD (1979) A high-order cnoidal wave theory. J Fluid Mech 94(1):129–161
38. Karabulut EA (2000) Higher-order approximation of Cnoidal-wave Theory. J Appl Mech Tech Phys 41(1):84–94
39. Sander J, Hutter K (1991) On the development of the theory of the solitary wave, a historical essay. Acta Mech 86:111–152
40. Craik ADD (2004) The origins of water wave theory. Annu Rev Fluid Mech 36:1–28
41. Kinsman B (1965) Wind waves, their generation and propagation on the ocean surface. Prentice-Hall, Englewood Cliffs
42. Longuet-Higgins MS (1957) The statistical analysis of a random, moving surface. Phil Trans Roy Soc Lond A 249:321–387
43. Longuet-Higgins MS (1984) Statistical properties of wave groups in a random sea state. Phil Trans Roy Soc Lond A 312:219–250
44. Ochi MK (1982) Stochastic analysis and probabilistic prediction of random seas. Adv Hydrosci 13:217–375
45. Goda Y (1990) Random waves and spectra. In: Herbich JB (ed) Handbook of coastal and ocean engineering, vol 1. Gulf Publishing, Houston, pp 175–212
46. Memos C, Tzanis K, Zographou K (2002) Stochastic description of sea waves. J Hydraulic Res 40(3):265–274
47. Pierson WJ, Moskowitz L (1964) A proposed spectral form for fully developed wind seas based on the similarity theory of S. A. Kitaigorodskii. J Geophys Res 69(24):5181–5190
48. Hasselmann K, Barnett TP, Bouws E, Carlson H, Cartwright DE, Enke K, Ewing JA, Gienapp H, Hasselmann DE, Kruseman P, Meerburg A, Müller P, Olbers DJ, Richter K, Sell W, Walden H (1973) Measurements of wind-wave growth and swell decay during the Joint North Sea Wave Project (JONSWAP). Dtsch Hydrogr Z A8(12):1–95
49. Komen GJ, Hasselmann K, Hasselmann K (1984) On the existence of a fully developed wind-sea spectrum. J Phys Oceanogr 14(8):1271–1285

50. Battjes JA, Zitman TJ, Holthuijsen LH (1987) Reanalysis of the spectra observed in JONSWAP. J Phys Oceanogr 17(8):1288–1295
51. DNV-RP-C205 (2007) Environmental conditions and environmental loads. recommended practice. Det Norske Veritas, Høvik
52. Le Méhauté B, Hanes DM (1990) Ocean engineering science, Part 1. Wiley, New York
53. Ochi MK, Hubble EN (1976) On six-parameter wave spectra. In: Proceedings of the 15th Coastal Engineering Conference, vol 1, pp 301–328
54. Torsethaugen K (1996) Model for a doubly peaked wave spectrum. SINTEF report STF22 A96204, Trondheim
55. Torsethaugen K, Haver S (2004) Simplified double peak spectral model for ocean waves. In: Proceedings of 14th international offshore and polar engineering conference. ISOPE-2004, vol 3, pp 76–84
56. Soares CG, Nolasco MC (1992) Spectral modelling of sea states with multiple wave systems. J Offshore Mech Arct Eng 114(4):278–284
57. Longuet-Higgins MS (1962) The directional spectrum of ocean waves, and processes of wave generation. Proc Roy Soc Lond A265:286–315
58. Borgman LE (1969) Directional spectra models for design. Proc Offshore Technol Conf OTC1069:721–746
59. Hasselmann DE, Dunckel M, Ewing JA (1980) Directional wave spectra observed during JONSWAP 1973. J Phys Oceanogr 10(8):1264–1280
60. Donelan MA, Hamilton J, Hui WH (1985) Directional spectra of wind-generated waves. Phil Trans Roy Soc Lond A 315(1534):509–562
61. Kumar VS, Deo MC, Anand NM, Chandramohan P (1999) Estimation of wave directional spreading in shallow water. Ocean Eng 26:83–98
62. Wen SC, Guo PF, Zhang DC (1993) Analytically derived wind-wave directional spectrum: Part 1. derivation of the spectrum. J Oceanogr 49:131–147
63. Chakrabarti SK (1990) Nonlinear methods in offshore engineering. Elsevier, Amsterdam
64. Peregrine DH (1976) Interaction of water waves and currents. Advan Appl Mech 16:9–117
65. Longuet-Higgins MS, Stewart RW (1961) The changes of amplitude of short gravity waves on steady non-uniform currents. J Fluid Mech 10:529–549
66. Thomas GP (1981) Wave-current interactions: an experimental and numerical study, Part 1, Linear waves. J Fluid Mech 110:457–474
67. Ismail NM (1983) Effects of wave-current interaction on the design of marine structures. Proc Offshore Technol Conf OTC4615:307–316
68. Ismail NM (1984) Wave-current models for design of marine structures. J Waterw Port Coastal Ocean Eng 110(4):432–447
69. Srokosz MA (1987) Models of wave-current interaction. Advances in underwater technology, ocean science and offshore engineering, vol 12: modelling the offshore environment. Graham and Trotman Ltd., London, pp 313–325
70. Baddour RE, Song S (1990) On the interaction between waves and currents. Ocean Eng 17(1–2):1–21
71. Hedges TS (1987) Combinations of waves and currents: an introduction. Proc Inst Civ Eng 82:567–585
72. Huang NE, Chen DT, Tung CC, Smith JR (1972) Interactions between steady non-uniform currents and gravity waves with applications for current measurements. J Phys Oceanogr 2:420–431
73. Tung CC, Huang NE (1973) Combined effects of current and waves on fluid force. Ocean Eng 2:183–193
74. Tayfun MA, Dalrymple RA, Yang CY (1976) Random wave-current interactions in water of varying depth. Ocean Eng 3:403–420
75. Hedges TS, Anastasion K, Gabriel D (1985) Interaction of random waves and currents. J Waterw Port Coastal Ocean Eng 111(2):275–288
76. Burrows R, Hedges TS (1985) The influence of currents on ocean wave climates. Coastal Eng 9:247–260

77. Li YC (1990) Wave-current interaction. In: Herbich JB (ed) Handbook of coastal and ocean engineering, vol 1. Gulf Publishing, Houston, pp 704–726
78. Masson D (1996) A case study of wave–current interaction in a strong tidal current. J Phys Oceanogr 26:359–372
79. Bretherton FP, Garrett CJR (1969) Wave trains in inhomogeneous moving media. Proc Roy Soc Lond A302:529–554
80. Bouwse E (1978). Wind and wave climate in the Netherlands sectors of the North Sea between 53° and 54° north latitude. Report W.R. 78-9, Koninklijk Netherlands Meterologisch Inst., De Build
81. Haver S (1987) On the joint distribution of heights and periods of sea waves. Ocean Eng 14(5):359–376
82. Ferreira JA, Soares CG (2002) Modelling bivariate distributions of significant wave height and mean wave period. Appl Ocean Res 24:31–45
83. Capitao R, Burrows R (1995) Wave predictions based on scatter diagram data. A review. Adv Eng Softw 23:37–47
84. Mathisen J, Bitner-Gregerse E (1990) Joint distributions for significant wave height and wave zero-up-crossing period. Appl Ocean Res 12(2):93–103
85. Kumar VS, Deo MC (2004) Design wave estimation considering directional distribution of waves. Ocean Eng 31:2343
86. Battjes JA (1972) Long-term wave height distributions at seven stations around the British Isles. Ocean Dyn 25(4):179–189
87. Haver S (1985) Wave climate off northern Norway. Appl Ocean Res 7:85–92
88. Nordenstrom N (1969) Long term distribution of wave heights and periods. DNV Rep. No. 69-21-S, Oslo
89. Bitner-Gregersen EM (2005) Joint probabilistic description for combined seas. In: Proceedings of OMAE 2005 conference of OMAE 2005-67382, Halkidiki, Greece
90. Nagai S (1973) Wave forces on structures. Adv Hydrosci 9:253–324
91. Sarpkaya T, Isaacson M (1981) Mechanics of wave forces on offshore structures. Van Nostrand Reinhold, New York
92. Clauss G, Lehmann E, Ostergaard C (1992) Offshore structures, vol. 1, Conceptual design and hydromechanics. Springer, London
93. API RP 2A WSD (1993) Recommended practice for planning, designing and constructing fixed offshore platforms, working stress design (20 edn). API, Dallas
94. Gudmestad OT, Moe G (1996) Hydrodynamic coefficients for calculation of hydrodynamic loads on offshore truss structures. Mar Struct 9:745–758
95. Sarpkaya T (1986) Force on a circular cylinder in viscous oscillatory flow at low Keulegan–Carpenter numbers. J Fluid Mech 165:61–71
96. Haritos N (2007) Introduction to the analysis and design of offshore structures—an overview. EJSE Special Issue Load Struct 7:55–65
97. Havelock TH (1940) The pressure of water waves upon a fixed obstacle. Proc Roy Soc Lond A175(963):409–421
98. MacCamy RC, Fuchs RA (1954) Wave forces on piles: a diffraction theory. Tech. Memo, 69, U.S. Army Corps of Engineers, Beach Erosion Board, Washington DC
99. Rahman M, Chakravartty IC (1981) Hydrodynamic loading calculations for offshore structures. SIAM J Appl Math 41(3):445–458
100. Rahman M, Chehil DS (1982) Second order wave force on a large cylinder. Acta Mech 44:127–136
101. Taylor RE, Kernot MP (1999) On second order wave loading and response in irregular seas. In: Liu, PL-F (ed) Advances in coastal and ocean engineering. vol 5. World Scientific, Singapore, pp 155–212
102. Abramowitz M, Stegun IA (eds) (1972) Handbook of mathematical functions with formulas, graphs and mathematical tables. Dover, New York
103. Atalik TS, Utku S (1976) Stochastic linearization of multi-degree-of-freedom non-linear systems. Earthq Eng Struct Dyn 4:411–420

104. Roberts JB, Spanos PD (1990) Random vibration and statistical linearization. Wiley, Chichester
105. Borgman LE (1967) Random hydrodynamic forces on objects. Ann Math Stat 38(1):37–51
106. Gudmestad OT, Connor JJ (1983) Linearization methods and the influence of current on the nonlinear hydrodynamic drag force. Appl Ocean Res 5(4):184–194
107. Langley RS (1984) The linearisation of three-dimensional drag force in random seas with current. Appl Ocean Res 6(3):126–131
108. Leira BJ (1987) Multidimensional stochastic linearisation of drag forces. Appl Ocean Res 9(3):150–162
109. Chen Y-H, Lin F-M (1989) General drag-force linearization for nonlinear analysis of marine risers. Ocean Eng 16(3):265–280
110. Karadeniz H (1993) Wave-current and fluid-structure interaction effects on the stochastic analysis of offshore structures. Int J Offshore Polar Eng 3(2):107–114
111. Kareem A, Hsieh CC, Tognarelli MA (1998) Frequency-domain analysis of offshore platform in non-Gaussian seas. J. Eng. Mech. 124(6):668–683
112. Hartnett M, Mullarkey T (1999) Statistical equivalent linearisation of drag forces on immersed slender members. Adv Eng Softw 30:657–662
113. Wolfram J (1999) On alternative approaches to linearization and Morison's equation for wave forces. Proc Roy Soc Lond A455:2957–2974
114. Karadeniz H (1992) Stochastic analysis of offshore structures under wave-current and fluid-structure interactions. In: Proceedings of 11th international conference on offshore mechanics and arctic engineering, OMAE, 1A, pp 241–248
115. Karadeniz H (1995) A stochastic analysis approach for the calculation of hydrodynamic damping. In: Proceedings of 14th international conference on offshore mechanics and arctic engineering, OMAE, 1B, pp 89–94
116. Karadeniz H (1999) Spectral analysis of offshore structures under combined wave and earthquake loadings. In: Proceedings of 9th international offshore and polar engineering conference, ISOPE, vol 4, pp 504–511
117. Karadeniz H (2009) SAPOS-spectral analysis program of structures. Report, Structural Mechanics Division, Faculty of Civil Engineering and Geosciences, Delft University of Technology, Delft

Chapter 4
Spectral Analysis of Offshore Structures Under Wave and Earthquake Loadings

4.1 Introduction

Structural response analyses under dynamic loading are well documented in general in many text books, see e.g. [1–10], especially for offshore applications, see e.g. [11–14]. The response analysis can be carried out either in the time domain or in the frequency domain as depending on the analysis and the loading type [15]. For a nonlinear analysis, the time domain approach is traditionally used [16] and, for the linear analysis, both time domain and frequency domain analyses may be used. For the spectral analysis, a frequency domain approach is more adequate and therefore it is used generally. Offshore structures are continuously subjected to random water waves, and therefore, a spectral analysis method is essentially used to determine response statistical quantities that needed in the calculation of fatigue damages and averages of extreme-value responses [17–20]. In order to apply a spectral analysis procedure to offshore jacket structures, the calculation of wave forces on structural members and their linearization techniques has been presented in details in Chap. 3. These forces have been formulated in terms of the random water elevation so that the corresponding responses will also depend on this random variable. Since the stochastic description of the random water elevation is determined as presented in Sect. 3.4, stochastic structural responses can be calculated in terms of the spectral values of the random water elevation through a spectral analysis procedure that will be presented in this chapter.

One other important loading category of offshore structures is introduced by seismic ground motions [21–27] which are commonly known as earthquakes that may cause catastrophic consequences in both economical and social terms. In order to avoid these unpleasant consequences of earthquakes, structures to be built in seismically active offshore areas must be designed against probable structural failures. Earthquakes occurring in ocean environments cause also *tsunamis* [23], which are huge long water waves traveling long distances without dissipating sufficient energy to become harmless. Since tsunamis are water waves created by

H. Karadeniz, *Stochastic Analysis of Offshore Steel Structures*,
Springer Series in Reliability Engineering, DOI: 10.1007/978-1-84996-190-5_4,

earthquakes their impacts on structural systems can be considered in the wave loading category. In this chapter, the earthquake loading due to random ground motion and related structural responses will be presented. Earthquakes are complex random phenomena and their sizes are used in practice to indicate their significances. The size of an earthquake can be measured in terms of magnitude and intensity scales [26, 27]. The magnitude of an earthquake is the amount of energy that released from its source, and the intensity is a measure of an earthquake hazard at a specific location, which differs from location to location. Earthquakes are recorded at different locations by seismometers as being accelerations of the ground motion in time domain, which are termed as time histories [28]. Time histories at different sites may differ considerably from each other in duration, frequency content, and amplitude. The average duration of strong-motion acceleration is about 45 s and, with increasing intensity, the duration decreases, and becomes on the average of 20–25 s [29]. The spatial variation of seismic ground motions has an important effect on the response of long structures [30]. Since they extend over long distances parallel to the ground, their supports undergo different motions during an earthquake. The spatial variation of the ground motions is described by a deterministic time delay function, which is required for the waveforms to reach faraway supports of the structures, and a stationary Gaussian process representing the random ground motion. The time varying property of the ground motion makes the process nonstationary and, in the spectral form, it is represented by a coherency function [31–37]. The detail of this subject will be presented in Sect. 4.5.

4.2 Dynamic Analysis of Structures in the Frequency Domain, the Transfer Function Approach

The calculation of response displacements of a linear structural system in the frequency domain is outlined in Sect. 1.7.2. The formulation of the global displacements in the frequency domain has been given by Eq. (1.229), which is rewritten below, in general for the convenience as,

$$\{D(\omega)\} = [H(\omega)]_{DP}\{P(\omega)\} \quad \rightarrow \quad [H(\omega)]_{DP} = \left([K] + i\omega[C] - \omega^2[M]\right)^{-1} \tag{4.1}$$

where $\{P(\omega)\}$ is the system load vector and $[H(\omega)]_{\mathrm{DP}}$ is defined as the structural transfer function matrix between the displacement and load vectors. For each frequency, Eq. (4.1) must be solved to find the frequency content of displacements. Since this procedure requires multiple matrix inversion, the solution of Eq. (4.1) is practically time-consuming, and therefore, the modal analysis procedure is usually applied, in practice, to obtain the displacements. By using the modified modal analysis, the displacements vector is written from Eq. (1.265) as

$$\{D(\omega)\} = [K]^{-1}\{P(\omega)\} + \left(\sum_{j=1}^{q} \alpha_j(\omega) f_j(\omega)\{\phi\}_j\right) \tag{4.2}$$

in which the first term is the contribution of the quasi-static response and the second term in brackets is the contribution of the dynamic response. The vector $\{\phi\}_j$ is the jth natural mode vector, $\alpha_j(\omega)$ and $f_j(\omega)$ are respectively the eigenmode participation factor and the generalized force, which are defined by Eq. (1.264a) and the generalized force is

$$f_j(\omega) = \{\phi\}_j^T \{P(\omega)\} \tag{4.3}$$

Once the load vector $\{P(\omega)\}$ and structural eigenmodes information are known, the response displacements in the global coordinates can be easily calculated from Eq. (4.2). The stiffness matrix $[K]$ needs to be inverted once for all frequency variations, and therefore, the calculation will be relatively very fast. In the case of wave loading, the structural load vector $\{P(\omega)\}$ is stated in terms of the random water elevation $\eta(\omega)$ and a transfer function vector $\{H(\omega)\}_{P\eta}$ in the frequency domain as explained in Sect. 3.7.3 for the consistent member forces. For the global system, it can be written as,

$$\{P(\omega)\}_{\text{wave}} = \{H(\omega)\}_{P\eta}\eta(\omega) \tag{4.4a}$$

where the subscript (wave) denotes loading due to waves. $\{H(\omega)\}_{P\eta}$ denotes the system transfer function vector between the wave load vector $\{P(\omega)\}$ and the random water elevation $\eta(\omega)$, which is obtained from the assembly process of member transfer functions. Having used Eq. (4.4a) in Eqs. (4.2) and (4.3), the global displacements can be stated as written by,

$$\{D(\omega)\}_{\text{wave}} = \{H(\omega)\}_{D\eta}\eta(\omega)$$
$$\rightarrow \begin{cases} \{H(\omega)\}_{D\eta} = [K]^{-1}\{H(\omega)\}_{P\eta} + \sum_{j=1}^{q} \{h(\omega)\}_{j\eta} \\ \{h(\omega)\}_{j\eta} = \alpha_j(\omega)\,\{\phi\}_j^T\{H(\omega)\}_{P\eta}\{\phi\}_j \end{cases} \tag{4.4b}$$

in which $\{H(\omega)\}_{D\eta}$ is the response transfer function vector for the global displacements under the wave loading. When the structure is subjected to the ground motion with a random base acceleration, $\ddot{u}_g$, like in the case of an earthquake, the corresponding system load vector can be stated in a similar way to Eq. (4.4a) as written in the frequency domain by,

$$\{P(\omega)\}_g = \{H(\omega)\}_{P\ddot{u}_g} u_g(\omega) \tag{4.5a}$$

where the subscript (g) denotes loading due to a ground motion, $\{H(\omega)\}_{P\ddot{u}_g}$ denotes the system transfer function vector between $\{P(\omega)\}_g$ and the random base acceleration $\ddot{u}_g(\omega)$. Having used Eqs. (4.5a) in (4.2) and (4.3), the global displacements under an earthquake loading can be stated as written by,

$$\{D(\omega)\}_g = \{H(\omega)\}_{D\ddot{u}_g} \ddot{u}_g(\omega)$$
$$\rightarrow \begin{cases} \{H(\omega)\}_{D\ddot{u}_g} = [K]^{-1}\{H(\omega)\}_{P\ddot{u}_g} + \sum_{j=1}^{q} \{h(\omega)\}_{j\ddot{u}_g} \\ \{h(\omega)\}_{j\ddot{u}_g} = \alpha_j(\omega)\,\{\phi\}_j^T \{H(\omega)\}_{P\ddot{u}_g} \{\phi\}_j \end{cases} \quad (4.5b)$$

in which $\{H(\omega)\}_{D\ddot{u}_g}$ is the response transfer function vector for the global displacements under an earthquake loading. Details of the calculations are presented in the following section.

4.3 Calculation of Response Transfer Functions of Offshore Structures

In the previous section, definitions of response transfer functions for the global displacements under wave and earthquake forces, $\{H(\omega)\}_{D\eta}$ and $\{H(\omega)\}_{D\ddot{u}_g}$, are presented. In this section, their calculations are explained in detail.

4.3.1 Response Transfer Functions for Wave Loads

Response transfer functions of global displacements under the wave loading, which are defined in the vectorial form in Eq. (3.42b), are calculated by using transfer functions of global wave forces $\{H(\omega)\}_{P\eta}$ and the eigenvalue information of the structural system. The transfer functions of global wave forces are obtained from the assembly process of the transfer functions of member wave forces. For a member, the transfer functions of wave forces in the global coordinates can be deducted from Eq. (3.60e) as written by,

$$\left.\begin{array}{l}\text{transfer functions of the wave}\\ \text{loading for a member}\end{array}\right\} \rightarrow \{h_w(\omega)\}_G = \omega \begin{Bmatrix} q_1[T_n] \\ q_3[T_\theta] \\ q_2[T_n] \\ q_4[T_\theta] \end{Bmatrix} \{\phi\} \quad (4.6a)$$

in which q_j ($j = 1,2,3,4$), $[T_n]$, $[T_\theta]$ and $\{\phi\}$ are explained in Sect. 3.7.3. For each loaded member, the vector of transfer functions $\{h_w(\omega)\}_G$ will be calculated from Eq. (4.6a) and then assembled to form the vector of transfer functions of the wave forces, $\{H(\omega)\}_{P\eta}$, in the global coordinates. Once the vector $\{H(\omega)\}_{P\eta}$ is calculated the transfer function vector $\{H(\omega)\}_{D\eta}$ will be calculated by using Eq. (4.4b) in which the quasi-static term (the first term) can be calculated easily since the system stiffness matrix $[K]$ is inverted only once. Calculation of the dynamic term (the second term) can also be carried out easily as stated below for a vibration mode (eigenmode).

$$\left.\begin{array}{l}\text{dynamic contribution of response}\\ \text{transfer functions for an eigenmode}\\ (j)\ \text{under wave loading}\end{array}\right\} \rightarrow \left\{\begin{array}{l}\{h(\omega)\}_{j\eta}=\beta_{j\eta}(\omega)\{\phi\}_j\\ \beta_{j\eta}(\omega)=\alpha_j(\omega)\,\{\phi\}_j^T\{H(\omega)\}_{P\eta}\end{array}\right. \tag{4.6b}$$

where $\beta_{j\eta}(\omega)$ is a frequency dependent scalar function which defines the participation of the eigenmode vector j to the transfer functions of the global displacements. $\alpha_j(\omega)$ is the natural frequency dependent eigenmode participation factor given by Eq. (1.264a), and $\{\phi\}_j$ is the eigenmode vector for the jth vibration mode. As stated in Eq. (4.4b), the total dynamic contribution to the transfer functions of the global displacements is obtained by superimposing the contributions of eigenmodes considered.

4.3.2 Response Transfer Functions for Earthquake Loading

For offshore structures, the earthquake loading is somewhat different than that for onshore structures. The existence of water surrounding offshore structures produces additional hydrodynamic forces on structures during a ground motion like in the case of an earthquake [38]. Besides this hydrodynamic force, earthquakes produce inertia forces on offshore structures due to the ground acceleration as they occur for onshore structures. In this section, these forces are explained separately in the following sub-sections.

4.3.2.1 Hydrodynamic Forces Produced by Earthquakes

The additional hydrodynamic force produced by earthquakes can be calculated using the Morison's equations which have been explained in detail in Chap. 3 for wave-current-structure interactions. In order to calculate this additional hydrodynamic force, the earthquake ground velocity and acceleration will be added to the relative water particle velocity and acceleration. For this purpose, it is assumed that the sea water is inviscid and do not move with the ground during the earthquake. It is also assumed that the structure is fixed at the bottom to make the same motion with the ground at the bottom. Under these conditions, the motion of the structure is like as shown in Fig. 4.1. The assumption of water at the sea bottom may be crude due to roughness of the bottom. In the reality, the water at the bottom moves with the ground. But, this motion decreases gradually toward the sea surface, and therefore, this assumption may be considered to be reasonable for deep water environments. Relative velocity and acceleration of water particles are used in the Morison's equation to calculate the wave force as written by Eq. (3.88b) in the absence of earthquakes. In the case of earthquakes, the relative velocity and acceleration of water particles that used in Eq. (3.88b) will be modified to include

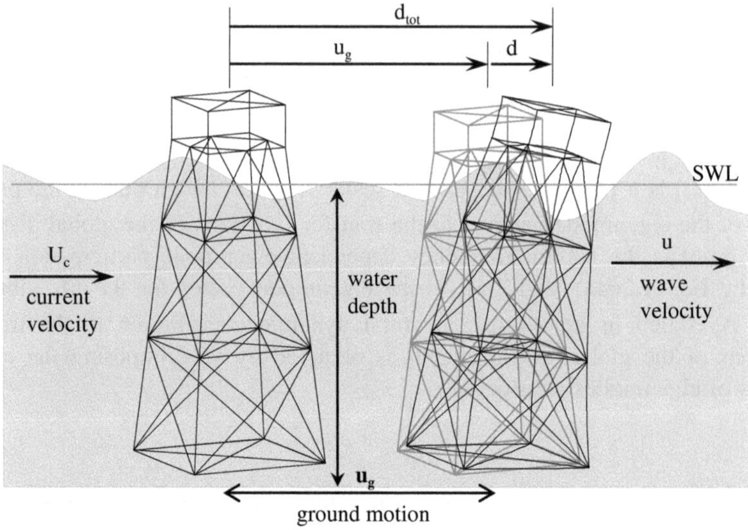

Fig. 4.1 Deformation of an offshore structure under wave and earthquake actions

the ground velocity and acceleration. These modified relative velocity and acceleration of water particles can be stated from Fig. 4.1 as written by, in the normal direction to the member axis,

$$\left.\begin{array}{l}\text{relative water particle velocity and}\\ \text{acceleration during an earthquake}\end{array}\right\} \rightarrow \left\{\begin{array}{l} u_r = u + U_c - (\dot{d} + \dot{u}_g) \\ \dot{u}_r = \dot{u} - (\ddot{d} + \ddot{u}_g) \end{array}\right. \tag{4.7a}$$

where u and U_c are respectively wave and current velocities, d and u_g are respectively structural and ground deformations as shown in Fig. 4.1. With these relative velocity and acceleration the Morison's equation can be written similarly to Eq. (3.88b) as,

$$f = C_D\left|u + U_c - (\dot{d} + \dot{u}_g)\right|\left(u + U_c - (\dot{d} + \dot{u}_g)\right) + C_M\dot{u} - C_A(\ddot{d} + \ddot{u}_g) \tag{4.7b}$$

In the vectorial form, it can be stated similarly to Eq.(3.88d) as written by,

$$\{f_n\}_G = \{f_n\}_{G,\text{wave}} + \{f_n\}_{G,\text{current}} - \{f_{dn}\}_{G,\text{struc.}} - \{f_{u_g n}\}_{G,\text{ground}} \tag{4.7c}$$

where the force vectors $\{f_n\}_{G,\text{wave}}$, $\{f_n\}_{G,\text{current}}$ and $\{f_{dn}\}_{G,\text{struc.}}$ have been defined in Eq. (3.88e). The vector $\{f_{u_g n}\}_{G,\text{ground}}$ denotes the hydrodynamic force vector in the global coordinates due to the ground motion. It is calculated from,

$$\{f_{u_g n}\}_{G,\text{ground}} = [T_n]\left(C_D A\,\{\dot{u}_g\} + C_A\{\ddot{u}_g\}\right) \tag{4.7d}$$

in which $\{\dot{u}_g\}$ and $\{\ddot{u}_g\}$ are the velocity and acceleration vectors of the ground in the global coordinates, C_D and C_A are respectively the drag and added mass constants defined in Eqs. (3.45b) and (3.88c) respectively, $[T_n]$ is the transformation matrix defined in Eq. (3.55b), and A is the linearization coefficient of the Morison's equation which has been explained in Sects. 3.8 and 3.9 in the absence of earthquakes. In the presence of earthquakes, it can be calculated as explained in Sect. 3.9 provided that the relative water velocity vector $\{u_{rn}\}$ in Eq.(3.84a) is taken to be $\{u_{rn}\}_g$ which is defined as,

$$\{u_{rn}\}_g = \{u_n\} - (\{\dot{d}_n\} + \{\dot{u}_{gn}\}) \rightarrow \begin{cases} \{u_{rn}\}_g = \{u_{rn}\} - \{\dot{u}_{gn}\} \\ \{u_{rn}\} = \{u_n\} - \{\dot{d}_n\} \end{cases} \tag{4.7e}$$

This relative velocity vector is a function of two independent random variables, one is the water elevation η and the other one is the earthquake ground acceleration $\ddot{u}_g$. Since the structural velocity vector $\{\dot{d}_n\}$ is dependent on both the water elevation η and the earthquake ground acceleration $\ddot{u}_g$, the components, $\{u_{rn}\}$ and $\{\dot{u}_{gn}\}$, of the relative velocity vector $\{u_{rn}\}_g$ are not independent. Since the linearization coefficient of the Morison's equation A is dependent on the standard deviation of this relative normal water velocity, a similar procedure presented in Sect. 3.9.1 can be applied to calculate it. If the structural velocity vector $\{\dot{d}_n\}$ is not taken into account in the calculation of hydrodynamic forces, the loading will be independent of the structural response velocities. In this case, calculation of the standard deviation of the relative normal water velocity can be simplified considerably. In the formulation presented here, it is assumed that the linearization coefficient A of the Morison's equation is available. By introducing Eq. (4.7d) into the general statement of the consistent member force vector given by Eq. (3.60d) the hydrodynamic consistent force vector of a submerged member, which is produced by an earthquake, can be obtained in the global coordinates as stated by,

$$\{p_{\ddot{u}_g}\}_{G,\text{hyro}} = \begin{cases} \{h_{\ddot{u}_g}(\omega)\}_{G,\text{hyro}}\dot{u}_g(\omega) \\ \{h_{\ddot{u}_g}(\omega)\}_{G,\text{hyro}} = \left(\int_0^{\ell} (C_D A + i\omega C_A) \begin{Bmatrix} N_1[T_n] \\ N_3[T_\theta] \\ N_2[T_n] \\ N_4[T_\theta] \end{Bmatrix} ds \right) \{\delta\} \end{cases} \tag{4.7f}$$

where $\{h_{\ddot{u}_g}(\omega)\}_{G,\text{hyro}}$ is the transfer function vector between $\{p_{\ddot{u}_g}\}_{G,hyro}$ and the absolute ground velocity $\dot{u}_g(\omega)$ and $\{\delta\}$ is a constant vector containing cosine directions (translational components) of the ground motion in the global coordinates. The integration in Eq. (4.7f) can be calculated in a similar way explained in Sect. 3.10.1. The drag term (the term containing C_DA) in Eq. (4.7f) is similar to the statement of the consistent force vector due to a constant current which is given by Eq. (3.89a). From this similarity, the integration of the drag term is readily available from Eq. (3.89c). The integration of the inertia term (the term containing $i\omega C_A$) in Eq. (4.7f) can be calculated easily using the shape functions N_j ($j = 1$ to 4) given by

Eq. (3.59c). Having carried out these operations, the transfer function vector of consistent hydrodynamic forces due to a ground deformation can be obtained as written by,

$$\{h_{\ddot{u}_g}(\omega)\}_{G,hyro}=\left(C_D A_1\begin{Bmatrix} q_{c1}[T_n] \\ q_{c3}[T_\theta] \\ q_{c2}[T_n] \\ q_{c4}[T_\theta] \end{Bmatrix}+\frac{\ell}{12}i\omega C_A\begin{Bmatrix} 6\,[T_n] \\ \ell\,[T_\theta] \\ 6\,[T_n] \\ -\ell\,[T_\theta] \end{Bmatrix}\right)\{\delta\} \tag{4.7g}$$

where q_{cj} ($j = 1$ to 4) are presented in Table 3.13 and A_1 is the value of A at the member end (1).

4.3.2.2 Inertia Forces Produced by Earthquakes

In addition to hydrodynamic forces, earthquakes produce also inertia forces on structural elements due to structural masses vibrating with the earthquake ground acceleration. These forces are calculated at member (element) levels from the dynamic equilibrium equation of a member in which the inertia term (term with the structural mass) is calculated by using the total acceleration vector of the member. For offshore structures subjected to wave, current, and earthquakes, the dynamic equilibrium equation of a member can be stated as similar to Eq. (1.84). It is written by,

$$[k]\{d\}+[c_s]\{\dot{d}\}+[m_s]\left(\{\ddot{d}\}+\{\Delta\}\ddot{u}_g\right)-\{p\}_G=\{f_{\mathrm{int.}}\}_G \tag{4.8a}$$

in which the vectors $\{\Delta\}$, $\{p\}_G$ and $\{f_{\mathrm{int.}}\}_G$ are respectively the earthquake direction vector, consistent applied load vector and the vector of internal forces at the ends of the member, which are defined in the global coordinates. The earthquake direction and consistent applied load vectors are defined as written by,

$$\begin{aligned} &\{\Delta\}^T=\{\ \{\delta\}^T\ \{0\}^T\ \{\delta\}^T\ \{0\}^T\} \\ &\{p\}_G=\{p_w\}_G+\{p_c\}_G-\{p_{\ddot{u}_g}\}_{G,hyro}-\{p_{\dot{d}}\}_G-\{p_{\ddot{d}}\}_G \end{aligned} \tag{4.8b}$$

where the vector $\{\delta\}$ is the same as defined in Eq. (4.7f) and the calculation of the vector $\{p\}_G$ is carried out by using Eq. (4.7c). The terms, $\{p_w\}_G$ and $\{p_c\}_G$, of this vector are calculated respectively from Eqs. (3.60e) and (3.89c). The term $\{p_{\ddot{u}_g}\}_{G,\mathrm{hyro}}$ is calculated from Eq. (4.7f) and the terms, $\{p_{\dot{d}}\}_G$ and $\{p_{\ddot{d}}\}_G$ are calculated respectively from Eqs. (3.90c) and (3.92b). Having substituted the statements of these force vectors in Eq. (4.8b) and rearranged Eq. (4.8a) it can be written that,

$$[k]\{d_G\}+[c]\{\dot{d}_G\}+[m]\{\ddot{d}_G\}-\left(\{p_w\}_G+\{p_c\}_G-\{p_{\ddot{u}_g}\}_G\right)=\{f_{\mathrm{int.}}\}_G \tag{4.8c}$$

in which $[c]$ and $[m]$ are the total damping and mass matrices of the member in the global coordinates defined as,

$$[c] = [c_s] + [c_h] \quad \text{and} \quad [m] = [m_s] + [m_a] \tag{4.8d}$$

The terms $[c_s]$ and $[m_s]$ in Eq. (4.8d) are the structural damping and mass matrices, $[c_h]$ and $[m_a]$ are the hydrodynamic damping and added mass matrices which are calculated respectively from Eqs. (3.90d) and (3.92c). The vector $\{p_{\ddot{u}_g}\}_G$ in Eq. (4.8c) is the total force vector of the member in the global coordinates produced by the earthquake. This force vector is defined as,

$$\{p_{\ddot{u}_g}\}_G = \{h_{\dot{u}_g}(\omega)\}_{G,\text{hyro}}\dot{u}_g + [m_s]\{\Delta\}\ddot{u}_g \tag{4.8e}$$

in which the first term is the hydrodynamic force vector as defined in Eq. (4.7f) and the second term is the inertia force vector produced by the structural mass vibrating with the ground acceleration. The response transfer functions for the total earthquake forces are presented in the following section.

4.3.2.3 Response Transfer Functions for Combined Earthquake Hydrodynamic and Inertia Forces

The combination of the earthquake hydrodynamic and inertia forces of a member is calculated from Eq. (4.8e). In order to find the transfer function of this combined earthquake force, the ground velocity $\dot{u}_g$ is stated in terms of the ground acceleration $\ddot{u}_g$ in the frequency domain. Having carried out this operation, the earthquake force vector can be written as,

$$\{p_{\ddot{u}_g}\}_G = \begin{cases} \{h_{\ddot{u}_g}\}_G \ddot{u}_g(\omega) \\ \{h_{\ddot{u}_g}\}_G = [m_s]\{\Delta\} - \dfrac{i}{\omega}\{h_{\dot{u}_g}(\omega)\}_{G,\text{hyro}} \end{cases} \tag{4.8f}$$

in which $\{h_{\ddot{u}_g}\}_G$ is the transfer function vector of the total earthquake forces of a member. For the whole system, the transfer function vector $\{H(\omega)\}_{P\ddot{u}_g}$ defined in Eq. (4.5a) will be obtained by using the assembly process of member transfer functions $\{h_{\ddot{u}_g}\}_G$. The contribution of concentrated masses at the deck of the platform will be added to this assembled transfer function, $\{H(\omega)\}_{P\ddot{u}_g}$. Having calculated the vector $\{H(\omega)\}_{P\ddot{u}_g}$ for the system, the associated response transfer function vector $\{H(\omega)\}_{D\ddot{u}_g}$ will be calculated by using Eq. (4.5b) in a similar way presented in Sect. 4.3.1 for the wave loading. Calculation of the quasi-static contribution from Eq. (4.5b) is straightforward and calculation of the dynamic contribution for an eigenmode is similar to Eq. (4.6b). It is written as,

$$\left.\begin{array}{l}\text{dynamic contribution of response} \\ \text{transfer functions for an eigenmode} \\ (j) \text{ under earthquake loading}\end{array}\right\} \rightarrow \begin{cases} \{h(\omega)\}_{j\ddot{u}_g} = \beta_{j\ddot{u}_g}(\omega)\{\phi\}_j \\ \beta_{j\ddot{u}_g}(\omega) = \alpha_j(\omega)\,\{\phi\}_j^T\{H(\omega)\}_{P\ddot{u}_g} \end{cases} \tag{4.8g}$$

where $\beta_{j\ddot{u}_g}(\omega)$ is a frequency dependent scalar function which is similar to $\beta_{j\eta}(\omega)$ defined in Eq.(4.6b). The only difference is, in $\beta_{j\ddot{u}_g}(\omega)$, the earthquake loading is used instead of wave loading.

4.4 Calculation of Response Spectra of Offshore Structures

In the preceding sections, response transfer functions for the global displacements have been presented for both wave and earthquake loadings. By using these transfer functions, response spectra of the structure that required, which may be for displacements, forces or stresses, will be presented in this section. The calculation is explained firstly for the stochastic wave and earthquake loadings separately, and then for the combination of these loading cases. In these calculations, current loads do not produce any dynamic response. Since a constant current profile is assumed, they are purely static as given by Eq. (3.89c).

4.4.1 Response Spectra Under Stochastic Wave Loads

Stochastic wave loads are formulated in terms of transfer functions and the water elevation η of random waves as stated vectorially in the frequency domain in Eq. (4.4a). The water elevation η is a random scalar quantity which characterizes the randomness of waves. It is taken to be the input variable of response spectra of offshore structures under stochastic wave loads. Its stochastic description and spectral representation are presented in Sect. 3.4. Knowing its spectral values the calculation of response spectra of structures, which are subjected to only wave loads for the time being, is presented in this section. The spectral analysis and input–output relations have been presented generally in Sects. 2.5 and 2.6. Their applications to random wave loads and corresponding structural responses are explained here. If we assume that the outputs are global displacements of the structure, the input–output relation has already been constructed as stated vectorially in Eq. (4.4b), which is rewritten below for the convenience.

$$\{D(\omega)\}_{\text{wave}} = \{H(\omega)\}_{D\eta}\eta(\omega) \tag{4.9a}$$

By using Eq. (2.136) and Eq. (4.9a), the spectral matrix of displacements can be readily obtained as written by,

$$\left.\begin{array}{l}\text{Spectrum of displacements}\\ \text{under random wave loads}\end{array}\right\} \rightarrow [S_{D_{wave}}(\omega)] = \{H(\omega)\}^*_{D\eta}\{H(\omega)\}^T_{D\eta}\, S_{\eta\eta}(\omega) \tag{4.9b}$$

in which $S_{\eta\eta}(\omega)$ is the spectral function of the random water elevation η, the transfer function vector $\{H(\omega)\}_{D\eta}$ is calculated from Eq. (4.4b), the superscripts, (*) and (*T*), denote respectively a complex conjugate and transposition. Any structural response output which is derived from displacements can be calculated in a similar way step-by-step. Since every step requires a matrix operation, the step-by-step calculation procedure is not efficient in terms of calculation time. An alternative and most powerful algorithm is to calculate firstly the transfer function of the response which is required, and then to apply the spectral calculation. This can be carried out only if the transfer function vector $\{H(\omega)\}_{D\eta}$ is evaluated at all frequency points considered. This evaluation can be carried out easily using Eq. (4.4b). Response transfer functions of member displacements are extracted from the transfer functions of the global displacement $\{H(\omega)\}_{D\eta}$, and consequently transfer functions of member internal forces and stresses are calculated for the frequencies considered. Then, the spectral values of the response required are calculated by using the procedure presented above in Eq. (4.9b). For example, if the response is assumed to be a stress in a member, its transfer function is calculated by using transfer functions of displacements. It is denoted by $h_{s\eta}(\omega)$. Since this stress transfer function is a scalar function, the spectrum of the stress is calculated similarly to Eq. (4.9b) from,

$$S_{ss}(\omega) = h^*_{s\eta}(\omega)h_{s\eta}(\omega)S_{\eta\eta}(\omega) \rightarrow S_{ss}(\omega) = \left|h_{s\eta}(\omega)\right|^2 S_{\eta\eta}(\omega) \tag{4.9c}$$

The calculation algorithm of a stress spectrum is summarized below in Fig. 4.2.

In the case of wave-current actions, the response will be nonzero-mean process due to static current loads which are given by Eq. (3.89c) for members. But, as explained in Sect. 3.5, the existence of current alters the spectral form of the water elevation η. In this case, the global displacement vector $\{D\}$ can be stated in two terms in the frequency domain as,

$$\{D(\omega)\} = \{D(\omega)\}_{\text{wave}} + \{\mu_D\} \tag{4.10a}$$

in which $\{D(\omega)\}_{wave}$ is given by Eq. (4.9a) and $\{\mu_D\}$ is a mean value displacement vector. $\{D(\omega)\}_{wave}$ is a stochastic process with zero mean and $\{\mu_D\}$ is calculated from the stiffness equation as written by,

$$[K]\{\mu_D\} = \{P_c\}_G \tag{4.10b}$$

where $\{P_c\}_G$ is the vector of system forces in the global coordinates due to a constant current. Similar to the displacements, any response quantity of the structure comprises a stochastic term with zero mean and a constant term defining the mean value response.

4.4.2 Response Spectra Under Stochastic Earthquake Loading

The response spectra of offshore structures under an earthquake loading are similar to those calculated from wave loads. They are calculated by using the response transfer functions defined in Eq. (4.5b). For the convenience, the system

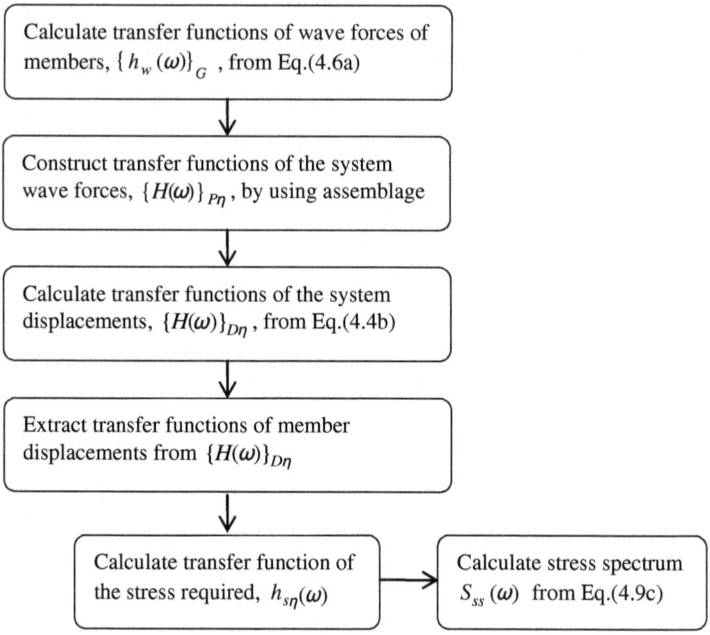

Fig. 4.2 Calculation algorithm of a stress spectrum under stochastic wave loads

displacement vector $\{D(\omega)\}_g$ due to an earthquake ground motion is rewritten in terms of transfer functions $\{H(\omega)\}_{D\ddot{u}_g}$, from Eq. (4.5b), as,

$$\{D(\omega)\}_g = \{H(\omega)\}_{D\ddot{u}_g}\ddot{u}_g(\omega) \tag{4.11a}$$

in which $\ddot{u}_g(\omega)$ is the acceleration of the ground in the frequency domain due to the earthquake motion. This acceleration is a random process and constitutes the input of the response spectra under earthquake loadings. Once a spectral function of the ground acceleration, $S_{\ddot{u}_g}(\omega)$, is known, the calculation of any response spectrum of the structure will be carried out in the same way as explained in the preceding section for the wave loads. Thus, as similar to Eq. (4.9b), the spectral matrix of system displacements is written by,

$$\left.\begin{array}{l}\text{Spectrum of displacements}\\ \text{under earthquake loadings}\end{array}\right\} \rightarrow [S_{D_g}(\omega)] = \{H(\omega)\}^*_{D\ddot{u}_g}\{H(\omega)\}^T_{D\ddot{u}_g}\, S_{\ddot{u}_g}(\omega) \tag{4.11b}$$

Any response spectrum can also be calculated by using the calculation algorithm presented in Fig. 4.2. As it can be realized from these spectral calculations, the determination of the earthquake spectrum $S_{\ddot{u}_g}(\omega)$ is the central issue in the spectral earthquake analysis of structures. In the following sections, the stochastic ground motion and its spectral representation under earthquakes are presented briefly.

4.4.2.1 Stochastic Earthquake Ground Motion

Earthquakes are random ground motions displaying a broadband character. They may result from various sources among which the tectonic-related earthquake motions are the largest and most important [23–28] for engineering, economical, and social points of view since their consequences are mostly catastrophic, especially in the near fault region. Earthquakes are measured in terms of *magnitudes* and *intensities*. The magnitude is the amount of energy that released during the earthquake. It is a unique measure for all locations, i.e. it is site independent. The best known measure of the earthquake magnitude is the Richter scale [39]. The intensity is a scale of the effect of an earthquake hazard at a specific location. It is based on observed human behavior and structural damages at a specific location so that the earthquake intensity scale is site dependent. There are numerous intensity scales that are in use currently in different parts of the world [26] such as the modified Mercalli (MMI) scale in the United States, the European Macroseismic scale (EMS-98) in Europe, the Shindo intensity scale in Japan, etc. The earthquake ground motion decreases with distance from the source of the earthquake and gradually dies away. This feature of the earthquake motion is termed as the *attenuation*. It is a function of not only the distance but also earthquake magnitude and geological site conditions [40–43]. The attenuation relations can be obtained by a statistical process of earthquake data (a regression analysis) measured at different locations. The basic data of earthquakes are recordings of ground accelerations at different sites during different earthquakes in time series, which contain valuable, and complete information that used in seismic analysis [23, 25]. These data vary significantly from site-to-site due to various factors. Depending on these factors, especially the magnitude and local site response, the recorded strong ground motions can display long duration that cause potential damages. The maximum amplitude of a recorded acceleration is defined as the peak ground acceleration (PGA), and similarly, the peak ground velocity (PGV) and peak ground displacement (PGD) are defined to indicate respectively the maximum amplitudes of the ground velocity and displacement [23]. The time histories of the ground velocity and displacement are obtained from integrations of the recorded time history of the ground acceleration. Based on the recorded time histories of the ground acceleration, which are mostly broadband random processes, spectral functions of the ground acceleration are determined. This spectral representation is presented briefly in the following section.

4.4.2.2 Spectral Representation of Stochastic Earthquake Ground Motion

The earthquake ground motion is provided by accelerograms in time domain that are recorded by accelerographs. For an earthquake occurrence, the earthquake accelerograms recorded at different locations are typically different from each other even the locations are within the range of dimensions of engineering structures and even there are similarities in accelerograms. These differences become larger for larger distances between recording stations (*separations distance*) due to different arrival

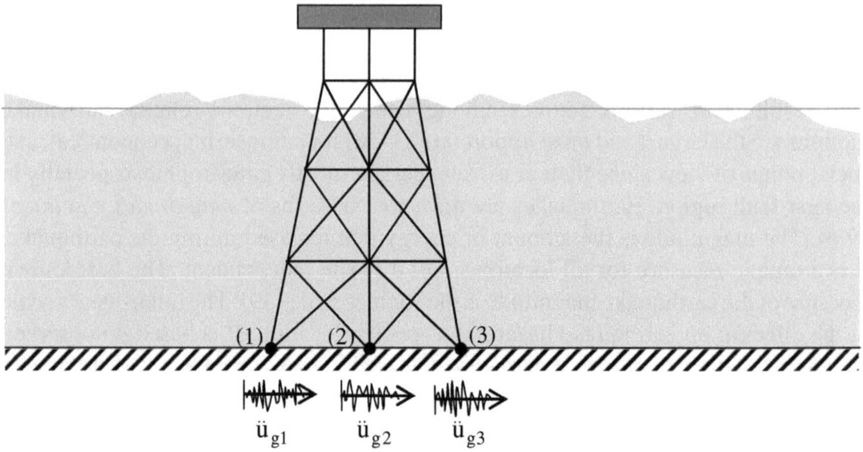

Fig. 4.3 Multiple support excitation of an offshore structure under a nonuniform earthquake ground motion

time of seismic waves at different stations (*wave passage effect*), different soil conditions at different locations (*site effect*). These phenomena of the earthquake motion are fully described by so-called the *coherency effect* and being considerable for lifeline and large-scale engineering structures, which can be properly modeled by using multi-support excitation input [44–50]. In the literature, there are numerous publications [30–33, 37, 44, 51–63] to address the coherency of the earthquake motion. In this case, for linear structures, a response value of the structure (e.g. displacement, force, stress) is calculated by superimposing relevant responses under multiple support excitations due to a nonuniform earthquake ground motion. Multiple support excitations are shown schematically in Fig. 4.3 for an example offshore structure with three supports which are numbered as (1), (2), and (3). Each support is subjected to a random ground acceleration, namely $\ddot{u}_{g1}$, $\ddot{u}_{g2}$ and $\ddot{u}_{g3}$. A response value (displacement, force, stress) at a location in the structure, which is denoted by r, can be calculated generally in the frequency domain as written by,

$$\left.\begin{array}{l}\text{Frequency response under}\\ \text{multiple-support excitations}\end{array}\right\} \rightarrow r(\omega) = \sum_{j=1}^{n} H_{r\ddot{u}_{gj}}(\omega)\, \ddot{u}_{gj}(\omega) \tag{4.12a}$$

where $H_{r\ddot{u}_{gj}}(\omega)$ is the transfer function between the response r and the ground acceleration at the support *j*, $\ddot{u}_{gj}(\omega)$, and *n* is the number of prescribed supports of the structure. The frequency response *r(ω)* given by Eq. (4.12a) can be conveniently written in the vector notation as,

$$r(\omega) = \{H_{r\ddot{u}_g}(\omega)\}^T \{\ddot{u}_g(\omega)\} \tag{4.12b}$$

in which the vector $\{H_{r\ddot{u}_g}(\omega)\}$ contains all transfer functions $H_{r\ddot{u}_{gj}}(\omega)$ for ($j = 1$ to n), and the vector $\{\ddot{u}_g(\omega)\}$ contains all support accelerations $\ddot{u}_{gj}(\omega)$ for ($j = 1$ to n).

From Eq. (4.12b), the spectrum of the response r, $S_{rr}(\omega)$, can be easily stated by using the input–output relation of a linear systems as similar to Eq. (2.136). It is written as,

$$\left.\begin{array}{l}\text{Response spectrum under}\\ \text{multi-support random excitations}\end{array}\right\} \rightarrow S_{rr}(\omega) = \{H_{r\ddot{u}_g}(\omega)\}^{*T}[S_{\ddot{u}_g}(\omega)]\{H_{r\ddot{u}_g}(\omega)\} \tag{4.13}$$

in which the superscript (*) denotes the complex conjugate of a row vector, and $[S_{\ddot{u}_g}(\omega)]$ is the spectral matrix of the ground accelerations at the supports, $\ddot{u}_{gj}(\omega)$ for ($j = 1$ to n). This spectral matrix is formed as written by,

$$[S_{\ddot{u}_g}(\omega)] = \begin{bmatrix} S_{11}(\omega) & . & . & S_{1j}(\omega) & . & . & S_{1n}(\omega) \\ . & . & . & . & . & . & . \\ . & . & . & . & . & . & . \\ S_{k1}(\omega) & . & . & S_{kj}(\omega) & . & . & S_{kn}(\omega) \\ . & . & . & . & . & . & . \\ . & . & . & . & . & . & . \\ S_{n1}(\omega) & . & . & S_{nj}(\omega) & . & . & S_{nn}(\omega) \end{bmatrix} \tag{4.14}$$

in which the term, in general, $S_{kj}(\omega)$ denotes the cross spectrum of the ground accelerations at the supports k and j respectively, i.e. $\ddot{u}_{gk}(\omega)$ and $\ddot{u}_{gj}(\omega)$. Based on the recorded earthquake data at different sites, the spectral matrix of the earthquake ground motion can be calculated in the smoothed forms. By using the spectral information of two recording stations of the earthquake, say k and j, a coherency function of the seismic motion is defined [33, 51–57] as written by,

$$\left.\begin{array}{l}\text{Coherency function of}\\ \text{a seismic motion}\end{array}\right\} \rightarrow \gamma_{kj}(\delta,\omega) = \frac{S_{kj}(\omega)}{\sqrt{S_{kk}(\omega)S_{jj}(\omega)}} \tag{4.15a}$$

in which δ denotes the separation distance between the locations k and j in the earthquake wave propagation direction. Its absolute value varies between zero and one, i.e. $0 \leq |\gamma_{kj}(\delta,\omega)| \leq 1$. The coherency function is conveniently stated in an alternative form [56] written as,

$$\left.\begin{array}{l}\text{Alternative statement of}\\ \text{the coherency function}\end{array}\right\} \rightarrow \gamma_{kj}(\delta,\omega) = |\gamma_{kj}(\delta,\omega)| \exp[i\theta_{kj}(\delta,\omega)] \tag{4.15b}$$

where $(i = \sqrt{-1})$, $|\gamma_{kj}(\delta,\omega)|$ is the modulus, which is also called as *loss of coherency* or *lagged coherency*, and $\theta_{kj}(\delta,\omega)$ is the phase of the coherency function describing the wave passage effect between the locations k and j. The real part of $\gamma_{kj}(\delta,\omega)$, i.e. Re $\gamma_{kj}(\delta,\omega)$, is called *the unlagged coherency*. The modulus $|\gamma_{kj}(\delta,\omega)|$ is a measure of the similarity in the seismic motion which indicates the degree of linearity between recorded data at the stations k and j. The phase of the coherency function is defined [33, 54, 56] as written by,

$$\left.\begin{array}{l}\text{Definition of phase of}\\ \text{the coherency function}\end{array}\right\} \rightarrow \theta_{kj}(\delta,\omega) = \arctan\left(\frac{\mathrm{Im}S_{kj}(\omega)}{\mathrm{Re}S_{kj}(\omega)}\right) \tag{4.16a}$$

For a seismic wave propagating with an approximately constant velocity c along the line between the stations k and j, the phase of the coherency function can be obtained [64] as written by,

$$\theta_{kj}(\delta,\omega) = -\frac{\delta\,\omega}{c} \tag{4.16b}$$

which is a function of the separation distance δ, frequency ω, and the apparent velocity of the wave propagation c. For the lagged (loss of) coherency, or the modulus $|\gamma_{kj}(\delta,\omega)|$, there have been numerous parametric expressions reported in the literature, see e.g. [30, 56] for a survey. The most popular one mentioned in the literature is introduced in [33]. It is written as,

$$\text{the lagged coherency} \rightarrow \left\{\begin{array}{l} |\gamma_{kj}(\delta,\omega)| = A\exp\left(-\frac{B(\omega)}{a}\right) + (1-A)\exp(-B(\omega)) \\ B(\omega) = \frac{2|\delta|}{\beta(\omega)}\,[1 + A(a-1)]\ \text{and} \\ \beta(\omega) = k\left[1 + \left(\frac{\omega}{\omega_0}\right)^b\right]^{-1/2} \end{array}\right. \tag{4.17a}$$

where A, a, k, b, and ω_0 are five empirical parameters. These parameters can be obtained from the minimization of the error function defined [54] as,

$$f(A,\,a,\,k,\,b,\,\omega_0) = \sum_{i=1}^{n_s}\sum_{j=1}^{n_\omega}\left[\arctan\left(\gamma(\delta_i,\omega_j)\right) - \arctan\left(|\gamma(\delta_i,\omega_j)|\right)\right]^2 \tag{4.17b}$$

in which n_s and n_ω are respectively the station pairs and discrete frequencies, and δ_i denotes the separation distance between the station pair i. It has been reported [54] that, for large separation distances and frequencies, Eq. (4.17a) produces erroneous values for a and k parameters. Therefore, for large δ and ω values, the coherency model given in Eq (4.17a) is degenerated to a simpler form [54] written as,

$$|\gamma_{kj}(\delta,\omega)| = A\ \exp\left(-\frac{2|\delta|}{\chi}\left[1 + \left(\frac{\omega}{\omega_0}\right)^b\right]^{1/2}\right) + (1-A) \tag{4.17c}$$

in which ($\chi = ak/(1-A)$). This model of $|\gamma_{kj}(\delta,\omega)|$ contains four empirical parameters as A, χ, b, and ω_0. In order to give an idea about the magnitudes of the empirical parameters, the values estimated from data [33] are written below.

$$A = 0.736, \quad a = 0.147, \quad k = 5210, \quad \omega_0 = 6.85, \quad b = 2.78 \tag{4.17d}$$

These values may be varied according to soil conditions and stations of the data that they are estimated from. It is, usually, assumed that the supports of the structure have the same site conditions with different separation distances and frequencies (homogeneous ground condition), and therefore, the processes of ground accelerations at the supports will be identical, i.e. $\left(S_{kk}(\omega) = S_{\ddot{u}_g}(\omega)\right)$ where ($k = 1$ to N_s) with N_s is the number of supports, and $S_{\ddot{u}_g}(\omega)$ is the spectrum of the ground acceleration at a point (point earthquake spectrum). In this case, the spectral matrix of ground accelerations at the supports given by Eq. (4.14) will be stated in terms of the coherency functions and the point earthquake spectrum as written by,

$$[S_{\ddot{u}_g}(\omega)] = \begin{bmatrix} \gamma_{11}(\delta,\omega) & . & . & \gamma_{1j}(\delta,\omega) & . & . & \gamma_{1n}(\delta,\omega) \\ . & . & . & . & . & . & . \\ . & . & . & . & . & . & . \\ \gamma_{k1}(\delta,\omega) & . & . & \gamma_{kj}(\delta,\omega) & . & . & \gamma_{kn}(\delta,\omega) \\ . & . & . & . & . & . & . \\ . & . & . & . & . & . & . \\ \gamma_{n1}(\delta,\omega) & . & . & \gamma_{nj}(\delta,\omega) & . & . & \gamma_{nn}(\delta,\omega) \end{bmatrix} S_{\ddot{u}_g}(\omega) \tag{4.18}$$

The point earthquake spectrum $S_{\ddot{u}_g}(\omega)$ is also defined as the homogeneous earthquake spectrum for random ground motion which is assumed to be a zero-mean ergodic-Gaussian process of finite duration represented by a filtered white noise. This spectrum is represented by the modified Kanai–Tajimi spectrum [4] which is stated as written by,

$$S_{\ddot{u}_g}(\omega) = \frac{1 + 4\xi_g^2 (\omega/\omega_g)^2}{\left[1 - (\omega/\omega_g)^2\right]^2 + 4\xi_g^2 (\omega/\omega_g)^2} S_f(\omega)\, S_0 \tag{4.19a}$$

where ξ_g is the characteristic ground damping ratio, ω_g is the characteristic ground frequency, $S_f(\omega)$ is the filter spectrum for low frequency region to avoid difficulties and singularity when the frequency approaches zero, and S_0 is an intensity factor. The filter spectrum is given [4] by

$$S_f(\omega) = \frac{(\omega/\omega_f)^4}{\left[1 - (\omega/\omega_f)^2\right]^2 + 4\xi_f^2 (\omega/\omega_f)^2} \tag{4.19b}$$

in which ω_f and ξ_f are respectively the characteristic frequency and damping ratio of the filter. The intensity factor S_0 is calculated from the integration of Eq. (4.19a) as it is obtained to be,

$$S_0 = \frac{4\xi_g \sigma_{\ddot{u}_g}^2}{\pi\left(1 + 4\xi_g^2\right)\omega_g} \tag{4.19c}$$

in which $\sigma_{\ddot{u}_g}$ is the standard deviation (rms) of the ground acceleration. The original Kanai–Tajimi spectrum, which does not include the filter spectrum $S_f(\omega)$, has a constant value equal to S_0 at the zero frequency, $(\omega = 0)$. When this acceleration spectrum is transformed to the spectrum of ground displacement, at the zero frequency point, there will be a singularity and, in the small frequency region, high values of the displacement spectrum are obtained. This occurrence may result in exaggerated response spectral values in the small frequency region, and consequently, wrong response statistical values. Using the filter spectrum in the original Kanai–Tajimi spectrum, as stated in Eq. (4.19a), prevents this undesirable situation. In order to demonstrate the effect of the filter spectrum, the modified Kanai–Tajimi spectrum is plotted for $\xi_f = 0.6$ and for different values of the characteristic filter frequency ω_f. In this demonstration, the standard deviation of the ground acceleration, the characteristic of ground frequency, and damping ratio are respectively assumed to be,

$$\sigma_{\ddot{u}_g} = 0.25\,\mathrm{m/s^2}, \quad \omega_g = 15\,\mathrm{rad/s}, \quad \xi_g = 0.6 \tag{4.19d}$$

The shapes of these spectra are shown in Fig. 4.4 where it is seen that the filter spectrum dominates the small frequency region while, in the high frequency region, it has inconsiderable effect. If all supports undergo the same random ground motion, i.e. $\left(\ddot{u}_{gk}(\omega) = \ddot{u}_g(\omega)\right)$ for $(k = 1$ to $N_s)$, then a unified ground motion is obtained with the homogeneous earthquake spectrum given by Eq. (4.19a). In this case, the response transfer functions will be calculated as explained in Sect. 4.3.2.3, and the response spectra will be calculated by using the response transfer functions explained in Sect. 4.4.2 and presented by Eq. (4.11b) for the displacements. Under multiple support excitations, as in the case of non-uniform ground motion, the calculation of response transfer functions is somewhat different than that explained in the Sect. 4.3.2.3. This calculation is explained briefly in the following section.

4.4.2.3 Calculation of Structural Response Transfer Functions Under NonUniform Earthquake Ground Motion

When the structure is subjected to a random multiple ground motion the calculation of response transfer functions differs from that under a uniform ground motion which is presented in the Sect. 4.3.2.3. In this case, a matrix partitioning is used in the dynamic equilibrium equation, for which the displacements of the structure are separated into two parts as,

1. prescribed displacements at the supports (base displacements of the structure) which are denoted by the vector $\{u_g\}$,
2. all other structural displacements excluding the supports (displacements of superstructure at the unconstrained degrees of freedom) which are denoted by the vector $\{D\}$ in the global coordinate system.

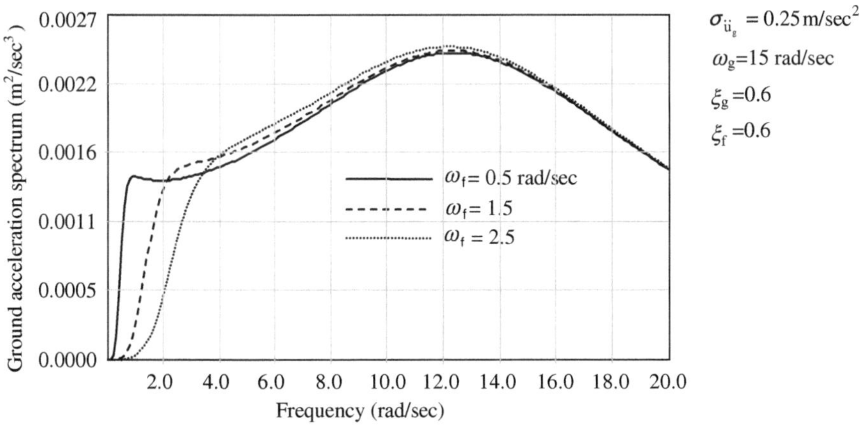

Fig. 4.4 The modified Kanai–Tajimi earthquake spectrum for different filter frequencies, ω_f

It is now assume that the structure is subjected to ground motion only. Thus, the forces acting on the structure are the support reactions which are denoted by the vector $\{P_g\}$, and forces at the superstructure are zero. Under this condition, the dynamic equilibrium equation of the structure can be written as,

$$\begin{bmatrix} \mathbf{K}_{\mathrm{dd}} & \mathbf{K}_{\mathrm{du}} \\ \mathbf{K}_{\mathrm{du}}^T & \mathbf{K}_{\mathrm{uu}} \end{bmatrix} \begin{Bmatrix} \mathbf{D} \\ \mathbf{u}_g \end{Bmatrix} + \begin{bmatrix} \mathbf{C}_{\mathrm{dd}} & \mathbf{C}_{\mathrm{du}} \\ \mathbf{C}_{\mathrm{du}}^T & \mathbf{C}_{\mathrm{uu}} \end{bmatrix} \begin{Bmatrix} \dot{\mathbf{D}} \\ \dot{\mathbf{u}}_g \end{Bmatrix} + \begin{bmatrix} \mathbf{M}_{\mathrm{dd}} & \mathbf{M}_{\mathrm{du}} \\ \mathbf{M}_{\mathrm{du}}^T & \mathbf{M}_{\mathrm{uu}} \end{bmatrix} \begin{Bmatrix} \ddot{\mathbf{D}} \\ \ddot{\mathbf{u}}_g \end{Bmatrix} = \begin{Bmatrix} \mathbf{0} \\ \mathbf{P}_g \end{Bmatrix} \tag{4.20}$$

in which the bold characters denote matrices and vectors, i.e. $\mathbf{D} = \{D\}$, $\mathbf{u}_g = \{u_g\}$, $\mathbf{P}_g = \{P_g\}$ and $\mathbf{K}_{\mathrm{dd}} = [\mathrm{K}_{\mathrm{dd}}]$, $\mathbf{K}_{\mathrm{du}} = [\mathrm{K}_{\mathrm{du}}]$, etc. Eq. (4.20) gives two sets of equations. The zero loading set can be stated explicitly in terms of the support motion as written by,

$$\mathbf{K}_{\mathrm{dd}}\mathbf{D} + \mathbf{C}_{\mathrm{dd}}\dot{\mathbf{D}} + \mathbf{M}_{\mathrm{dd}}\ddot{\mathbf{D}} = -\left(\mathbf{K}_{\mathrm{du}}\mathbf{u}_g + \mathbf{C}_{\mathrm{du}}\dot{\mathbf{u}}_g + \mathbf{M}_{\mathrm{du}}\ddot{\mathbf{u}}_g\right) \tag{4.21a}$$

At this time, we assume that the displacement vector **D** is decomposed into a pseudo-static and a dynamic components as written by,

$$\mathbf{D} = \mathbf{D}_s + \mathbf{D}_d \rightarrow \begin{cases} \mathbf{D}_s\text{: pseudo-static component} \\ \mathbf{D}_d\text{: dynamic component} \end{cases} \tag{4.21b}$$

The pseudo-static component is calculated from Eq.(4.21a) by setting all dynamic terms to zero which leads to the following equation.

$$\mathbf{D}_s = -\mathbf{R}\mathbf{u}_g \quad \text{where} \quad \mathbf{R} = \mathbf{K}_{dd}^{-1}\mathbf{K}_{du} \tag{4.21c}$$

Having substituted Eqs. (4.21c) and (4.21b) into (4.21a) it can be obtained that,

$$\mathbf{K}_{\mathrm{dd}}\mathbf{D}_d + \mathbf{C}_{\mathrm{dd}}\dot{\mathbf{D}}_d + \mathbf{M}_{\mathrm{dd}}\ddot{\mathbf{D}}_d = -(\mathbf{C}_{\mathrm{du}} - \mathbf{C}_{\mathrm{dd}}\mathbf{R})\dot{\mathbf{u}}_g - (\mathbf{M}_{\mathrm{du}} - \mathbf{M}_{\mathrm{dd}}\mathbf{R})\ddot{\mathbf{u}}_g \quad (4.21\text{d})$$

In Eq. (4.21d), the damping force term is usually small in comparison with the inertia force term [45], and therefore, it is neglected. Then, the differential equation of the dynamic component $\mathbf{D}_{\mathrm{d}}$ of the displacement vector becomes as written by,

$$\mathbf{K}_{\mathrm{dd}}\mathbf{D}_d + \mathbf{C}_{\mathrm{dd}}\dot{\mathbf{D}}_d + \mathbf{M}_{\mathrm{dd}}\ddot{\mathbf{D}}_d = -(\mathbf{M}_{\mathrm{du}} - \mathbf{M}_{\mathrm{dd}}\mathbf{R})\ddot{\mathbf{u}}_g \quad (4.22\text{a})$$

which can be solved as explained in Sect. 4.2. The response transfer function matrix of this displacement vector, $\mathbf{H}_{D_d\ddot{u}_g}(\omega)$, can be stated implicitly from Eq. (4.1) as written by,

$$\begin{aligned} \mathbf{D}_d(\omega) &= \mathbf{H}_{D_d\ddot{u}_g}(\omega)\,\ddot{\mathbf{u}}_g(\omega) \\ \mathbf{H}_{D_d\ddot{u}_g}(\omega) &= \left(\mathbf{K}_{\mathrm{dd}} + i\omega\mathbf{C}_{\mathrm{dd}} - \omega^2\mathbf{M}_{\mathrm{dd}}\right)^{-1}\left(\mathbf{M}_{\mathrm{dd}}\mathbf{R} - \mathbf{M}_{\mathrm{du}}\right) \end{aligned} \quad (4.22\text{b})$$

By using the modal analysis method the response transfer function matrix $\mathbf{H}_{D_d\ddot{u}_g}(\omega)$ can be obtained explicitly from Eq.(4.2) as stated by,

$$\mathbf{H}_{D_d\ddot{u}_g}(\omega) = \left(\mathbf{K}_{\mathrm{dd}}^{-1} + \sum_{j=1}^{q}\alpha_j(\omega)\,\boldsymbol{\varphi}_j\boldsymbol{\varphi}_j^T\right)(\mathbf{M}_{\mathrm{dd}}\mathbf{R} - \mathbf{M}_{\mathrm{du}}) \quad (4.22\text{c})$$

in which $\boldsymbol{\varphi}_j$ denotes the eigenmode vector of the jth natural mode, i.e. $\boldsymbol{\varphi}_j = \{\phi\}_j$, $\alpha_j(\omega)$ is the modal participation factor, see Eq. (4.2), and q is the number of eigenmodes included. The response transfer function matrix of the quasi-static displacements can be readily written from Eq. (4.21c) as,

$$\mathbf{D}_s(\omega) = -\mathbf{R}\mathbf{u}_g(\omega) = \mathbf{H}_{D_s\ddot{u}_g}(\omega)\ddot{\mathbf{u}}_g(\omega) \rightarrow \mathbf{H}_{D_s\ddot{u}_g}(\omega) = \frac{1}{\omega^2}\mathbf{R} \quad (4.22\text{d})$$

Since the total displacements, $\mathbf{D}$, are obtained from the superimposition of the quasi-static and dynamic components of the displacements as written in Eq. (4.21b), the transfer function matrix of the total displacements can be stated as,

$$\mathbf{D}(\omega) = \mathbf{H}_{D\ddot{u}_g}(\omega)\,\ddot{\mathbf{u}}_g(\omega) \rightarrow \mathbf{H}_{D\ddot{u}_g}(\omega) = \left(\mathbf{H}_{D_s\ddot{u}_g}(\omega) + \mathbf{H}_{D_d\ddot{u}_g}(\omega)\right) \quad (4.23\text{a})$$

Having substituted the statements of $\mathbf{H}_{D_s\ddot{u}_g}(\omega)$ and $\mathbf{H}_{D_d\ddot{u}_g}(\omega)$ from Eqs. (4.22d) and (4.22c) into (4.23a) it can be obtained that,

$$\mathbf{H}_{D\ddot{u}_g}(\omega) = \frac{1}{\omega^2}\mathbf{R} + \left(\mathbf{K}_{\mathrm{dd}}^{-1} + \sum_{j=1}^{q}\alpha_j(\omega)\,\boldsymbol{\varphi}_j\boldsymbol{\varphi}_j^T\right)(\mathbf{M}_{\mathrm{dd}}\mathbf{R} - \mathbf{M}_{\mathrm{du}}) \quad (4.23\text{b})$$

When a lumped mass matrix is used, the off diagonal term will be zero, i.e. $\mathbf{M}_{du} = \mathbf{0}$. In this case, Eq. (4.23b) will be simplified as written by,

$$\mathbf{H}_{D\ddot{u}_g}(\omega) = \left[\frac{1}{\omega^2}\mathbf{I} + \left(\mathbf{K}_{\mathrm{dd}}^{-1} + \sum_{j=1}^{q} \alpha_j(\omega)\, \boldsymbol{\varphi}_j \boldsymbol{\varphi}_j^T\right)\mathbf{M}_{\mathrm{dd}}\right]\mathbf{R} \tag{4.23c}$$

in which **I** denotes a unit matrix. The calculation of the matrix **R** can be carried out easily from Eq. (4.21c) by applying unit displacements at supports and solving to the structural displacements $\mathbf{D}_{\mathrm{s}}$.

4.4.2.4 Calculation of Spectral Responses Under Non-Stationary Multi-Support Ground Excitation

In the spectral earthquake analysis, it is mostly assumed that the random earthquake ground acceleration is a filtered white noise stationary Gaussian process. Its spectral representation is given by the Kanai–Tajimi spectrum as shown in Fig. 4.4. This is an adequate simplifying assumption to ease the calculation of spectral responses of structures. However, it is realized from the study of recorded accelerograms that the earthquake ground excitation is generally a non-stationary random process [65], and therefore, the due response spectra depend on not only the frequency content but also a function of time. Non-stationary responses of structures have been studied in the past and reported in the literature, see i.e. [34–36, 58] and [65–89]. In this section, calculation of response spectra of structures under non-stationary multi-support earthquake excitation are explained briefly.

A non-stationary unified earthquake ground motion is represented by a uniformly modulated random process [71, 76] which is defined as,

$$\ddot{x}_g(t) = \alpha(t)\, \ddot{u}_g(t) \tag{4.24a}$$

In Eq.(4.24), $\ddot{x}_g(t)$ is the non-stationary earthquake ground acceleration, $\alpha(t)$ is a deterministic envelope time function and $\ddot{u}_g(t)$ is the unified stationary earthquake ground acceleration for which the spectral function is defined by Eq. (4.19a). Similar to the stationary multiple support excitations presented in the previous sections, the non-stationary multiple support excitations can be written from Eq. (4.24a) as,

$$\{\ddot{x}_g(t)\} = [\alpha(t)]\{\ddot{u}_g(t)\} \tag{4.24b}$$

where $[\alpha(t)]$ is a diagonal matrix allowing different deterministic time functions at different supports. The calculation procedure of non-stationary random vector processes is explained below in general. Then, the spectral formulation of multi-support earthquake random excitations is obtained from the general formulation. For this purpose, two non-stationary input processes and the corresponding response output processes are used. The assumed non-stationary input vector processes are denoted by $\{y_1(t)\}$ and $\{y_2(t)\}$, and the corresponding response output vector processes are denoted by $\{z_1(t)\}$ and $\{z_2(t)\}$, respectively. As similar to Eq. (4.24b), the non-stationary input processes are defined as,

$$\text{non-stationary input processes} \rightarrow \begin{cases} \{y_1(t)\} = [\alpha_1(t)]\{x_1(t)\} \\ \{y_2(t)\} = [\alpha_2(t)]\{x_2(t)\} \end{cases} \tag{4.25a}$$

whrere $\{x_1(t)\}$ and $\{x_2(t)\}$ are stationary input vector processes with zero mean, $[\alpha_1(t)]$ and $[\alpha_2(t)]$ are diagonal matrices containing deterministic time functions for the elements of the stationary $\{x_1(t)\}$ and $\{x_2(t)\}$ processes. The corresponding output processes for a linear system at the time stations, t_1 and t_2, are obtained from the convolution integral explained in Sect. 2.6 as written by,

$$\begin{aligned} \{z_1(t_1)\} &= \int_{-\infty}^{t_1} [h(t_1 - \tau_1)]\{y_1(\tau_1)\}d\tau_1 \\ \{z_2(t_2)\} &= \int_{-\infty}^{t_2} [h(t_2 - \tau_2)]\{y_2(\tau_2)\}d\tau_2 \end{aligned} \tag{4.25b}$$

in which $[h(t)]$ is a matrix of system impulse response functions which are shown in Fig. 4.5 with non-stationary input processes. Having introduced Eqs. (4.25a) into (4.25b) the response output processes at time stations, t_1 and t_2, can be obtained as written by,

$$\text{non-stationary output processes} \rightarrow \begin{cases} \{z_1(t_1)\} = \int_{-\infty}^{t_1} [h(t_1 - \tau_1)][\alpha_1(\tau_1)]\{x_1(\tau_1)\}d\tau_1 \\ \{z_2(t_2)\} = \int_{-\infty}^{t_2} [h(t_2 - \tau_2)][\alpha_2(\tau_2)]\{x_2(\tau_2)\}d\tau_2 \end{cases} \tag{4.25c}$$

The cross-correlation matrix of the non-stationary output processes, $\{z_1(t_1)\}$ and $\{z_2(t_2)\}$, can be written from Eq. (2.131a) as,

$$[R_{z_1 z_2}(t_1, t_2)] = E\left[\{z_1(t_1)\}\{z_2(t_2)\}^T\right] \tag{4.26a}$$

Since $\{x_1(t_1)\}$ and $\{x_2(t_2)\}$ are stationary processes Eq. (4.26a) can be stated as written by,

$$\begin{aligned} [R_{z_1 z_2}(t_1, t_2)] &= \int_{-\infty}^{t_1} \int_{-\infty}^{t_2} [h(t_1 - \tau_1)][\alpha_1(\tau_1)][R_{x_1 x_2}(\tau)]\,[\alpha_2(\tau_2)]^T [h(t_2 - \tau_2)]^T d\tau_1 d\tau_2 \\ &\text{where} \quad \tau = (\tau_2 - \tau_1) \end{aligned} \tag{4.26b}$$

in which $[R_{x_1 x_2}(\tau)]$ is the cross-correlation matrix of the stationary $\{x_1(t)\}$ and $\{x_2(t)\}$ processes. It is obtained from the Fourier transform of their cross-spectral matrix as written by,

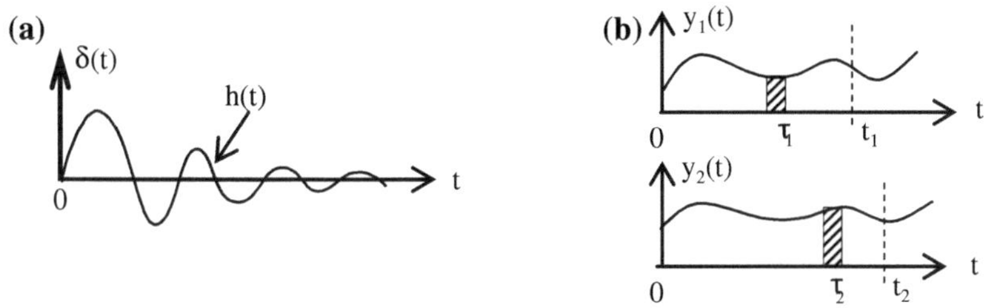

Fig. 4.5 An impulse response function and non-stationary input functions. **a** an impulse response function, **b** non-stationary input functions

$$[R_{x_1x_2}(\tau)] = \int_{-\infty}^{\infty} [S_{x_1x_2}(\omega)]\, e^{i\omega\tau} d\omega \quad \text{with} \quad (\tau = \tau_2 - \tau_1) \tag{4.26c}$$

Having introduced Eqs. (4.26c) into (4.26b), the cross-correlation matrix of the response outputs $[R_{z_1z_2}(t_1, t_2)]$ can be expressed as written by,

$$\left.\begin{array}{l}\text{Cross-correlation of}\\ \text{non-stationary functions}\end{array}\right\} \rightarrow [R_{z_1z_2}(t_1, t_2)] = \int_{-\infty}^{\infty} [A_1^*(t_1, \omega]\, [S_{x_1x_2}(\omega)]\, [A_2(t_2, \omega]^T\, d\omega \tag{4.27a}$$

in which $[A_j(t,\omega)]$, where $(j = 1,2)$, is a matrix of modulating functions as depending on the time t and frequency ω, the superscript (*) denotes a complex conjugate. The matrix of modulating functions is defined as stated by,

$$[A_j(t, \omega] = \int_{-\infty}^{t} [h(t - \tau)][\alpha_j(\tau)]e^{i\omega\tau} d\tau \tag{4.27b}$$

When $(t_1 = t_2 = t)$ the integrant of $[R_{z_1z_2}(t_1, t_2)]$ in Eq. (4.27a) is defined as the evolutionary power cross-spectral matrix. It is written as,

$$\left.\begin{array}{l}\text{Evolutionary cross-}\\ \text{specral matrix}\end{array}\right\} \rightarrow [S_{z_1z_2}(t, \omega)] = [A_1^*(t, \omega][S_{x_1x_2}(\omega)]\, [A_2(t, \omega]^T \tag{4.28a}$$

For this special case, since $\{z_1(t)\}$ and $\{z_2(t)\}$ are zero mean value non-stationary processes, the cross-correlation matrix $[R_{z_1z_2}(t_1, t_2)]$ becomes as to be the cross-covariance matrix. It is stated from Eqs. (4.27a) and (4.28a) as written by,

$$\left.\begin{array}{l}\text{Cross-covariances of}\\ \text{non-stationary processes}\end{array}\right\} \rightarrow [R_{z_1z_2}(t, t)] = [\sigma_{z_1z_2}(t)] = \int_{-\infty}^{\infty} [S_{z_1z_2}(t, \omega)]\, d\omega \tag{4.28b}$$

When $(\{z_1(t)\} = \{z_2(t)\} = \{z(t)\})$, which is the case of multi-support non-stationary earthquake excitations, the cross-covariance matrix given by Eq. (4.28b) becomes as the auto-covariance matrix. When it is compared with the stationary processes, the statements given by Eqs. (4.28a) and (2.133) become similar, i.e.,

$$\begin{aligned} \text{stationary to non-stationary:} \quad & [H_z(\omega)] \rightarrow [A_z(t,\omega] \\ \text{non-stationary to stationary:} \quad & [A_z(t,\omega] \rightarrow [H_z(\omega)] \end{aligned} \tag{4.29}$$

The stationary case is obtained from the non-stationary case when the matrix of deterministic time functions in Eq. (4.27b) becomes a unit matrix, i.e. $[\alpha(t)] = \mathbf{I}$ where $\mathbf{I}$ is a unit matrix. The calculation of the evolutionary power cross-spectral matrix from Eq. (4.28a) requires the calculation of the matrix of modulating functions, $[A_j(t,\omega)]$. Under assumed deterministic time functions, i.e. $[\alpha_j(t)]$, it is calculated by using Eq. (4.27b). For different deterministic time functions, the corresponding modulating functions have been calculated and reported in the literature [90].

4.5 Calculation of Response Statistical Quantities

In the previous sections, calculation of response spectra of offshore structures under wave and earthquake loadings has been presented. Since the input random wave elevation $\eta(t)$ for the wave loading, and the earthquake ground acceleration $\ddot{u}_g(t)$ for the earthquake loading, are zero mean stationary processes, the corresponding responses become also zero mean random processes. If the non-stationary earthquake motion is considered, then the response becomes also non-stationary with zero mean. Statistical quantities of a response random variable are defined as the mean value, variance and probability distribution information such as skewness, kurtosis, and spectral bandwidth which are explained in Chap. 2. For stationary processes, these statistical quantities are calculated from spectral moments of the response process considered using Eq. (2.96) in Chap. 2. The spectrum of the response considered, i.e. a stress process due to wave loading, will be calculated from Eq. (4.9c). In the case of stationary multi-support earthquake excitations, a response spectrum will be calculated from Eq. (4.13). In the case of non-stationary earthquake excitations, variances, and covariances of the derived processes cannot be calculated directly from spectral moments of responses which are given by Eq. (2.100). Calculation of the variance $\sigma_{zz}(t)$ of a response non-stationary process z, which is a time function, is explained in the previous section. The variances and covariances of the derived processes of z, which may be $\sigma_{z\dot{z}}(t)$, $\sigma_{\dot{z}\dot{z}}(t)$, $\sigma_{\dot{z}\ddot{z}}(t)$, $\sigma_{z\ddot{z}}(t)$ and $\sigma_{\ddot{z}\ddot{z}}(t)$, can be calculated as similar to the calculation of $\sigma_{zz}(t)$ by using the evolutionary spectra of the derived processes. This subject has been studied and reported in the literature, see e.g. [91–98]. Statistical characteristics of a non-stationary response process are defined in terms of the standard deviations of z and its derivative processes. These characteristics are:

$$\begin{aligned} &\text{Up-crossing frequency} \ldots\ldots\ldots :\rightarrow \omega_0(t) = \sigma_{\dot{z}}(t)/\sigma_z(t) \\ &\text{Average frequency of maxima} \ :\rightarrow \omega_m(t) = \sigma_{\ddot{z}}(t)/\sigma_{\dot{z}}(t) \\ &\text{Spectral Bandwidth}\ldots\ldots\ldots\ldots :\rightarrow \varepsilon(t) = \sqrt{1 - \omega_0^2(t)/\omega_m^2(t)} \end{aligned} \tag{4.30}$$

where $\sigma_z(t)$, $\sigma_{\dot{z}}(t)$ and $\sigma_{\ddot{z}}(t)$ are respectively standard deviations of z, $\dot{z}$ and $\ddot{z}$. It is worth noting that the process z can be a narrow banded at a time and a broad banded at another time instances.

4.6 Example

For the demonstration purposes, a jacket type offshore structure is analyzed by using the SAPOS program [99]. The structure, geometrical data and the 3D calculation model are shown in Fig. 4.6. It is supported on piles of 25 m depth in the soil as shown in Fig. 4.6. The soil properties are given in Table 4.1. Member dimensions and material properties of the structure are given in Table 4.2. It is assumed that the members in water are empty. Added masses of surrounding water of submerged members are taken into account by increasing the structural mass density according to the following statement:

$$\text{Total mass density of submerged members} :\rightarrow \rho_{\text{tot}} = \left(\rho_s + \frac{\rho_w \gamma}{4} \frac{D}{h}\right) \tag{4.31}$$

where ρ_s is the mass density of structural material, ρ_w is the mass density of water, γ is the added mass coefficient, D and h are respectively diameter and wall thickness of the structural member, which are given in Table 4.2. It is assumed that the water depth is 75.0 m and the jacket is subjected to uni-directional random waves in the global X direction as shown in Fig. 4.6. The Pierson–Moskowitz (PM) sea spectrum, which is given by Eq. (3.31a) in Chap. 3, for an assumed sea state of $H_s = 9.0$ m is used in the analysis. It is shown in Fig. 4.7 where ω_z is the zero crossings frequency and ω_p is the peak frequency of waves. It is also used a long-term Weibull probability distribution function of sea states given by Eq. (3.41) as shown in Fig. 4.8. In the analysis, the water structure interaction is taken into account with one iteration. The wave force data, marine growth, density of water and parameters of the Weibull probability distribution are given in Table 4.3 where c_d and c_m are respectively drag and inertia force coefficients, A, B and C are the parameters of the Weibull distribution. In the spectral analysis of the structure, both the quasi-static and dynamic contributions of the response are considered.

Two natural mode shapes are used for the calculation of the dynamic response contribution. The mode shapes with corresponding natural frequencies are shown in Fig. 4.9. These two mode shapes are the same in orthogonal directions with the same natural frequencies. The spectrum and spectral moments of the hot-spot normal-stress at the bottom of a leg of the jacket are calculated for the assumed sea state of $H_s = 9.0$ m. For the calculation of the hot-spot normal-stress, a stress

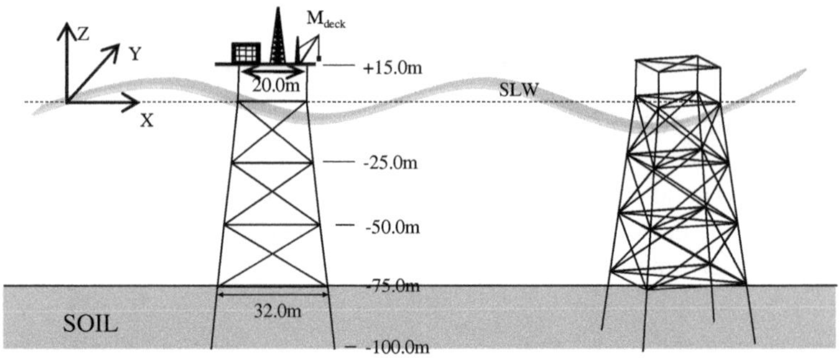

Fig. 4.6 An example jacket type offshore structure, geometrical data and 3D calculation model

Table 4.1 Properties of the soil under the example jacket structure

Axial skin friction, k_x	Subgrade modulus, k_y	Subgrade modulus, k_z
100 MPa	100 MPa	100 MPa

Table 4.2 Member dimensions, material properties and mass of the deck of the example jacket

Members	Diameter (mm)	Thickness(mm)
Bracings at top (+15.0 m)	Rigid	Rigid
Horizontal bracings and diagonals	1,200	16
Vertical diagonals	1,600	20
Top legs (above SWL)	2,000	35
Piles and Legs in water (below SWL)	1,750	30
Material properties (steel): E = 205 GPa, ρ_s = 7,800.0 kg/m^3, Poisson's ratio = 0.30		
Mass of the deck....... : M_{deck} = 4,800 ton, Added mass coefficient = 0.9		

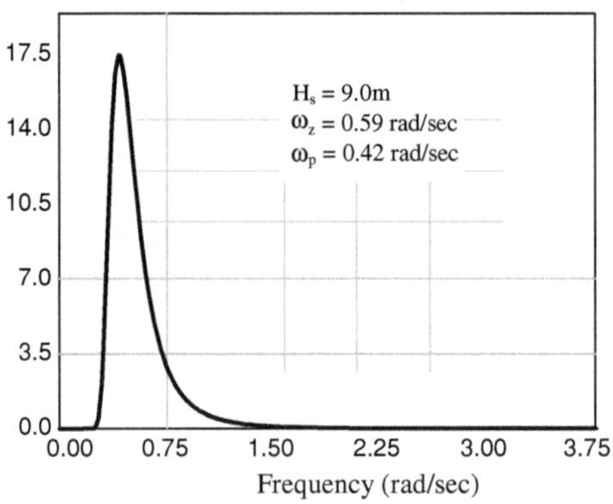

Fig. 4.7 Pierson–Moskowitz sea spectrum

concentration factor of (SCF = 2.0) is assumed. The calculated hot-spot stress transfer function is illustrated in Fig. 4.10 where the peak corresponds to the natural frequency of (ω_1 = 2.98 rad/s). The stress spectrum is illustrated in Fig. 4.11

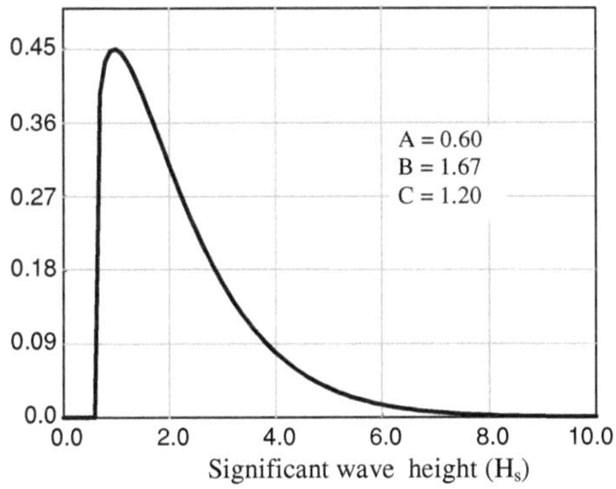

Fig. 4.8 Weibull probability distribution

Table 4.3 Wave force data and parameters of the Weibull probability function

c_d	c_m	Marine growth thick., t_{marine}	Density of water, ρ_w	A	B	C
1.0	1.6	20 cm	1,024 km/m^3	0.60	1.67	1.20

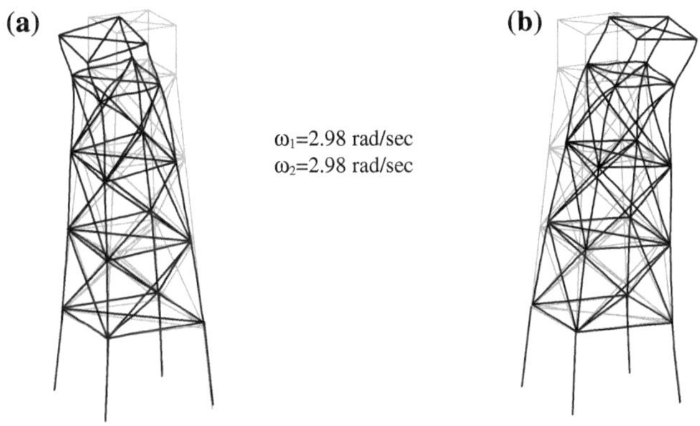

Fig. 4.9 Natural mode shapes and frequencies of the example jacket structure. **a** First natural mode shape. **b** Second natural mode shape

where the first peak corresponds to the peak frequency of waves at $\omega_p = 0.42$ rad/s and the second peak corresponds to the lowest natural frequency of the structure at $\omega_1 = 2.98$ rad/s. The stress spectral moments calculated from this spectrum are $m_0 = 0.345 \times 10^{15}$, $m_2 = 1.626 \times 10^{15}$ and $m_4 = 13.72 \times 10^{15}$. By using Eqs. (2.116) and (2.117) in Chap. 2, the mean frequencies of zero crossings and maxima of the hot-spot stress are calculated to be $\omega_0 = 2.172$ rad/s and $\omega_m = 2.905$ rad/s. From Eq. (2.102) the spectral bandwidth of the hot-spot stress process is calculated to be $\varepsilon = 0.664$ which indicates that, for the assumed sea state, the stress process is not narrow banded. It is a stochastic process between narrow and broadband.

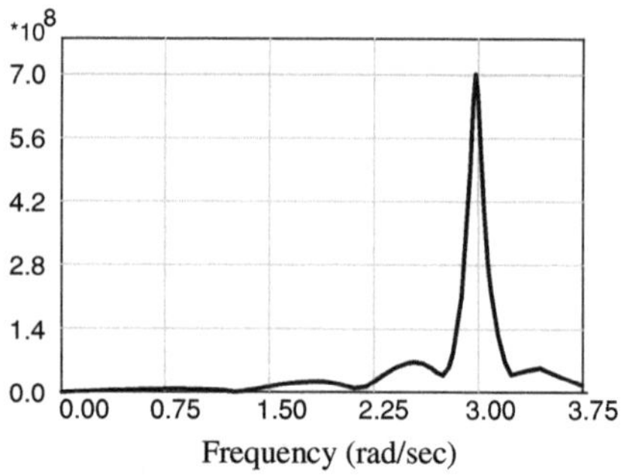

Fig. 4.10 Transfer function of hot-spot normal-stress

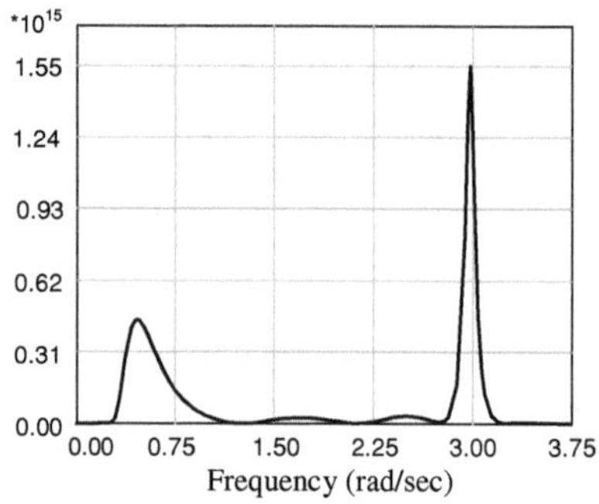

Fig. 4.11 Spectrum of the hot-spot stress

Exercise 1

A monopod tower shown in Fig. 4.12 is subjected to uni-directional random waves. The tower is fixed at the bottom. It is made of steel and has a length of $h_s = 120.0$ m, a diameter of $D = 15.0$ m with a wall thickness of $t = 0.08$ m. The mass of the deck is $M_{\text{dec}} = 300.0$ ton. The water depth is $d = 100.0$ m, the drag and inertia force coefficients are respectively $c_d = 1.3$ and $c_m = 2.0$. The structural damping ratio is assumed to be $\xi = 0.01$. The random waves are represented by the Pierson–Moskowitz sea spectrum given by Eq. (3.31a) in Chap. 3 with a sea state of $H_s = 9.0$ m. The mass density of water is assumed to be $\rho_w = 1{,}024$ kg/m^3 and added mass coefficient is $\gamma = 0.9$. The deep water condition is used in the analysis. The stress concentration factor at the bottom of the tower is assumed to be SCF $= 2.0$. The following items are required:

1. Calculate natural frequency of the tower with and without containing water up to still water level (SWL)
2. Calculate the added damping ratio due to surrounding water of the tower
3. Plot transfer function of the normal hot-spot stress at the bottom of the tower

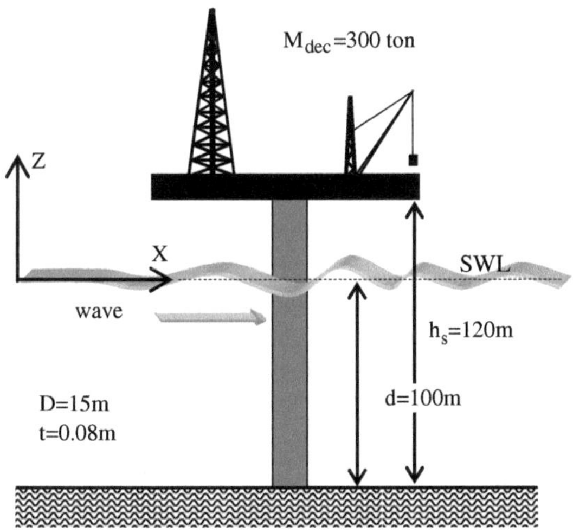

Fig. 4.12 A monopod tower under random wave loading

4. Calculate and plot the spectrum of the hot-spot stress
5. Calculate stress spectral moments, variance, frequencies of zero-crossings and maxima, and the bandwidth parameter of the hot-spot stress process at the bottom of the tower.

Exercise 2

The monopod tower shown in Fig. 4.12 is now subjected to a stationary earthquake random ground motion which is represented by the modified Kanai–Tajimi spectrum given by Eq. (4.19a). It is assumed that the characteristic ground damping ratio and frequency are respectively $\xi_g = 0.6$ and $\omega_g = 15.0$ rad/s. The characteristic damping ratio and frequency of the filter are assumed respectively to be $\omega_f = 1.5$ rad/s, $\xi_f = 0.6$. The standard deviation of the ground acceleration is assumed to be $\sigma_{\ddot{u}_g} = 0.25$ m/s^2. The followings are required to be calculated.

1. Calculate and plot the transfer function of the absolute horizontal displacement at the top of the tower
2. Calculate and plot the spectra of the displacement at the top and hot-spot normal stress at the bottom
3. Calculate stress spectral moments, variances, frequencies of zero-crossings and maxima, and the bandwidth parameters of the displacement at the top and the hot-spot stress at the bottom of the tower.

References

1. Biggs JM (1964) Introduction to structural dynamics. McGraw-Hill, New York
2. Bishop RED, Gladwell GML, Michaelson S (1965) The matrix analysis of vibration. Cambridge University Press, London

3. Meirovitch L (1967) Analytical methods in vibrations. Collier-McMillan, London
4. Clough RW, Penzien J (1993) Dynamics of structures, 2nd edn. McGraw-Hill, New York
5. Chopra AK (2001) Dynamics of structures: theory and applications to earthquake engineering. Prentice-Hall, Upper Saddle River
6. Cheng FY (2001) Matrix analysis of structural dynamics: applications and earthquake engineering. Dekker, New-York
7. Humar JL (2002) Dynamics of structures, 2nd edn. Swets & Zeitlinger, Lisse
8. Wunderlich W, Pilkey WD (2003) Mechanics of structures: variational and computational methods. CRC, Boca Raton
9. Paz M, Leigh W (2004) Structural dynamics: theory and computation, 5th edn. Kluwer, Dordrecht
10. Craig RR, Kurdila AJ (2006) Fundamentals of structural dynamics. Wiley, Hoboken
11. Brebbia CA, Walker S (1979) Dynamic analysis of offshore structures. Butterworths, London
12. Barltrop NDP, Adams AJ (1991) Dynamics of fixed marine structures, 3rd edn. Butterworth-Heinemann, Oxford
13. Han SM, Benaroya H (2002) Nonlinear and stochastic dynamics of compliant offshore structures. Kluwer, Dordrecht
14. Wilson JF (2003) Dynamics of offshore structures. Wiley, New-York
15. Soares D, Mansur WJ (2003) An efficient time/frequency domain algorithm for modal analysis of non-linear models discretized by the FEM. Comput Methods Appl Mech Eng 192:3731–3745
16. Shinozuka M, Yun C, Vaicaitis R (1977) Dynamic analysis of fixed offshore structures subjected to wind generated waves. Mech Based Des Struct Mach 5(2):135–146
17. Penzien J, Kaul MK, Berge B (1972) Stochastic response of offshore towers to random sea waves and strong motion earthquakes. Comput Struct 2:733–756
18. Kirk CL (1985) Dynamic response of marine risers by single wave and spectral analysis methods. Appl Ocean Res 7(1):2–13
19. Ahmad S, Datta TK (1989) Dynamic response of marine risers. Eng Struct 11:179–188
20. Hartnett M (2000) The application of a spectral response model to fixed offshore structures. Comput Struct 78:355–364
21. Bolt BA (1999) Earthquakes, 4th edn. Freeman, New-York
22. Manolis GD, Koliopoulos PK (2001) Stochastic structural dynamics in earthquake engineering. WIT Press, Southampton
23. Chen WF, Scawthorn C (2003) Earthquake engineering handbook. CRC, Boca Raton
24. Bolt BA (2003) Earthquakes. W.H. Freeman, New York
25. Bozorgnia Y, Bertero VV (2004) Earthquake engineering—from engineering seismology to performance-based engineering. CRC, Boca Raton
26. Chen WF, Lui EM (2006) Earthquake engineering for structural design. CRC, Boca Raton
27. Sen TK (2009) Fundamentals of seismic loading on structures. Wiley, Chichester
28. Chen WF, Lui EM (2006) Earthquake engineering for structural design. CRC, Boca Raton
29. Trifunac D, Brady AG (1975) A study on the duration of strong earthquake ground motion. Bull Seismol Soc Am 65(3):581–626
30. Zerva A (2009) Spatial variation of seismic ground motions-modeling and engineering applications. CRC, Boca Raton
31. Laouami N, Labbe P (2001) Analytical approach for evaluation of the seismic ground motion coherency function. Soil Dyn Earthq Eng 21:727–733
32. Nadim F, Vanmarcke EH, Gudmestad OT, Hetland S (1991) Influence of spatial variation of earthquake motion on response of gravity base platforms. Struct Saf 10:113–128
33. Harichandran RS, Vanmarcke EH (1986) Stochastic variation of earthquake ground motion in space and time. J Eng Mech ASCE 112(2):154–174
34. Ryu CS, Yun CB (1997) Nonstationary response analysis of offshore guyed tower subjected to earthquake loading. Eng Struct 19(1):63–70
35. Gao W (2007) Random seismic response analysis of truss structures with uncertain parameters. Eng Struct 29:1487–1498

36. Takewaki I (2004) Critical envelope functions for non-stationary random earthquake input. Comput Struct 82(20–21):1671–1683
37. Shinozuka M, Deodatis G (1988) Stochastic process models for earthquake ground motion. Probab Eng Mech 3(3):114–123
38. Karadeniz H (1999) Spectral analysis of offshore structures under combined wave and earthquake loadings. Proceedings of the 9th Internataional Offshore and Polar Engineering Conference, ISOPE, vol 4. pp 504–511
39. Richter CF (1935) An instrumental earthquake scale. Bull Seism Soc Am 25:1–32
40. Hwang H, Huo JR (1997) Attenuation relations of ground motion for rock and soil sites in eastern United States. Soil Dyn Earthq Eng 16:363–372
41. Campbell KW (1981) Near-source attenuation of peak horizontal acceleration. Bull Seism Soc Am 71(6):2039–2070
42. Campbell KW, Bozorgnia Y (2003) Updated near-source ground motion (attenuation) relations for the horizontal and vertical components of peak ground acceleration and acceleration response spectra. Bull Seism Soc Am 93(1):314–331
43. Campbell KW, Bozorgnia Y (2006) Campbell-Bozorgnia next generation attenuation (NGA) relations for PGA, PGV and Spectral acceleration: A progress report. Proceedings of the 8th US National Conference on Earthquake Engineering, Paper No.906
44. Oliveira CS, Hao H, Penzien J (1991) Ground motion modeling for multiple input structural analysis. Struct Saf 10:79–93
45. Der Kiureghian A, Neuenhofer A (1992) Response spectrum method for multi-support seismic excitations. Earthq Eng Struct Dyn 21:713–740
46. Heredia-Zavoni E, Vanmarcke EH (1994) Seismic random-vibration analysis of multi-support-structural systems. J Eng Mech 120:1107–1128
47. Bai FL, Hao H, Li HN (2010) Seismic response of a steel trussed arch structure to spatially varying earthquake ground motions including site effect. Adv Struct Eng 13(6):1089–1103
48. Chun-guang L, Hui-jun L (2010) A novel method to calculate the dynamic reliability of space structures subjected to multi-dimensional multi-support excitations. Int J Space Struct 25(1):25–34
49. Zhang YH, Li QS, Lin JH, Williams FW (2009) Random vibration analysis of long-span structures subjected to spatially varying ground motions. Soil Dyn Earthq Eng 29:620–629
50. Rambabu KV, Allam MM (2007) Response of an open-plane frame to multiple support horizontal seismic excitations with soil–structure interaction. J Sound Vib 299:388–396
51. Boissieres HP, Vanmarcke EH (1995) Spatial correlation of earthquake ground motion: non-parametric estimation. Soil Dyn Earthq Eng 14:23–31
52. Goda K, Hong HP (2008) Spatial correlation of peak ground motions and response spectra. Bull Seism Soc Am 98(1):354–365
53. Luco JE, Wong HL (1986) Response of a rigid foundation to a spatially random ground motion. Earthq Eng Struct Dyn 14:891–908
54. Harichandran RS (1991) Estimating the special variation of earthquake ground motion from dense array recordings. Struct Saf 10:219–233
55. Zerva A (1994) On the spatial variation of seismic ground motions and its effects on lifelines. Eng Struct 16(7):534–546
56. Zerva A (2002) Spatial variation of seismic ground motions: an overview. Appl Mech Rev 55(3):271–297
57. Zembaty Z, Rutenberg A (2002) Spatial response spectra and site amplification effects. Eng Struct 24:1485–1496
58. Cacciola P, Deodatis G (2011) A method for generating fully non-stationary and spectrum-compatible ground motion vector processes. Soil Dyn Earthq Eng 31:351–360
59. Hong HP, Zhang Y, Goda K (2009) Effect of spatial correlation on estimated ground-motion prediction equations. Bull Seism Soc Am 99(2A):928–934
60. Price TE, Eberhard MO (1998) Effects of spatially varying ground motions on short bridges. J Struct Eng 124(8):948–955

61. Dumanoglu AA, Soyluk K (2003) A stochastic analysis of long span structures subjected to spatially varying ground motions including the site-response effect. Eng Struct 25:1301–1310
62. Bayraktar A, Bilici Y, Akkose M (2010) The effect of the spatially varying earthquake ground motion on random hydrodynamic pressures. Adv Struct Eng 13(6):1153–1165
63. Sun J, Ye J, Cheng W, Zhang Q (2007) Application of the pseudo-excitation method with spatial coherence in random vibration analysis of long-span space structures. Adv Struct Eng 10:135–151
64. Vanmarcke E (1983) Random fields: analysis and synthesis. MIT Press, Cambridge
65. Yeh CH, Wen YK (1990) Modeling of nonstationary ground motion and analysis of inelastic structural response. Struct Saf 8:281–298
66. Priestley MB (1965) Evolutionary spectra and non-stationary processes. J Royal Stat Soc Ser B (Methodol) 27(2):204–237
67. Priestley MB (1967) Power spectral analysis of non-stationary random processes. J. Sound Vib 6(I):86–97
68. Hammond JK (1968) On the response of single and multidegree of freedom systems to non-stationary random excitations. J Sound Vib 7(3):393–416
69. Zhixin X (1988) On the stochastic modeling of earthquake strong ground motion. Nucl Eng Des 110:171–175
70. Langley RS (1986) Structural response to non-stationary non-white stochastic ground motions. Earthq Eng Struct Dyn 14:909–924
71. Der Kiureghian A, Crempien J (1989) An evolutionary model for earthquake ground motion. Struct Saf 6(1989):235–246
72. Iwan WD, Hou ZK (1989) Explicit solution for the response of simple systems subjected to nonstationary random excitation. Struct Saf 6:77–86
73. Quek ST, Teo YP, Balendra T (1990) Non-stationary structural response with evaluationary spectra using seismological input model. Earthq Eng Struct Dyn 19:275–288
74. Lin J, Zhang W, Williams FW (1994) Pseudo-excitation algorithm for nonstationary random seismic responses. Eng Struct 16(4):270–276
75. Deodatis G (1996) Non-stationary stochastic vector processes: seismic ground motion applications. Probab Eng Mech 11:149–168
76. Conte JP, Peng BF (1997) Fully nonstationary analytical earthquake ground-motion model. J Eng Mech 123(1):15–24
77. Fang T, Sun M (1997) A unified approach to two types of evolutionary random response problems in engineering. Arch Appl Mech 67:496–506
78. Deodatis J, Shinozuka M (1988) Auto-regressive model for nonstationary stochastic processes. J Eng Mech 114(11):1995–2012
79. Zembaty Z (1988) A note on non-stationary stochastic response and strong motion duration. Earthq Eng Struct Dyn 16:1189–1200
80. Gupta ID, Trifunac MD (2000) A note on the nonstationarity of seismic response of structures. Eng Struct 22(11):1567–1577
81. Rofooei FR, Mobarake A, Ahmadi G (2001) Generation of artificial earthquake records with a nonstationary Kanai–Tajimi model. Eng Struct 23:827–837
82. Takewaki I (2001) Nonstationary random critical excitation for nonproportionally damped structural systems. Comput Methods Appl Mech Eng 190(31):3927–3943
83. Gao W, Chen JJ, Ma J, Liang ZT (2004) Dynamic response analysis of stochastic frame structures under nonstationary random excitation. AIAA 42(9):1818–1822
84. Xue SD, Wang S, Cao Z (2004) Multi-dimensional pseudo excitation method for nonstationary random seismic analysis of spatial lattice shells. Int J Space Struct 19(3):129–136
85. Pousse G, Bonilla LF, Cotton F, Margerin L (2006) Nonstationary stochastic simulation of strong ground motion time histories including natural variability: application to the K-net Japanese Database. Bull Seism Soc Am 96(6):2103–2117
86. Chaudhuri A, Chakraborty S (2006) Reliability of linear structures with parameter uncertainty under non-stationary earthquake. Struct Saf 28:231–246

87. Rezaeian S, Der Kiureghian A (2008) A stochastic ground motion model with separable temporal and spectral nonstationarities. Earthq Eng Struct Dyn 37:1565–1584
88. Bougioukou AP, Leros AP, Papakonstantinou V (2008) Modeling of non-stationary ground motion using the mean reverting stochastic process. Appl Math Modell 32:1912–1932
89. Cacciola P (2010) A stochastic approach for generating spectrum compatible fully nonstationary earthquakes. Comput Struct 88:889–901
90. Jangid RS (2004) Response of SDOF system to non-stationary earthquake excitation. Earthq Eng Struct Dyn 33:1417–1428
91. Senthilnathan A, Lutes LD (1991) Nonstationary maximum response statistics for linear structures. J Eng Mech ASCE 117(2):294–311
92. Michaelov G, Sarkani S, Lutes LD (1999) Spectral characteristics of nonstationary random processes—a critical review. Struct Saf 21:223–244
93. Michaelov G, Sarkani S, Lutes LD (1999) Spectral characteristics of nonstationary random processes - response of a simple oscillator. Struct Saf 21:245–267
94. Sarkani S, Lutes LD, Michaelov G (2000) Extreme response of linear structures to nonstationary base excitation. Paper No.1293, Proceedings of the 12th WCEE, Auckland, New Zealand
95. Naprstek J, Fischer C (2000) Analysis of non-stationary response of structures due to seismic random processes of evolutionary type. Paper No.1110, Proceedings of the 12th. WCEE, Auckland, New Zealand
96. Fuentes M (2002) Spectral methods for nonstationary spatial processes. Biometrika 98(1):197–210
97. Barbato M, Conte JP (2008) Spectral characteristics of non-stationary random processes: theory and applications to linear structural models. Probab Eng Mech 23:416–426
98. He J (2009) An approximation of the first passage probability of systems under nonstationary random excitation. Appl Math Mech Engl Ed 30(2):255–262
99. Karadeniz H (2009) SAPOS, Spectral analysis program of structures. Report, Structural Mechanics Division, Faculty of Civil Engineering and Geosciene, TUDelft, Delft, The Netherlands

Chapter 5
Fatigue Analysis of Offshore Structures

5.1 Introduction

Fatigue is an important phenomenon that causes damages and failure around the weld toe at member connections of offshore structures. Fatigue occurs due to high stress concentrations in the long term since loading conditions are time-dependent and continuous in sea and ocean environments. Because of severe corrosion and uncertain cyclic loading in the sea environment, fatigue becomes potentially one of the main problems causing degradation in the long-term structural integrity. Structures to be built in such environments are imposed on wind, wave, current, and earthquake exposures. Among them, waves play a major role in fatigue failure due to their continuity in time in random sequences, as being tiny, moderate, and sometimes catastrophic, which produce random fluctuating stress responses in structural components. Explorations and productions of offshore energy resources in more hostile and deeper ocean environments require structures that are relatively flexible and dynamic sensitive. For such structures, fatigue damage accumulations and consequently the fatigue failure become more important due to stress fluctuations with large amplitudes that cause failure in a shorter time period. Since fatigue is one of the main design criteria for offshore structures, it requires a sophisticated and advanced analysis procedure to be formulated and determined in a proper way to maintain the long-term integrity of welded joints of the structure for a safe and reliable operation. However, fatigue is a complex phenomenon that a little is known about its cumulative excursion in time under random loading. It depends on loading types, structural material, joint configuration, initial conditions and imperfections of the geometry, etc.

In the theoretical point of view, methods of the fracture mechanics suffer from lack of completeness to solve this complicated problem in a general sense. Fatigue damage occurs in two main stages as being crack initiation and crack growth including propagation of one dominant crack and the final fracture. The fracture mechanics approach can be performed straightforwardly if the critical value of fracture toughness at the service conditions [1] is known. The amount of fracture

H. Karadeniz, *Stochastic Analysis of Offshore Steel Structures*,
Springer Series in Reliability Engineering, DOI: 10.1007/978-1-84996-190-5_5,

toughness involved in the development of fracture criterion depends on some factors. These are the knowledge of service conditions of the structure such as temperature and loading, the desired level of performance in the structure such as plane-strain, elastic–plastic, or plastic behavior, and the consequences of structural component failure. Owing to these complexities, the fracture mechanics approach can be applied to the fatigue damage estimate for very simple cases of structural components that are mostly far from the realistic welded joint configurations of offshore structures. However, in general, it provides some useful explanations and results in the fatigue damage formulation in the uncertainty space. Due to considerable variations in the controlling parameters of uncertainties, a reliability-based fatigue [2–12] and fracture mechanics assessments [13–17] need to be essentially carried out.

In the experimental point of view, our knowledge of fatigue is mainly based on experimental observations. Therefore, fatigue testing and interpretation of test results are of utmost importance. Fatigue test results constitute the basis of design and assessment procedures for the fatigue resistance of offshore structural components to cyclic loadings. As a phenomenon, fatigue strength cannot be theoretically formulated from fundamental engineering principles, but is rather analytically modeled from experimental observations. For offshore structures, laboratory tests have been carried out to obtain fatigue data for design purposes, see for example [18]. The analysis of existing experimental data and corresponding fatigue damage formulation is crucial to successful design of offshore structures. However, most of the data have been produced from constant amplitude tests, and few from variable amplitude tests (random loading). In all loading conditions, some specific joint types have been tested, and therefore the data available are subjected to some limitations of the joint types as well as the loading conditions. In practical applications, no limitation does exist concerning loading and other environmental factors that cannot be represented in a laboratory setup. The joint configurations in practice are also much more complex than those used in experimental tests. Therefore, owing to these limitations and uncertainties in loading and environment, a probabilistic fatigue analysis will be more eligible than a deterministic one for offshore steel structures. The probabilistic fatigue damage estimate of offshore structures can be carried out by using a fatigue model, which is based on available fatigue data obtained from experiments, and theoretical tools that have been developed so far. This chapter outlines the procedure of analytical calculation for probabilistic fatigue damages and fatigue life of offshore steel structures assuming that the experimental fatigue model (*S–N* curve) is provided.

5.2 Fatigue Phenomenon and Fatigue Damages

In general, fatigue is defined as a process of damage accumulation in a material under fluctuating stress histories. Eventually, it results in failure and even the maximum working stress remains below the ultimate elastic limit of the stress. It reduces local strengths of structural materials. The phenomenological details of the fatigue process may be different for different materials [19], but it is common

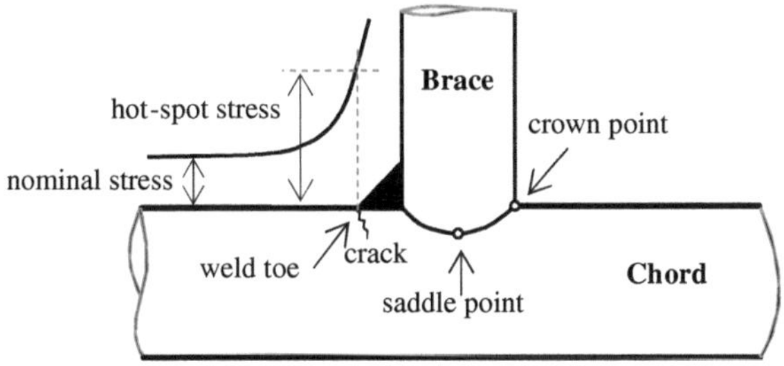

Fig. 5.1 A tubular T joint, definition of the hot-spot stress in the chord, and a possible crack region due to the fatigue damage

for all materials that the process of progressive change in a material under fluctuating stresses may be ended by development of cracks or complete fracture after a sufficient number of stress cycles. For tubular welded joints, the fatigue crack starts mainly at the weld toe where stress concentration occurs as shown in Fig. 5.1 for a T joint assuming that any initial imperfection does not exist to cause high stress concentrations. Then, the microscopic crack grows gradually through the thickness and along the circumference of the stressed member, e.g., chord in Fig. 5.1, to become visible, and finally to a degree of fully damaged cross-section. The stress concentration at the weld toe is defined as the hot-spot stress.

In the fatigue process, the damage develops gradually in early stages and then accelerates very rapidly in late stages toward the end at the location of the hot-spot stresses. The first stage consists of a threshold, crack initiation, and formation as schematically shown in Fig. 5.2. The second stage belongs to the crack-propagation process which follows a linear propagation rule in the logarithmic scale. The third stage belongs to unstable crack growth and fracture which finalizes with the failure as shown in Fig. 5.1. In the threshold region of Stage I, i.e., $\Delta K \leq \Delta K_{th}$ where ΔK and ΔK_{th} are, respectively, stress- and threshold stress-intensity factors, cracks do not propagate [1]. This point defines an endurance fatigue limit of the (*S–N*) curve. In this region an infinite fatigue life is obtained. Stage II represents the crack propagation process according to the Paris–Erdogan crack propagation Law [20] which dominates the process of fatigue damage accumulation. In Stage III, the acceleration of the crack propagation process increases and, at a maximum (critical) constant stress-intensity factor, ΔK_{max}, the crack grows to failure as an unstable fracture when the remaining cross-sectional area is too small to take a stress value. There are two main methods applied practically to predict the fatigue damage growth in offshore structural members. These are the fracture mechanics approach and the (*S–N*) curve approach as presented briefly in the following sections.

5.2.1 Fracture Mechanics Approach to Predict Fatigue Damages

Fracture mechanics approach is used at the design stage of offshore structures as it provides the basis for fatigue life prediction, and also during the operational stage

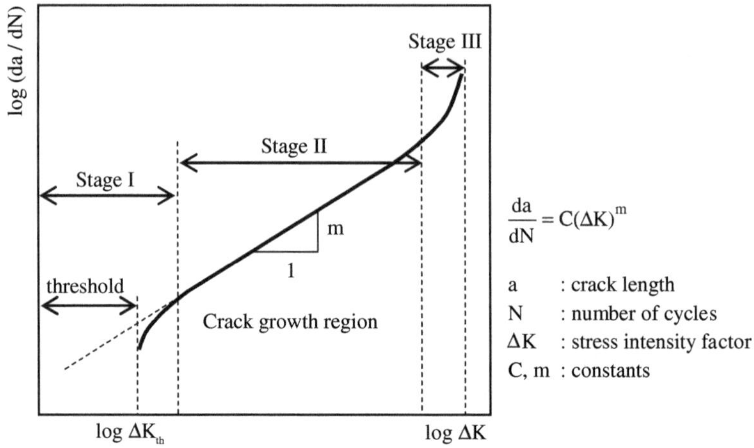

Fig. 5.2 Crack developments in materials under cyclic stress processes

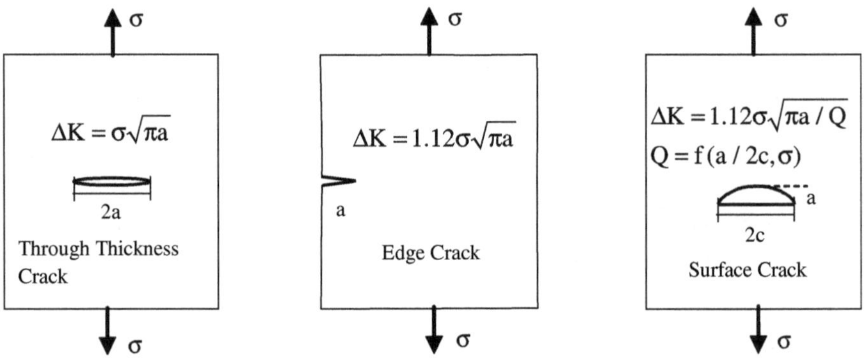

Fig. 5.3 Different crack geometries and related stress intensity factors for an infinite plate

to make decisions on inspection scheduling and repair strategies [18]. Like other large and complex structures, offshore tubular structures can also have crack-like imperfections, notches, or other kind of discontinuities. The fundamental concept of the fracture mechanics is the characterization of the stress field in terms of a single parameter, ΔK, which is called as the *stress intensity factor*, in the vicinity of a crack [1]. This parameter is related to both the stress σ and crack size a. For an infinite plate with different crack geometries under a uniform stress σ shown in Fig. 5.3, the stress intensity factors are written in Fig. 5.3 [21]. Welded tubular joints are much more complicated than a plate shown in Fig. 5.3 regarding the geometry and the loading case in which cracks are usually in a complex stress field. Therefore, the stress intensity factor in tubular-welded joints should contain the effect of the geometrical and loading complexities. This is achieved by defining the range of the stress intensity factor as written [15, 18, 22, 23] by,

$$\Delta K = \Delta S_h Y \sqrt{\pi a} \tag{5.1}$$

in which Y is a correction function accounting for a free front surface, finite plate width, crack geometry, non-uniform stress field, geometrical discontinuity, and changes in structural restraint [18]. In Eq. (5.1), the parameter a is the crack size and ΔS_h is the range of stress amplitude at the hot spot. For cracked tubular welded joints, different analysis methods can be applied to determine the correction function Y using different solutions of the stress intensity factor for semi-elliptical surface cracks. Empirical and semi-empirical solutions which are based on experimental results and those obtained from the finite element analysis are used for this purpose, see e.g. [18, 22, 23]. As far as experimental method is concerned an analytic expression of Y can be obtained from curve fitting to experimental data as being a function of crack size a, member thickness and the applied stress range [18, 23]. The finite element method is widely used to calculate the *stress intensity factor* (SIF) of tubular joints since it includes all leading parameters and produces accurate results.

Having defined the range of stress intensity factor, a crack growth law must be established to estimate the cumulative fatigue damage. For this purpose, the Paris–Erdogan crack growth law [20] is commonly used in practice as given by,

$$\frac{da}{dN} = C(\Delta K)^{\mathrm{m}} \tag{5.2}$$

where C and m are material constants and N is the number of stress cycles. Having introduced Eq. (5.1) into (5.2), for a constant hot-spot stress range ΔS_h, the expression of the fatigue life (number of stress cycles to failure, N) can be calculated from the integration of Eq. (5.2) as written by,

$$N = \frac{\left(\Delta S_h \sqrt{\pi}\right)^{-\mathrm{m}}}{\mathrm{C}} \int_{a_i}^{a_f} \left(Y\sqrt{a}\right)^{-\mathrm{m}} da \tag{5.3}$$

where a_i and a_f are, respectively, the initial and final crack sizes.

For variable amplitude stress ranges, an incremental procedure can be applied to estimate the crack growth with an equivalent hot-spot stress range denoted by ΔS_{eq}. For a hot-spot stress range ΔS_j, an infinitesimal crack size increment $d(\Delta a_j)$ can be expressed by using Eqs. (5.1) and (5.2) as written by,

$$d(\Delta a_j) = \mathrm{C}\left(Y\sqrt{\pi a}\right)^{\mathrm{m}} \left(\Delta S_j\right)^{\mathrm{m}} d(\Delta N_j) \tag{5.4}$$

in which $d(\Delta N_j)$ is the corresponding infinitesimal increment of the number of stress cycles to failure for the stress range ΔS_j. The differential crack size, da, is obtained by summation of infinitesimal crack size increments, $d(\Delta a_j)$, for all different stress ranges as written by,

$$da = \sum_{j=1}^{k} d(\Delta a_j) \quad \rightarrow \quad da = \mathrm{C}\left(Y\sqrt{\pi a}\right)^{\mathrm{m}} \sum_{j=1}^{k} \left(\Delta S_j\right)^{\mathrm{m}} d(\Delta N_j) \tag{5.5a}$$

where k is the number of different applied stress ranges. Having divided Eq. (5.5a) by differential number of stress cycles, dN, the crack growth law given by Eq. (5.2) can be expressed as,

$$\frac{da}{dN} = C\left(Y\sqrt{\pi a}\right)^{m}\sum_{j=1}^{k}\left(\Delta S_j\right)^{m}\left(\frac{d(\Delta N_j)}{dN}\right) \quad \text{with} \quad dN = \sum_{j=1}^{k} d(\Delta N_j) \tag{5.5b}$$

The ratio, $d(\Delta N_j)/dN$, in Eq. (5.5b) is simply the probability, $P(\Delta S_j)$, that the stress range ΔS_j occurs in the life time. Therefore, Eq. (5.5b) can be rewritten as,

$$\frac{da}{dN} = C\left(Y\sqrt{\pi a}\right)^{m}\sum_{j=1}^{k}\left(\Delta S_j\right)^{m}P(\Delta S_j) = C\left(\Delta S_{eq}Y\sqrt{\pi a}\right)^{m} \tag{5.6a}$$

in which ΔS_{eq} is an equivalent hot-spot stress range [24] defined as,

$$\Delta S_{eq} = \left(\sum_{j=1}^{k}\left(\Delta S_j\right)^{m}P(\Delta S_j)\right)^{\frac{1}{m}} = \left(\int_{0}^{\infty} s^{m} f_{\Delta S_h}(s)\, ds\right)^{\frac{1}{m}} \tag{5.6b}$$

The function $f_{\Delta S_h}(s)$ in Eq. (5.6b) is the probability density function of the variable (random) hot-spot stress range ΔS_h. This technique of equivalent stress range assumes that the material constants C and m are same for all different stress ranges, i.e., a single linear function is used for all stress ranges to represent the crack growth rate in the logarithmic scale. If, however, different linear functions are applied for different stress ranges to represent the crack growth rate (multiple segment representation of the crack growth), then an equivalent crack growth concept is defined [25]. The multiple segmented crack growth occurs with the corrosion fatigue crack growth [16]. In this case, for the segment j, the material constants are denoted by C_j and m_j. For the hot-spot stress range ΔS_h in the region of segment j, with lower and upper bounds, i.e., $(\Delta S_{j-1} \le \Delta S_h \le \Delta S_j)$, the contribution of the crack growth rate can be stated from Eqs. (5.1) and (5.2) as written by,

$$\left(\frac{da}{dN}\right)_j = C_j\left(Y\sqrt{\pi a}\right)^{m_j}\left(\Delta S_h\right)^{m_j}P\left(\Delta S_{j-1} \le \Delta S_h \le \Delta S_j\right) \tag{5.7a}$$

where $P(.)$ is the probability that the stress range ΔS_h is in the region of $(\Delta S_{j-1} \le \Delta S_h \le \Delta S_j)$. For all stress ranges, i.e., for all segments, the total crack growth rate is calculated by superimposing contributions of individual segments. It can be written from Eq. (3.7a) as,

$$\left(\frac{da}{dN}\right)_{AV} = \sum_{j=1}^{k}\left(\frac{da}{dN}\right)_j \rightarrow \left(\frac{da}{dN}\right)_{AV} = \sum_{j=1}^{k} C_j\left(Y\sqrt{\pi a}\right)^{m_j}\int_{\Delta S_{j-1}}^{\Delta S_j} s^{m_j} f_{\Delta S_h}(s)\,ds \tag{5.7b}$$

where $f_{\Delta S_h}(s)$ is the probability density function of the hot-spot stress range ΔS_h, k is the number of segments and $(\mathrm{d}a/\mathrm{d}N)_{\mathrm{AV}}$ denotes the average value of the crack growth rate for all stress ranges applied. By using this average crack growth rate, the fatigue life is calculated from the integration,

$$N = \int_{a_i}^{a_f} \frac{\mathrm{d}a}{(\mathrm{d}a/\mathrm{d}N)_{\mathrm{AV}}} \tag{5.8}$$

in which a_i and a_f are the initial and final crack sizes, respectively. By introducing $(\mathrm{d}a/\mathrm{d}N)_{\mathrm{AV}}$ from Eq. (5.7b) into (5.8), the fatigue life N can be numerically calculated.

5.2.2 S–N Curve Approach to Predict Fatigue Damages

The *S–N* curve approach is based on experimental fatigue-test data. In this approach, the fatigue life N, which is the number of stress cycles to failure, is related to the hot-spot stress range S_h with the constant amplitude. In the light of fracture mechanics formulation as given by Eq. (5.3), it can be written that,

$$N = \mathrm{C} S_h^{-\mathrm{k}} \tag{5.9}$$

where C and k are constants which are determined from the regression of the fatigue-test data [26] in the logarithmic scale as shown in Fig. 5.4. In the logarithmic scale, Eq. (5.9) is a linear function as written in the natural logarithmic scale by,

$$\ln N = \ln \mathrm{C} - \mathrm{k} \ln S_h \tag{5.10}$$

which is shown in Fig. 5.4. The *S–N* curve approach is commonly used in practice and well described in Recommended Practices and Guide for fatigue design of offshore structures [27–31]. A multi-segmented *S–N* line approach [32–34] which is shown in Fig. 5.4 can also be used for a better curve fitting to the fatigue-test data. The multi-segmented *S–N* fatigue model allows for an endurance limit under which stress ranges do not produce fatigue damages practically. The *S–N* line of a segment, e.g., segment j is defined as written by,

$$N_j(S_h) = \mathrm{C}_j S_h^{-\mathrm{k}_j} \quad \rightarrow \quad (S_h)_j \leq S_h \leq (S_h)_{j+1} \tag{5.11}$$

which is valid only in the hot-spot stress range defined in Eq. (5.11). Each segment has its own fatigue constants, C_j and k_j, and accordingly a fatigue life N_j for a constant amplitude stress range S_h. In practice, the hot-spot stress range is defined in terms of the nominal stress range S_j and a *stress concentration factor* (SCF) as stated by,

$$S_h = (\mathrm{SCF} \times S_j) \tag{5.12}$$

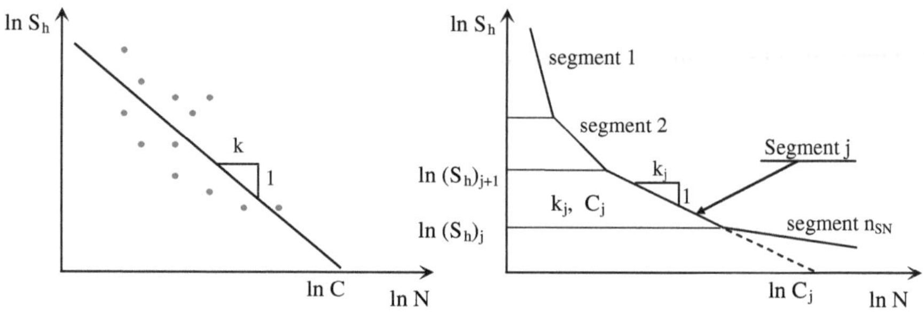

Fig. 5.4 A regression line (*S–N* Line) of fatigue-test data and a multi-segmented *S–N* model

In the fatigue damage calculation, the hot-spot and nominal stress ranges are calculated as being normal to the weld toe. The SCF can be accurately calculated by using a FE analysis of tubular joints. But, in practice for different joint configuration and loading types, approximate empirical parametric formulae [23, 29, 35] may be used to estimate the SCF. A number of semi-empirical relations are also available in the literature [27] as depending on the joint type and member dimensions.

5.2.3 Cumulative Fatigue Damage Rule

Having defined a relation between the stress range and the number of stress cycles to failure as presented above, a damage accumulation rule is used to predict the cumulative damage under different stress ranges. For this purpose, the Palmgren–Miner's rule [36–38] is commonly used in practice and it is implemented in design codes [28–31]. This rule assumes a linear fatigue damage accumulation for different stress ranges. It is stated for a number of constant amplitude stress blocks as written by,

$$D = \sum_{j=1}^{q} \frac{n(S_j)}{N(S_j)} \tag{5.13}$$

in which q is the number of stress blocks applied, $n(S_j)$ is the number of constant amplitude stress cycles for the block j with the stress range S_j, $N(S_j)$ is the number of stress cycles to failure for the stress range S_j, and D is the cumulative damage rate in the lifetime. The stress blocks and corresponding fatigue lives are shown in Fig. 5.5. The Palmgren–Miner's rule states that the total fatigue damage for all applied stress blocks is obtained by linear summation of damages produced by individual stress blocks. Since the total damage D is a fraction between working stress cycles and those to failure, it is assumed that fatigue failure occurs when this fraction reaches unity, i.e., ($D = 1.0$), theoretically. In general, if we denote the ultimate value of the damage ratio as D_f, then fatigue failure occurs when the total damage ratio D exceeds this reference damage value D_f, i.e., $(D \geq D_f)$.

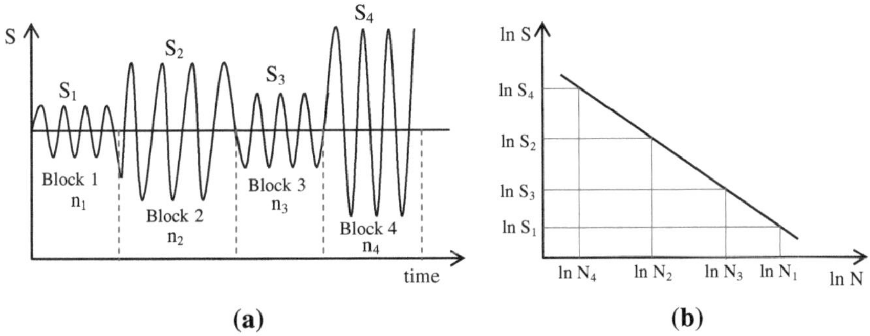

Fig. 5.5 Example of different applied stress blocks and corresponding fatigue lives, N_i. **a** Applied stress blocks **b** S-N line

Investigations to check the validity of the theoretical value of D_f, which is unity, have revealed that it varies considerably under realistic circumstances, e.g., for welded structures a range of $D_f = (0.5 - 2.0)$ has been considered to be appropriate [39, 40]. In DNV [29] and NORSOK [35] codes, this range is taken to be $D_f = (0.1 - 1.0)$ which produces rather conservative results [18]. It has been reported [41, 42] that the Palmgren–Miner's damage sum (D_f) is greater than unity for low–high loading sequence and less than unity for high–low loading sequence tests. For deterministic fatigue analysis, a mean value of D_f equals unity can be used in practical applications. For probabilistic analysis, the reference damage value D_f should be taken as random with mean value of unity

5.3 Cycle Counting Procedure for Random Stresses

In the case of variable amplitude stress variation, the stress range and cycle cannot be defined as simple as those for constant amplitude stress variation. In practice, stress time histories of offshore structures under wave and other random loading sequences are random, and therefore, stress cycles and corresponding ranges can only be determined by counting algorithms. The main issue of a counting procedure is that the stress history must be available in the time domain by recording, simulated numerically or obtained from a time domain dynamic analysis. In the stochastic analysis of offshore structures, stress spectra are calculated usually by applying a spectral analysis method. In order to apply a cycle counting algorithm the stress time history must be generated first from the corresponding spectrum. Then the cycle counting is processed. A number of methods have been proposed for stress cycle counting [43]. The methods most commonly accepted for use in connection with Codes and Standards are the Reservoir and Rainflow methods [44, 45]. The reservoir cycle counting is employed for short stress histories while the rainflow counting is employed for longer and more complex stress histories [44].

It has been reported [43] that the rainflow method is generally regarded as the best estimator of the fatigue damage in actual lives, and therefore, it is used largely in practice. As the rainflow cycle counting method is the most commonly adopted for use in practice it is explained briefly in the following section.

5.3.1 Rainflow Cycle Counting

When the stress-time history is random, determination of the stress cycles and ranges are not possible analytically. In this case, a cycle counting algorithm is applied to determine stress cycles and ranges from the time series of the stress process. For this purpose, the rainflow cycle counting method is widely used in practice and implemented in design codes. The rainflow method has obtained its name from an analogy of rain falling down a pagoda roof. It was developed by Matsuishi and Endo in 1968 [46], and since then it has been studied and well documented in the literature, e.g., [43–45, 47–50]. Rychlik presented a mathematical definition for the rainflow cycle counting method [51–56], which enables closed-form computations from the statistical properties of the load signal. Since the cumulative damage is affected by the loading sequence, in random loading, the loading sequence is also random. Thus, the classical rainflow cycle counting method does not take into account the loading sequence to reduce the Palmgren–Miner's damage sum. A modification of the classical rainflow cycle counting has been proposed to eliminate this drawback [57]. For long stress time series, an alternate rainflow algorithm was proposed to count cycles by dividing the long series into smaller parts [50]. In this section, the principle of the classical rainflow cycle counting method is presented.

The rainflow cycle counting method is based on visualization of rainflow over a sequence of pagoda roofs. Essentially, it counts half cycles. To simulate a pagoda roof from a stress time series, the peaks and troughs of the stress time series are connected by linear lines as shown in Fig. 5.6b. Then the linearized model is rotated clock wise with 90° such that the time axis is in vertical direction with the origin at the top and positive downward, as shown in Fig. 5.7a in which example stress values of peaks and troughs are written relative to mean values. In this figure, the odd numbers denote the troughs and even numbers denote the peaks. Rainflow is assumed to begin from a peak or a trough and keeps falling on the roof until it stops according to the following conditions.

1. A drop begins to flow right from a trough or left from a peak onto subsequent roofs.
2. When a drop starts from a trough, it stops if it meets an equal or deeper trough than that it is started from, e.g., path (1–2–2′) in Fig. 5.7b. The drop falls on another roof until a stop condition arises, e.g., path (3–4–4′–6–6′–10–10′) in Fig. 5.7b. The drop stops flowing when it meets a flow from an earlier path, e.g., path (5–4′) in Fig. 5.7b.

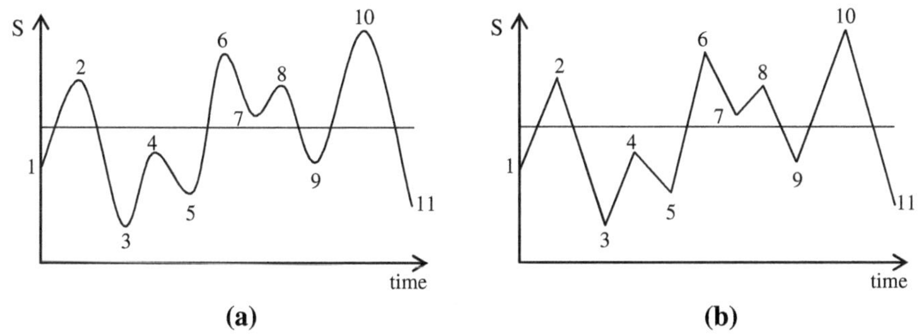

Fig. 5.6 An example stress time history and the equivalent linearized model. **a** An actual stress time histrory **b** the equivalent linearized model

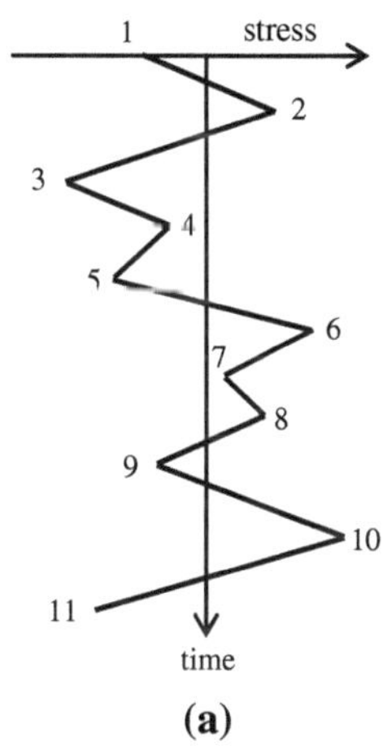

No.	stress	No.	stress
1	-34.0	6	60.0
2	40.0	7	8.0
3	-66.0	8	36.0
4	-20.0	9	-24.0
5	-48.0	10	78.0
		11	-54.0

path	stress range	path	stress range
1-2'	74.0	6-9'	84.0
2-3'	106.0	7-8'	28.0
3-10	144.0	8-7'	28.0
4-5'	28.0	9-6'	84.0
5-4'	28.0	10-11	132.0

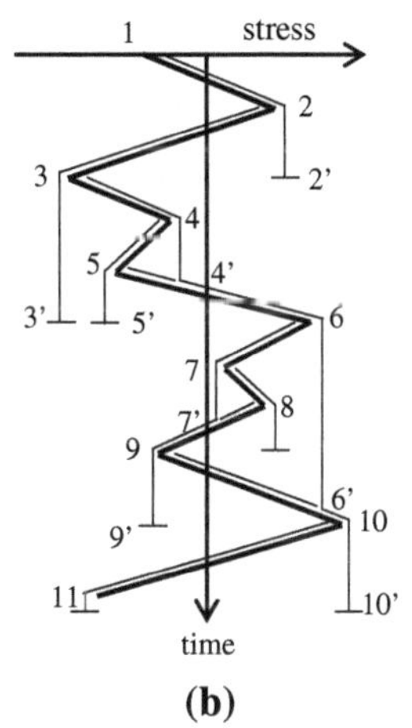

Fig. 5.7 An example pagoda roof and corresponding rainflow cycle counting patterns. **a** The pagoda roof model of stress history **b** the rainflow patterns

3. When a drop starts from a peak, it stops if it meets an equal or larger peak than that it is started from, e.g., path (2–3–3′) in Fig. 5.7b. The drop falls on another roof until a stop condition arises, e.g., path (6–7–7′–9–9′) in Fig. 5.7b. The drop stops flowing when it meets a flow from an earlier path, e.g., path (8–7′) in Fig. 5.7b.
4. A drop stops flowing when it reaches the tip of the roof (end of stress history record).
5. The stress range of a half cycle, which is equivalent to that of a constant amplitude load, is defined as the projection of a rainflow path between the start and stop points.

As an example, the rainflow paths and corresponding stress ranges of half cycles for the stress history given in Fig. 5.6a are written in Fig. 5.7.

5.4 Probability Distribution of Random Stress Ranges

As it occurs in offshore structural analysis, the stress process of a structural system in practice is usually stochastic due to environmental random loading or other phenomenal loading condition. For narrow-banded stress processes with zero means, stress ranges are defined as double of stress peaks, and for Gaussian stress processes, the probability distribution of stress peaks follows the Rayleigh distribution as given by Eq. (2.121). For broad-banded stress processes with Gaussian distribution, probability distribution of stress peaks (maxima) approaches Gaussian when spectral bandwidth parameter approaches unity, i.e., $(\varepsilon \rightarrow 1)$. In practice, for offshore structures, the peak distribution of the stress process will neither be Rayleigh nor be Gaussian. In addition, stress ranges identified by the rainflow counting algorithm cannot be related to stress peaks as they are also random. Therefore, for broad-banded processes, a probability distribution of stress ranges cannot be deducted using the probability distribution of stress peaks. The subject of fatigue damages under broad-banded processes has been studied by various investigators and presented in different forms [34, 58–65]. The simplest and one of earliest form, for a broad-banded process, is the concept of a damage correction factor [58] of the damage based on the narrow-banded stress assumption. Later, this simple damage correction factor has been extended to a stress process with two separated spectral peaks [60]. In some other studies, different types of peak distributions are used to calculate damages of broad-banded stress processes [34, 61, 63, 64]. Based on rainflow data, semi-empirical formula of the stress range distribution have also been introduced [39, 65], which are assumed to be superior to those of peak distributions. In this section, an alternative probability distribution of stress ranges is presented.

Since the probability distribution of stress maxima is the combination of Gaussian and Rayleigh types, see Eqs. (2.127a) and (2.127b), a similar formulation can also be used for the distribution of stress ranges. As the Rayleigh distribution is a special case of the two parameters Weibull distribution, for the generality, we assume that the probability distribution of stress ranges can be defined parametrically as the sum of the Gaussian and Weibull types. Thus, the distribution function is written as,

$$F_s(x) = F_{gs}(x) + F_{ws}(x) \quad \rightarrow \quad (x = S/\sqrt{m_0}) \tag{5.14a}$$

where $F_{gs}(x)$ is the Gaussian-type distribution and $F_{ws}(x)$ is the Weibull-type distribution, x denotes the normalized stress range, or stress range variable, S denotes the stress range and m_0 is the zeroth spectral moment (variance) of the stress process. They are defined as written by,

$$F_{gs}(x) = \varepsilon^2 \mathrm{erf}\left(\frac{x}{A\varepsilon\sqrt{2}}\right) \quad \rightarrow \quad (0 \leq x \leq +\infty) \quad \text{and}$$
$$F_{ws}(x) = \alpha\left[\alpha\,\mathrm{erf}\left(\frac{1}{\varepsilon}\left(\frac{x}{B}\right)^{C/2}\right) - \exp\left(-\left(\frac{x}{B}\right)^{C}\right)\mathrm{erf}\left(\frac{\alpha}{\varepsilon}\left(\frac{x}{B}\right)^{C/2}\right)\right] \rightarrow \quad \alpha = \sqrt{1-\varepsilon^2} \tag{5.14b}$$

in which ε is the bandwidth parameter, m_0 is the zeroth spectral moment (variance) of the stress process, A, B, and C are parameters to be determined, and erf(.) denotes the error function. In the extreme cases, when $\varepsilon = 1$ and $\varepsilon = 0$, the density function $F_s(x)$ becomes, respectively, Gaussian and Weibull distributions which are:

$$F_s(x) = \mathrm{erf}\left(\frac{x}{A\sqrt{2}}\right) = 2\Phi\left(\frac{x}{A}\right) - 1 \quad \rightarrow \quad \text{for } (\varepsilon = 1) \quad : \quad \text{Gaussian}$$
$$F_s(x) = \left[1 - \exp\left(-\left(\frac{x}{B}\right)^{C}\right)\right] \quad \rightarrow \quad \text{for } (\varepsilon = 0) \quad : \quad \text{Weibull} \tag{5.15}$$

in which $\Phi(.)$ denotes the standard normal cumulative distribution function. In the case of $\varepsilon = 0$ and $C = 2$, the function $F_s(x)$ becomes a Rayleigh distribution. The probability density function (PDF) of the normalized stress range x is obtained from the derivative of Eq. (5.14a) as written by,

$$f_s(x) = \frac{\partial F_{gs}(x)}{\partial x} + \frac{\partial F_{ws}(x)}{\partial x} = \left(f_{gs}(x) + f_{ws}(x)\right) \tag{5.16a}$$

where $f_{gs}(x)$ and $f_{ws}(x)$ denote, respectively, Gaussian- and Weibull-type density functions which are obtained as written by,

$$f_{gs}(x) = \frac{2\varepsilon}{A\sqrt{2\pi}}\exp\left(-\frac{x^2}{2A^2\varepsilon^2}\right) \quad \rightarrow \quad (0 \leq x \leq +\infty) \quad \text{and}$$
$$f_{ws}(x) = \alpha\frac{C}{B}\left(\frac{x}{B}\right)^{C-1}\exp\left(-\left(\frac{x}{B}\right)^{C}\right)\mathrm{erf}\left(\frac{\alpha}{\varepsilon}\left(\frac{x}{B}\right)^{C/2}\right) \rightarrow \quad \alpha = \sqrt{1-\varepsilon^2} \tag{5.16b}$$

which are defined in the positive region of x, i.e., $(0 \leq x \leq +\infty)$. For extreme values of ε, the PDF of the normalized stress range x becomes as written by,

$$f_s(x) = \frac{2}{A\sqrt{2\pi}}\exp\left(-\frac{x^2}{2A^2}\right) \rightarrow \quad \text{for } (\varepsilon = 1) \quad : \quad \text{Gaussian} \tag{5.17a}$$

$$f_s(x) = \frac{C}{B}\left(\frac{x}{B}\right)^{C-1}\exp\left(-\left(\frac{x}{B}\right)^{C}\right) \rightarrow \quad \text{for } (\varepsilon = 0) \quad : \quad \text{Weibull} \tag{5.17b}$$

In practical application, e.g., calculation of fatigue damages with integer slopes of S–N lines, the moments of stress-range probability distribution may be useful. They are calculated from,

$$M_n = \int_0^\infty x^n f_s(x)\,dx = \int_0^\infty x^n \left(f_{gs}(x) + f_{ws}(x)\right) dx \tag{5.18a}$$

Having introduced $f_{gs}(x)$ and $f_{ws}(x)$ from Eq. (5.16b) into Eq. (5.18a, b), the moments for ($n = 0, 1, 2, \ldots, m$) can be obtained as stated by,

$$\begin{aligned} M_n &= \frac{1}{\sqrt{\pi}} \left(\sqrt{2}\,A\right)^n \varepsilon^{n+2}\, \Gamma\left(\frac{1}{2}(n+1)\right) + J_n \\ J_n &= B^n \alpha \int_0^\infty y^{\frac{n}{C}} \operatorname{erf}\left(\frac{\alpha}{\varepsilon}\sqrt{y}\right) e^{-y}\,\mathrm{d}y \quad \rightarrow \quad \alpha = \sqrt{1-\varepsilon^2} \end{aligned} \tag{5.18b}$$

in which $\Gamma(.)$ denotes Gamma function. The first part is the contribution of the Gaussian distribution and the second part (J_n) is the contribution of the Weibull distribution. All these formulations are dependent on the parameters A, B and C. In order to calculate the probability distribution and moments of the stress range x the parameters A, B and C must be determined. This subject is presented in the following section.

5.4.1 Parameters of the Probability Distribution of Stress Ranges

For a general stress process, the probability distribution of the stress range has been defined by Gaussian and Weibull distributions in terms of parametric constants as it is presented in the previous section. The Gaussian part has only one parameter, A, and the Weibull part has two parameters, B and C, as they are clearly seen in Eq. (5.16b). For a narrow-banded stress process, the Gaussian part disappears and the Weibull part becomes the Rayleigh distribution with the parameters of B and C equal to:

$$\text{narrow-banded stress process } (\varepsilon = 0) \rightarrow \begin{cases} B = 2\sqrt{2} \\ C = 2 \end{cases} \tag{5.19}$$

For a general process, all parameters (A, B, C) must be determined from experimental stress range data. For this purpose, the following steps are carried out for a spectral shape.

- Artificial stress time histories are generated from a stress spectrum with a random phase angle by using Monte Carlo simulation.
- Stress ranges are identified by using the rainflow cycle counting algorithm for all Monte Carlo simulations.
- Based on these simulations, mean values of the probability density histogram are calculated.

- Spectral moments of the stress range are calculated using the mean value histogram, and accordingly the parameters A, B, and C.

This procedure is repeated for a large number of spectral shapes of the stress process to obtain population data with respect to spectral bandwidth parameter, ε.

There are alternative techniques to calculate the parameters A, B, C. The most probable techniques may be:

1. Using the minimization of the mean square error of the probability density function with respect to the parameters A, B and C.
2. The Parameter A is calculated from the minimization of the mean square error. The parameter B is calculated from the third spectral moment, and the parameter C is calculated from the fourth spectral moment.

In the first technique, the mean square error is defined as the sum of errors at all intervals of the probability density histogram. It is written as,

$$E = \sum_{i=1}^{n_s} (y_i - f_s(x_i))^2 \tag{5.19a}$$

where n_s is the number of intervals of the histogram, y_i is the value of the histogram at the interval i and $f_s(x_i)$ is the value of the PDF given by Eq. (5.16a) at the same interval i. The minimization of this error requires that,

$$\frac{\partial E}{\partial A} = 0, \qquad \frac{\partial E}{\partial B} = 0, \qquad \frac{\partial E}{\partial C} = 0 \tag{5.19b}$$

from which the following statements can be obtained.

$$\sum_{i=1}^{n_s} y_i \frac{\partial f_s(x_i)}{\partial A} - \sum_{i=1}^{n_s} f_s(x_i) \frac{\partial f_s(x_i)}{\partial A} = 0 \tag{5.20a}$$

$$\sum_{i=1}^{n_s} y_i \frac{\partial f_s(x_i)}{\partial B} - \sum_{i=1}^{n_s} f_s(x_i) \frac{\partial f_s(x_i)}{\partial B} = 0 \tag{5.20b}$$

$$\sum_{i=1}^{n_s} y_i \frac{\partial f_s(x_i)}{\partial C} - \sum_{i=1}^{n_s} f_s(x_i) \frac{\partial f_s(x_i)}{\partial C} = 0 \tag{5.20c}$$

Since these equations are highly nonlinear an iterative solution may be appropriate. The solution algorithm can be:

1. Chose an initial value of A between $(0.17 \le A \le 2.0)$. For $(\varepsilon = 1)$, the PDF of the stress range will be Gaussian and A is calculated by using the minimization of the mean square error. It is obtained approximately as $A = 0.17$. In this case, the stress range is not equal to double of the stress amplitude. Instead, it equals roughly 0.17 times amplitude, i.e., $(x = 0.17|\hat{s}|)$. From a numerical investigation it is obtained that the maximum value of A equals approximately 2.0

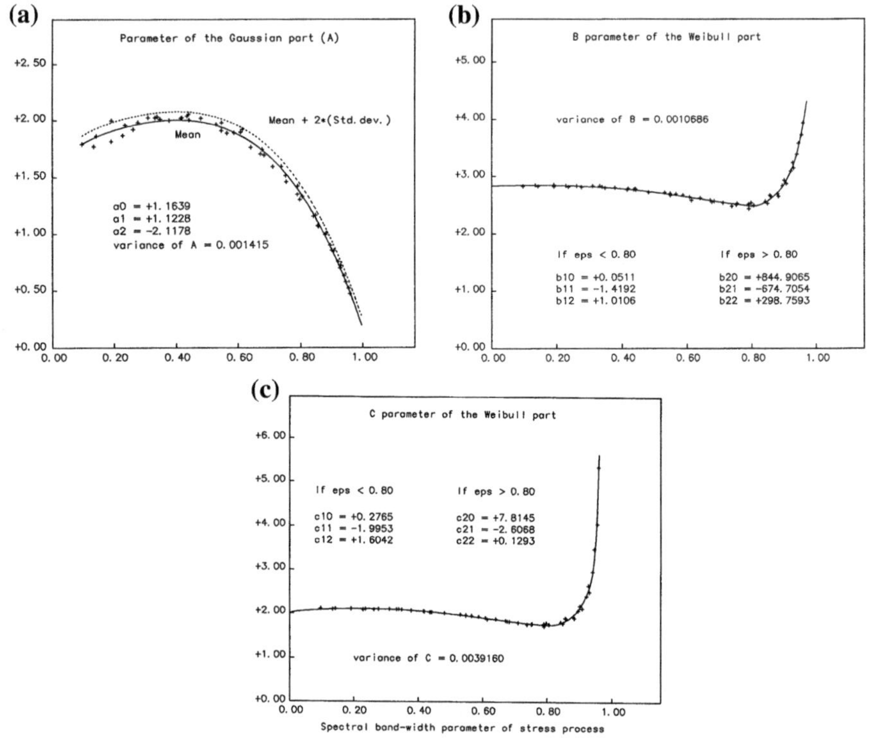

Fig. 5.8 **a** Population of parameter A of the Gaussian part of the PDF and its mean curve, **b** population of parameter B of the Weibull part of the PDF and its mean curve, **c** population of parameter C of the Weibull part of the PDF and its mean curve

around ($\varepsilon = 0.45$) as shown in Fig. 5.8a. Therefore, an initial value of A between (0.17–2.0) is appropriate.
2. Chose the value of $C = 2.0$.
3. Calculate B from Eq. (5.20b).
4. Calculate C from Eq. (5.20c).
5. Repeat steps (3–4) until both B and C converge.
6. Using B and C calculated from previous steps, calculate A from Eq. (5.20a).
7. Repeat steps (3–6) until A converges.

The convergence of this algorithm is very fast and provides good results for all spectral bandwidth parameter, ε.

In the second algorithm of the determination of A, B and C mentioned above, the third and fourth equal spectral moments calculated from PDF and histograms are used together with the minimization of the mean square error. It is an iterative calculation as presented below.

1. Choose A between (0.17–2.0) and $C = 2.0$.

2. Since A and C are known, calculate B using the third probability moment which is obtained as $\left(M_3 = \sqrt{8/\pi}\ \varepsilon^5 A^3 + J_3\right)$ where J_3 is calculated from Eq. (5.18b).
3. Since A and B are known from the previous steps, calculate C using the fourth probability moment which is obtained as $\left(M_4 = 3\,\varepsilon^6 A^4 + J_4\right)$ where J_4 is calculated from Eq. (5.18b).
4. Since B and C are known from the previous steps, calculate A by using the minimization of the mean square error, i.e., by using Eq. (5.20a).
5. Repeat steps 2 through 4 until a required convergence is obtained.

This iteration algorithm is quite fast and convergent. It produces better results for higher probability moments than other methods. Therefore, it is adopted in this book. For each ε value, the parameters A, B, and C are calculated and their populations are generated with respect to ε. Then, by using nonlinear regression analysis, analytical statements of their mean values are obtained as functions of the spectral bandwidth parameter, ε. Since the population data of A, B, and C are scattered, their variances are estimated from,

$$\sigma_X^2 = \frac{1}{(N-1)} \sum_{i=1}^{N} \left[x_i - \mu_X(\varepsilon_i)\right]^2 \tag{5.21}$$

in which N is the number of population, X denotes one of the A, B, or C parameter, and μ_X is its mean curve as function of ε. The mean curves and variances of the parameter A is obtained for all ε values as shown in Fig. 5.8a, and written by,

$$\begin{aligned} \mu_A &= 1.1639 + 1.1228\,\varepsilon^{1/4} - 2.1178\,\varepsilon^4 \\ \sigma_A^2 &= 0.001415 \end{aligned} \tag{5.22}$$

The mean curves of the parameters of the Weibull part, B and C, are defined in two regions of ε as being one for $(0 \le \varepsilon \le 0.8)$ and one for $(0.8 \le \varepsilon \le 1.0)$. Their populations and mean curves are shown in Fig. 5.8b and c, respectively. The analytical statements of the mean curves are presented below.

$$\left.\begin{aligned} \mu_B &= 2\sqrt{2} + 0.0511\,\varepsilon^{1/2} - 1.4192\,\varepsilon^3 + 1.0106\,\varepsilon^5 \\ \mu_C &= 2 + 0.2765\,\varepsilon^{1/2} - 1.9953\,\varepsilon^3 + 1.6042\,\varepsilon^5 \end{aligned}\right\} \to \text{ for } (\varepsilon \le 0.8)$$

$$\left.\begin{aligned} \mu_B &= b_0 + b_1\varepsilon + b_2 g_2 + b_3 g_3 + b_4 g_4 \\ \mu_C &= c_0 + c_1\varepsilon + c_2 h_2 + c_3 h_3 + c_4 h_4 \end{aligned}\right\} \to \text{ for } (\varepsilon \ge 0.8) \tag{5.23a}$$

in which b_i and c_i ($i = 0, 1, 2, 3, 4$) are constants obtained from the regression analysis, and g_i and h_i ($i = 2, 3, 4$) are some functions of ε. The constants are presented in Table 5.1. The functions g_i and h_i ($i = 2, 3, 4$) are obtained as they are presented in Eq. (5.23b).

Table 5.1 Constants of the parameters B and C of the Weibull part of the PDF for $(\varepsilon \geq 0.8)$

Constants of the parameter B	Constants of the parameter C
$b_0 = 2.979926$	$c_0 = 2.064185$
$b_1 = -0.626589$	$c_1 = -0.391006$
$b_2 = 844.9065$	$c_2 = 7.8145$
$b_3 = -674.7054$	$c_3 = -2.6068$
$b_4 = 298.7593$	$c_4 = 0.1293$

$$\left\{\begin{array}{l} g_2 = \varepsilon^3 - 1.920000\varepsilon + 1.024000 \\ g_3 = \varepsilon^6 - 1.966080\varepsilon + 1.310720 \\ g_4 = \varepsilon^8 - 1.677722\varepsilon + 1.174405 \end{array}\right\} \text{ and } \left\{\begin{array}{l} h_2 = \delta - 1.684500\varepsilon + 0.227310 \\ h_3 = \delta^2 - 3.774258\varepsilon + 1.764356 \\ h_4 = \delta^4 - 9.473766\varepsilon + 6.003862 \\ \delta = 1/\left(1 - \varepsilon^{10}\right) \end{array}\right\} \tag{5.23b}$$

The value of μ_B at $\varepsilon = 1$ is approximately $\mu_B = 6.11$ while μ_C is infinity. The variances of these parameters are calculated to be:

$$\begin{aligned} \sigma_B^2 &= 0.0010686 \\ \sigma_C^2 &= 0.0039160 \end{aligned} \tag{5.23c}$$

The populations of the parameters A, B, and C are created by using bimodal stress spectra. For this purpose, a monopod column is used in a water depth of $d = 100.0$ m and a wall thickness of $h = 0.20$ m. A Pierson–Moskowitz sea spectrum is assumed. By using only the inertia force term of the Morison's equation the simplified stress spectrum at the bottom of the column is obtained as,

$$S_{ss}(\omega) \simeq \frac{7.57x10^{11}(10.194 - 1/\omega^2)^2 \exp\left(-\dfrac{3.118}{H_s^2\omega^4}\right)}{\omega^5\left[\left(1 - \dfrac{\omega^2}{\omega_k^2}\right)^2 + 0.0004\dfrac{\omega^2}{\omega_k^2}\right]} \tag{5.24}$$

in which ω_k is the natural frequency of the column and H_s is the significant wave height. This function displays two peaks, one at the fundamental wave frequency and one at the natural frequency ω_k. By changing ω_k and H_s different shapes are produced to obtain ε values between $(0.0 < \varepsilon < 1.0)$ which covers the range of narrow- and broad-banded stress processes.

5.5 Spectral Fatigue Damages Based on Multilinear (*S–N*) Model

In previous sections, all ingredients of the fatigue damage calculation have been presented. These are:

- *S–N* fatigue model given by Eq. (5.9)
- Cumulative damage model given by Eq. (5.13)
- The PDF of stochastic stress ranges given by Eq. (5.16a) and (5.16b)

In the case of random stress histories, the damage of one stress cycle with the hot-spot range S_h is calculated from Eq. (5.13) as given by,

$$dD = \frac{1}{N(S_h)} \tag{5.25a}$$

For a multi-segmented *S–N* fatigue model, by using Eqs. (5.11) and (5.12) in (5.25a) the damage of one stress cycle for the segment j can be written as,

$$dD_j = \frac{1}{C_j}(\text{SCF } S)^{k_j} \rightarrow \quad \left(S_j \le S \le S_{j+1}\right) \tag{5.25b}$$

or, by using the normalized stress range variable x from Eq. (5.14a), it can be stated as,

$$dD_j = \frac{1}{C_j}(\text{SCF } \sigma_S)^{k_j} x^{k_j} \rightarrow \begin{cases} x_j \le x \le x_{j+1} \\ x = S/\sigma_S \end{cases} \tag{5.25c}$$

in which σ_S is the standard deviation of the nominal stress process. The mean value of dD_j is calculated by using the PDF of x given by Eqs.(5.16a) and (5.16b). It can be stated as,

$$E[dD_j] = \frac{1}{C_j}(\text{SCF } \sigma_S)^{k_j} \int_0^{\infty} x^{k_j} f_s(x)\, dx = \frac{1}{C_j}(\text{SCF } \sigma_S)^{k_j} M_{kj} \tag{5.26}$$

where M_{kj} is a spectral bandwidth factor of the stress process calculated similarly to Eq. (5.18b) from,

$$M_{kj} = \frac{\varepsilon^2}{\sqrt{\pi}}\left(\sqrt{2}\,\varepsilon A\right)^{k_j} \Gamma\left(\frac{1+k_j}{2}\right) + \alpha B^{k_j} \int_0^{\infty} x^{\frac{k_j}{C}} \operatorname{erf}\left(\frac{\alpha}{\varepsilon}\sqrt{x}\right) e^{-x} dx$$
$$\alpha = \sqrt{1-\varepsilon^2} \tag{5.27a}$$

In this statement, the parameters A, B and C are calculated from Eqs. (5.22) and (5.23a). For the extreme cases, ($\varepsilon = 0.0$) and ($\varepsilon = 1.0$), i.e., for narrow- and broad-banded stress processes, the M_{kj} factor becomes as,

$$M_{kj} = \left(\sqrt{2}\right)^{k_j}\Gamma\left(1+\frac{k_j}{2}\right) \rightarrow \text{ narrow band (Rayleigh)}$$
$$M_{kj} = \frac{(0.23886)^{k_j}}{\sqrt{\pi}}\Gamma\left(\frac{1+k_j}{2}\right) \rightarrow \text{ broad band (Gaussian)} \tag{5.27b}$$

The total mean damage of one stress cycle, $E[\mathrm{d}D]$, for the complete (*S–N*) model is calculated from a probabilistic summation of mean damages of all segments as written by,

$$E[\mathrm{d}D] = \sum_{j=1}^{NL} P_j(S)\,E[\mathrm{d}D_j] \tag{5.28a}$$

in which *NL* denotes the number of segments of the (*S–N*) fatigue model shown in Fig. 5.4, $E[\mathrm{d}D_j]$ is calculated from Eq. (5.26) and $P_j(S)$ is the probability that a stress range *S* falls into the region of the segment *j*. It is calculated from Eq. (2.15) as written by,

$$P_j(S) = \int_{x_j}^{x_{j+1}} f_s(x)\,\mathrm{d}x \quad \text{with} \;\; (x = S/\sigma_S) \tag{5.28b}$$

where the PDF $f_s(x)$ is given by Eq. (5.16a).

Having calculated the mean fatigue damage of one stress cycle $E[\mathrm{d}D]$, the fatigue damage that occurs during a sea state is calculated by multiplying $E[\mathrm{d}D]$ with the number of stress cycles that occur during the sea state. This is stated as,

$$D_{Hs} = n_{sHs}E[\mathrm{d}D] = (T_{Hs}/T_m)E[\mathrm{d}D] \tag{5.29a}$$

in which n_{sHs} is the number of total stress cycles in a sea state denoted by H_s, T_{Hs} is the period of the sea state H_s and T_m is the mean period of stress maxima in the sea state H_s. The cumulative damage for all identical sea states (sea states with the same period of T_{Hs}) is calculated as,

$$(D_{\text{tot}})_i = (N_{Hs})_i D_{Hs} = (N_{Hs})_i (T_{Hs}/T_m)E[\mathrm{d}D] \tag{5.29b}$$

where $(N_{Hs})_i$ denotes the number of identical sea states. For a lifetime *T*, the total damage is obtained by accumulation of the damages of identical sea states as stated by,

$$D_{\text{tot}} = \sum_{i=1}^{N_{HS}} (D_{\text{tot}})_i = \sum_{i=1}^{N_{HS}} (N_{Hs})_i\,(T_{Hs}/T_m)E[\mathrm{d}D] \tag{5.29c}$$

in which N_{Hs} is the total number of different identical sea states in the lifetime *T*. Having multiplied and divided Eq. (5.29c) by the total number of sea states in the lifetime *T*, the cumulative damage, D_{tot}, can be conveniently written as,

$$D_{\rm tot} = \sum_{i=1}^{N_{HS}} \frac{(N_{Hs})_i}{N} N \frac{T_{Hs}}{T_m} E[\mathrm{d}D] = \sum_{i=1}^{N_{HS}} P_i(H_s) N \frac{T_{Hs}}{T_m} E[\mathrm{d}D] \tag{5.29d}$$

where N is the total number of sea states occur in the lifetime T and $P_i(H_s)$ is the probability that a sea state occurs in T. Having stated N in terms of periods T and T_{Hs}, the total damage can be written as,

$$D_{\rm tot} = \sum_{i=1}^{N_{HS}} P_i(H_s) \frac{T}{T_{Hs}} \frac{T_{Hs}}{T_m} E[\mathrm{d}D] = T \sum_{i=1}^{N_{HS}} \frac{1}{T_m} P_i(H_s)\; E[\mathrm{d}D] \tag{5.30a}$$

Equation (5.30a) is the formulation of the total fatigue damage when the sea states are discrete, i.e., scatter diagrams are used for the long-term probability distribution of sea states. $D_{\rm tot}$ in Eq. (5.30a) can also be interpreted as the probabilistic damage accumulation for all sea states in the lifetime T. The ratio, (T/T_m), is the total number of stress cycles in the lifetime T and the total damage of one sea state in T, which is denoted by $(D_{\rm tot})_{Hs}$, can be expressed as,

$$(D_{\rm tot})_{Hs} = \frac{T}{T_m} E[\mathrm{d}D] \tag{5.30b}$$

When it is multiplied by the probability that the sea state occurs in T, i.e., by $P_i(H_s)$, the damage contribution to the total damage is obtained. Then, the total damage can be expressed as the summation of contributions of all sea states. It is written as,

$$D_{\rm tot} = \sum_{i=1}^{N_{HS}} (D_{\rm tot})_{Hs} P_i(H_s) \tag{5.30c}$$

Equation (5.30c) is the same statement as that written in Eq. (5.30a). If, however, a continuous sea state is used instead of discrete sea states, then the total damage in the long term is calculated as being the mean value of $(D_{\rm tot})_{Hs}$ given in Eq. (5.30b). It is stated from Eq. (5.30b) as written by,

$$D_{\rm tot} = T \int_0^\infty \int_0^\infty \frac{1}{T_m} E[\mathrm{d}D] f_{Hs,Tz}(h,t)\; \mathrm{d}h\, \mathrm{d}t \tag{5.31}$$

in which $f_{Hs,Tz}(h,t)$ is the density function of the joint probability distribution of the significant wave height and zero-crossings period of waves, H_s and T_z, in the long term. It has been explained in Sect. 3.6. In this statement, T_m and $E[\mathrm{d}D]$ are dependent on the sea state parameters, namely on the significant wave height H_s and the zero-crossing period of waves T_z

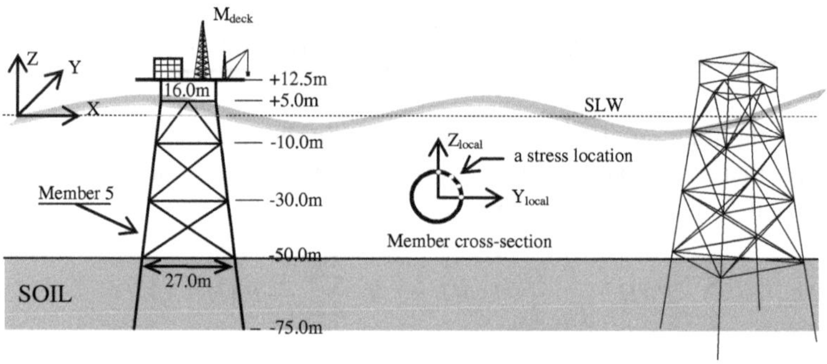

Fig. 5.9 An example jacket-type offshore structure for the fatigue damage calculation by SAPOS

Table 5.2 Member dimensions, material properties, and mass of the deck of the example jacket

Members	Diameter (mm)	Thickness (mm)
Bracings and diagonals at top (+12.5 m)	Rigid	Rigid
Bracings and diagonals at level +5.0 m	800	8
Bracings and diagonals at level −10.0 m and bracings at level −30.0 m	1,200	14
Jacket legs, Vertical diagonals and diagonals at level −30.0 m	1,200	16
Top legs (above +5.0 m)	2,000	50
Piles	1,200	16
Material properties (steel): $E = 205$ GPa, $\rho_s = 7{,}800.0$ kg/m^3, Poisson's ratio = 0.25		
Mass of the deck: $M_{\text{deck}} = 7{,}000$ ton, added mass coefficient = 0.9		

Table 5.3 Wave force data and parameters of the Weibull probability function

c_d	c_m	Marine growth thick., t_{marine}	Density of water, ρ_w	A	B	C
1.3	1.6	20 cm	1,024 km/m^3	0.60	1.67	1.20

5.6 Examples

For the demonstration of the fatigue damage and fatigue lifetime calculations, an example jacket type structure is used. The fatigue analysis is carried out by the SAPOS program [66]. The structure, stress locations on member cross-sections, and the 3D analysis model are as shown in Fig. 5.9. The structure is supported on piles of 25 m depth in the soil and the data of soil are given in Table 4.1 in Chap. 4. Member dimensions and material properties of the structure are given in Table 5.2. It is assumed that the members in water are empty. Added masses of surrounding water of submerged members are taken into account by increasing the structural mass density according to the Eq. (4.31) given in Chap. 4. In this analysis, it is assumed that the water depth is 50 m and the jacket is subjected to random waves in

Table 5.4 Results of the eigenvalue solution of the example jacket structure

Mode	Eigenvalue	Natural frequency (rad/sec)	Natural period (sec)
1	7.3282	2.707	2.321
2	7.3282	2.707	2.321
3	33.1930	5.761	10.906

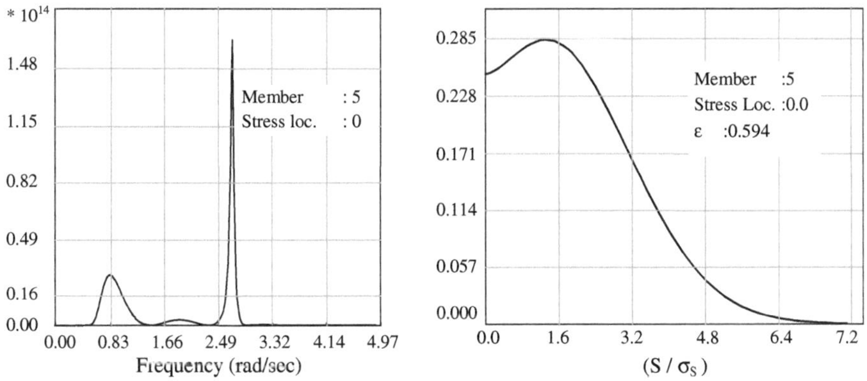

Fig. 5.10 Stress spectrum (**a**) and PDF (**b**) of the stress range at the bottom of member 5 shown in Fig. 5.9 of the example structure for $H_s = 2.5$ m (**a**) and $\varepsilon = 0.594$ (**b**)

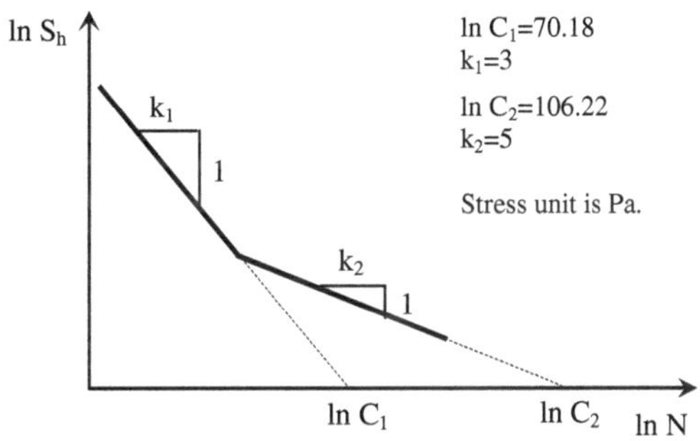

Fig. 5.11 API design basic *S–N* curve for welded joints of tubular connections

the global X direction as shown in Fig. 5.9. In the short term, the Pierson–Moskowitz (PM) sea spectrum is used and, in the long term, a Weibull probability distribution function for the significant wave height H_s is used with the parameters A, B, and C given in Table 5.3. It is shown in Fig. 4.7. In the analysis, the water-structure interaction is taken into account. The wave force data, marine growth, and density of water are given in Table 5.3 where c_d and c_m are, respectively, drag and inertia force coefficients. In the analysis of the structure, both the quasi-static and dynamic contributions of the response are taken into account. Three natural mode shapes are used for the calculation of the dynamic response contribution. The first two mode shapes are obtained as being flexural in orthogonal directions with the

Table 5.5 Occurrence probabilities of sea states and corresponding fatigue damages per year at the given stress location

Sea state (H_s)	Occurrence probability	Stress ω_0 (rad/sec)	Stress ω_m (rad/sec)	Damage/year
1.150	0.521756	2.577	2.663	0.000561
2.828	0.398667	1.948	2.549	0.001844
5.452	0.075942	1.300	2.351	0.010820
9.138	0.003612	0.927	2.126	0.05533
14.431	0.000023	0.717	1.932	0.23390
Probabilistic cumulative damage				0.002055

Table 5.6 Probabilistic cumulative damages and fatigue lives at different stress locations

Stress location	Probabilistic cumulative damage/year	Fatigue life (years)
0	0.00205481	486.662
15	0.00162059	617.061
30	0.00130057	768.891
45	0.00108663	920.280
60	0.00095309	1049.220
90	0.00085966	1163.240

same natural frequency and the third mode shape is torsional. The results of the eigenvalue solution of the jacket are presented in Table 5.4. For the calculation of the hot-spot normal stress, a stress concentration factor of (SCF $=$ 2.0) is assumed. The stress spectrum and probability density function of the stress range, for $H_s = 2.5$ m and $\varepsilon = 0.594$, at the bottom of the member 5 and at the stress location in the local Y direction shown in Fig. 5.9 are illustrated in Fig. 5.10. For the fatigue damage calculation, API Basic Design S–N curve for welded joints [28] is used. This S–N curve consists of two segments as shown in Fig. 5.11. The same stress concentration factor, SCF $=$ 2.0, for the normal force and bending moments, is used for demonstration purposes. Fatigue damages on the bottom cross-section of the member 5, which is shown in Fig. 5.9, at the stress location in the local Y axis are calculated for a number of H_s values by using the SAPOS program. The results are presented in Table 5.5 together with the occurrence probabilities of the sea states. In this table, ω_0 and ω_m are, respectively, mean frequencies of the zero-crossings and maxima of the stress process at the location specified above. As it is seen from Table 5.5, for small sea states, fatigue damages are considerably small and occurrence probabilities of small sea states are much higher than those of higher sea states. As it is also seen from Table 5.5 the probabilistic cumulative damage for all sea states is almost equivalent to the fatigue damage caused by a single sea state with ($H_s = 3.0$ m) approximately. For different stress locations, the probabilistic cumulative damages and fatigue lives in years are presented in Table 5.6. The given stress locations are measured from the local Y axis in degrees. As it is expected the maximum damage occurs at the stress location in the wave direction which coincides with the local Y axis shown in Fig. 5.9.

Exercise 1

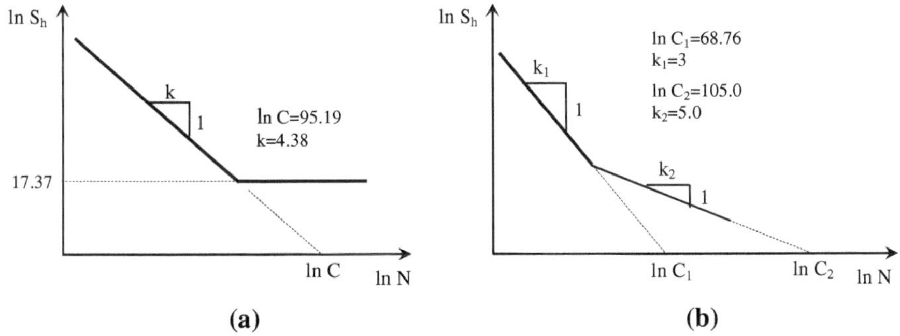

Fig. 5.12 API X and ABS T(CP) *S–N* curves [30]

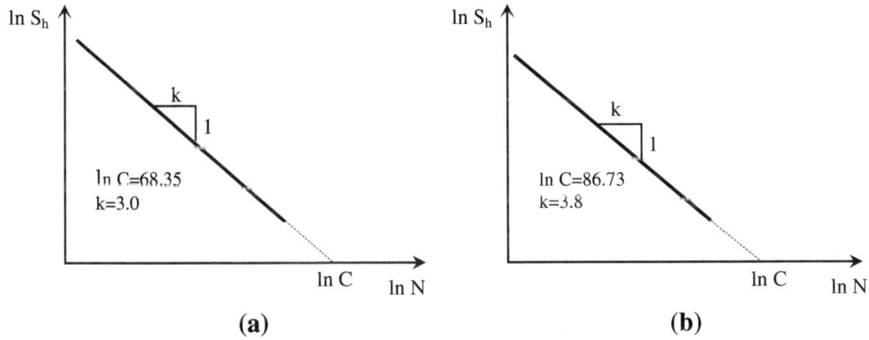

Fig. 5.13 ABS T(FC) [30] and experimentally determined [26, 67] *S–N* curves

The fatigue analysis of the monopod tower, which is described in Sect. 4.6 for the Exercise 1, is required. The tower is shown in Fig. 4.11 and the data are explained in the exercise 1. The fatigue analysis will be carried out after the calculation step 5 in the Exercise 1 is completed. The following items are required. Calculate mean fatigue damage for one stress cycle for the sea state of $H_s = 9.0$ m by using Eq. (5.26) for a single *S–N* curve, and Eq. (5.28a) for multi segmented *S–N* curve. The *S–N* curves to be used are API X and Basic Design *S–N* curves [28, 30], ABS T(CP and T(FC) *S–N* curves [30] and that based on experimental fatigue data [26, 67]. These *S–N* curves are shown in Figs. 5.11, 5.12 and 5.13.

1. Plot probability density function of stress ranges given by Eq. (5.16a) and (5.16b), for the sea state of $H_s = 9.0$ m.
2. Calculate probabilistic cumulative fatigue damages and lifetimes in the long-term sea states by using Weibull probability distribution with the parameters given in Table 5.3.
3. Repeat the calculations in items 1 and 2 assuming that stress ranges are narrow-banded, and thus, their probability distributions are Rayleigh type.

4. Compare the fatigue damages calculated from the narrow- and broad-banded stress assumptions

References

1. Barsom JM, Rolfe ST (2000) Fracture and fatigue control in structures: applications of fracture mechanics, 3rd edn. Butterworth-Heinemann, Guildford
2. Kirkegaard PH, Sorensen JD, Brincker R (1991) Fatigue reliability analysis of a mono-tower platform. Mar Struct 4:413–434
3. Kam JCP (1988) Fatigue reliability assessment of offshore structures. Qual Reliab Eng Int 4:41–48
4. Wirsching PH (1980) Fatigue reliability in welded joints of offshore structures. Int J Fatigue 2(2):77–83
5. Ang AHS, Cheung MC, Shugar TA, Fernie JD (2001) Reliability-based fatigue analysis and design of floating structures. Mar Struct 14:25–36
6. Karamchandani A, Dalane JI, Bjerager P (1991) Systems reliability of offshore structures including fatigue and extreme wave loading. Mar Struct 4:353–379
7. Chryssanthopoulos MK, Righiniotis TD (2006) Fatigue reliability of welded steel structures. J Constr Steel Res 62:1199–1209
8. Ayala-Uraga E, Moan T (2007) Fatigue reliability-based assessment of welded joints applying consistent fracture mechanics formulations. Int J Fatigue 29:444–456
9. Dalane JI (1997) Fatigue reliability-measured response of the heidrun TLP tethers. Mar Struct 10:611–628
10. Tovo R (2001) On the fatigue reliability evaluation of structural components under service loading. Int J Fatigue 23:587–598
11. Ramsamooja DV, Shugar TA (2002) Reliability analysis of fatigue life of the connectors—the US mobile offshore base. Mar Struct 15:233–250
12. Kam JCP, Birkinshaw M (1994) Reliability-based fatigue and fracture mechanics assessment methodology for offshore structural components. Int J Fatigue 16(3):183–192
13. Siddiqui NA, Ahmad S (2001) Fatigue and fracture reliability of TLP tethers under random loading. Mar Struct 14:331–352
14. Schijve J (2003) Fatigue of structures and materials in the 20th century and the state of the art. Int J Fatigue 25:679–702
15. Dover WD (1981) Fatigue fracture mechanics analysis of offshore structures. Int J Fatigue, April, pp 52–60
16. Kam JCP (1990) Recent development in the fast corrosion fatigue analysis of offshore structures subject to random wave loading. Int J Fatigue 12(6):458–468
17. Yao JTP, Kozin F, Wen Y-K, Yang J-N, Schueller GI, Ditlevsen O (1986) Stochastic fatigue fracture and damage analysis. Struct Saf 3:231–267
18. Etube LS (2001) Fatigue and fracture mechanics of offshore structures. Professional Engineering, London
19. Boyer HE, ASM, Gall TL (1985) Metals handbook. ASM International, Ohio
20. Paris PC, Erdogan F (1963) A critical analysis of crack propagation laws. J Basic Eng Trans ASME 85(Series D):528–534
21. Kanniwen MF, Popelar CH (1985) Advanced fracture mechanics. Oxford University Press, Oxford
22. Dover WD, Dharmavasan S, Brennan FB, Marsh KJ (eds) (1995) Fatigue crack growth in offshore structures. Engineering Materials Advisory Services (EMAS), Warley
23. Dover WD, Rao AGM (1996) fatigue in offshore structures, vol 1–2. Balkema, Rotterdam
24. Dover WD (1979) Variable amplitude fatigue of welded structures. In: Proceedings of international conference on fracture mechanics, current status, future prospects, Cambridge

25. Kam JCP, Dover WD (1988) Fatigue crack growth in offshore welded tubular joints under real life variable amplitude loading. In: Proceedings of international conference on fatigue crack growth under variable amplitude loading, Paris
26. de Back J, Vaessen GHG (1981) Fatigue and corrosion fatigue behaviour of offshore structures. Final report, foundation for materials research in the sea, Delft
27. Gupta A, Singh RP (1986) Fatigue Behaviour of Offshore Structures. Springer-Verlag, Berlin
28. API RP 2A WSD (2000) Recommended practice for planning, designing and constructing fixed offshore platforms. American Petroleum Institute, Washington, DC
29. DNV-RP-C203 (2005) Fatigue design of offshore steel structures. Recommended practice, Det Norske Veritas, Høvik
30. ABS (2003) Guide for the fatigue assessment of offshore structures. American Bureau of Shipping, Houston
31. ABS (2010) Guide for spectral based fatigue analysis for FPSO installations. American Bureau of Shipping, Houston
32. Karadeniz H (1991) An improved fatigue analysis for offshore structures. Mar Struct 4:333–352
33. Karadeniz H (2001) Uncertainty modelling in the fatigue reliability calculation of offshore structures. Reliab Eng Syst Saf 74(3):323–335
34. Karadeniz H (1990) Fatigue analysis of offshore structures under non-narrow banded stress processes. In: Proceedings of 1st european offshore mechanics. symposium (EUROMS-90), pp 221–228
35. NORSOK Standards (1998) Design of steel structures, N-004. Norwegian Technology Standards Institution, Oslo
36. Miner MA (1945) Cumulative damage in fatigue. J Appl Mech 12(3):A159–A164
37. Fatemi A, Yangt L (1998) Cumulative fatigue damage and life prediction theories: a survey of the state of the art for homogeneous materials. Int J Fatigue 20(1):9–34
38. Fisher JW, Kulak GL, Smith IFC (1998) A fatigue primer for structural engineers. National Steel Bridge Alliance, Chicago
39. Karadeniz H (1991) An improved fatigue analysis for offshore structures. Mar Struct 4:333–352
40. Schutz W (1981) Procedures for the prediction of fatigue life of tubular joints. In: Proceedings of international conference on steel in marine structures, Paris, pp 254–308
41. Hwang W, Han KS (1986) Cumulative damage models and multi-stress fatigue life prediction. J Compos Mater 20(2):125–153
42. van Paepegem W, Degrieck J (2002) Effects of load sequence and block loading on the fatigue response of fibre-reinforced composites. Mech Adv Mater Struct 9(1):19–35
43. Dowling NE (1972) Fatigue failure predictions for complicated stress-strain histories. J Mater 7(1):71–87
44. Research Designs & Standards Organization (2008) Guidelines for assessment of residual fatigue life of steel girder bridges, BS-91, Government of India Ministry of Railways, Lucknow
45. Ship Structure Committee (200) Fatigue of aluminum structural weldments, SSC-410, Washington DC
46. Matsuishi M, Endo T (1968) Fatigue of metals subjected to varying stress. In: Proceedings of Kyushi Branch JSME, pp 37–40
47. Olagnon M, Guede Z (2008) Rainflow fatigue analysis for loads with multimodal power spectral densities. Mar Struct 21:160–176
48. Bishop WM, Sherratt F (1990) A theoretical solution for the estimation of rainflow ranges from power spectral density data. Fatigue Fract Eng Mater Struct 13(4):311–326
49. Sunder R, Seetharam SA, Bhaskaran TA (1984) Cycle counting for fatigue crack growth analysis. Int J Fatigue 6(3):147–156
50. Glinka G, Kam JCP (1987) Rainflow counting algorithm for very long stress histories. Int J Fatigue 9(3):223–228

51. Rychlik I, Gupta S (2007) Rain-flow fatigue damage for transformed Gaussian loads. Int J Fatigue 29:406–420
52. Rychlik I (1987) A new definition of the rainflow cycle counting method. Int J Fatigue 2:119–121
53. Rychlik I (1996) Extremes, rainflow cycles and damage functionals in continuous random processes. Stoch Process Appl 63:97–116
54. Rychlik I (1993) Note on cycle counts in irregular loads. Fatigue Fract Eng Mater Struct 16(4):377–390
55. Lindgren G, Rychlik I (1987) Rain flow cycle distributions for fatigue life prediction under Gaussian load processes. Fatigue Fract Eng Mater Struct 10(3):251–260
56. Rychlik I (1988) Rain-flow-cycle distributions for ergodic load processes. SIAM J Appl Math 48(3):662–679
57. Anthes RJ (1997) Modified rainflow counting keeping the load sequence. Int J Fatigue 19(7):529–535
58. Wirsching PH, Light MC (1980) Fatigue under wide band random stresses. J Struct Div ASCE 106(7):1593–1606
59. Larsen CE, Lutes LD (1991) Predicting the fatigue life of offshore structures by the single-moment spectral method. Probab Eng Mech 6(2):96–108
60. Jiao G, Moan T (1990) Probabilistic analysis of fatigue due to Gaussian load processes. Probab Eng Mech 5(2):76–83
61. Bouyssy V, Naboishikov SM, Rackwitz R (1993) Comparison of analytical counting methods for Gaussian processes. Struct Saf 12:35–57
62. Petrucci G, Zuccarello B (1999) On the estimation of the fatigue cycle distribution from spectral density data. Proc Inst Mech Eng Part C J Mech Eng Sci 213:819–831
63. Zhao W, Baker MJ (1992) On the probability density function of rainflow stress range for stationary Gaussian processes. Int J Fatigue 14(2):121–135
64. Chaudhury GK, Dover WD (1985) Fatigue analysis of offshore platforms subject to sea wave loadings. Int J Fatigue 7(1):13–19
65. Dirlik T (1985) Application of computers in fatigue. Ph.D. thesis, University of Warwick, Coventry
66. Karadeniz H (2009) SAPOS, spectral analysis program of structures. Report, structural mechanical division, faculty of civil engineering and geoscience, TUDelft, Delft
67. Karadeniz H (1982) Thoughts on determination of fatigue parameters for offshore steel structures. Report, structural mechanical division, faculty of civil engineering and geoscience, Faculty of civil engineering and geoscience, TUDelft, Delft

Chapter 6
Reliability Analysis of Offshore Structures

6.1 Introduction

Design of structures is carried out to satisfy the condition that the structures remain safely in operational circumstances and withstand possible extreme loading sequence. This safety condition is assured in the deterministic analysis by reducing the resistance, or increasing the strength, of the structure by a factor. Generally, the safety factor is understood to be the ratio between the expected strength of the structure and the expected applied load. Alternatively, the safety condition can also be satisfied by increasing the applied load by a load factor. A combination of resistance, strength, and load factors can also be used. This analysis procedure is known as the partial safety factor method and explained in detail in the literature, see e.g., [1, 2]. It has been used in practice extensively [3–19] to allow a safe margin in the analysis. In practice, structures, applied loads, and environmental conditions contain uncertainties that are not taken into account in a deterministic base analysis procedure which uses only the mean or expected values of design parameters. Actually, both the strengths and loads are variables and their values are scattered about their respective mean values. When these scatters are considered in the analysis, the safety factor could potentially be less than unity, and therefore the traditional safety factor design would fail. This drawback of the safety factor design is overcome by using probabilistic and other non-deterministic methods for design under uncertainty. Probabilistic methods in structural design are steadily increasing in latest design philosophy, especially in the aerospace, automotive, and the ocean engineering industries [20] since they provide more confidence about the safety of the structure during an operational lifetime under uncertainty. Some codes of practice have already included provisions for reliability analysis either in the calibration of partial safety factors or for the use of reliability methods in design and analysis [2, 21–26]. The reliability method provides a systematic procedure for evaluating influences of uncertainties in the various design parameters, which affects the safety factor. Consequently, it can

H. Karadeniz, *Stochastic Analysis of Offshore Steel Structures*,
Springer Series in Reliability Engineering, DOI: 10.1007/978-1-84996-190-5_6,

assist design engineers in judging the acceptability level of the risk. The reliability method deals with the probability evaluation of a system that it performs its function over an assumed lifetime under specified service conditions. The structural response, consequently the structural strength, depends on many factors such as loads, boundary conditions, stiffness, and mass properties that include uncertainties. On the other hand, different resistance categories of the structure to be compared with corresponding strengths also include uncertainties that are incorporated in the probabilistic structural analysis. The latter uncertainties concern mostly experimental data and they arise due to the lack of knowledge and scarcity of data available. This kind of uncertainties is grouped into the epistemic uncertainty [27]. In general, uncertainties are considered in two main categories [28] as:

- Epistemic uncertainty
- Aleatory uncertainty

The epistemic uncertainty is due to lack of knowledge about the behavior of the system and can be reduced by increasing the knowledge through research, or other similar activities, and collecting more relevant data. In principle, it can be eliminated with sufficient study, and therefore, it can be affected by human activities. The aleatory uncertainty, or variability, is the natural randomness in a process. It is inherent in the behavior of the system under study, e.g., wind speed, intensity of earthquake ground motion, and fatigue phenomenon in structures. The aleatory uncertainty is irreducible and can be characterized in the analysis by frequency distributions of the randomness, such as by probability distributions of representing parameters of the randomness. These two categories of uncertainties can be encountered in most engineering problems, and the analyst should decide their degrees in the analysis by using available information through a study, or collected data, and also his experience. Whatever uncertainty types are concerned the uncertainties, in general, should be taken into account in the design of structures. This subject involves a probabilistic analysis. It is performed by a reliability analysis that this chapter deals with as it is presented in the following sections.

6.2 Structural Reliability Methods

In a safe design of structures, there are various failure modes to be checked and satisfy the prescribed safety conditions. The response is considered satisfactory or safe when the design requirements imposed on the structural behavior are met. Each design requirement is termed as the limit state or constraint. There are different types of limit states as:

- Ultimate limit states
- Conditional limit states
- Serviceability limit states

The ultimate limit states correspond to the maximum load carrying capacity of the structure due to e.g., plastic, fatigue, and buckling deformations. The conditional limit states correspond to the load carrying capacity of the structure if a local failure occurs in the structural system due to e.g., exceeding the material strength or instability. The serviceability limit states are related to the normal use of the structure, but some constraints such as e.g., limitation of deflection, vibration, or local damages are imposed for some reasons. The study of structural reliability concerns the calculation and prediction of the probability of limit state violation at any stage during its life. The probability of occurrence of a limit state violation, which is termed as a failure state, is a numerical measure of the chance that the failure state occurs in the structural lifetime. The objective of the reliability method is to find this numerical measure of the probability of a failure occurrence. In general, reliability methods are categorized in three levels as:

- Level-I reliability methods
- Level-II reliability methods
- Level-III reliability methods

The simplest method is Level-I, which is the classical design procedure. In this method, the safety of structures is achieved by using safety factors on applied loads and material resistances. The uncertainties that may occur in a structural design are incorporated in these safety factors. Then, the analysis is carried out deterministically. The safety factors can be different for different load combinations and material resistances. This type of analysis is known as the partial safety factor procedure that is commonly used in practice. The Level-I method is not further discussed in this book and attention is laid on more realistic reliability analysis methods. The Level-II reliability method is still approximate, but more precise. In this method, any type uncertainties are incorporated in the design by representative random parameters with assumed or prescribed probability models, i.e., by probability distributions, mean values, standard deviations, and correlation coefficients. This method uses approximations of the failure surface between failure and safe domains of the prescribed limit state function to a hyperplane or quadratic surfaces at design points. The Level-III method is the most accurate reliability analysis method, in which uncertain quantities in the design are modeled by their joint distribution functions. The probability of failure is directly calculated from the probability integral. These methods of the reliability analysis are well documented and can be found in a number of textbooks, see e.g., [1, 2] and [29–33]. In the reliability analysis, all possible failure modes of the system are taken into account to calculate the system reliability. Failure of a structural component may cause hazards to some other components, and even it may or may not lead directly to the system failure. The system reliability methods are already been developed and well documented, see e.g., [30, 34–36] The system reliability for both parallel and series systems can be calculated in terms of the component reliability that is calculated from the first-order approximation of the failure function. This highlights the importance of the component reliability in the estimation of global system reliability. In many cases, reliability aspects are time dependent. The reason may, for instance, be fluctuating loads or some kind or degrading mechanism on the resistance side. The instantaneous, or point in time, failure probability at time t in those cases should also be formulated as in the same manner of formulating failure probability for time invariant system [1, 37–39] so that the failure probability will be a function of time. The interest herein often lies in estimating the probability of failure over a time interval, say from 0 to t_L. For both cases, time-independent and time-dependent reliability methods, the basis of the analysis is the same. Time-independent reliability methods receive considerable attention in practice. They are also encountered in more realistic time-variant reliability analyses. In some cases, time-dependent failure modes are transferred into corresponding time-independent modes. Therefore, time-independent reliability methods for component structural failure modes are presented in the following sections.

6.2.1 Design Variables, Limit State Functions, and Failure Probability in the Reliability Analysis

As we have mentioned in the introduction, a structural design contains some uncertainties that may arise partly from structural material used, from different applied loads, from structural configuration, dimensions, and modeling, from the environment where the structure will be built, etc. All these uncertainties are taken into account in the analysis by representative random parameters or variables with assumed or determined probability distributions. Each uncertain category is represented by an individual random variable Y_i so that the design will be a function of multiple random variables, Y_j $(j = 1,2,..,n)$, representing uncertainties that occur in the design. These random variables are called as design, or basic, variables of the reliability analysis. The collection of design variables in a vector form is the design variables vector denoted by $\mathbf{Y} = (Y_1, Y_2,\ldots,Y_n)$. The design variables may be independent, correlated, or partly independent, and partly correlated with assumed or known probability distributions. For a safe design, their values will be calculated from the reliability analysis together with the probability of a failure mode, and these values of variables are called design values. The mathematical representation of a failure mode, or a serviceability condition, is defined as the limit state, or the failure, function of the reliability analysis. A failure mode can be any event that causes a structural danger or unfavorable situation in the structural behavior. Examples of failure modes may be those are stress condition that exceeds the ultimate limit stress capacity, excessive buckling or plastic stresses, fatigue phenomenon, etc. of a structural member or members, or the complete system collapse mechanism. There may also be some other conditions that even structural members and the structure do not fail due to serviceability or operational restrictions, the structural behavior is taken to be unfavorable. In such cases, the excess of the structural behavior from the specified limit is the serviceability failure mode. The mathematical statement of a failure mode is written as,

$$g(Y) = 0 \tag{6.1}$$

which can be any function of the design variables, linear or nonlinear. We define the following regions as depending on the value of the failure function.

$$\begin{aligned} g(Y) &\leq 0 \rightarrow \text{ Failure (unsafe) region} \\ g(Y) &> 0 \rightarrow \text{ Safe region} \\ g(Y) &= 0 \rightarrow \text{ Failure (limit state) surface} \end{aligned} \tag{6.2}$$

When the failure (limit state) function is zero, then we obtain a surface between the safe and unsafe regions. By definition, we call this surface as the failure (limit state) surface. The negative value of the failure function defines the failure region and the positive value defines the safe region. The simplest case of the failure function can be written as,

$$Z = g(Y) = (R - S) \rightarrow \begin{cases} S \geq R : \text{Failure region} \\ S < R : \text{Safe region} \\ S = R : \text{Failure surface} \end{cases} \tag{6.3}$$

where R denotes the resistance and S denotes the loading (stress) terms. When the stress S is greater than or equal to the resistance R, then failure occurs. Thus, we can write the probability of failure as,

$$P_F = P(R \leq S) = P(g(Y) \leq 0) = P(Z \leq 0) = F_Z(0) \tag{6.4a}$$

where $F_Z(.)$ is the probability distribution (cumulative distribution) function, CDF, of the failure function Z. It can also be calculated from the joint probability density function of R and S, $f_{RS}(r,s)$, as written by,

$$P_F = P(R \leq S) = \int_D f_{RS}(r, s)\, \mathrm{d}r\, \mathrm{d}s \tag{6.4b}$$

in which D denotes the failure domain. If we assume that R and S are independent, then the joint probability density function $f_{RS}(r,s)$ becomes to be:

$$f_{RS}(r, s) = f_R(r) f_S(s) \tag{6.4c}$$

By using this statement in Eq. (6.4b) it can be written that,

$$\begin{aligned} &P_F = P(R \leq S) = \int_{-\infty}^{\infty} \left(\int_{-\infty}^{s} f_R(r)\, \mathrm{d}r \right) f_S(s)\, \mathrm{d}s = \int_{-\infty}^{\infty} F_R(s) f_S(s)\, \mathrm{d}s \\ &\text{with } F_R(s) = \int_{-\infty}^{s} f_R(r)\, \mathrm{d}r \end{aligned} \tag{6.4d}$$

from which the probability of failure is obtained as the mean value of the probability distribution function of the resistance R evaluated at any stress variable s, $F_R(s)$. If we assume that the probability distribution of the failure function, $Z = g(Y)$, is Normal (Gaussian), from Eq. (6.4a) we can write the failure probability as,

$$F_Z(z) = \Phi\left(\frac{z - \mu_Z}{\sigma_Z}\right) \rightarrow P_F = F_Z(0) = \Phi\left(-\frac{\mu_Z}{\sigma_Z}\right) \tag{6.5a}$$

where $\Phi(.)$ is the Standard Normal Distribution Function which has zero mean and unit variance. Equation (6.5a) can be written in a slightly different form as,

$$\left(\beta = \frac{\mu_Z}{\sigma_Z}\right) \rightarrow P_F = \Phi(-\beta) \tag{6.5b}$$

in which β is defined as the *Reliability Index*, and the probability of failure is calculated from the Standard Normal Distribution Function at $(u = -\beta)$ provided that the failure function Z is Normally distributed [40]. This means that the resistance and loading functions, R and S, are independent and Normally distributed. The β index and the failure probability are shown in Fig. 6.1 where the shaded area indicates the failure probability. In the Level-II reliability method, all we need is to calculate the mean value μ_Z and the standard deviation σ_Z of the Normally distributed failure function $Z = g(Y)$, and consequently to calculate the reliability index β. In some situations, it is very simple and, in some other situations, it is relatively complicated. For example, if failure functions Z can be stated as a linear superposition of independent Normally distributed design variables $\mathbf{Y}$, i.e.,

$$Z = a_0 + \sum_{i=1}^{n} a_i Y_i \tag{6.6a}$$

then the calculation of μ_Z, σ_Z and β are straightforward as written below.

$$\left(\mu_Z = a_0 + \sum_{i=1}^{n} a_i \mu_{Y_i} \text{ and } \sigma_Z = \sqrt{\sum_{i=1}^{n} a_i^2 \sigma_{Y_i}^2}\right) \rightarrow \left(\beta = \frac{\mu_Z}{\sigma_Z}\right) \tag{6.6b}$$

In general, if the failure function is defined as given by Eq. (6.3) with Normally distributed independent resistance and loading, R and S, then the reliability index β is calculated from,

$$\beta = \frac{\mu_R - \mu_S}{\sqrt{\sigma_R^2 + \sigma_S^2}} \tag{6.6c}$$

For nonlinear failure functions, correlated design variables and/or non-Normally distributed design variables, the calculation of β is not as simple as explained above. For different cases of design variables and the failure function, the calculation of β is presented briefly in the following sections.

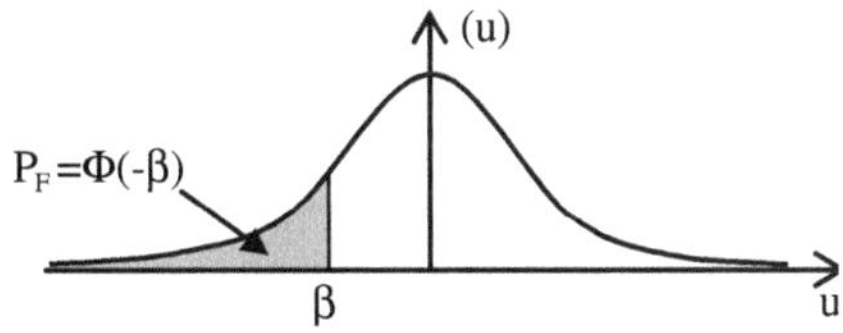

Fig. 6.1 Standard normal probability density function and failure probability

6.2.2 *First-Order Reliability Methods*

In the aforementioned section, we have assumed that the design variables are independent Normally distributed random variables and the failure function is a linear combination of design variables. Consequently, the failure function becomes a Normally distributed random variable. On this basis, the reliability index β is defined and calculated. However, in most practical applications, failure functions are nonlinear and, in some situations, the design variables are correlated with non-Normal distributions. First of all, the correlated and non-Normal design variables are transformed to independent Standard Normal variables space, **U**, with zero means and unit variances. Then, the nonlinear failure function $g(Y)$ in the original design variables space, **Y**, is also transformed to a failure function $g(U)$ in the **U** space, which is also a nonlinear function of the independent Standard Normal variables. A first order approximation of the failure surface $g(U) = 0$ at some design points is made. This is equivalent to representing the failure surface by tangent hyperplanes at the design points in the **U** space [1, 2, 41]. The hyperplane at a design point is obtained by using the Taylor series solution of the nonlinear failure function $g(U)$. This method of calculating β is called as FORM. This is exact for linear failure functions and approximate for nonlinear failure functions. Its approximation depends on the degree of the nonlinearity in the failure function. Different methods of FORM are available depending on the condition of design variables. In the following sections, these methods are briefly explained.

6.2.2.1 Calculation of β for Nonlinear Failure Functions of Independent Normal Random Variables

Calculation of the reliability index β for linear failure function of independent Normal random variables is presented above in Sect. 6.2.1. In this section, we assume that the failure function is nonlinear and design variables are independent and Normal. In order to calculate β, as it is mentioned above, the original problem will be transformed to the Standard Normal variables space **U** first. Then the linearization process is performed [41]. The transformation from the original **Y** space to **U** space, which is the space of independent normalized Normal variables, can be written as,

$$\mathbf{Y}=\boldsymbol{\mu}_{\mathbf{Y}}+\boldsymbol{\sigma}_{\mathbf{Y}}\mathbf{U} \tag{6.7}$$

where $\boldsymbol{\mu}_{\mathbf{Y}}$ is the vector of mean values of **Y** and $\boldsymbol{\sigma}_{\mathbf{Y}}$ is a diagonal matrix containing the variances of **Y**. Having used **Y** from Eq. (6.7) in the failure function $g(Y)$, the failure function $g(U)$ in the **U** space can be obtained. Next step is to perform the linearization of $g(U)$ at design points, say at $\mathbf{U}=\mathbf{U}^*$where $\mathbf{U}^*$denotes a design point. This is written as,

$$Z = g(U) \approx g(U^*) + \nabla g_{U^*}^T(\mathbf{U} - \mathbf{U}^*) \tag{6.8a}$$

in which ∇g_{U^*} is the gradient vector of $g(U)$ with respect to **U** evaluated at $\mathbf{U} = \mathbf{U}^*$, i.e.,

$$\left(\nabla g_{U^*}^T = \nabla g_U^T\big|_{U=U^*}\right) \rightarrow \nabla g_U^T = \left\{\frac{\partial g(U)}{\partial u_1}, \frac{\partial g(U)}{\partial u_2}, \ldots, \frac{\partial g(U)}{\partial u_n}\right\} \tag{6.8b}$$

The mean value μ_Z and standard deviation σ_Z of the linearized failure function $Z = g(U)$, and consequently, the reliability index β can be written from Eq. (6.8a) as,

$$\left.\begin{array}{l}\mu_Z \approx g(U^*) - \nabla g_{U^*}^T\mathbf{U}^* \\ \sigma_Z \approx |\nabla g_{U^*}|\end{array}\right\} \rightarrow \beta = \frac{\mu_Z}{\sigma_Z} = \frac{g(U^*) - \nabla g_{U^*}^T\mathbf{U}^*}{|\nabla g_{U^*}|} \tag{6.9a}$$

in which $|\nabla g_{U^*}|$ is the absolute value of the gradient vector ∇g_{U^*}, i.e.,

$$|\nabla g_{U^*}| = \sqrt{\nabla g_{U^*}^T \nabla g_{U^*}} \tag{6.9b}$$

With the help of Eq. (6.9a) the failure function $g(U)$ at the failure surface can be stated in terms of β as written by,

$$\beta|\nabla g_{U^*}| = g(U^*) - \nabla g_{U^*}^T\mathbf{U}^* \rightarrow g(U) \approx \beta|\nabla g_{U^*}| + \nabla g_{U^*}^T\mathbf{U} = 0 \tag{6.9c}$$

from which the reliability index β is stated alternatively as,

$$\left(\beta = \boldsymbol{\alpha}^{\mathbf{T}}\mathbf{U}\right) \rightarrow \boldsymbol{\alpha} = -\frac{\nabla g_{\mathbf{U}^*}}{|\nabla g_{\mathbf{U}^*}|} \tag{6.10a}$$

In Eq. (6.10a), $\boldsymbol{\alpha}$ denotes a unit vector being normal to the tangent hyperplane (unit direction vector) at the linearization point $\mathbf{U}^*$. Since $\boldsymbol{\alpha}$ is a unit vector its absolute value is equal to one, i.e.,($\boldsymbol{\alpha}^{\mathbf{T}}\boldsymbol{\alpha} = 1$). In Eq. (6.10a), having multiplied β by ($\boldsymbol{\alpha}^{\mathbf{T}}\boldsymbol{\alpha}$) from both sides and rearranged the statement, it can be obtained that,

$$\boldsymbol{\alpha}^{\mathbf{T}}(\beta\boldsymbol{\alpha} - \mathbf{U}) = 0 \rightarrow \mathbf{U} = \beta\boldsymbol{\alpha} \tag{6.10b}$$

If the the design variables vector **U** is known, then the unit direction vector $\boldsymbol{\alpha}$ and consequently the reliability index β can be calculated from Eq. (6.10a). Conversely, if β and $\boldsymbol{\alpha}$ are known, then the **U** vector will be calculated from Eq. (6.10b). Calculation of β and the final design point $\mathbf{U}^*$ is carried by an iterative procedure starting from an initial vector $\mathbf{U} = \mathbf{U}_0$ as shown in Fig. 6.2. The reliability index β is the shortest distance from the origin to the tangent hyperplane at the final design point on

Fig. 6.2 Tangent hyperplane, β, unit direction vector $\boldsymbol{\alpha}$ and the final design point **U***

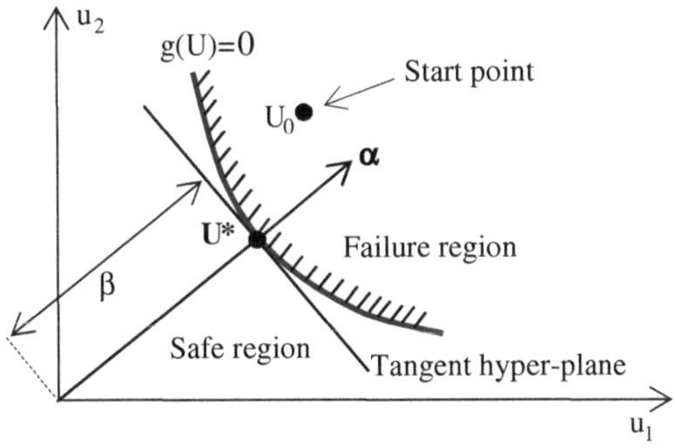

the failure surface $g(U) = 0$. Thus, the calculation of β in **U** space becomes as an optimization problem to find the minimum value. It is stated as,

$$\beta = \min_{g(U)=0} \sqrt{\sum_{i=1}^{n} U_i^2} \tag{6.11}$$

The failure function and design variables are given or specified in the original variables space **Y**. The transformation from **U** to **Y** and vice versa is given by Eq. (6.7). By using this transformation, the gradient vector, ∇g_U, with respect to **U** can be calculated in terms of the gradient vector, ∇g_Y, with respect to the original design variables **Y** as written by,

$$\nabla g_U = \frac{\partial g}{\partial \mathbf{U}} = \left(\frac{\partial \mathbf{Y}}{\partial \mathbf{U}}\right)^{\mathbf{T}} \frac{\partial g}{\partial \mathbf{Y}} \rightarrow \nabla g_U = \boldsymbol{\sigma}_Y^T \nabla g_Y \tag{6.12}$$

The iterative calculation algorithm is presented below in Fig. 6.3. At the final calculation step, the failure function $g(U^*)$ will be obtained to be zero, i.e., $g(U^*) = 0$, and β will be ($\beta = \boldsymbol{\alpha}_i^T$ **U***) as it is given in Eq. (6.10a). In this algorithm, the convergence of iteration is quite fast and after few iteration cycles the correct result is obtained. As the start vector, $\mathbf{U}_0 = 0$, can be used which corresponds to the mean values μ_Y of the original design variables vector **Y**.

6.2.2.2 Calculation of β for Correlated Normal Random Variables

It is assumed in this section that the original design variables **Y** are correlated and Normally distributed. The failure function $g(Y)$ can be linear or nonlinear. The design variables **Y** are given with mean values, μ_Y, and correlation coefficient matrix $\boldsymbol{\rho}_Y$ as,

$$\boldsymbol{\mu}_Y^T = \{\mu_{Y1}, \mu_{Y2}, \ldots, \mu_{Yn}\} \text{ and } \boldsymbol{\rho}_Y = \begin{bmatrix} 1 & \rho_{12} & & \rho_{1n} \\ \rho_{12} & 1 & . & \rho_{2n} \\ . & . & . & . \\ \rho_{1n} & \rho_{2n} & . & 1 \end{bmatrix} \tag{6.13a}$$

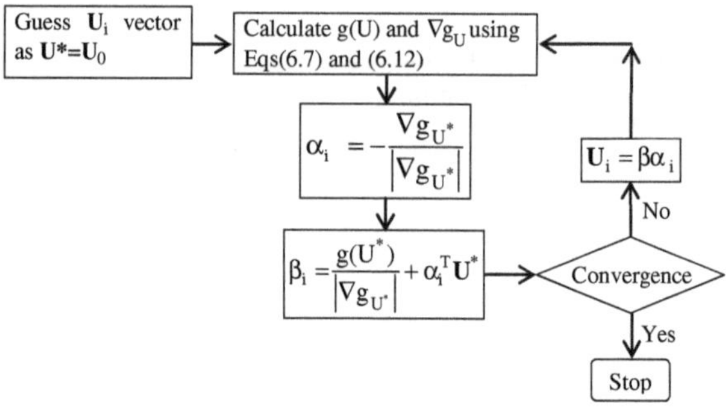

Fig. 6.3 Calculation algorithm of the reliability index β for nonlinear failure functions

In this statement, the correlation coefficient ρ_{ij} of the design variables Y_i and Y_j is defined as,

$$\rho_{ij} = \frac{\sigma_{YiYj}}{\sigma_{Yi}\sigma_{Yj}} \tag{6.13b}$$

where σ_{YiYj} is the covariance and σ_{Yi} and σ_{Yj} are the standard deviations of Y_i and Y_j. The first step is to transform **Y** vector to a correlated Normal variables vector **X** with zero means and unit standard deviations (normalized Normal Variables), i.e., $\mu_{Xi} = 0$ and $\sigma_{Xi} = 1$ ($i = 1,2,..,n$). This is done by the transformation,

$$\mathbf{Y} = \boldsymbol{\mu}_{Y} + \boldsymbol{\sigma}_{Y}\mathbf{X} \tag{6.14}$$

in which $\boldsymbol{\sigma}_Y$ denotes a diagonal matrix containing standard deviations of **Y**, i.e., σ_{Yi} ($i = 1,2,..,n$). It can easily be verified from Eq. (6.14) that the correlation coefficient matrices of **Y** and **X** are equal. Thus, $(\boldsymbol{\rho}_X = \boldsymbol{\rho}_Y)$. The second step is to transform the correlated normalized Normal variables vector **X** to the **U** variables space (normalized uncorrelated Normal variables space). This transformation can be performed by two methods as:

- Choleski Triangulation Method
- Orthogonal Transformation (Eigenvalue Solution) Method

These two methods are briefly outlined below. First, the Choleski triangulation method is explained.

In the Choleski triangulation method, the correlated Normal random variables, **X**, are transformed to the Standard Normal Variables, **U**, space. This transformation is written as,

$$\mathbf{X} = \mathbf{TU} \tag{6.15a}$$

where **T** is an unknown transformation matrix that will be calculated. For this purpose, we use the correlation coefficient matrix of **X**, which is stated from Eq. (6.15) as written by,

$$\boldsymbol{\rho}_X = \mathbf{T}E\left[\mathbf{U}\mathbf{U}^T\right]\mathbf{T}^T \rightarrow \boldsymbol{\rho}_X = \mathbf{T}\mathbf{T}^T \tag{6.15b}$$

Since, ($\rho_X = \rho_Y$), the transformation matrix **T** can be calculated from Eq. (6.15b) by the Choleski decomposition method from which it is clear that **T** will be a lower triangular matrix. Having introduced Eq. (6.15a) into Eq. (6.14), the original design variables vector **Y** can be transformed to the standard normal variables space, **U,** by the transformation:

$$\mathbf{Y} = \boldsymbol{\mu}_{\mathrm{Y}} + \mathbf{DU} \rightarrow (\mathbf{D} = \boldsymbol{\sigma}_{\mathrm{Y}}\mathbf{T}) \tag{6.16a}$$

Having introduced this transformation into the failure function $g(Y)$ it is calculated in the **U** space, i.e., $g(U)$. For the iteration of the reliability index, the gradient vector with respect to the standard normal variables, ∇g_U, can be easily calculated in terms of the gradient vector with respect to the original design variables **Y**, ∇g_Y, as written by,

$$\nabla g_U = \mathbf{D}^T \frac{\partial g}{\partial \mathbf{Y}} \rightarrow \left(\nabla g_U = \mathbf{D}^T \nabla g_Y\right) \tag{6.16b}$$

Having obtained the failure function $g(U)$ and its gradient vector ∇g_U, The reliability index β is calculated as explained in the previous Sect. 6.2.2.1.

An alternative transformation method from the correlated normalized Normal variables vector **X** to the **U** variables is to use the orthogonal transformation which forms an eigenvalue problem. For this purpose, we use auxiliary independent Normal random variables with zero means, i.e., $\mu_{\mathrm{Vi}} = 0$ (i = 1,2,...,n), which are denoted by the vector **V** as it is defined by,

$$\mathbf{V} = \mathbf{C}^{\mathbf{T}}\mathbf{X} \tag{6.17a}$$

Eq. (6.17a) is an orthogonal transformation. Since **V** is an independent variables vector the covariance matrix of these variables will be diagonal and it is denoted by $\boldsymbol{\sigma}_{\mathrm{V}}$. It is calculated from,

$$\boldsymbol{\sigma}_V = E\left[\mathbf{V}\mathbf{V}^T\right] = \mathbf{C}^T E\left[\mathbf{X}\mathbf{X}^T\right]\mathbf{C} \rightarrow \left(\boldsymbol{\sigma}_V = \mathbf{C}^T \rho_X \mathbf{C}\right) \tag{6.17b}$$

This equation is an eigenvalue problem and can be solved by an eigenvalue solver. The variances of **V** are the eigenvalues of ρ_X, i.e., $\left(\sigma_{V_i}^2 = \lambda_i\right)$ with (i = 1,2,...,n). The eigenvalue λ_i and the eigenvector $\mathbf{C}_i$, which is the ith. column of the matrix **C** are calculated from the eigenvalue solution,

$$\begin{aligned} &\rho_{\mathrm{X}}\mathbf{C}_i = \lambda_i\mathbf{C}_i \text{ with } \mathbf{C} = [\mathbf{C}_1, \mathbf{C}_2, \ldots, \mathbf{C}_i, \ldots \mathbf{C}_\mathbf{n}] \\ &\sigma_{V_i}^2 = \lambda_i \end{aligned} \tag{6.17c}$$

Having calculated the transformation matrix **C** from Eq. (6.17c) in which, it is transformed to the **U** variables space as written by,

$$\mathbf{V} = \boldsymbol{\sigma}_{\mathrm{V}}\mathbf{U} \tag{6.18a}$$

Since the transpose of an orthogonal matrix is equal to its inverse, from Eq. (6.17a) it can be stated that,

$$\mathbf{V} = \mathbf{C}^{\mathrm{T}}\mathbf{X} = \boldsymbol{\sigma}_{\mathbf{V}}\mathbf{U} \rightarrow (\mathbf{X} = \mathbf{C}\sigma_{\mathbf{V}}\mathbf{U}) \tag{6.18b}$$

Having introduced **X** from Eq. (6.18b) into Eq. (6.14), the original design variables vector **Y** can be stated similar to that given by Eq. (6.16a). This is written as,

$$\mathbf{Y} = \boldsymbol{\mu}_{\mathbf{Y}} + \mathbf{DU} \rightarrow (\mathbf{D} = \boldsymbol{\sigma}_{\mathbf{Y}}\mathbf{C}\sigma_{\mathbf{V}}) \tag{6.18c}$$

with this transformation the reliability index β can be calculated as explained above in the case of the Choleski triangulation method.

6.2.2.3 Calculation of β for Non-Normal Independent Random Variables

In practice, probability distributions of design variables are not always Normal. Usually, they are non-Normal although they may be independent which is the case considered in this section. For such case, in order to calculate the β index, the non-Normal design variables are transformed to equivalent Normal random variables at design points. For this purpose, probability distributions and probability density functions of non-Normal and Normal variables are equalized at design points of the reliability iteration [42]. For original non-Normal design variables Y_i and equivalent standard Normal variables U_i ($i = 1, 2,\ldots,n$), this is given at any point as,

$$F_Y(Y_i){=}\Phi(U_i) \rightarrow Y_i = F_Y^{-1}(\Phi(U_i)) \tag{6.19a}$$

and from the derivative of both sides, it can be stated that,

$$\frac{\partial\Phi(U_i)}{\partial U_i} = \frac{\partial F_Y(Y_i)}{\partial Y_i}\frac{\partial Y_i}{\partial U_i} \rightarrow \varphi(U_i){=}f_Y(Y_i)\frac{\partial Y_i}{\partial U_i} \rightarrow \frac{\partial Y_i}{\partial U_i} = \frac{\varphi(U_i)}{f_Y(Y_i)} \tag{6.19b}$$

where $\varphi(.)$ and $f_Y(.)$ denote the probability density functions of the standard Normal random variable and the original design variable. The gradient of the failure function, ∇g_{U_i}, with respect to the standard Normal variable U_i can be written by using the chain rule as,

$$\frac{\partial g}{\partial U_i} = \frac{\partial g}{\partial Y_i}\frac{\partial Y_i}{\partial U_i} \rightarrow \nabla g_{U_i}{=}\frac{\varphi(U_i)}{f_Y(Y_i)}\nabla g_{Y_i} \tag{6.19c}$$

This transformation is equal to the transformation defined as (with unknown standard deviation σ_i and mean value μ_i to be determined),

$$Y_i = \mu_i + \sigma_i U_i \tag{6.20a}$$

The standard deviation is calculated from Eqs. (6.19b) and (6.20a) and the mean value μ_i is calculated from Eqs. (6.19a) and (6.20a) as written by,

$$\sigma_i = \frac{\partial \mathbf{Y}_i}{\partial U_i} = \frac{\varphi(U_i)}{f_Y(Y_i)} \text{ and } \mu_i{=}\ F_Y^{-1}(\Phi(U_i)) - \sigma_i U_i \tag{6.20b}$$

For a given or updated iteration point in the **U** space, **U***, the corresponding original design variables vector **Y*** is calculated by using Eq. (6.19a), and consequently, the failure function $g(Y_i^*) = g\left(F_Y^{-1}\left(\Phi(U_i^*)\right)\right)$ is calculated. For the reliability iteration in the **U** space, the gradient ∇g_U is calculated by using Eq. (6.19c). Having calculated the failure function and its gradient in the **U** space, the rest of the reliability iteration is carried out by using the algorithm given in Fig. 6.3.

6.2.2.4 Calculation of *β* for Non-Normal Correlated Random Variables

If the failure function $g(Y)$ is a function of non-Normal correlated random variables, which are denoted by the vector **Y**, in order to calculate the reliability index β, these variables are transformed to uncorrelated standard Normal variables **U** For this purpose, two transformation methods are used [2, 43, 44]. These methods are:

Rosenblatt transformation
Nataf transformation

The Rosenblatt transformation is used if the joint probability distribution function of the random variables **Y** is known or given as a sequence of conditional probability distribution functions, i.e., if $F_Y(Y_1, Y_2, \ldots, Y_n)$ or $f_Y(Y_1, Y_2, \ldots, Y_n)$ is known [2, 43]. This transformation uses equal probability distribution functions of uncorrelated standard Normal variables **U** and the conditional distribution functions of the correlated non-Normal variable **Y**. It is stated as,

$$\begin{aligned} \Phi(U_1) &= F_{Y_1}(Y_1) \\ \Phi(U_2) &= F_{Y_2}(Y_2|Y_1) \\ &\vdots \\ \Phi(U_n) &= F_{Y_n}(Y_n|Y_1, Y_2, \ldots, Y_{n-1}) \end{aligned} \tag{6.21}$$

where $F_{Y_i}(Y_i|Y_1, Y_2, \ldots, Y_{i-1})$ is the distribution function of Y_i given $(Y_1, Y_2, \ldots, Y_{i-1})$ for $(i = 1, 2, \ldots, n)$. This conditional probability distribution function is calculated from the integration of the corresponding probability density function (PDF) as written by,

$$F_{Y_i}(Y_i|Y_1, Y_2, \ldots, Y_{i-1}) = \int_{-\infty}^{Y_i} f_{Y_i}(z|Y_1, Y_2, \ldots, Y_{i-1})\, dz \tag{6.22a}$$

in which $f_{Y_i}(z|Y_1, Y_2, \ldots, Y_{i-1})$ is the conditional probability density function of Y_i defined by,

$$f_{Y_i}(z|Y_1, Y_2, \ldots, Y_{i-1}) = \frac{f_Y(Y_1, Y_2, \ldots, z)}{f_Y(Y_1, Y_2, \ldots, Y_{i-1})} \tag{6.22b}$$

where z is a variable for Y_i, i.e., ($z = -\infty$ to Y_i), and $f_Y(Y_1, Y_2, \ldots, Y_i)$ is the joint probability density function of the random variables (Y_1, Y_2, …, Y_i). Having introduced Eq. (6.22b) into Eq. (6.22a) the conditional distribution function can be calculated from,

$$F_{Y_i}(Y_i|Y_1, Y_2, \ldots, Y_{i-1}) = \frac{\int_{-\infty}^{Y_i} f_Y(Y_1, Y_2, \ldots, Y_{i-1}, z)\mathrm{d}z}{f_Y(Y_1, Y_2, \ldots, Y_{i-1})} \tag{6.22c}$$

The joint PDF $f_Y(Y_1, Y_2, \ldots, Y_i)$ is calculated from the integration,

$$f_Y(Y_1, Y_2, \ldots, Y_i) = \int_{-\infty}^{\infty} \cdots \int_{-\infty}^{\infty} f_Y(Y_1, Y_2, \ldots, Y_i, Y_{i+1}, \ldots, Y_n)\mathrm{d}Y_{i+1} \ldots \mathrm{d}Y_n \tag{6.22d}$$

In the reliability iteration as given in Fig. 6.3, the initial standard normal variables, U, are chosen and updated at each iteration cycle. In order to calculate the failure function, $g(Y)$, in the original design variables space, the variables Y_i are calculated from the transformation given by Eq. (6.21). This is written as,

$$Y_i = F_{Y_i}^{-1}(\Phi(U_i)|Y_1, Y_2, \ldots, Y_{i-1}) \text{ where } (i = 1, 2, \ldots, n) \tag{6.23a}$$

In the iteration, the gradient vector $\nabla g_{\mathbf{U}}$ with respect to the standard normal variables **U** is also required. It is calculated in terms of the gradient vector ∇g_Y with respect to the original design variables **Y**. Since **Y** is a function of **U** the gradient vectors, ∇g_Y and ∇g_U, are calculated from the following relations.

$$\mathbf{Y} = f(\mathbf{U}) \rightarrow \nabla g_Y = \mathbf{J}^T \frac{\partial g}{\partial U} = \mathbf{J}^T \nabla g_U \rightarrow \nabla g_U = \left(\mathbf{J}^T\right)^{-1} \nabla g_Y \tag{6.23b}$$

where **J** is the Jacobian matrix defined as,

$$\mathbf{J} = \begin{bmatrix} \frac{\partial U_1}{\partial Y_1} & \cdot & \cdot & \frac{\partial U_1}{\partial Y_n} \\ \cdot & \cdot & \cdot & \cdot \\ \cdot & \cdot & \cdot & \cdot \\ \frac{\partial U_n}{\partial Y_1} & \cdot & \cdot & \frac{\partial U_n}{\partial Y_n} \end{bmatrix} \tag{6.23c}$$

For the Rosenblatt transformation, $\Phi(U_i) = F_{Y_i}(Y_i|Y_1, Y_2, \ldots, Y_{i-1})$, the derivatives can be written as,

$$\frac{\partial \Phi(U_i)}{\partial Y_j} = \frac{\partial \Phi(U_i)}{\partial U_i}\frac{\partial U_i}{\partial Y_j} = \frac{\partial F_{Y_i}(Y_i|Y_1, Y_2, \ldots, Y_{i-1})}{\partial Y_j} \tag{6.24a}$$

from which a term of the Jacobian matrix can be obtained as,

$$\frac{\partial U_i}{\partial Y_j} = \frac{1}{\varphi(U_i)} \frac{\partial F_{Y_i}(Y_i|Y_1, Y_2, \ldots, Y_{i-1})}{\partial Y_j} \rightarrow \frac{\partial U_i}{\partial Y_j} = 0 \text{ for } (i<j) \tag{6.24b}$$

From Eq. (6.24b), it can be intuitively seen that the Jacobian matrix **J** for Rosenblatt transformation is a lower triangle matrix, i.e.,

$$\mathbf{J} = \begin{bmatrix} \frac{\partial U_1}{\partial Y_1} & \cdot & \cdot & 0 \\ & \cdot & \cdot & \\ & \cdot & \cdot & \\ \frac{\partial U_n}{\partial Y_1} & \cdot & \cdot & \frac{\partial U_n}{\partial Y_n} \end{bmatrix} \tag{6.24c}$$

As it is mentioned above, if the joint probability density function of the original design variables **Y** is known, then the transformation from **Y** to **U**, and vice versa, can be carried out by using the Rosenblatt transformation. The gradient vectors are calculated from Eq. (6.23b). In practice, the joint probability distribution is not always known. In this case, the Nataf transformation is used.

When the complete probability information of correlated random variables is not available, i.e., the joint probability distribution is not known, the Rosenblatt transformation cannot be applied. In practice, usually the marginal probability distribution functions of random variables, $F_Y(Y_i)$ where (i = 1,2,..,n), with correlation coefficients ρ_{ij} are available. In order to calculate the reliability index β using such probability information, the Nataf transformation method is used [2, 44–46]. The essence of this method is to transform first the original design variables **Y** to jointly normalized normal random variables **X**, i.e., zero mean values and unit variances ($\mu_{\text{X}} = 0$ and $\sigma_{\text{Xi}} = 1$), by using equal marginal and joint probability distributions. In this context, the correlation coefficients ρ'_{ij} of two jointly normal variables, X_i and X_j, and the joint probability distribution function $F_Y(Y_i,Y_j)$ of two original random variables, Y_i and Y_j, are unknown. Since **X** is jointly normal with zero mean and unit variances, the joint distribution function of X_i and X_j is readily available. From the definition of the Nataf transformation it is written that,

$$\begin{aligned} F_{Y_i}(Y_i) &= \Phi_X(X_i) \rightarrow Y_i = F_{Y_i}^{-1}(\Phi_X(X_i)) \\ F_Y(Y_i, Y_j) &= \Phi_X(X_i, X_j) \rightarrow f_Y(Y_i, Y_j)\text{d}Y_i\text{d}Y_j = \varphi_X(X_i, X_j)\text{d}X_i\text{d}X_j \end{aligned} \tag{6.25}$$

in which $\Phi_X(.)$ and $\varphi_X(.)$ are probability distribution and density functions (joint and marginal) of normalized normal variables. The $\varphi_X(X_i,X_j)$ is defined as,

$$\varphi_X(X_i, X_j) = \frac{1}{2\pi\sqrt{1-\rho'^2_{ij}}} \exp\left[-\frac{\left(X_i^2 + X_j^2 - 2\rho'_{ij}X_iX_j\right)}{2\left(1-\rho'^2_{ij}\right)}\right] \tag{6.26}$$

The correlation coefficient ρ_{ij} of Y_i and Y_j, which is known, is calculated from,

$$\rho_{ij} = \frac{1}{\sigma_{Y_i}\sigma_{Y_j}} \int_{-\infty}^{\infty} \int_{-\infty}^{\infty} (Y_i - \mu_{Yi})(Y_j - \mu_{Yj}) f_Y(Y_i, Y_j) \mathrm{d}Y_i \mathrm{d}Y_j \tag{6.27a}$$

or from Eq. (6.25) it is stated that,

$$\rho_{ij} = \frac{1}{\sigma_{Y_i}\sigma_{Y_j}} \int_{-\infty}^{\infty} \int_{-\infty}^{\infty} (Y_i - \mu_{Yi})(Y_j - \mu_{Yj}) \varphi_X(X_i, X_j) \mathrm{d}X_i \mathrm{d}X_j \tag{6.27b}$$

which will be a function of ρ'_{ij} though $\varphi_X(X_i, X_j)$ from Eq. (6.26). The variables Y_i and Y_j are calculated from Eq. (6.25) dependently on X_i and X_j so that the integration given by Eq. (6.27b) can be carried out iteratively to solve the correlation coefficient ρ'_{ij} of X_i and X_j. For practical calculations, the ratio $\left(R = \rho'_{ij}/\rho_{ij}\right)$ is calculated and presented in [45] for a number of probability distribution functions. Thus, the correlation coefficient of two jointly normalized normal variables X_i and X_j is calculated from,

$$\rho'_{ij} = R\,\rho_{ij} \tag{6.28}$$

and X_i is calculated from the transformation given by Eq. (6.25). With this information the reliability index β can be easily calculated as explained previously for correlated normal random variables in Sect. 6.2.2.2. The solution algorithm is:

1. Calculate correlation coefficient of X_i and X_j from $\left(\rho'_{ij} = R\,\rho_{ij}\right)$.
2. Select **U** vector.
3. Calculate **X** vector from $\mathbf{X} = \mathbf{TU}$ (**T** is obtained from the Choleski decomposition of $\boldsymbol{\rho}_X$ with elements ρ'_{ij}, see Sect. 6.2.2.2).
4. Calculate **Y** vector from $\left(Y_i = F_Y^{-1}(\Phi_X(X_i))\right)$.
5. Calculate failure function $g(Y)$ and the gradient ∇g_U.
6. Calculate the reliability index β as given in Fig. 6.3.
7. Update **U** vector and go to step 3 until a required convergence is obtained.

6.2.3 Second-Order Reliability Method

The FORM methods use first-order (linear) approximation of the failure surface at the design point. It will be accurate if the failure function is linear or weakly nonlinear (relatively flat). For heavily nonlinear failure function, the FORM methods may not be adequate to find a correct failure probability. In such cases, a better approximation of the failure surface at the design point is required. For this purpose, a second-order approximation of the failure surface at the design point is used as shown in Fig. 6.4. The failure probability calculated on basis of the second-order approximation is known as the Second-order reliability methods (SORM) [29, 47–51]. It is a relatively complicated process and computationally more time consuming. By using the second-order Taylor series solution of the failure surface in the **U** space at the design point **U*** the approximate failure function can be stated [47, 50] as written by,

$$Z = g(U) \approx g(U^*) + \nabla g_{U^*}^T(\mathbf{U} - \mathbf{U}^*) + \frac{1}{2}(\mathbf{U} - \mathbf{U}^*)^T \nabla g_{U^*}^2(\mathbf{U} - \mathbf{U}^*) \quad (6.29a)$$

in which $\nabla g_{U^*}^2$ is the second derivative matrix of the failure function, which is known as the Hessian matrix, at the design point **U***. It is defined as,

$$\nabla g_{U^*}^2 = \begin{bmatrix} \frac{\partial^2 g(U)}{\partial U_1 \partial U_1} \cdots \frac{\partial^2 g(U)}{\partial U_1 \partial U_n} \\ \cdots \\ \frac{\partial^2 g(U)}{\partial U_n \partial U_1} \cdots \frac{\partial^2 g(U)}{\partial U_n \partial U_n} \end{bmatrix}_{U=U*} \quad (6.29b)$$

Having denoted FORM reliability index by β_F and using Eqs. (6.9c) and (6.10a), the second-order approximation of the failure function given by Eq. (6.29a) can be stated at the failure surface $(g(U) = 0)$ as,

$$g(U) \approx \beta_F - \boldsymbol{\alpha}^T\mathbf{U} + \frac{1}{2}(\mathbf{U} - \mathbf{U}^*)^T\mathbf{H}(\mathbf{U} - \mathbf{U}^*) = 0 \rightarrow \mathbf{H} = \frac{\nabla g_{U^*}^2}{|\nabla g_{U^*}|} \quad (6.29c)$$

in which **H** is the normalized Hessian matrix. In the second-order approximation of the failure surface as given by Eq. (6.29c), the failure function is stated in the form of two independent random functions, one is linear and other one is quadratic in the **U** space. Due to limited knowledge on the probability distribution of the quadratic term, an exact analytical calculation of the failure probability cannot be performed. Therefore, some approximations are used in the calculation. Breitung [48] has used an asymptotic approximation and obtained the failure probability in terms of the FORM reliability index β_F as written [51] by,

$$P_F \approx \Phi(-\beta_F) \prod_{j=1}^{n-1} \left(1 - \beta_F\, \kappa_j\right)^{-1/2} \quad (6.30a)$$

in which n is the dimension of the **U** space and κ_j, where $j = 1$ to $(n-1)$, are the principal curvatures of the failure surface at the design point. Breitung [48] has shown that Eq. (6.30a) asymptotically approaches the exact failure probability as β_F approaches infinity while $(\beta_F \kappa_j)$ $(j = 1,2,\ldots n-1)$ remain fixed. Therefore, the asymptotic approximation gives good results for large values of β_F. This approximation of the failure probability, which is given by Eq. (6.30a), can be used for $(1 \leq \beta_F)$ and $(\beta_F \kappa_j < 1)$ $(j = 1,2,..n-1)$ [51]. For smaller values of β_F, the results are not reliable. The drawback of the asymptotic approximation is that it produces erroneous results when the curvatures are in the vicinity of $(1/\beta_F)$, i.e.,$(\kappa_j \approx 1/\beta_F)$, due to the singularity at this point. For small β_F values,

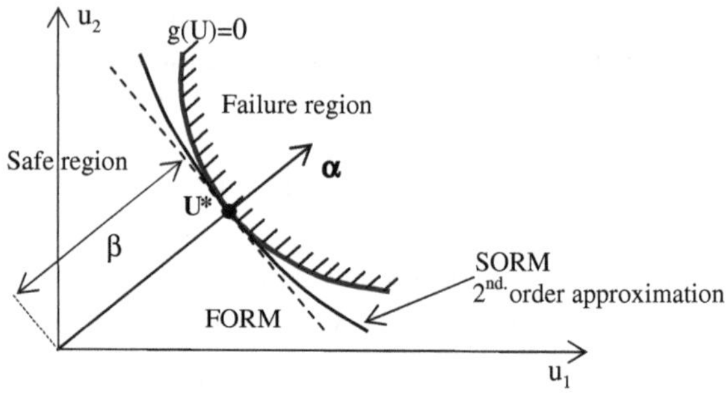

Fig. 6.4 FORM (first-order) and SORM (second-order) approximations, β and unit direction vector $\boldsymbol{\alpha}$

especially for $(\beta_F < 1)$, Tvedt [50] introduced an exact result in terms of a one-dimensional integral given [51] as,

$$P_F = \frac{1}{2} - \frac{1}{\pi}\int_0^{\infty} \sin\left(\beta t + \frac{1}{2}\sum_{j=1}^{n-1} \tan^{-1}(-\kappa_j t)\right) \frac{\exp(-\frac{t^2}{2})}{t\prod_{j=1}^{n-1}\left(1+\kappa_j^2 t^2\right)^{1/4}} \mathrm{d}t \qquad (6.30\mathrm{b})$$

which may be used for small β_F values. The Tvedt's result is exact for a fitting parabola to the failure surface so that it is considered as a second-order approximation. For moderate β_F values, an interpolation between the results of Tvedt and the asymptotic solutions, i.e., Eqs. (6.30a) and (6.30b), can be used. In the literature, some variations of the second-order reliability method are available [52–56]. In order to calculate the failure probability from SORM the principle curvatures κ_j, $j = 1$ to $(n-1)$, are required. Their calculation is outlined in the following section.

6.2.3.1 Calculation of Principle Curvatures

It can be shown [55] that the failure function $g(U)$ can be expressed in terms of the principal curvatures of the failure surface at the design point by using orthogonal transformations. This transformation is defined as written by,

$$\mathbf{U} = \mathbf{RV} \qquad (6.31)$$

where $\mathbf{R}$ is the orthogonal transformation matrix to be determined and $\mathbf{V}$ is the vector of rotated coordinates in the independent standard normal variables space. The first axis, V_1, of the vector $\mathbf{V}$ is chosen to coincide with the direction of the $\boldsymbol{\alpha}$ vector shown in Figs.6.2 and 6.4. The orthogonal transformation matrix $\mathbf{R}$ is obtained by using the Gram–Schmidt orthogonalization [57]. In order to satisfy the condition that the first axis, V_1 of the vector $\mathbf{V}$ is in the direction of the $\boldsymbol{\alpha}$ vector, the first column of $\mathbf{R}$, i.e., $\mathbf{R}_1$, is taken to equal the $\boldsymbol{\alpha}$ vector. The other columns of $\mathbf{R}$ are calculated by using the Gram–Schmidt orthogonalization starting from the matrix,

$$\mathbf{B} = \begin{bmatrix} \alpha_1 & 0 & . & 0 \\ \alpha_2 & 1 & . & 0 \\ . & . & . & . \\ \alpha_n & 0 & . & 1 \end{bmatrix} = [\mathbf{B}_1, \mathbf{B}_2, \ldots, \mathbf{B}_n] \tag{6.32a}$$

where the column ($\mathbf{B}_1 = \boldsymbol{\alpha}$) and other columns of $\mathbf{B}$ are all zero, except the element b_i of the column $\mathbf{B}_i$ which is equal to 1, i.e., ($b_i = 1$) with ($i = 2,3,\ldots,n$). With this starting matrix $\mathbf{B}$, the columns of the orthogonal transformation matrix $\mathbf{R}$, which are vectors, are calculated from,

$$\mathbf{R}_i = \mathbf{B}_i - \sum_{j=1}^{i-1} (\mathbf{B}_i^T \mathbf{r}_j)\, \mathbf{r}_j \rightarrow \left(\mathbf{r}_j = \frac{\mathbf{R}_j}{|\mathbf{R}_j|} \right) \text{ where } (i = 2 \text{ to } n) \tag{6.32b}$$

Then, the transformation matrix $\mathbf{R}$ is obtained as written by,

$$\mathbf{R} = [\mathbf{R}_1, \mathbf{R}_2, \ldots, \mathbf{R}_n] \rightarrow (\mathbf{R}_1 = \boldsymbol{\alpha}) \tag{6.32c}$$

Having calculated the matrix $\mathbf{R}$ from Eqs. (6.32b) and (6.32c) the transformation given by Eq. (6.31) can be carried out, and by introducing it into Eq. (6.29c), the equation of the failure surface is obtained as written by,

$$g(U) \approx \beta_F - \boldsymbol{\alpha}^T \mathbf{R} \mathbf{V} + \frac{1}{2} (\mathbf{V} - \mathbf{V}^*)^T \mathbf{R}^T \mathbf{H} \mathbf{R} (\mathbf{V} - \mathbf{V}^*) = 0 \tag{6.33a}$$

The vectors of the rotated coordinates at any point and at the design point on the failure surface, $\mathbf{V}$ and $\mathbf{V}^*$, are defined as written by,

$$\mathbf{V} = \begin{Bmatrix} V_1 \\ \mathrm{v} \end{Bmatrix} \rightarrow \mathrm{v} = \{V_2, \ldots, V_n\}^T \text{ and } \mathbf{V}^* = \{\beta_F, 0, \ldots, 0\}^T \tag{6.33b}$$

Since the matrix $\mathbf{R}$ is orthogonal, i.e., $\left(\mathbf{R}_i^T \mathbf{R}_j = 0\right)$ for ($i \neq j$), and having introduced Eq. (6.33b) into Eq. (6.33a) the equation of the failure surface becomes as written by,

$$\beta_F - V_1 + \frac{1}{2} \begin{Bmatrix} (V_1 - \beta_F) \\ \mathbf{v} \end{Bmatrix}^T \mathbf{A} \begin{Bmatrix} (V_1 - \beta_F) \\ \mathbf{v} \end{Bmatrix} = 0 \rightarrow \left(\mathbf{A} = \mathbf{R}^{\mathbf{T}} \mathbf{H} \mathbf{R}\right) \tag{6.33c}$$

in which $\mathbf{A}$ is the rotated Hessian matrix. By using the following partition,

$$\mathbf{A} = \begin{bmatrix} \mathbf{A}_{11} & \mathbf{A}_{12} \\ \mathbf{A}_{12}^T & \mathbf{G} \end{bmatrix} \tag{6.33d}$$

and using it in Eq. (6.33c) it can be obtained that,

$$V_1 = \beta_F + \frac{1}{2} \left[\mathbf{A}_{11} (V_1 - \beta_F)^2 + 2 (V_1 - \beta_F) \mathbf{A}_{12} \mathbf{v} + \mathbf{v}^T \mathbf{G} \mathbf{v} \right] \tag{6.33e}$$

Since, near the design point $\mathbf{V}^*$, the value of V_1 is approximately equal to β_F, i.e., $(V_1 \approx \beta_F)$, Eq. (6.33e) can be stated approximately as written by,

$$V_1 = \beta_F + \frac{1}{2}\mathbf{v}^T\mathbf{G}\mathbf{v} \tag{6.34a}$$

where the vector $\mathbf{v}$ has a dimension of $(n-1)$ and the matrix $\mathbf{G}$ has dimensions of $(n-1)$ by $(n-1)$. At this point, we use another orthogonal transformation by rotating the axes of the vector $\mathbf{v}$ around V_1 and the new rotated vector is denoted by $\mathbf{y}$ defined as,

$$\mathbf{v} = \mathbf{Q}\,\mathbf{y} \tag{6.34b}$$

in which $\mathbf{Q}$ is a normalized orthogonal matrix with dimensions of $(n-1)$ by $(n-1)$. Having introduced Eq. (6.34b) into Eq. (6.34a) it can be stated that,

$$V_1 = \beta_F + \frac{1}{2}\mathbf{y}^T\mathbf{B}\mathbf{y} \rightarrow \left(\mathbf{B} = \mathbf{Q}^T\mathbf{G}\mathbf{Q}\right) \tag{6.34c}$$

which can be written in a quadratic form as,

$$V_1 = \beta_F + \frac{1}{2}\sum_{i=1}^{n-1}\kappa_i y_i^2 \tag{6.35}$$

where κ_i and y_i $(i = 1,2,\ldots,n-1)$ are the principal curvatures of the failure surface at the design point and the elements of the rotated vector $\mathbf{y}$. Eq. (6.35) can be written only if the matrix $\mathbf{B}$ is diagonal with the terms of κ_i on diagonals. From Eq. (6.34c) it may be realized that $\mathbf{B}$ can be diagonal only if the orthogonal matrix $\mathbf{Q}$ is the eigen matrix of $\mathbf{G}$ and κ_i $(i = 1,2,\ldots,n-1)$ are the corresponding eigenvalues which are calculated from,

$$|\mathbf{G} - \kappa\mathbf{I}| = 0 \tag{6.36a}$$

in which $\mathbf{I}$ is the unit matrix with $(n-1)\times(n-1)$ and κ is an eigenvalue. Alternatively, it can also be written for the ith. column, $\mathbf{Q}_i$, of orthogonal transformation matrix $\mathbf{Q}$ as,

$$\mathbf{G}\mathbf{Q}_i = \kappa_i\mathbf{Q}_i \text{ and } \mathbf{Q} = [\mathbf{Q}_1, \mathbf{Q}_2, \ldots, \mathbf{Q}_{n-1}] \tag{6.36b}$$

Having calculated the principal curvatures κ_i of the failure surface at the design point from Eq. (3.36a) or (3.36b), the failure probability P_F can be calculated by using Eq. (6.30a) or (6.30b). Since $\mathbf{Q}_i$ is a normalized eigen vector it can be written from Eq. (6.36b) that $\left(\mathbf{Q}_i^T\mathbf{G}\mathbf{Q}_i = \kappa_i\right)$ and $\left(\mathbf{Q}_j^T\mathbf{G}\mathbf{Q}_i = 0 \text{ for } j \neq i\right)$.

6.2.4 Level-III (Exact) Reliability Methods

The failure probability of a structural component with respect to a single failure mode has been given previously in Eq. (6.4b) as rewritten by,

$$P_F = \int_{g(Y) \leq 0} f_Y(y)\,\mathrm{d}y \tag{6.37}$$

where Y is the vector of basic random variables (design variables). Calculation of this integration requires that the failure surface, $g(Y) = 0$, and the joint probability density function of Y, $f_Y(y)$, are known. The direct calculations of the failure probability from the integration, given by Eq. (6.37), are called as Level-III reliability methods. They are considered as the most accurate methods. The well-known Level-III methods are:

- Analytical Integration (AI),
- Numerical Integration (NI)
- Monte-Carlo Simulation (MCS).

Unlike the Level-II methods, unfortunately all Level-III methods may be rather time consuming. In general, the integral given by Eq. (6.37) cannot be computed analytically. The Analytical Integration can only be eligible if a few basic design variables are concerned and the failure surface has a simple form. For example, if the failure function, $Z = g(Y)$, is stated in terms of independent functions of the stress S and the resistance R as written by,

$$Z = R - S \tag{6.38a}$$

their joint probability density function will be the product of PDF of each term as given by Eq. (6.4c), which is rewritten below for convenience.

$$f_Y(r, s) = f_R(r) f_S(s) \tag{6.38b}$$

Since the failure domain is defined as $(R < S)$, the failure probability, P_F, will be calculated as the mean value of the probability distribution function $F_R(s)$ of the resistance R, i.e.,

$$P_F = \int_{-\infty}^{\infty} F_R(s) f_S(s)\,\mathrm{d}s = E[F_R(s)] \tag{6.38c}$$

where $E[.]$ denotes expectation operator and $F_R(s)$ is calculated from Eq. (6.4d). In some cases, the analytical integration given by Eq. (6.38c) can be carried out. However, for complicated distributions, complex limit state functions and multiple basic variables, as well as a complex failure surface, the analytical integration is not practically possible. In this case, a numerical integration procedure is applied as explained in the following section.

6.2.4.1 Numerical Integration Method

The analytical integration is not always possible due to complexities in the failure functions, probability distributions of design variables or due to multiple design variables, etc. In such cases, the numerical integration procedure is applied to calculate the failure probability P_F. For this purpose, all basic random variables in the original **Y** space are transformed to the independent standard normal variables, **U**, space. Then, the integration given by Eq. (6.37) is carried out numerically, which can be stated [58] in the multiple summation as written by,

$$P_F = \sum_{i=-\infty}^{\infty} \sum_{j=-\infty}^{\infty} \cdots\cdots \sum_{k=-\infty}^{\infty} I(g(u)) f_U(u)\, \Delta u \tag{6.39a}$$

where $I(.)$ is a failure domain identifier, which equals 1 in the failure domain and 0 in the safe domain, i.e.,

$$\begin{aligned} I(g(u)) &= 1 \text{ if } g(u) \leq 0 \\ I(g(u)) &= 0 \text{ if } g(u) > 0 \end{aligned} \tag{6.39b}$$

where u is the variable of the vector **U**. The infinite integration (summation) limits in Eq. (6.39a) can be removed by appropriate upper and lower bounds if they are available. The numerical integration is feasible only for a limited number of random variables [59], e.g., maximum ($n = 10$). For more dimensional ($n > 10$) problems, the computation time of the numerical integration increases considerably, and therefore, in most applications the numerical integration procedure is applied only for validation of other methods with a small number of variables.

6.2.4.2 Monte Carlo Simulation Methods

As stated in the previous sections, calculations of the failure probability from Eq. (6.37) by using the analytical and numerical integrations suffer from the inefficiencies due to difficult conditions in the failure state and excessive time-consuming characters in the calculation process. Consequently, these methods are not feasible in practice. This drawback of the analytical and numerical integrations is overcome by applying the Monte Carlo Simulation technique, which is always feasible to apply. However, it can become inefficient when the limit state, of failure condition, requires heavy calculations by calling a computer code such as a finite element analysis suite. In this case, the Monte Carlo Simulation technique may not be used also in practical applications. The basic idea in the Monte Carlo Simulation technique is to sample random values of the design variables (vector **Y**), which are generated on basis of their probability distributions. Then, the number of samples falling into the failure domain (N_F) is identified, i.e., N_F is the number of samples satisfying the condition of $g(Y) \leq 0$. The failure probability, P_F, is then estimated from,

$$\bar{P}_F = \frac{N_F}{N} \approx P_F \tag{6.40a}$$

where N is the total number of samples. This estimator $\bar{P}_F$ of the failure probability P_F can also be written as,

$$\bar{P}_F = \frac{1}{N}\sum_{i=1}^{N} I(g(Y)) \tag{6.40b}$$

where $I(.)$ is as defined in Eq. (6.39b). This sample mean converges to the true mean, P_F, as the number of samples increases. Based on the law of large numbers, it can be shown [60, 61] that the variance of the estimator $\bar{P}_F$ is,

$$\sigma^2_{\bar{P}_F} = \frac{P_F(1 - P_F)}{N} \tag{6.41}$$

which is inversely proportional to N. Thus, for a small value of N, the estimation of failure probability may be subject to a considerable uncertainty. To reduce this uncertainty to an acceptable level, a large number of simulations are required. For a given target coefficient of variation of the failure probability, $V(P_F)$, i.e., $V(P_F) = \sigma_{\bar{P}_F}/P_F$, the minimum required number of samples N can be calculated from Eq. (6.41) as stated by,

$$N > \frac{1}{V^2(P_F)}\left(\frac{1}{P_F} - 1\right) \tag{6.42a}$$

In this statement, since P_F is unknown, the worst case $(P_F = 1/2)$ obtained from the maximization of the variance of estimator $\sigma^2_{\bar{P}_F}$ may be used. Having introduced this worst case in Eq. (6.42a) the following condition of the total number of samples can be obtained.

$$N > \frac{1}{V^2(P_F)} \tag{6.42b}$$

The efficiency of the direct MCS method depends on the magnitude of the failure probability, i.e., smaller failure probabilities, which usually appear in structural reliability, require larger numbers of samples. Since the probability of failure in practice is small, most of sample values are wasted, and therefore the direct MCS method becomes inefficient in the reliability methods. This drawback of the MCS method can be overcome by generating samples on the base of a different probability density function so that a large number of outcomes fall in the failure domain. This technique is called as *Importance Sampling* and the related probability density function, which is a selected *Importance Sampling Density*, is indicated by $h_Y(y)$. With this information, the failure probability given by Eq. (6.37) can be rewritten as, without loosing the generality,

$$P_F = \int\limits_{g(Y)\leq 0} \left(\frac{f_Y(y)}{h_Y(y)}\right) h_Y(y)\, dy \tag{6.43a}$$

The estimator of the failure probability can then be given as, in the light of Eq. (6.43a),

$$\bar{P}_F = \frac{1}{N}\sum_{i=1}^{N} I(g(Y)) \left(\frac{f_Y(y)}{h_Y(y)}\right) \tag{6.43b}$$

in which the samples are based on the density function $h_Y(y)$, instead of the density function of basic design variables $f_Y(y)$. The efficiency of this technique depends on the selection of $h_Y(y)$. Importance sampling is generally recognized as the most efficient variance reduction technique. A successful choice of $h_Y(y)$ produces reliable results and reduces the number of simulations significantly, while an inappropriate choice of $h_Y(y)$ produces inaccurate results.

Other variance reduction techniques are also available as:

Adaptive sampling technique
Conditional expectation technique

Adaptive sampling technique updates sampling density dynamically as the simulation proceeds [62, 63]. Conditional expectation technique consists of *Directional Simulation*, which is recommended for convex safe sets [64] and *Axis-orthogonal Simulation* technique, which is recommended for convex failure sets [65, 66].

6.3 Inverse Reliability Method

In the standard reliability analysis of structures, the uncertainties in limit state functions are modeled by random design variables with known or prescribed probability distributions. Then, a reliability measure is calculated iteratively in the form of the reliability index for a prescribed limit state function as explained in previous sections. In some cases, the target reliability measure is given and the unknown design variables are required with a deterministic design parameter or one of statistical parameters (mean value or standard deviation) of the design variables in the limit state function. This problem involves the inverse reliability method [67, 68]. In a structural design, more than one parameter for a given reliability measure may be required to determine. In this case, a unique solution can be obtained only if sufficient numbers of constraints are provided as being geometric or reliability related. Under insufficient constraints an optimum solution may be obtained by using optimization which is not considered in this section. In the case of reliability related constrains, different target reliability measures may be used for different limit state functions. The solution of the inverse reliability problem is carried out by using the FORM reliability methods as explained in [67–70] and used in different engineering applications [71–76]. In this section, the solution of the inverse reliability problem is explained briefly for multiple parameters to be determined. In the inverse reliability method for multiple parameters, the number of limit state functions must be equal to the number of unknown parameters to be determined. The multiple limit state functions in the correlated non-Normal variables space (Y-space) is written as,

$$\left.\begin{array}{r} g_i(\mathbf{Y},\mathbf{P}) = 0 \\ \text{given } \beta_i = a_i \end{array}\right\} \text{for } (i = 1, 2, \ldots, k) \tag{6.44a}$$

in which **Y** is the design variables vector representing uncertainties in the limit state function, **P** is the vector of parameters to be determined using the reliability constraints (β_i = a$_i$) with (i = 1,2,...,k), k is the number of parameters. The vectors **Y** and **P** are defined as,

$$\begin{array}{l} \mathbf{P} = \{p_1, p_2, \ldots\ldots, p_k\}^T \\ \mathbf{Y} = \{y_1, y_2, \ldots\ldots, y_n\}^T \end{array} \tag{6.44b}$$

In order to solve Eq. (6.44a), the limit state functions with unknown vector **P** are stated in the standard normal independent variables space (U-space) as written by,

$$\left.\begin{array}{r} g_i(\mathbf{U},\mathbf{P}) = 0 \\ \text{given } \beta_i = a_i \end{array}\right\} \text{for } (i = 1, 2, \ldots, k) \tag{6.45}$$

For known **P** vector and unknown β_i values, the solution methods have been presented in the previous sections. For a prescribed **U** vector, the solution of the limit state functions to the **P** vector can be obtained without any constraint such as ($\beta_i = a_i$). Since the limit state functions, $g_i(\mathbf{U},\mathbf{P})$, are usually nonlinear functions of the unknown **P** vector, the solution can be performed numerically in an iterative sequence. For this purpose, the Newton–Raphson method is used, which requires linearization of the limit state functions at some points of **P**, say **P***. The linearization of $g_i(\mathbf{U},\mathbf{P})$ at (**P** = **P***) is stated as,

$$g_i(\mathbf{U},\mathbf{P}) = g_i(\mathbf{U},\mathbf{P}^*) + \nabla g_{iP*}^T(\mathbf{P} - \mathbf{P}^*) = 0 \text{ where } (i = 1, 2, \ldots, k) \tag{6.46a}$$

in which ∇g_{iP*} is the gradient vector of $g_i(\mathbf{U},\mathbf{P})$ with respect to **P** evaluated at (**P** = **P***), i.e.,

$$\left(\nabla g_{iP*}^T = \nabla g_{iP}^T\big|_{P=P^*}\right) \rightarrow \nabla g_{iP} = \left\{\frac{\partial g_i(\mathbf{U},\mathbf{P})}{\partial p_1}, \frac{\partial g_i(\mathbf{U},\mathbf{P})}{\partial p_2}, \ldots, \frac{\partial g_i(\mathbf{U},\mathbf{P})}{\partial p_k}\right\}^T \tag{6.46b}$$

where (i = 1, 2,...,k). The linearized limit state functions given by Eq. (6.46a) can be stated for (i = 1, 2,...,k) in the vector form as written by,

$$g(\mathbf{U},\mathbf{P}^*) + \nabla g_{\mathbf{P}*}(\mathbf{P}-\mathbf{P}^*) = 0 \rightarrow \left(\mathbf{P} = \mathbf{P}^* - \nabla g_{\mathbf{P}*}^{-1}\, g(\mathbf{U},\mathbf{P}^*)\right) \qquad (6.46c)$$

where $g(\mathbf{U},\mathbf{P}^*)$ is a vector consisted of $g_i(\mathbf{U},\mathbf{P}*)$ for (i = 1,2,…,k) and $\nabla g_{\mathbf{P}*}$ is a matrix consisted of the gradient vectors ∇g^T_{iP*}. They are defined as,

$$\begin{aligned} g(\mathbf{U},\mathbf{P}^*) &= \{g_1(\mathbf{U},\mathbf{P}*),\, g_2(\mathbf{U},\mathbf{P}*),\, \ldots,\, g_k(\mathbf{U},\mathbf{P}*)\}^T \\ \nabla g_{\mathbf{P}*} &= [\nabla g_{1P*},\, \nabla g_{2P*}, \ldots,\, \nabla g_{kP*}]^T \end{aligned} \qquad (6.46d)$$

For a given initial vector **P*** and assumed **U** vector, the vector **P** will be updated by using Eq. (6.46c). This update will continue until the vector $g(\mathbf{U},\mathbf{P}^*)$ becomes zero, i.e., $g(\mathbf{U},\mathbf{P}^*) = 0$. This process will be repeated for each variation of the **U** vector. For given **P** and **U** vectors of the limit state function $g_i(\mathbf{U},\mathbf{P}^*)$, the corresponding unit direction vector $\boldsymbol{\alpha}_i$ will be calculated by using Eq. (6.10a) as written by,

$$\boldsymbol{\alpha}_\mathbf{i} = -\frac{\nabla g_{iU^*}}{|\nabla g_{iU^*}|} \qquad (6.47a)$$

in which ∇g_{iU*} is the gradient vector of $g_i(\mathbf{U},\mathbf{P})$with respect to **U** evaluated at (**U** = **U***) given by Eq. (6.8b). Then, the **U** vector for the limit state function $g_i(\mathbf{U},\mathbf{P})$ will be updated by using Eq. (6.10a) as written by,

$$\mathbf{U} = \beta_i \boldsymbol{\alpha}_i \qquad (6.47b)$$

Since the target reliability measure is different for different limit state functions, it is realized from Eq. (6.47b) that the values of the **U** vector are different for different limit state functions. The solution of the inverse reliability problem requires a double iteration procedure as one is for the **U** vector and one is the **P** vector which is shown in Fig. 6.5. This is a straightforward and normal calculation procedure.

Due to double iteration processes this calculation procedure may be somewhat time consuming. However, it proves from some numerical investigations that an accelerated calculation algorithm can be obtained. In this algorithm, the updates of **U** and **P** are performed within a single iteration loop that reduces the computation time. This accelerated calculation algorithm is shown in Fig. 6.6, in which the inner loop (the Newton–Raphson iteration loop) of the straightforward calculation algorithm shown in Fig. 6.5 may be disregarded. At the end of iterations, when $|\Delta U| \approx 0$, the condition of $g_i(\mathbf{U},\mathbf{P}) = 0$ is automatically satisfied. In this way, a considerable shortcut of the execution time is obtained. Since the limit state functions, $g_i(\mathbf{U},\mathbf{P})$, are nonlinear, there may be multiple solutions for **P** vector.Therefore, the solution obtained from the calculation algorithms given in Figs.6.5 and 6.6 must be checked if it is feasible or not. If the solution is not feasible, then another solution must be searched by using a different trial (start) vector **P***. The degree of the nonlinearity with respect to **P** vector can be visualized by the graph of $g_i(\mathbf{U},\mathbf{P}) = 0$for an assumed **U** vector. Only one element of the **P** vector, e.g., p_j, is kept as being variable and the rest are assumed to be known at a time. In this way, a single variable function,

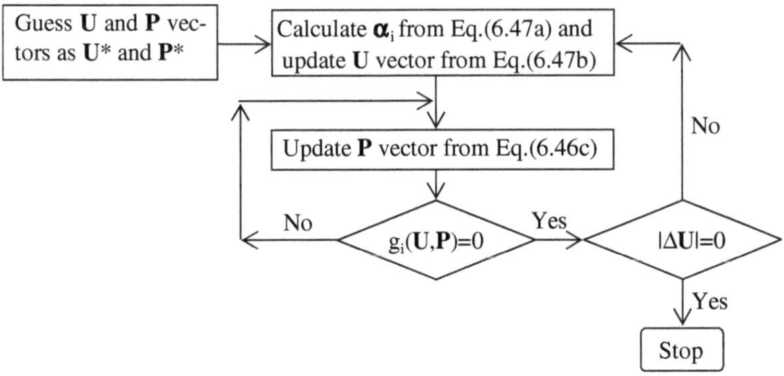

Fig. 6.5 Normal calculation algorithm of the inverse reliability method

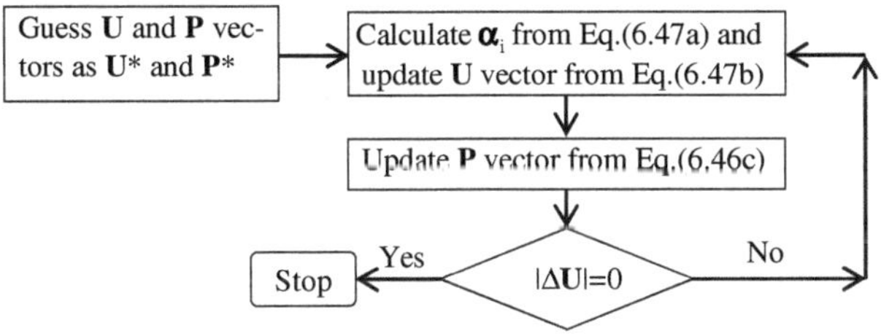

Fig. 6.6 An alternative calculation algorithm of the inverse reliability method

$g_i(p_j) = 0$, is obtained and the conditional graph of $g_i(\mathbf{U}, \mathbf{P}) = 0$ can be plotted easily. It is a conditional graph, because all variables and parameters are assumed to be known except the parameter p_j. By changing j from 1 to k, i.e., ($j = 1, 2,\ldots,k$), the conditional graphs of all elements of the vector **P** can be plotted. By changing the values of the assumed variables and parameters, except p_j, multiple conditional graphs can be obtained. Although it may be computationally time consuming, this process provides more information about the nonlinearity of the limit state functions with respect to the vector **P**, which helps to find a feasible solution for the vector **P**.

6.4 Uncertainties in Spectral Stresses and Fatigue Damages of Offshore Structures

Offshore structures are subjected to many uncertainties [77–82] that should be taken into account in the design process. These uncertainties arise partly from environmental conditions that offshore structures are subjected such as wave, current, wind and earthquakes [77], and arise partly from underlying foundation, structural materials and imperfections, damping phenomenon [83], degradations, and fatigue strength data [84, 85]. Since the environmental conditions (among which waves are the most important phenomena) are random in time, corresponding response behaviors become also random and can be determined by a spectral analysis procedure as it is presented in Chap. 4. By using a spectral analysis procedure the statistical values of the response stresses can only be calculated, and any limit state function [86–88] must be defined in terms of the stress statistical values such as in the case of the fatigue damage formulation presented in Chap. 5. The fatigue damage is one of the most important design criteria for the offshore structures in the long term since the sea waves exist always in the offshore environment with randomly varying sea conditions, smooth and harsh. Since spectral fatigue damages include stress statistical values, in this section, the uncertainties occurring in the fatigue damages are presented firstly. Then, fatigue and stress based reliability calculations will be explained.

Uncertainties in the fatigue damage are generally embedded in response characteristics of the stress process and the fatigue damage model used. Besides commonly accepted uncertainties in offshore structural analysis, which are associated with the modeling of structures and the random wave environment, there are also uncertainties arising from joint flexibility and degradation of structural members that occur during the response. They all appear in the response statistical characteristics. In the cumulative fatigue damage, some other uncertainties arise from the fatigue damage modeling and fatigue data obtained from laboratory experiments as well as hot-spot stress modeling of structural joint configurations. The uncertainties of the cumulative fatigue damage can be classified into two main categories in general as:

Uncertainties in stress statistical characteristics.
Uncertainties in fatigue model parameters.

Most of uncertainties fall into the stress statistical characteristics which are calculated from a spectral analysis. Uncertainties in the fatigue model are considered to be in stress concentration factors (SCF) and parameters of experimentally determined damage model which can be an *S–N* line or a fracture mechanics model. These uncertainties are briefly explained in the following sections.

6.4.1 Uncertainties in Stress Statistical Characteristics

The uncertainties occurring in the stress statistical characteristics, which are zero-crossings period, period of maxima, and the standard deviation of stresses, involve a spectral analysis. They can be further considered in two groups as one group is from structural origin and the other group is from loading and environmental origin. These two group uncertainties are presented below.

6.4.1.1 Uncertainties Arising From Structural Origin

Uncertainties, which affect natural frequencies and structural transfer functions, are considered in this category. These uncertainties are generally encountered in the system stiffness and mass matrices and also in damping ratios, which are discussed below separately.

Uncertainties in Stiffness Matrix

Major uncertainty sources in the stiffness matrix are mostly due to imperfections of member thicknesses, joint flexibility, foundation parameters, and ovalization of member cross-sections during the response in general. Imperfections in member thicknesses occur during the fabrication. Since stiffness properties of tubular members are linear functions of member thicknesses, a single variable X_t, having a mean value of $\mu(X_t) = 1$, represents the inherent uncertainties of all member thicknesses assuming that they are fully correlated. Uncertainties in a joint flexibility are more complicated than the uncertainties in member thicknesses. Members are welded at joints of the structure so that original connections are always rigid as it is assumed in the analysis. During the response of the structure, actual joint behaviors display some flexibility in the vicinity of connections due to local deformations of members [89], plastic deformations under an extreme loading case or a fatigue crack accumulation in the long term. Apart from joint flexibility, ovalization of member cross-sections changes stiffness of member during the response process [90]. This is closely related to flexural deformation of members. In the case of damage, the load carrying capacity of members is lost considerably, but the member can still perform its functionality to some degree until it is fully collapsed. This functionality of a flexible joint can be taken into account in the analysis [91] by using a computation model that always includes uncertainties, which can be represented by a random variable X_r. The stiffness matrix of a damaged member becomes a nonlinear function of the uncertainty variables X_t and X_r. The linearized form of this stiffness matrix can be stated in terms of the uncertainty variables, X_t and X_r, as written by,

$$\mathbf{k} = X_t \boldsymbol{\mu}_{kt} + X_r \boldsymbol{\mu}_{kr} \tag{6.48a}$$

where $\boldsymbol{\mu}_{kt}$ and $\boldsymbol{\mu}_{kr}$ are the mean value matrices related to member thicknesses and joint flexibilities.

Some additional uncertainties in the stiffness matrix are also introduced by the foundation due to the lack of proper determination of soil parameters. These are spring coefficients if a spring system represents the foundation, or subgrade and shear moduli if a structure–foundation interaction is used. In any case, all uncertainties related to the foundation are represented by a single uncertainty parameter, X_g, having a mean value of $\mu(X_g) = 1$. Then, the global stiffness matrix of the structural system can be stated in the uncertainty space as written by,

$$X_K \boldsymbol{\mu}_K = X_t \boldsymbol{\mu}_{Kt} + X_r \boldsymbol{\mu}_{Kr} + X_g \boldsymbol{\mu}_{Kg} \tag{6.48b}$$

where X_K is the uncertainty parameter of the system stiffness matrix with a mean value of $\boldsymbol{\mu}(X_K) = 1$, $\boldsymbol{\mu}_K$ is the mean value of the system stiffness matrix, $\boldsymbol{\mu}_{Kt}$, $\boldsymbol{\mu}_{Kr}$ and $\boldsymbol{\mu}_{Kg}$ are respectively contributions of mean values of the system stiffness matrix due to member thicknesses, joint flexibilities, and the foundation. By this uncertainty modeling of the stiffness matrix, the single variable X_K is assumed to represent the total uncertainties related to the system stiffness matrix. The variance of X_K can be estimated from,

$$X_K = X_t \left(\frac{\mu_{Kt}}{\mu_K} \right) + X_r \left(\frac{\mu_{Kr}}{\mu_K} \right) + X_g \left(\frac{\mu_{Kg}}{\mu_K} \right) \tag{6.48c}$$

where $(\mu_K = \mu_{Kt} + \mu_{Kr} + \mu_{Kg})$ and the scalars μ_{Kt}, μ_{Kr} and μ_{Kg} are the mean value generalized stiffnesses calculated by using natural modes of the structural system.

Uncertainties in Mass Matrix

The uncertainty modeling of the system mass matrix is similar to that of the system stiffness matrix in general. The mass matrix contains uncertainties that arise mainly from added masses due to structure–water interactions and the structural mass. The mass of the deck is also assumed to be uncertain because of its variation during the service lifetime. Member thicknesses and the added mass are the major uncertainty sources in the structural mass. As similar to the stiffness matrix, the mass matrix can also be stated in the uncertainty space as,

$$X_M \boldsymbol{\mu}_M = X_t \boldsymbol{\mu}_{Mt} + X_a \boldsymbol{\mu}_{Ma} + X_d \boldsymbol{\mu}_{Md} + \mathbf{M}_w \tag{6.49a}$$

where X_M, X_t, X_a and X_d are respectively uncertainty variables of the system, member thicknesses, added and the deck mass matrices with unit mean values whereas $\boldsymbol{\mu}_M$, $\boldsymbol{\mu}_{Mt}$, $\boldsymbol{\mu}_{Ma}$, $\boldsymbol{\mu}_{Md}$ are the corresponding mean value matrices and $\mathbf{M}_w$ is the mass matrix due to water inside members, which is assumed to be fully deterministic. The variable, X_t, is defined in the stiffness matrix and the variable X_a is assumed to be fully correlated with the inertia force coefficient of the Morison's equation. The variable, X_M, is assumed to represent total uncertainties in the system mass matrix. Its variance can be estimated from,

$$X_M = X_t \left(\frac{\mu_{Mt}}{\mu_M} \right) + X_a \left(\frac{\mu_{Ma}}{\mu_M} \right) + X_d \left(\frac{\mu_{Md}}{\mu_M} \right) + m_w \tag{6.49b}$$

where $(\mu_M = \mu_{Mt} + \mu_{Ma} + \mu_{Md} + m_w)$ and the scalars μ_{Mt}, μ_{Ma}, μ_{Md} and m_w are the mean value generalized masses calculated by using natural modes of the structural system. Since both the stiffness and mass matrices are functions of the same uncertainty variable, X_t, of the member thicknesses, the representative uncertainty variables, X_K and X_M, will be correlated to some degree. The correlation coefficient can be calculated from Eqs. (6.48c) and (6.49b).

Uncertainties in Damping Ratios

Since the modal analysis technique is used in the spectral analysis of structures, damping ratios are considered to be uncertain. For offshore structures, the damping ratios consist of two parts as being structural and hydrodynamic contributions. Uncertainties in the structural damping ratios are represented by an independent uncertainty variable while uncertainties in the hydrodynamic damping ratios are more complicated. Because of structure–water interactions, they are dependent on the response itself and the linearization constant of the drag force term of the Morison's equation, in general. For the lack of simplicity, uncertainties introduced by the linearization constant of the drag force term are ignored and a mean value linearization constant is used. Hydrodynamic damping ratios depend on the drag force coefficient, c_d, thickness of the marine growth and also the mean value of the linearization constant of the drag force term. It can be shown that the contribution of the uncertainty introduced by marine growths to the hydrodynamic damping is considerably small compared to the uncertainty introduced by the drag force coefficient, c_d. Therefore, the uncertainty introduced by marine growths is disregarded. The hydrodynamic damping ratio for a natural mode, n, can be stated in the uncertainty space [92] as written by,

$$\xi_n = X_\xi \mu(\xi_{sn}) + \frac{X_{cd}}{\sqrt{X_K X_M}} \mu(\xi_{hn}) \tag{6.50}$$

where X_ξ is the uncertainty variable of the structural damping ratios with a mean value of $\mu(X_\xi) = 1$, $\mu(\xi_{sn})$ is the mean structural damping ratio, X_{cd} is the uncertainty variable of c_d with a mean value of $\mu(X_{cd}) = 1$, $\mu(\xi_{hn})$ is the mean value of the hydrodynamic damping ratio for the natural mode, n. The uncertainty variables, X_K and X_M, are defined in Eqs. (6.48c) and (6.49b). Uncertainties in the damping ratios affect the peaks of structural transfer functions so that they may be considerable for dynamic sensitive structure. However, for dynamic insensitive structures, a quasi-static analysis may be carried out to calculate responses, in which case, uncertainties in the damping ratios and natural frequencies do not affect any response value of the structure.

6.4.1.2 Uncertainties Arising from Loading and Environmental Origin

Other group of uncertainties in stress statistical characteristics arises from the loading and environmental origins. These uncertainties are introduced by the wave loading and modeling of random waves. The parameters of wave loads and wave spectra are major uncertainty sources. They are briefly presented below.

Uncertainties in Wave Loading

As it is presented in Chap. 3, the wave load is calculated by using the Morison's equation which is a nonlinear function of the wave velocity. In order to apply a spectral analysis procedure in the response calculation of offshore tubular structures, the nonlinear wave load calculated from the Morison's equation will be linearized. The linearized wave load can be written in the vector form as,

$$\mathbf{q} = A_D C_D \mathbf{w}_n + C_M \dot{\mathbf{u}}_n \tag{6.51}$$

where $\mathbf{w}_n$ is the relative water velocity vector and $\dot{\mathbf{u}}_n$ is the water acceleration vector being normal to the member axis. A_D is the linearization constant which is calculated as explained in Chap. 3, C_D and C_M are respectively the drag and inertia force parameters calculated from,

$$\begin{aligned} C_D &= \rho_w c_d D_h/2 \\ C_M &= \pi \rho_w c_m D_h^2/4 \end{aligned} \tag{6.52a}$$

in which, ρ_w is the water density c_d and c_m are respectively the drag and inertia force coefficients, and D_h is the increased member diameter due to marine growths. The uncertainties in the wave loading arise from the force coefficients, c_d and c_m, which are assumed to be independent, and marine growths. For a submerged member, the increased member diameter due to marine growths can be expressed as written by,

$$D_h = D + 2h \tag{6.52b}$$

where D is the original member diameter, h is the thickness of marine growths, which includes uncertainties. It is assumed that marine growths on all submerged members are fully correlated. The statistical values of the marine growth thickness may be estimated from data of marine growth measurements in the offshore environment where the structure to be built. The uncertainty in the member diameter D_h can be represented by a random variable X_D with a mean value of $\mu(X_D) = 1$ so that it can be written as,

$$D_h = X_D \mu_{Dh} \tag{6.52c}$$

where μ_{Dh} is the mean value of the increased diameter of a submerged member. In the whole system, the drag and inertia force parameters C_D and C_M, can be stated in the uncertainty space as written by,

$$C_D = X_{CD} \mu_{CD} \quad \text{and} \quad C_M = X_{CM} \mu_{CM} \tag{6.53a}$$

where X_{CD} and X_{CM} are random variables representing uncertainties in C_D and C_M while μ_{CD} and μ_{CM} are the values of C_D and C_M for the whole system, which are calculated at the mean values of the drag force coefficient c_d and c_m, and also marine growth thickness h, i.e., $X_{cd} = 1$, $X_{cm} = 1$ and $X_D = 1$. For the whole system, the mean values and variances of X_{CD} and X_{CM} can be calculated from,

$$X_{CD} = (X_{cd}X_D) \quad \text{and} \quad X_{CM} = \left(X_{cm}X_D^2\right) \tag{6.53b}$$

since X_{cd}, X_{cm} and X_D are assumed to be the same for all submerged members.

Uncertainties in the Modeling of Random Waves

Random waves are described by spectral functions of the wave elevation, η, based on observations so that they contain always uncertainties. In a wave-current field, wave spectral functions can be determined in terms of the current velocity and the sea spectrum in a zero-current field, $S_{\eta\eta}(\omega)$, as it is presented in Sect. 3.5. The sea spectrum in a wave-current field has been given by Eq. (3.38a) as it is rewritten below for the convenience.

$$S_{\eta\eta}(\omega_a, U) = \frac{4}{\zeta(1+\zeta)^2} S_{\eta\eta}(\omega_a) \tag{6.54a}$$

where ζ is a dimensionless parameter defined in Eq. (3.36c) as rewritten by,

$$\zeta = \sqrt{1 + 4\omega_a U/g} \tag{6.54b}$$

In Eqs. (6.54a) and (6.54b), ω_a is the angular frequency in a stationary reference frame (absolute frequency), U is current velocity in the wave direction, and g is the gravitational acceleration. Current affects not only the sea spectrum but also transfer functions of water velocity and acceleration. The uncertainty introduced by the current, which is represented by a random current velocity U, is reflected into the analysis via the dimensionless parameter ζ defined in Eq. (6.54b).

In a zero-current field, the Pierson–Moskowitz (PM) and JONSWAP spectral functions describe commonly random waves in a short-term sea state. Uncertainties arising from the modeling of these spectral functions are dependent on representation of the severity of sea states. The PM spectrum is a special case of the JONSWAP spectrum. It is presented in Sect. 3.4.2.2. Its function can be written as,

$$S_{\eta\eta}(\omega) = \frac{A}{\omega^5} \exp\left(-\frac{5}{4}\frac{\omega_p^4}{\omega^4}\right) \gamma^{g(\omega)} \rightarrow g(\omega) = \exp\left[-\frac{1}{2}\left(\frac{\omega - \omega_p}{\omega_p \sigma}\right)^2\right] \tag{6.55a}$$

where A is a constant of the sea state, ω_p is the peak frequency of waves, γ is a peak enhancement factor, and σ is defined as $\sigma = \sigma_a$ if $\omega < \omega_p$, and $\sigma = \sigma_b$ if $\omega > \omega_p$. Since they are obtained from measured data, the parameters γ, σ_a and σ_b contain some uncertainties. Their mean values are given as $\gamma = 3.3$, $\sigma_a = 0.07$, and $\sigma_b = 0.09$. If $\gamma = 1.0$ then the PM sea spectrum is obtained. The constant A in Eq. (6.55a) depends on the modeling of the sea state. If only the significant wave height H_s represents the severity of the sea state, then the constant A is calculated from,

$$A = \alpha_\eta g^2 \tag{6.55b}$$

where g is the gravitational acceleration, α_η is the shape parameter obtained from wave data so that it contains always uncertainties. Therefore, it is used as an uncertainty variable. If both H_s and T_z (zero crossings period of waves) represent together the severity of the sea state, then the constant A and the peak wave frequency ω_p is calculated from Eq. (3.33c) in terms of γ. Uncertainties in A arise from either α_η or γ as depending on the representation of sea states. The peak enhancement function $g(\omega)$ of the JONSWAP spectrum written in Eq. (6.55a) is defined in a narrow frequency region around ω_p. It is assumed that variations in σ_a and σ_b have insignificant effects on response spectra. Uncertainties in these parameters may be included in γ. Therefore, σ_a and σ_b may be considered as deterministic. As a result, in the short-term sea state, α_η and γ represent the inherent uncertainties if the severity of the sea state is described by H_s only. If the severity of the sea state is described by both H_s and T_z, then the inherent uncertainties in the sea spectrum are represented by γ. In the long term, using a wave scatter diagram does not introduce additional uncertainties assuming that the measured wave scatter data are correct. However, if H_s is used only to describe the short-term sea state, then a probability distribution function of H_s represents the long-term sea state. More information about this subject has been given in Sect. 3.6. In this section, it is emphasized that a three parameters Weibull distribution function is generally used for the probabilistic model of the long-term significant wave height as written by Eq. (3.41). The parameters of this Weibull distribution function, A, B and C, contain some uncertainties since they are calculated from wave data which are location dependent. These uncertainties are also taken into account in the reliability analysis.

6.4.2 *Uncertainties in Fatigue Model Parameters*

Uncertainties in the fatigue damage, which are not related to the spectral analysis, are considered in this category. They are encountered mainly in stress concentration factors (SCF) at joints, in the experimentally determined fatigue parameters (parameters of *S–N* lines), in the probability distribution of broad-banded stress ranges and in the reference damage at which the fatigue failure occurs. Using a multilinear *S–N* fatigue model, formulation of the cumulative fatigue damage in offshore tubular structures has been presented by Eqs. (5.30a) and (5.31) in Sect. 5.5 in terms of the mean damage, $E[dD]$, for one stress cycle. By using Eqs. (5.26) and (5.28a) it can be stated as,

$$E[\mathrm{d}D] = \sum_{j=1}^{NL} \frac{1}{C_j} P_j \,(\mathrm{SCF}\,\sigma_S)^{k_j} M_{kj} \tag{6.56}$$

in which NL denotes the total number of segments of the (S–N) fatigue model shown in Fig. 5.4, C_j and k_j are respectively the constant and slope of the segment, j, σ_s is the standard deviation of the nominal normal stress, P_j is the probability that a stress range lies between the bounds of the segment j, which is calculated from Eq. (5.28b), M_{kj} is a factor of non-narrowness of the stress process calculated from Eq. (5.27a). In the non-narrowness factor M_{kj}, there are three parameters (A, B and C) which are determined from numerical data based on Rainflow cycle counting algorithm as presented in Sect. 5.4.1. These parameters contain some uncertainties due to scattering of numerical data. Their mean values and variances are calculated in Sect. 5.4.1 and given in Eqs. (5.22), (5.23a) and (5.23c).

Fatigue data are obtained from laboratory tests. The laboratory condition cannot reflect the actual environmental condition in the sea. The test loading in the laboratory cannot represent the actual loading in the sea. Therefore, the use of experimental data in the fatigue damage calculation of real structures is indicative only and it introduces inevitable uncertainties. Besides, the parameters of the S–N model are calculated from statistical analysis of the experimental data that display a wide range of scatter. This feature of the data leads to deviations from mean lines so that extra uncertainties occur due to data scarcity and scattering. In the case of a multilinear S–N model, it is assumed that two variables, X_{fk} and X_{fC}, represent uncertainties of all segments so that $k_j = X_{fk}\,\mu_{kj}$ and $\ln C_j = X_{fC}\,\mu_{\ln Cj}$ where μ_{kj} and $\mu_{\ln Cj}$ are the mean values of k_j and $\ln C_j$. It is also assumed that these parameters are correlated [93]. In practice, as it is explained in Sect. 5.2.2, stress concentration factors (SCF) are calculated from some empirical formula, and therefore they include uncertainties. Different SCF may be used for different stress components (axial force and bending moments). It is assumed that a single variable X_{SCF} represents uncertainties in all SCF.

The fatigue failure occurs when the cumulative fatigue damage D_{tot} reaches a value D_f. This limit value of the damage is considered as reference damage. It displays a large variation in practice [94], and therefore, it should be treated as an independent random variable. For steel structures a mean value of $\mu_{Df} = 1$ is commonly used.

6.5 Stress Spectrum and Stress Statistical Values in the Uncertainty Space

As it is presented in Sect. 6.5 the total mean fatigue damage, D_{tot}, is a function of stress statistical characteristics, which are calculated from stress spectral moments. In the reliability assessment, using a standard formulation of the spectral analysis may be excessively time consuming and it becomes impractical for large offshore structures since, for each variation of uncertainty variables in the stress, the spectral analysis procedure needs to be carried out. Therefore, an alternative formulation is required to reduce the calculation time. The essence of the alternative formulation is to split the stress spectrum to become a function of uncertainty variables in a way that the spectral analysis is carried out only once. In general, the spectral analysis of offshore structures is presented in Chap. 4 and the matrix of spectra of system displacements is given by Eq. (4.9b) as it is rewritten below for the convenience.

$$[S_{D_{\text{wave}}}(\omega)] = \{H(\omega)\}^{*}_{D\eta}\{H(\omega)\}^{T}_{D\eta}\, S_{\eta\eta}(\omega) \tag{6.57a}$$

in which $\{H(\omega)\}_{D\eta}$ is the vector of transfer functions between the system displacements and the water elevation η, $S_{\eta\eta}(\omega)$ is the sea spectrum (spectrum of water elevation). For offshore structures, using Eq. (1.265), the transfer function vector $\{H(\omega)\}_{D\eta}$ can be stated as written by,

$$\{H(\omega)\}_{D\eta} = [K]^{-1}\{H(\omega)\}_{P\eta} + \sum_{r=1}^{n} h_r(\omega) f_r(\omega)\{\phi\}_r \tag{6.57b}$$

where $[K]$ is the system stiffness matrix, $\{H(\omega)\}_{P\eta}$ is the wave force transfer function, n is the number of eigenmodes to be included in the analysis, $h_r(\omega)$ is the modal transfer function for the eigenmode r, $f_r(\omega)$ is the transfer function of the generalized force, i.e., $f_r(\omega) = \{\phi\}^T_r\{H(\omega)\}_{P\eta}$, and $\{\phi\}_r$ is the corresponding eigenmode vector. The modal transfer function, $h_r(\omega)$, is calculated from Eq. (1.264a) as written by,

$$h_r(\omega) = \frac{1}{k_r}\left[\frac{\omega_r^2}{(\omega_r^2 - \omega^2) + 2i\xi_r\omega_r\omega} - 1\right] \tag{6.57c}$$

where ω_r, k_r and ξ_r are respectively the natural frequency, generalized stiffness, and damping ratio for the eigenmode r. These parameters contain uncertainties. Member stresses are calculated from member internal forces. The transfer functions between member internal forces and the water elevation η can be written in vector notation as,

$$\{h(\omega)\}_{F\eta} = [k]\{h(\omega)\}_{d\eta} + \{h(\omega)\}_{p\eta} \tag{6.58a}$$

where $[k]$ is the member stiffness matrix, $\{h(\omega)\}_{d\eta}$ is the transfer function vector of member displacements which is extracted from Eq. (6.57b) and $\{h(\omega)\}_{p\eta}$ is the transfer function vector of member fixed-end forces under the wave loading. Since transfer functions of member internal forces are calculated as written in Eq. (6.58a) the transfer function of the normal stress σ_x at a location of a tubular member can also be calculated by using the stress–force relation,

$$\sigma_x = \frac{M_y}{I_y} z - \frac{M_z}{I_z} y + \frac{N}{A} \tag{6.58b}$$

in which M_y, M_z are the bending moments about the local coordinates y and z, I_y, I_z are the corresponding inertia moments, y an z are the coordinates of the stress location on the member cross-section, N is the normal force, and A is the member cross-sectional area. By using Eq. (6.58b) the transfer function, which is scalar, between the normal stress σ_x and the water elevation η can be stated as written by,

$$h_{s\eta}(\omega) = \frac{1}{X_t} \left[X_K \left(h_{sq}(\omega) + X_t \sum_{r=1}^{n} h_r(\omega)\, h_{sr}(\omega) \right) + h_{sf}(\omega) \right] \tag{6.58c}$$

where X_t is a random variable representing uncertainties in member thicknesses, X_K is a random variable representing uncertainties in both system and member stiffness matrices, $h_{sq}(\omega)$ is the transfer function of the quasi-static normal stress, $h_r(\omega)$ is given by Eq. (6.57c), $h_{sr}(\omega)$ is modal normal-stress and $h_{sf}(\omega)$ is the normal stress due to member fixed-end forces. Since wave forces are calculated from the Morison's equation in terms of the drag and inertia forces, Eq. (6.58c) can be further written in terms of contributions of these forces as,

$$h_{s\eta}(\omega) = \frac{1}{X_t} \left[C_D\, h_{\text{scd}}(\omega) + C_M\, h_{\text{scm}}(\omega) \right] \tag{6.59a}$$

where $h_{\text{scd}}(\omega)$ and $h_{\text{scm}}(\omega)$ are respectively stress contribution factors of the drag and inertia forces. C_D and C_M are the drag and inertia force parameters defined in Eq. (6.53a) in terms of the uncertainty parameters X_{CD} and X_{CM}. The stress contribution factors $h_{\text{scd}}(\omega)$ and $h_{\text{scm}}(\omega)$ can be calculated by using Eq. (6.58c) as written by,

$$\begin{aligned} h_{\text{scd}}(\omega) &= \left[X_K \left(h_{\text{sqd}}(\omega) + X_t \sum_{r=1}^{n} h_r(\omega)\, h_{\text{sdr}}(\omega) \right) + h_{\text{sfd}}(\omega) \right] \\ h_{\text{scm}}(\omega) &= \left[X_K \left(h_{\text{sqm}}(\omega) + X_t \sum_{r=1}^{n} h_r(\omega)\, h_{\text{smr}}(\omega) \right) + h_{\text{sfm}}(\omega) \right] \end{aligned} \tag{6.59b}$$

where $h_{\text{sqd}}(\omega)$, $h_{\text{sdr}}(\omega)$, $h_{\text{sfd}}(\omega)$ are functions related to drag forces, and $h_{\text{sqm}}(\omega)$, $h_{\text{smr}}(\omega)$, $h_{\text{sfm}}(\omega)$ are functions related to inertia forces only. They are obtained from the explicit manipulation of Eq. (6.58c). Having determined the stress transfer function as given by Eq. (6.59a) the stress spectrum $S_{ss}(\omega)$ is calculated from,

$$S_{ss}(\omega) = \left| h_{s\eta}(\omega) \right|^2 S_{\eta\eta}(\omega) \tag{6.60a}$$

where $|\cdot|$ denotes modulus of a complex function and $S_{\eta\eta}(\omega)$ is the short-term sea spectrum which is explained in Chap. 3. By using Eq. (6.59a) in Eq. (6.60a) the stress spectrum can be stated explicitly in terms of drag and inertia force components as written by,

$$S_{ss}(\omega) = \frac{1}{X_t^2}\left(C_D^2\, S_{cd}(\omega) + C_M^2\, S_{cm}(\omega) + C_D C_M\, S_{dm}(\omega)\right) \tag{6.60b}$$

where $S_{cd}(\omega)$, $S_{cm}(\omega)$ and $S_{dm}(\omega)$ are respectively auto- and cross-spectral functions defined by,

$$\begin{aligned} S_{cd}(\omega) &= |h_{\text{scd}}(\omega)|^2 S_{\eta\eta}(\omega) \\ S_{cm}(\omega) &= |h_{\text{scm}}(\omega)|^2 S_{\eta\eta}(\omega) \\ S_{dm}(\omega) &= \left(h_{\text{scd}}(\omega)\, h_{\text{scm}}^*(\omega) + h_{\text{scm}}(\omega)\, h_{\text{scd}}^*(\omega)\right) S_{\eta\eta}(\omega) \end{aligned} \tag{6.60c}$$

In Eq. (6.60c), the superscript (*) denotes a complex conjugate. It is obvious that these spectral functions contain uncertainties due to X_K, X_t, $h_r(\omega)$ see Eq. (6.59b), and the sea spectrum $S_{\eta\eta}(\omega)$ so that they affect the shape of the stress spectrum. For an approximate calculation, it is assumed that all these uncertainties embedded in stress spectrum are represented by a single uncertainty parameter X_S. With this approximation, the stress spectrum $S_{ss}(\omega)$ can be stated in the uncertainty space as,

$$S_{ss}(\omega) = \frac{X_S}{X_t^2}\left(C_D^2\, \mu_{S_{cd}}(\omega) + C_M^2\, \mu_{S_{cm}}(\omega) + C_D C_M\, \mu_{S_{dm}}(\omega)\right) \tag{6.61}$$

in which C_D and C_M are given by Eq. (6.53a) in the uncertainty space of X_{CD} and X_{CM}, the functions $\mu_{S_{cd}}(\omega)$, $\mu_{S_{cm}}(\omega)$ and $\mu_{S_{dm}}(\omega)$ are the mean values of the spectral functions $S_{cd}(\omega)$, $S_{cm}(\omega)$ and $S_{dm}(\omega)$. Once the stress spectrum is determined the stress spectral moments are calculated in the uncertainty space by using Eq. (6.61) as written by,

$$\begin{aligned} m_n &= \int_0^\infty \omega^n S_{ss}(\omega)\, d\omega \quad \text{or} \\ m_n &= \frac{X_S}{X_t^2}\left(C_D^2\, (m_{cd})_n + C_M^2\, (m_{cm})_n + C_D C_M (m_{dm})_n\right) \end{aligned} \tag{6.62}$$

in which $(m_{cd})_n$, $(m_{cm})_n$ and $(m_{dm})_n$ are spectral moments calculated from the mean value spectral functions $\mu_{S_{cd}}(\omega)$, $\mu_{S_{cm}}(\omega)$ and $\mu_{S_{dm}}(\omega)$ respectively. Then, stress statistical characteristics are calculated from the spectral moments as written by,

$$\sigma_s = \sqrt{m_0}, \quad T_0 = 2\pi\sqrt{\frac{m_0}{m_2}} \quad \text{and} \quad T_m = 2\pi\sqrt{\frac{m_2}{m_4}} \tag{6.63a}$$

where σ_s is the standard deviation, T_0 and T_m are respectively mean periods of zero crossings and maxima of the stress process. The irregularity parameter, α, and the bandwidth parameter, ε, of the stress process are calculated from,

$$\alpha = \frac{T_m}{T_0} \quad \text{and} \quad \varepsilon = \sqrt{1 - \alpha^2} \tag{6.63b}$$

With this uncertainty definition it is clear from Eqs. (6.62), (6.63a) and (6.63b) that T_0, T_m, α, and ε are not functions of the uncertainty parameters X_S and X_t unlike standard deviation σ_s which is also dependent on X_S and X_t. Further it can be seen from Eq. (6.62) that, for only drag or inertia dominant structures, uncertainties in C_D and C_M do not affect the mean periods T_0 and T_m considerably.

6.6 Fatigue and Stress Based Reliability Calculations of Offshore Structures

The methods of the reliability calculation is presented in Sect. 6.2. In this section, its application to the fatigue and stress reliability calculations of offshore structures are explained. In the analysis presented herein, the FORM reliability method is used to calculate reliability indices. The limit state functions of the reliability calculation are based on the total fatigue damage and the mean value of the normal stress-amplitudes. For the fatigue damage, it is assumed that failure occurs when the total fatigue damage D_{tot} reaches a reference damage D_f. In other words, damage occurs when the ratio (D_f/D_{tot}) equals one. Thus, the limit state function for the fatigue reliability calculation can be defined as written by,

$$g_{\text{fat}}(\mathbf{Y}) = \ln D_f - \ln D_{\text{tot}} \tag{6.64}$$

in which $\mathbf{Y}$ denotes the vector of uncertainty variables. This function satisfies the failure condition of $(g_{\text{fat}}(\mathbf{Y}) \le 0)$. As presented in previous sections, D_f and D_{tot} contain uncertainties. For the stress based reliability calculation, the limit state function can be defined on the mean value of stress-amplitudes as written by,

$$g_{\text{str}}(\mathbf{Y}) = s_d - \mu_S/2 \tag{6.65a}$$

in which $\mathbf{Y}$ denotes the vector of uncertainty variables, s_{d} is the threshold (design) stress value and μ_S is the mean value of the stress range. Here, it is assumed that the stress amplitude is half of the stress range. The mean value of the stress range is calculated from the first probability moment of the stress range process. It can be written from Eqs. (5.18a) and (5.18b) as,

$$\mu_S = \sigma_s \left(\sqrt{\frac{2}{\pi}} \varepsilon^3 A + \alpha B \int_0^\infty x^{\frac{1}{C}} \operatorname{erf}\left(\frac{\alpha}{\varepsilon}\sqrt{x}\right) e^{-x} \mathrm{d}x \right) \tag{6.65b}$$

where σ_s is given in Eq. (6.63a), α and ε are given in Eq. (6.63b), A, B and C are the parameters of the probability density function of the stress range process presented in Sect. 5.4.1. The limit state functions $g_{\text{fat}}(\mathbf{Y})$ and $g_{\text{str}}(\mathbf{Y})$ are continuously differentiable so that their gradients can be analytically calculated. For the fatigue limit state function, the gradients are calculated from the statement,

$$\frac{\partial g_{\text{fat}}(\mathbf{Y})}{\partial Y_i} = \frac{1}{D_f}\frac{\partial D_f}{\partial Y_i} - \frac{1}{D_{\text{tot}}}\frac{\partial D_{\text{tot}}}{\partial Y_i} \tag{6.66}$$

In Eq. (6.66), calculation of the first term is straightforward. The second term is more complicated and, usually, a numerical differentiation is applied for its calculation. For the stress limit state function, the gradients can be calculated from Eq. (6.65a) by numerical differentiation. The calculation algorithms of the fatigue reliability for both full uncertainty and reduced uncertainty spaces are shown in Fig. 6.7. As it is seen from Fig. 6.7 that, for the calculation with reduced uncertainties, the spectral analysis are not repeated for the numerical calculation of gradients of the failure function. This makes a considerable short cut in the total calculation time. It may be very useful for a preliminary calculation of the reliability index. If a small reliability index is obtained, then a precise calculation can be carried out by considering full uncertainties to be sure that the structure remains in the safe margin. In the following section, the reliability calculation is demonstrated by a couple of examples.

6.7 Examples

In order to demonstrate the reliability calculation a couple of examples are presented in this section. The first example is a tubular steel member subjected to a compression axial force and a bending moment. The second example is a jacket offshore structure presented as shown by Fig. 5.9.

6.7.1 Reliability Calculation of a Tubular Steel Member

A tubular steel member like a jacket leg is assumed to be subjected to a compressive axial force and a bending moment. The ultimate load carrying capacity of the member is calculated by forming a full plastic hinge. In this calculation, the effects of the shear force and torsion are neglected. For the ultimate load capacity, the cross-section of the member is fully yielded as shown in Fig. 6.8.

In this figure, the notations are defined as,

N_{m} Axial force couples which produces bending moment M
N Compressive axial force
x_g Gravity center of the cross-section above x (application point of the force N_{m})
σ_y Yield stress of the steel tube
h Wall thickness of the steel tube
R Radius of the steel tube

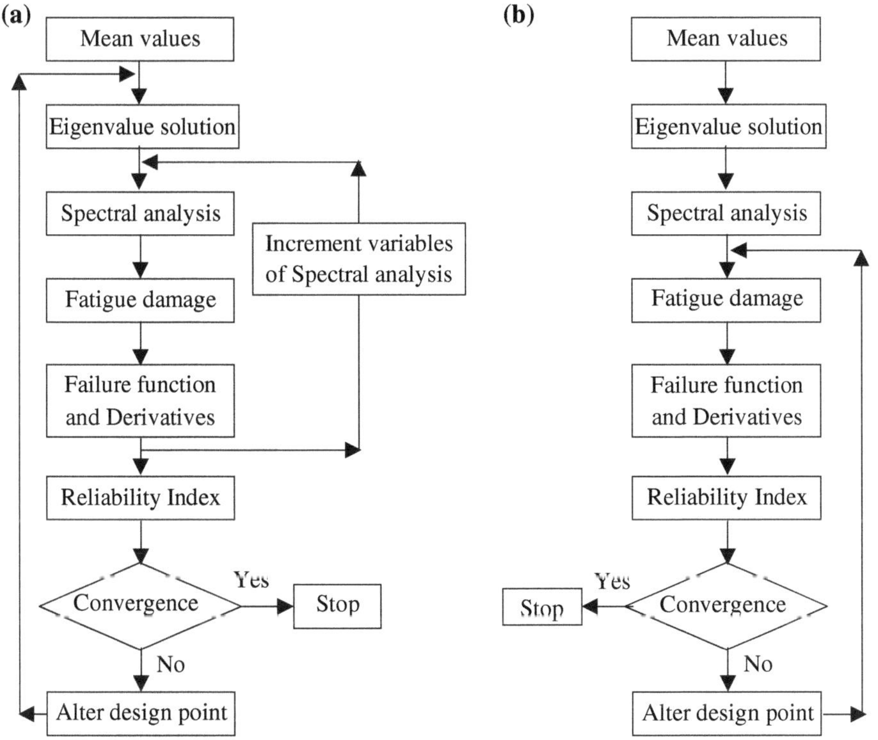

Fig. 6.7 Calculation flow diagrams of the fatigue reliability analysis of offshore structures. **a** For full uncertainties. **b** For reduced uncertainties

Fig. 6.8 A tubular steel member under compressive axial force and bending moment

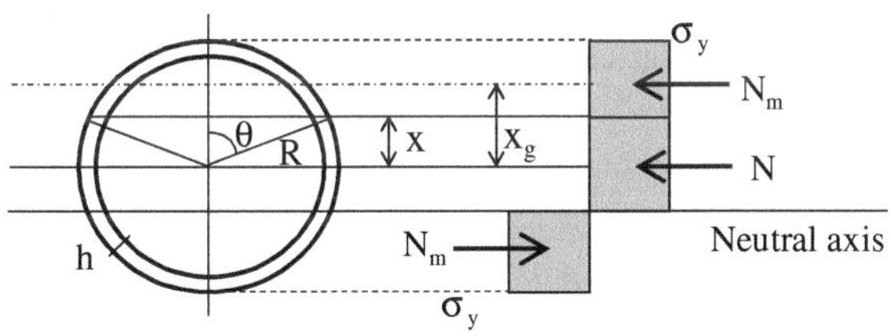

The cross-sectional areas of the axial force couple, N_m, and the axial force N are calculated as written by,

$$\begin{aligned}&\text{for } N_m \rightarrow A_m = 2\,R\,h\,\theta\\&\text{for } N \rightarrow A_n = 2Rh(\pi - 2\,\theta)\end{aligned} \tag{6.67}$$

The axial force N and plastic axial force N_p of the whole tube are calculated from,

$$\left.\begin{array}{l} N = A_n\sigma_y \rightarrow N = 2Rh\,\sigma_y(\pi - 2\theta) \\ N_p = A\sigma_y \rightarrow N_p = 2\,\pi R\,h\,\sigma_y \end{array}\right\} \rightarrow \frac{N}{N_p} = \left(1 - \frac{2\theta}{\pi}\right) \tag{6.68a}$$

The application point x_g of the force couple N_{m}, the bending moment M and the plastic bending moment M_p are calculated as written by,

$$\int_{-\theta}^{\theta} x\,dA = x_g A_m \rightarrow x_g = \frac{R\,\sin\theta}{\theta}$$

$$\left.\begin{array}{l} M = 2N_m x_g = 2\,A_m\sigma_y\,x_g \;\rightarrow\; M = 4\,R^2\,h\,\sigma_y\sin\theta \\ \text{for } \left(\theta = \frac{\pi}{2}\right),\; M = M_p \rightarrow M_p = 4\,R^2\,h\,\sigma_y \end{array}\right\} \rightarrow \frac{M}{M_p} = \sin\theta \tag{6.68b}$$

By using Eqs. (6.68a) and (6.68b) the yield surface of the tubular steel member can be formulated as written by,

$$\frac{M}{M_p} = \sin\left[\frac{\pi}{2}\left(1 - \frac{N}{N_p}\right)\right] \rightarrow \frac{M}{M_p} = \cos\left(\frac{\pi}{2}\frac{N}{N_p}\right) \tag{6.69}$$

which states that, if $(M/M_p \geq \cos(\pi N/2N_p))$, then failure occurs due to exceedance of the load carrying capacity of the section. From this condition, the failure function of the reliability analysis is defined as,

$$g(Y) = \cos\left(\frac{\pi}{2}\frac{N}{N_p}\right) - \frac{M}{M_p} \rightarrow \begin{cases} g(Y) = 0 \rightarrow \text{Failure surface} \\ g(Y) < 0 \rightarrow \text{Failure region} \\ g(Y) > 0 \rightarrow \text{Safe region} \end{cases} \tag{6.70a}$$

Having introduced N_p and M_p from Eqs. (6.68a) and (6.68b) into Eq. (6.70a) the failure function can be obtained as written by,

$$g(Y) = \cos\left(\frac{1}{4R}\frac{N}{h\sigma_y}\right) - \frac{1}{4R^2}\frac{M}{h\sigma_y} \tag{6.70b}$$

Eq. (6.70b) is used to calculate the reliability index of the tubular member for exceeding load carrying capacity. In this failure function, it is assumed that the axial force N, bending moment M, wall thickness h of the tube and the yield stress σ_y are uncertain. The radius R is considered as deterministic. Two reliability calculations are carried out as,

- All uncertainty variables (Y_1, Y_2, Y_3, Y_4) are assumed to be independent and normally distributed.
- The axial force N and bending moment M (Y_1 and Y_2) are assumed to be correlated and Normally distributed. Other variables are independent.

Deterministic and design variables of the reliability calculation with their statistical data are presented in Table 6.1. For the first calculation it is assumed that

Table 6.1 Design variables and deterministic data for the reliability calculation of the tubular steel member with respect to yielding load capacity

Designation	Design variables	Mean value	COV	Standard deviation	Probabiliy distribution
Axial force (*N*)	Y_1	2000 kN	0.20	400 kN	Normal
Bending moment (M)	Y_2	250 kNm	0.25	62.5 kNm	Normal
Wall thickness (h)	Y_3	0.02 m	0.10	0.002 m	Normal
Yield stress (σ_y)	Y_4	40000 kPa	0.10	4000 kPa	Normal
Radius (R)	–	0.80 m	–	–	Deterministic

all design variables are independent. The FORM methods are used to calculate the reliability index β. The failure function in the original design variables space (Y_1, Y_2, Y_3, Y_4) is written as,

$$g(Y) = \cos\left(\frac{1}{4R}\frac{Y_1}{Y_3Y_4}\right) - \frac{1}{4R^2}\frac{Y_2}{Y_3Y_4} \tag{6.71}$$

The gradient vector with respect to the original design variables can be calculated from Eq. (6.71) as written by,

$$\nabla g_Y = \begin{Bmatrix} -\dfrac{1}{4R}\dfrac{1}{Y_3Y_4}\sin\left(\dfrac{1}{4R}\dfrac{Y_1}{Y_3Y_4}\right) \\ -\dfrac{1}{4R^2}\dfrac{1}{Y_3Y_4} \\ \dfrac{1}{4R}\dfrac{1}{Y_3^2Y_4}\left[Y_1\sin\left(\dfrac{1}{4R}\dfrac{Y_1}{Y_3Y_4}\right)+\dfrac{Y_2}{R}\right] \\ \dfrac{1}{4R}\dfrac{1}{Y_3Y_4^2}\left[Y_1\sin\left(\dfrac{1}{4R}\dfrac{Y_1}{Y_3Y_4}\right)+\dfrac{Y_2}{R}\right] \end{Bmatrix} \tag{6.72}$$

The gradient vector with respect to the Standard Normal variables space **U** can be calculated from Eq. (6.12). The reliability index β is calculated by using the algorithm given in Fig. 6.3. For a given Standard Normal variables vector **U**, the original variables vector **Y** is calculated by using Eq. (6.7), and then introducing it to Eq. (6.71) the failure function in the **U** space can be calculated. The results of the 4th reliability iteration are presented in Table 6.2.

In the second calculation, it is assumed that the design variables Y_1 and Y_2 (axial force and bending moment) are correlated with the correlation coefficient of $\rho_{12} = 0.40$ and their probability distributions are Normal. Thus, the correlation coefficient matrix $\boldsymbol{\rho}_Y$ is written as,

$$\boldsymbol{\rho}_Y = \begin{bmatrix} 1 & 0.4 & 0 & 0 \\ 0.4 & 1 & 0 & 0 \\ 0 & 0 & 1 & 0 \\ 0 & 0 & 0 & 1 \end{bmatrix} \tag{6.73a}$$

The correlation of the axial force N and bending moment M to some degree is realistic since both N and M are subject to the same loading uncertainties. Therefore, a correlation between these variables is assumed. In this case, for the calculation of the transformation from the correlated normalized Normal variables to the Standard Normal variables (from **X** to **U**), the Choleski Triangulation Method is used. The transformation matrix **T** is calculated from Eq. (6.15b) and the transformation matrix **D** from **U** to **Y** (original design variables vector) vectors is calculated from Eq. (6.16a). These transformation matrices are written as,

$$\mathbf{T} = \begin{bmatrix} 1 & 0 & 0 & 0 \\ 0.4 & 0.91652 & 0 & 0 \\ 0 & 0 & 1 & 0 \\ 0 & 0 & 0 & 1 \end{bmatrix}, \quad \mathbf{D} = \begin{bmatrix} 400.0 & 0 & 0 & 0 \\ 25.0 & 57.2825 & 0 & 0 \\ 0 & 0 & 0.002 & 0 \\ 0 & 0 & 0 & 4000 \end{bmatrix} \tag{6.73b}$$

By using the transformation matrix **D** the reliability index β can be calculated as explained above in the case of the calculation of β for independent design variables. The results of the 4th reliability iteration are presented in Table 6.3. As it is seen from the comparison of Tables 6.2 and 6.3 that the reliability index β for correlated variables is somewhat reduced.

6.7.2 *Fatigue Reliability Calculation of an Example Offshore Jacket Structure*

In this section, the fatigue reliability calculation of offshore structures, which is explained above in Sect. 6.6, is demonstrated. For this purpose, an example jacket type structure, which is given in Sect. 5.6 and shown in Fig. 5.9, is used. The structural data are shown in Table 5.2. For the assumed data presented in Sect. 5.6, the result of the eigenvalue calculation, which is carried out by the SAPOS program [95], is presented in Table 5.4. For the fatigue damage calculation, API Design Basic *S–N* curve shown in Fig. 5.11 is used. In this calculation, it is assumed that the marine growth thickness is same for all submerged members with a mean value of $\mu_h = 0.20$ m. The stress concentration factors (SCF) for all stress components due to bending moments and axial force are also same with a mean value of $\mu_{\mathrm{SCF}} = 2.0$. For the fatigue reliability calculation, the lifetime of the structure is assumed as $T_{\mathrm{life}} = 25$ years. The short-term sea state is represented by the Pierson–Moskowitz (PM) sea spectrum with the significant wave height H_s. In the long term, the Weibull distribution given by Eq. (3.41) represents the probability distribution of H_s. The parameters, A, B and C, of this probability distribution are assumed uncertain with mean values of $\mu_A = 0.60$, $\mu_B = 1.67$ and $\mu_C = 1.20$. The mean values of the drag and inertia force coefficients are assumed as $\mu_{cd} = 1.3$ and $\mu_{cm} = 1.6$. The mean value of the reference damage at which fatigue failure occurs is assumed as $\mu_{Df} = 1.0$. Design variables, designations, and

Table 6.2 Results of the 4th reliability iteration of the tubular steel member under yielding load capacity. Design variables are independent and $\boldsymbol{\alpha}$ is the unit direction vector

Vectors	1	2	3	4	g(U) and β
U	1.612244	0.319437	−1.385563	−1.385563	
Δg_U	−0.207208	−0.041124	0.179668	0.179668	g(U) = 0.00003
$\boldsymbol{\alpha}$	0.627075	0.124454	−0.543732	−0.543732	$\beta = 2.55742$
Y	2644.8975	269.9648	0.017229	34457.7466	

Table 6.3 Results of the 4th reliability iteration of the tubular steel member under yielding load capacity. Axial force N and bending moment M are correlated with $\rho_{12} = 0.40$. Other design variables are independent and $\boldsymbol{\alpha}$ is the unit direction vector

Vectors	1	2	3	4	g(U) and β
U	1.629965	0.275298	−1.308351	−1.308351	
Δg_U	−0.218887	−0.037023	0.177434	0.177434	g(U) = 0.00008
$\boldsymbol{\alpha}$	0.653326	0.110506	−0.529600	−0.529600	$\beta = 2.48135$
Y	2651.9858	306.5184	0.017383	34766.5957	

the statistical data that were used in the reliability calculation are shown in Table 6.4. The reliability calculation is carried by the SAPOS program using the reduced uncertainty variables as explained in Sect. 6.6. The FORM reliability methods are used to calculate the reliability indices at different stress locations on cross sections of the member 5 at elevations of (−30 m) and (−50 m) shown in Fig. 5.9. It is observed that a maximum of four reliability iterations produces accurate results. The results of the reliability indices after four iterations and the corresponding failure functions are presented in Table 6.5. The stress locations θ are in degrees and measured from the local Y_{local} axis shown in Fig. 5.9.

Exercise

A monopod tower is described in Sect. 4.6 for the exercise 1 as shown in Fig. 4.11. By using the data used for the exercise 1 in Chap. 4 the fatigue reliability calculation of the monopod tower will be carried out. For this calculation the narrow-banded stress process will be used. The natural frequency of the tower is assumed to be deterministic and the damping ratio to the critical is assumed uncertain with $\mu_\xi = 0.015$ and COV = 0.15. The maximum dynamic stress is calculated by multiplying the maximum static stress with the dynamic amplification factor (DAF) which is given as,

$$\text{DAF} = \frac{\omega_n^2}{\sqrt{\left(\omega_n^2 - \omega^2\right)^2 + 4\xi^2\omega^2\omega_n^2}} \tag{6.74}$$

where ω is the frequency, ω_n denotes the natural frequency and ξ is the damping ratio. The probability distribution of the significant wave height H_s in the long term is Weibull distribution given by Eq. (3.41) with the mean value of parameters,

Table 6.4 Input of design variables for the fatigue reliability calculation of the example jacket structure. Design variables, probability distributions, means and coefficient of variations (COV)

Designation		Design Variable	Type	Mean	COV
Stress concentration factors (SCF)		Y_1	Log Normal	1.00	0.1000
Stress spectrum		Y_2	Log Normal	1.00	0.1500
Correlated fatigue parameters	lnC	Y_3	Normal	1.00	0.0240
(*S-N*) lines	k	Y_4	Normal	1.00	0.0260
Spectral bandwidth factor M_{kj}	A	Y_5	Normal	1.00	0.0376
of the stress process, see	B	Y_6	Normal	1.00	0.0326
Eq. (5.27a)	C	Y_7	Normal	1.00	0.0625
Coefficients in the long term	A	Y_8	Normal	1.00	0.1000
probability distribution of	B	Y_9	Normal	1.00	0.1000
H_s (Weibull distribution given by Eq. (3.41)	C	Y_{10}	Normal	1.00	0.1000
Reference damage, D_f		Y_{11}	Log Normal	1.00	0.2000
Drag force coefficient, c_d		Y_{12}	Log Normal	1.00	0.2000
Inertia force coefficient, c_m		Y_{13}	Log Normal	1.00	0.2000
Thickness of marine growth		Y_{14}	Log Normal	1.00	0.1000
Member thicknesses		Y_{15}	Log Normal	1.00	0.0500

Mean value of marine growth thickness : μ_h=0.20m, Service lifetime :T_{life} =25 years
Correlation coefficient of fatigue parameters : $\rho_{lnC,k}$ =0.96

Table 6.5 Results of the reliability calculation of the example jacket structure. Reliability indices β and failure functions

Stress location θ^c	Member 5, Elevation :-50 m		Member 5, Elevation :-30 m	
	β	Failure function	β	Failure function
0.0	1.8596	0.000084	1.9294	0.000086
15.0	1.9917	0.000084	1.7627	0.000086
30.0	2.1112	0.000084	1.6040	0.000086
45.0	2.2063	0.000086	1.4641	0.000085
60.0	2.2743	0.000086	1.3516	0.000086
90.0	2.3226	−0.000002	1.2334	−0.000008

$\mu_A = 0.60$, $\mu_B = 1.67$ and $\mu_C = 1.20$, and coefficient of variations are assumed as COV = 0.10 for all parameters. The marine growth thickness is not taken into account in the calculation. The stress concentration factor is assumed as $\mu_{SCF} = 1.5$ and COV = 0.10. Since the diameter of the tower is large, only the inertia force term is considered in response calculation with the values of $\mu_{cm} = 2.0$ and COV = 0.20 assuming that it is not a function of the frequency, ω. Thus, the frequency reduction function of the inertia force term is equal to 1. The wall thickness of the tower is also assumed as uncertain with the values of $\mu_t = 0.08$ m and COV = 0.05. A single *S–N* fatigue model, which is shown in Fig. 5.13b, is used in damage calculation with the values of $\mu_{\text{lnC}} = 86.73$, COV(lnC) = 0.0024, $\mu_k = 3.8$, COV(k) = 0.0026 and the correlation coefficient of $\rho_{\text{lnC},k} = 0.96$. The reference damage is assumed as $\mu_{Df} = 1.0$ and COV $= 0.20$. For reliability calculation, the lifetime of the tower is assumed to be $T_{\text{life}} = 25$ years and the FORM reliability methods are used to calculate the reliability index, β.

References

1. Melchers RE (1999) Structural reliability analysis and prediction, 2nd edn. Wiley, New York
2. Ditlevsen O, Madsen HO (1996) Structural reliability methods. Wiley, Chichester
3. Chabrolin B (2001) Partial safety factors for resistance of steel elements to EC3& EC4, calibration for various steel products and failure Criteria. Final Report, CTICM, Saint-Aubin
4. Burdekin FM (2007) General principles of the use of safety factors in design and assessment. Eng Fail Anal 14:420–433
5. Sedlacek G, Kraus O (2007) Use of safety factors for the design of steel structures according to the Eurocodes. Eng Fail Anal 14:434–441
6. Jonas Clausen J, Hansson SO, Nilsson F (2006) Generalizing the safety factor approach. Reliab Eng Syst Saf 91:964–973
7. Darlaston J, Wintle J (2007) Safety factors in the design and use of pressure equipment. Eng Fail Anal 14:471–480
8. Stacey A, Sharp JV (2007) Safety factor requirements for the offshore industry. Eng Fail Anal 14:442–458
9. Shama MA (2009) Basic concept of the factor of safety in marine structures. Ships Offshore Struct 4(4):307–314
10. Tarp-Johansen NJ (2005) Partial safety factors and characteristic values for combined extreme wind and wave load effects. Trans ASME 127:242–252
11. Wilson R (2007) A comparison of the simplified probabilistic method in R6 with the partial safety factor approach. Eng Fail Anal 14:450–489
12. Kala Z (2007) Influence of partial safety factors on design reliability of steel structures—probability and fuzzy assessment. J Civ Eng Manag 8(4):291–296
13. Wang SS, Hong HP (2004) Partial safety factors for designing and assessing flexible pavement performance. Can J Civ Eng 31:397–406
14. Mohamed A, Soares R, Venturini WS (2001) Partial safety factors for homogeneous reliability of nonlinear reinforced concrete columns. Struct Saf 23:137–156
15. Roos E, Wackenhut G, Lammert R, Schuler X (2011) Probabilistic safety assessment of components. Int J Press Vessels Pip 88:19–25
16. Val DV, Stewart MG (2002) Safety factors for assessment of existing structures. J Struct Eng ASCE 128(2):258–265
17. Mrazik A, Krizma M (1997) Probability-based design standards of structures. Struct Saf 19(2):219–234
18. Muhammed A (2007) Background to the derivation of partial safety factors for BS 7910 and API 579. Eng Fail Anal 14:481–488
19. Goh ATC, Phoon KK, Kulhawy FH (2009) Reliability analysis of partial safety factor design method for cantilever retaining walls in granular soils. J Geotech Geoenviron Eng 135(5) ASCE, 616–622
20. Nikolaidis E, Ghiocel DM, Singhal S (2005) Engineering design reliability handbook. CRC Press LLC, Boca Raton
21. ISO 2394 (1998) General principles on reliability for structures. ISO, Geneva
22. EN, 1990 Eurocode (2002) Basis of structural design. CEN, Brussels
23. JCSS (2001) Probabilistic model code. Joint Committee on Structure Safety
24. Sadovsky Z, Pales D (2008) Probabilistic optimization of partial safety factors for the design of industrial buildinghs. Int J Reliab Qual Saf Eng 15(5):411–424

25. Faber MH, Sorensen JD (2002) Reliability based code calibration. JCSS Workshop on Reliability Based Code Calibration, Zurich
26. Hansen PF, Sorensen JD (2002) Reliability-based code calibration of partial safety factors. JCSS Workshop on Reliability Based Code Calibration, Zurich
27. Swiler LP, Paez TM, Mayes RL (2009) Epistemic uncertainty quantification tutorial. In: Proceeding of the IMAC-XXVII, Orlando
28. Swiler LP, Giunta A (2007 Aleatory and epistemic uncertainty quantification for engineering applications. Technical Report, SAND2007-2670C, Sandia National Laboratories, Albuquerque
29. Madsen HO, Krenk S, Lind NC (1986) Methods of structural safety. Prentice-Hall, Englewood Cliffs
30. Thoft-Christensen P, Murotsu Y (1986) Application of structural system reliability theory. Springer, Berlin
31. AH-S, Tang WH (1984) Probability concepts in engineering planning and design. Volume II—decision, risk and reliability. Wiley, New York
32. Thoft-Christensen P, Baker MJ (1982) Structural reliability theory and its applications. Springer, Berlin
33. Ditlevsen O (1981) Uncertainty modeling with application to multidimensional civil engineering system. McGraw-Hill, New York
34. Thoft-Christensen P (1987) Recent advances in the application of structural system reliability methods. In: Proceeding 5th International Conference on Applications of Statistics and Probability in Soil and Structural Engineering, ICASP5, University of British Columbia, Vancouver
35. Karamchandani A (1987) Structural system reliability analysis methods. Report No.83, The John A. Blume Earthquake Engineering Center, Stanford
36. Shinozuka M (1983) Basic analysis of structural safety. J Struct Eng, ASCE, 109(3):721–740
37. Karadeniz H, Vrouwenvelder T (2006) Overview reliability methods. report: SAF-R5-1-TUD-01(10). Task 5.1.SAFERELNET
38. Vrouwenvelder T, Karadeniz H (2011) Overview of structural reliability methods. safety and reliability of industrial products, systems and structures, Edited by Soares CG, CRC Press, Boca Raton
39. Sun HH, Bai Y (2003) Time-variant reliability assessment of FPSO hull girders. Mar struct 16:219–253
40. Cornell CA (1969) A probability based structural code. J Am Concr Inst 66(12):974–985
41. Hasofer AM, Lind LC (1974) An exact and invariant first-order reliability format. J Eng Mech ASCE 100:111–121
42. Rackwitz R, Fiessler B (1978) Structural reliability under combined random load sequences. Comput Struct 9:489–494
43. Rosenblatt M (1952) Remarks on a multivariate transformation. Ann Math Stat 23:470–472
44. Nataf A (1962) Determination des distribution dont les marges sont données. C.R. Acad Sci 225:42–43
45. Der Kiureghian A, Liu PL (1986) Structural reliability under incomplete probability information. J. Eng. Mech., ASCE, 112(1):85–104
46. Liu PL, Der Kiureghian A (1986) Multivariate distribution models with prescribed marginals and covariances. Probab Eng Mech 1(2):105–112
47. Fiessler B, Neumann H-J, Rackwitz R (1979) Quadratic limit states in structural reliability. J Eng Mech, ASCE, 105(4):661–676
48. Breitung, K (1984) Asymptotic approximations for multi-normal integrals. J. Eng. Mech., ASCE, 110(3):357-366
49. Der Kiureghian A, Lin, H-Z, Hwang S-J (1987) Second-order reliability approximations. J Eng Mech, ASCE, 113(8):1208–1225
50. Tvedt, L (1990) Distribution of quadratic forms in normal space—Approximation to structural reliability. J Eng Mech, ASCE, 116(6):1183–1197

51. Rackwitz R (2001) Reliability analysis-a review and some perspectives. Struct Saf 23:365–395
52. Koyluoglu HU, Nielsen SRK (1994) New approximations for SORM integrals. Struct Saf 13:235–246
53. Cai GQ, Elishakoff I (1994) Refined second-order reliability analysis. Struct Saf 14:267–276
54. Zhao Y-G, Ono T (1999) A general procedure for first/second-order reliability method (FORM/SORM). Struct Saf 21:95–112
55. Zhao Y-G, Ono T (1999) New approximation for SORM: Part 1. J Eng Mech, ASCE, 125(1):79–85
56. Adhikari S (2005) Asymptotic distribution method for structural reliability analysis in high dimensions. Proc. R. Soc. A 461, 3141–3158
57. Golub GH, Van Loan CF (1996) Matrix computations. 3rd (edn) Johns Hopkins, Maryland
58. Ouypornprasert W (1988) Adaptive numerical integration for reliability analysis. Inst Eng Mech, Univ Innsbruck, Int. Rep. No. 12-87, Innsbruck
59. Waarts P (2000) Structural reliability using finite element analysis, Ph.D. thesis, Delft University Press, Delft
60. Hines WW, Montgomery DC (1980) Probability and statistics in engineering and management science. Wiley, New York
61. Ayyub BM, Haldar A. (1985) Improved simulation techniques as structural reliability models. 4th International Conference on Structure Saf. and Rel.,ICOSSAR'85, Vol. 1, 17–26
62. Karamchandani A (1987) Structural system reliability analysis methods. Report No.83, John A. Blume Earthquake Eng. Center, Stanford University, Stanford
63. Bucher CG (1988) Adaptive sampling: an iterative fast monte carlo procedure. Struct Saf 5(2):119–126
64. Bjerager P (1988) Probability integration by directional simulation. J Eng Mech, ASCE, 114 (8):1285–1302
65. Bjerager P (1989) Probability computation methods in structural and mechanical reliability. Computational Mechanics of Probabilistic and Reliability Analysis, eds. by Liu WK, Belytschko T, Elme Press Int., Lausanne
66. Hohenbichler M, Rackwitz R (1988) Improvement of second-order reliability estimates by importance sampling. J Eng Mech, ASCE, 114(12): 2195–2199
67. Der Kiureghian A, Zhang Y, Li CC (1994) Inverse reliability problem. J Eng Mech, ASCE, 120(5):1154–1159
68. Li H, Foschi RO (1998) An inverse reliability measure and its application. Struct Saf 20:257–270
69. Sadovsky Z (2000) Discussion on: An inverse reliability method and its application. Struct Saf 22:97–102
70. Li H, Foschi RO (200) Response: an inverse reliability method and its application. Struct Saf, 22:103–106
71. Nakamura Y, Nakamura T (2000) Inverse reliability-based design of shear buildings supported by springs with stochastic stiffnesses. Probab Eng Mech 15:295–303
72. Ramu P, Qu X, Youn BD, Haftka RT, Choi KK (2004) Safety factor and inverse reliability measures. 45th AIAA/ASME/ASCE/AHS/ASC Struct., Struct Dyn & Mater Conf, 1–11
73. Du X, Sudjianto A, Chen W (2004) An integrated framework for optimization under uncertainty using inverse reliability strategy. Trans ASME 126:562–570
74. Minguez R, Castillo E, Hadi AS (2005) Solving the inverse reliability problem using decomposition techniques. Struct Saf 27:1–23
75. Saha S, Manohar CS (2005) Inverse reliability based structural design for system dependent critical earthquake loads. Probab Eng Mech 20:19–31
76. Cheng J, Zhang J, Cai CS, Xiao R-C (2007) A new approach for solving inverse reliability problems with implicit response functions. Eng Struct 29:71–79
77. Bitner-Gregersen EM, Hagen O (1990) Uncertainties in data fort he offshore environment. Struct Saf 7:11–34
78. Olufsen A, Bea RG (1990) Loading uncertainties in extreme waves. Mar struct 3:237–260

79. Olufsen A, Leira BJ, Moan T (1992) uncertainty and reliability analysis of jacket platform. J Struct Eng 118(10):2699–2715
80. Golafshani AA, Ebrahimian H, Bagheri V, Holmas T (2011) Assessment of offshore platforms under extreme waves by probabilistic incremental wave analysis. J Constr Steel Res 67:759–769
81. Nikolaidis E, Kaplan P (1991) Uncertainties in stress analyses on marine structures. Report No. SSC-363, Ship Structure Committee, Washington
82. Karadeniz H (2001) Uncertainty modeling in the fatigue reliability calculation of offshore structures. Reliab Eng Syst Saf 74:323–335
83. Kareem A, Gurley K (1996) Damping in structures: its evaluation and treatment of uncertainty. J Wind Eng Ind Aerodyn 59:131–157
84. Pittaluga A, Cazzulo R, Romeo P (1991) Uncertainties in the fatigue design of offshore steel structures. Mar struct 4:317–332
85. Pillaia TMM, Prasad AM (2000) Fatigue reliability analysis in time domain for inspection strategy of fixed offshore structures. Ocean Eng 27:167–186
86. Skjong R, Gregersen EB, Cramer E, Croker A, Hagen O, Korneliussen G, Lacasse S, Lotsberg I, Nadim F, Ronold KO (1995) Guideline for offshore structural reliability analysis-general. DNV Report No.95-2018
87. Sigurdsson G, Cramer E, Lotsberg I, Berge B (1996) Guideline for offshore structural reliability: application to jacket platforms. DNV Report No. 95-3203, Hovik
88. Sigurdsson G, Cramer E (1996) Guideline for offshore structural reliability-examples for jacket platforms. DNV Report No. 95-3204, Hovik
89. Barltrop NDP, Adams AJ (1991) Dynamics of fixed marine structures, 3rd edn. Butterworth-Heinemann, Oxford
90. Karadeniz H (2001) A method for including ovalization effects of tubular member on cross-section properties. Proceeding 11th International Offshore and Polar Engineering Conference, ISOPE, 4:426–432
91. Karadeniz H (1994) An algorithm for member releases and partly connected members in offshore structural analysis. Proc. 13th.international Conference Offshore Mechanical and Arctic Engineering OMAE, 1, 471–476, Houston
92. Karadeniz H (1992) Stochastic analysis of offshore structures under wave-current and fluid-structure interactions. In:Proceeding 11th Internaztional Conference Offshore Mechanical and Arctic Engineering OMAE, 1-A, 241–248, Calgary
93. Karadeniz H, Vrouwenvelder A, Bouma AL (1983) Stochastic fatigue reliability analysis of jacked type offshore structures. In: Proceeding NATO Advanced Study Institute on Reliability Theory and Its Application in Structural and Soil Mechanics, Edited by Thoft-Christensen P, Martinus Nijhoff, The Hague
94. Schutz W (1981) Procedures for the prediction of fatigue life of tubular joints. In: Proceeding International Conference on Steel in Marine Structures, 254–308, Paris
95. Karadeniz H (2009) SAPOS, spectral analysis program of structures. Report, Structure Mechanical Div., Faculty of Civ. Eng. and Geosci., TUDelft, Delft, the Netherlands

Chapter 7
Optimization of Offshore Structures

7.1 Introduction

Structural optimization is relatively new branch of structural engineering compared to structural analysis and structural mechanics. This division of structural engineering is developed by applying the optimization techniques to structural design problems. Optimization which is a branch of applied mathematics, computational mathematics, and operations research deals in finding solution of problems where it is necessary to maximize or minimize a real function within a domain which contains the acceptable values of variables while some restrictions are to be satisfied. The domain naturally holds real or integer values for the variables. The set of variables that maximizes or minimizes the real function while satisfying the described restrictions are called optimum solution of the problem. This solution is the best solution among the large amount of acceptable solutions that satisfy constrains. The function that is to be required to be maximized or minimized is called objective function and the restriction functions that are to be satisfied in the solution are called constraints. Variables in an optimization problem are parameters that describe a particular entity. Since optimization problems are originated for finding solutions to decision making problems, the variables in optimization problems are called decision variables. Financial decisions of a bank or insurance company can be formulated as an optimization problem. One typical example is making the right decision about the optimal allocation of funds in such companies so that their profit can be maximized. Similar decisions related to the optimization of stocks, cash, accounts receivable are further examples of optimization problems. In these problems variables are selected such that their values represent decisions to be made regarding the policy of firms. This is why variables in optimization problems are called decision variables. Mathematical model of an optimization problem can be expressed as in the following.

H. Karadeniz, *Stochastic Analysis of Offshore Steel Structures*,
Springer Series in Reliability Engineering, DOI: 10.1007/978-1-84996-190-5_7,

$$\begin{array}{lll} \text{minimize} & W(\mathbf{d}) & \\ \text{subject to} & h_j(\mathbf{d}) = 0 & j = 1,\ldots,ne \\ & g_k(\mathbf{d}) \leq 0 & k = 1,\ldots,ni \\ & \mathbf{d}_L \leq \mathbf{d} \leq \mathbf{d}_U & \end{array} \tag{7.1}$$

where $\mathbf{d} = \{d_1,\ldots,d_n\}^T$ is the vector of decision variables, $W(\mathbf{d})$ is the objective function, $h_j(\mathbf{d})$ is the equality, and $g_k(\mathbf{d})$ is the inequality constraints of the optimization problem under consideration. $\mathbf{d}_L$ and $\mathbf{d}_U$ are lower and upper bounds vectors of variables. n represents the total number of variables, ne is the total number of equality constraints, and ni is the total number of inequalities in the optimization problem. It is possible that while both types of these constraints may exist in some optimization problems, in some others either equality or inequality type of constraints might be present. In fact there are optimization problems where there may be no constraints to be satisfied at all. However, in most of practical engineering problems constraints do exist. Optimization problems described in Eq. (7.1) is also called mathematical programming problems. Here the word *programming* should not be mixed with computer programming. It rather implies finding a *program* or *schedule* in terms of training or logistics for the decision making problem under consideration. Optimization techniques determine the values of variables such that constraints given in Eq. (7.1) are satisfied and the objective function shown in Eq. (7.1) attains its minimum or maximum value depending on the formulation of the problem.

7.2 Structural Optimization

In structural design problems decision variables are called design variables. They are the parameters that define the design problem. The solution of the design problem intends to find the numerical values of these parameters. In the design of a simple beam with a rectangular cross-section, the width and the depth of the rectangular cross-section can be design variables. In a truss design problem they can be taken as the cross-sectional areas of members. In a frame design problem the second moment of areas of frame members such as its beams and columns can be treated as design variables. The joint coordinates of a truss or a frame are required to be considered as design variables in addition to the cross-sectional dimensions of members if a designer intends to determine the optimum geometry of these structures. Accordingly, the design variables are those parameters that quantify the structural systems.

Design variables can have continuous, discrete, and integer values. If the value of a design variable in a structural design problem can have any value, such design variables are called continuous design variables as it is the case in the optimum design of steel plate girders. The width of flange plates and web plates may have any real value provided that no architectural limitations are present. The thickness of steel plate from which the flanges and the web are to be cut is usually selected since they are only produced with certain thickness in the practice. In some other structural design problems designer may not have this flexibility. In the design of steel frames the steel sections are required to be selected from steel profiles list available in the practice where the design variables have to have one of the fixed values within this table. Such design variables are called discrete design variables. In addition to these two there are certain cases where the value of a design variable must be integer. If a design variable represents the total number of bolts required in beam column connection it is apparent that the value of design variable cannot be real, it should be integer. Similarly if a design variable represents the total number of beams in longitudinal direction in a grillage system, it is required that it should have an integer value in the solution of the design problem.

7.2.1 Objective Function

Objective function represents the measure which is used to evaluate the goodness of acceptable solutions. It is expressed in terms of design variables. Designers in structural design aim at finding the design solution among the all possible designs which can be constructed economically. This necessitates taking the cost of a structure under consideration as objective function to be minimized. In reinforced concrete structures this is really the case because reinforced concrete structures involve different materials. The unit cost of these different materials used in the construction influence the total cost of the reinforced concrete structures. Hence the objective function is usually taken as the total cost of concrete, reinforcing bars, and the formwork used in the construction of the frame. Consequently, in the optimum design of reinforced concrete structures the design solution which gives the least cost among all other solutions is taken as the optimum solution. In steel structures steel profiles are connected to each other to construct a steel frame. In this case, there is only one material which is steel and cost of a steel structure is somewhat related with its weight. The transportation of steel sections is priced according to their weight. Cost of erection of the members is also function of the weight of beam and column sections. The connection of members which involves bolting and welding is not a function of their weight though the weight is also an important factor. This is why in the optimum design of steel structures generally the weight is taken as objective function to be minimized. However, it should be emphasized that the minimum weight is not the minimum cost in the optimum design of steel structures. This clearly indicates the fact that selection of objective function affects the optimum solution to be obtained. If the aim is the minimum cost then the cost function is to be written for the steel structure which is to be minimized within the optimization process if the correct optimum solution is desired.

In most of the practical optimum design problems there is only one objective function in the design problem. The cost or weight is required to be minimized. In some cases the stiffness of a structure is maximized. Such design problems are called single-criterion optimum design problems. However, there are certain cases there may be more than one objective function. In the design of satellite communication dishes it is desired that the dish has the minimum weight and in the meantime displacements of certain joint or joints are also minimum. In the design of tall steel frames both the cost of the frame and its top story sway are required to be minimized. Such structural design problems are referred to as multiobjective structural optimization problems. However, in some other type of design problems the objectives may be conflicting. It may be necessary that while certain mode of natural frequency of a structure is required to be maximized while the compliance of the structure is needed to be minimized. Certainly finding solution of such multicriteria structural optimization problems is much more complicated than single-criterion structural optimization ones.

7.2.2 Design Constraints

Structural designer is required to consider many restrictions during the design process. Design of a structure should abide by provisions of a design code that is adopted for the design. Basically the design codes make sure that structure to be designed has sufficient strength to withstand the external loads that are expected to act on the structure during its lifetime and it satisfies serviceability requirements. Satisfying serviceability limitations mean there are no excessive displacements in the structure which prevents the structure functioning properly during its service life. Both of these limitations are clearly defined in design codes and what all designer is supposed to do is to include these restrictions in the mathematical formulation of the design problem. In the design of steel frames the design constraints have different forms depending on the assumed structural behavior in the design process. If linear elastic behavior is adopted in the design process then the stresses develop in structural members under the combined axial and bending moments are required to be less than the allowable stresses of the steel material from which the members are produced. However, if ultimate state design is implemented in the design process then the strength constraint of a beam-column member necessitates satisfaction of an inequality which is to be <1. This inequality consists of combined axial and flexural strength of a beam column such that the required axial and flexural strength of the member is less than its nominal axial and flexural strength. In addition to strength constraints beam deflections and lateral displacements of the frame are required to be less than certain value specified in the code. Lateral deflections are of two kinds. One makes sure that the top story sway is less than its upper bound and the other is restricting the inter-story drift of the frame. In addition to these it may be necessary depending on the type of the design problem to impose lower and upper bound on the cross-sectional properties if they are treated as design variables. Because steel profiles are produced in certain dimensions such that cross-sectional variables cannot have values that are larger or smaller than those available in practice.

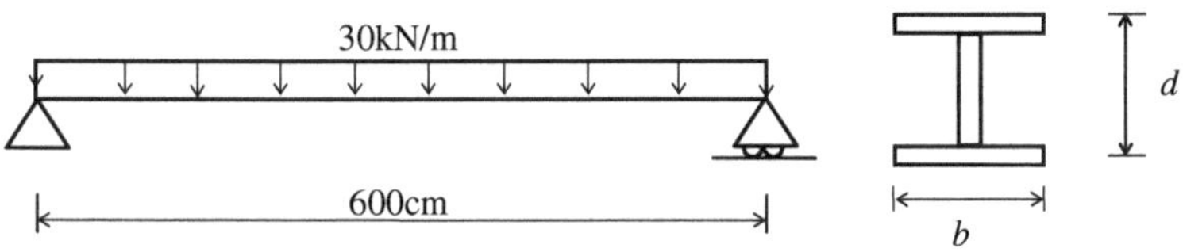

Fig. 7.1 Optimum design of welded beam

7.2.3 Design Example

Steel build up I section is required to span 6 m. The beam is expected to carry 30 kN/m uniformly distributed load. The flanges and the web of I beam section are decided to be cut from a steel plate that already exist in workshop which has 8 mm thickness. These flange and web pieces are welded to each other to make the build up section shown in Fig. 7.1. The modulus of elasticity of mild steel is 20,500 kN/cm^2, yield stress in bending is 25 kN/cm^2, and in shear is 15 kN/cm^2. It is desired that the bending stress and shear stress are not to exceed their upper bound of yield stresses while the maximum deflection of the beam is to be less than span/360 which is 1.67 cm. It is also necessary that the width and the depth of the welded beam should not be <5 cm and 10 cm, respectively. Determine the optimum values of the depth and the width of the beam so that it can be constructed by using the least amount of steel.

It is apparent from Fig. 7.1 that the design variables are the width and depth of the beam. Noticing that the thickness of the steel plate is 0.8 cm the area and the moment of inertia of the steel beam can be expressed in terms of design variables as

$$A = 1.6b + 0.8d, \quad I = 0.06667d^3 + 0.4bd^2 \tag{7.2}$$

The maximum bending moment in the mid-span is $M_{max} = w\ell^2/8 = 33{,}750$ kN cm and the maximum shear force occurs at the supports as $V_{max} = w\ell/2 = 225$ kN. The normal stress due to bending is $\sigma = M(0.5d + 0.8)/I$ and the maximum shear stress is calculated as $\tau = V/A_{web}$ where $A_{web} = 0.8d$. The mid-span deflection is $\delta = 5w\ell^4/(384EI)$. The cost of the beam that includes the material, production, and welding expenses is considered as the objective function which is related to the design variables as $C = 480b + 4{,}800d$. Hence the optimum design problem of the welded beam can be formulated as follows:

$$\begin{aligned}
&\text{minimize} && \text{Cost} = 480(b + 10d) \\
&\text{subject to} && \sigma = \frac{33{,}750}{0.06667d^3 + 0.4bd^2}(0.5d + 0.8) \le 25 \\
& && \tau = \frac{225}{0.8d} \le 15 \\
& && \delta = \frac{5}{384}\frac{0.75 \times 600^4}{20{,}500(0.06667d^3 + 0.4bd^2)} \le 1.67
\end{aligned} \tag{7.3}$$

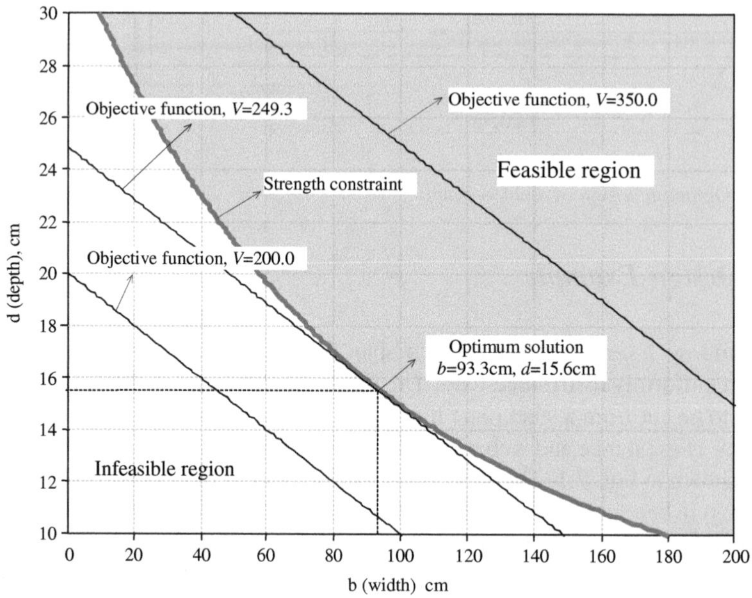

Fig. 7.2 Graphical solution of the optimum design problem of steel build up beam with strength constraint only

When these expressions are simplified, the following programming problem is obtained.

$$\begin{aligned} \text{minimize} \quad & \text{Cost} = 480(b + 10d) \\ \text{subject to} \quad & 1.6667d^3 + 10bd^2 - 16,875d - 27,000 \geq 0 \\ & d - 18.75 \geq 0 \\ & 0.11112d^3 + 0.6667bd^2 - 61737.8 \geq 0 \\ & b \geq 25, \quad d \geq 10 \end{aligned} \tag{7.4}$$

The inequalities, $b \geq 25$ and $d \geq 10$, given in Eq. (7.4) represent the lower bounds imposed on the design variables which might be necessary from the practical point of view. Finding the optimum solution of the above problem requires determining the optimum values of the design variables; namely b and d such that the value of the objective functions, Cost, given in Eq. (7.4) is the minimum and the design constraints; $1.6667d^3 + 10bd^2 - 16{,}875d - 27{,}000 \geq 0$, $d - 18.75 \geq 0$, and $0.11112d^3 + 0.6667bd^2 - 61737.8 \geq 0$ are satisfied. First, the graphical solution of the design problem is sought by only considering the strength constraint, $1.6667d^3 + 10bd^2 - 16{,}875d - 27{,}000 \geq 0$, for simplicity. The graphical solution of this problem is shown in Fig. 7.2.

The curve in Fig. 7.2 represents the strength constraint. The values of the design variables b and d that are taken from the upper part of this curve satisfy the strength constraint ($1.6667d^3 + 10bd^2 - 16{,}875d - 27{,}000 \geq 0$). This region is called as feasible region shown in Fig. 7.2 as shaded area. The values of the design variables

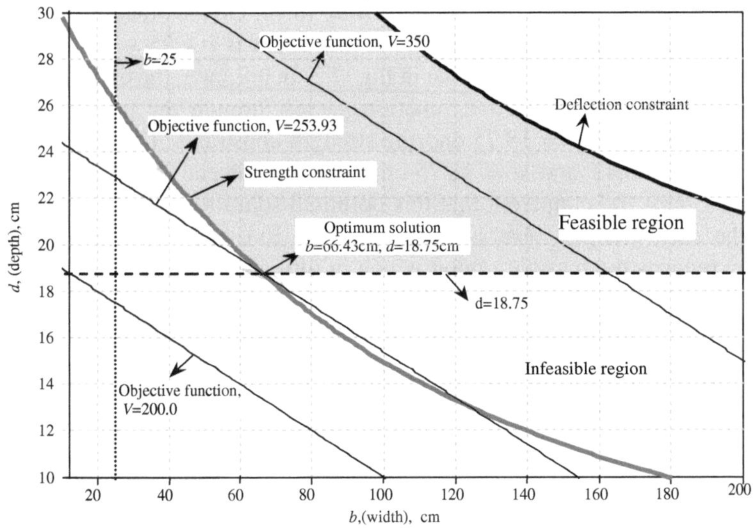

Fig. 7.3 Graphical solution of the optimum design problem of steel build up beam with all constraints

taken below this curve do not satisfy the strength constraints. This region is called as infeasible region. The optimum solution is the one which satisfies the constraint and in the meantime it makes the value of the objective function the minimum. To determine this particular couple of values of the design variables we need to draw the objective function. By keeping the constant out which does not have any effect in specifying the slope of the objective function and only considering the terms in the bracket as $V = b + 10d$, we can plot the objective function which is the linear function in the same graph by assigning values to V. For example if V is taken as 350, one gets the line located in the top part of the figure. $V = 200$ gives the line which is shown in the bottom part of Fig. 7.2. The one that makes the objective function the minimum is the one which is tangent to the feasible region.

This is obtained by letting $V = 249.3$ which corresponds to the optimum solution. The values of the width and depth of the beam can be read from the graph which gives the optimum solution as $b = 93.3$ and $d = 15.6$ cm with the objective function value of 249.3 cm^3.

The graphical solution of the design problem where all the constraints given in Eq. (7.4) are considered is shown in Fig. 7.3. It is apparent from the figure that deflection constraint is not active in the design problem compare to strength constraint. This means that the values of width and depth variables which satisfy the strength constraint satisfy the deflection constraint. However, the boundaries of the feasible region in this case are not only defined by strength constraint but also by the lower bound applied to the width and depth variables. The lower bound on the depth variable required because of shear strength constraint is 18.75 cm which is larger than the practical limit included in the problem as $d \geq 10$. Hence the feasible region is bounded horizontally by the line $d = 18.75$. On the other hand, the lower bound limitation imposed on the width of the beam is $b \geq 25$ which bounds the feasible region from the left as shown in Fig. 7.3. In this case, the optimum solution is the point where the objective function passes through the intersection point of the horizontal line $d = 18.75$ and the strength constraint. This point gives the values of $b = 66.43$ and $d = 18.75$ cm with the objective function value of $V = 253.93$ cm^3. It is apparent that this optimum solution is heavier than the one where the only strength constraint is considered. Existence of further constraints naturally reduces the feasible region which in turn affects the optimum solution.

7.3 Deterministic and Stochastic Solution Techniques of Optimization

There are various classifications for the optimization techniques available in the literature [1–3]. Among these probably the most general one is the one which divides the algorithms into deterministic and stochastic ones. Deterministic optimization techniques make use of derivatives of the objective function and constraints in the search of the optimum solution. They start the search at a preselected initial point and compute the gradients of the objective function and constraints at this point and take a step in the negative direction of the gradient of the objective function in the case of minimization problems to determine the next point. They continue the iterations until there is significant change in the values of design variables within 2 consecutive iterations. All the mathematical programming techniques fall into this classification. Among these linear programming, integer programming, and nonlinear programming techniques are widely used in solving engineering optimization problem [4–7]. Although these techniques are successful in obtaining the solution of small size optimum design problems, they present convergence difficulties in the design of real world problems. Furthermore, in some cases the objective function and constraints may have irregular peaks for which the gradient search can be quite difficult [8]. Among the constrained nonlinear programming methods, penalty function methods, feasible directions method, reduced gradient method, sequential linear programming method, and sequential quadratic programming method are widely used to find the optimum solution of engineering optimization problems.

Computational drawbacks of existing derivative-based numerical methods have forced researchers all over the world to rely on stochastic algorithms founded on simulations of nature for solving computationally intractable engineering optimization problems since the past 2 decades. The basic idea behind these techniques is to simulate the natural phenomena, such as survival of the fittest, immune system, swarm intelligence, and the cooling process of molten metals through annealing into a numerical algorithm. These methods are non-traditional stochastic search and optimization methods and they are very suitable and efficient in finding the solution of combinatorial optimization problems. They do not require the gradient information of the objective function and constraints and they use probabilistic transition rules not deterministic ones [9–17]. These techniques are also known as metaheuristics as they use heuristics to search the design space to attain a better solution than the current one. Metaheuristic algorithms do not guarantee finding the optimal solution but may end up reaching near optimal solution. They initiate the search either generating a population randomly which consists of candidate solutions of the optimization problem under consideration or they start with a randomly selected single candidate solution and try to improve this solution during the search process. Among these evolutionary algorithms are based on the Darwinian theory of evolution and survival of the fittest. Immune system algorithm simulates the body's immune system into a numerical algorithm. Simulated annealing is an iterative search technique inspired by annealing process of metals. Particle swarm optimizer is based on the social behavior of animals, such as fish schooling, insect swarming, and bird flocking. Ant colony optimization technique is inspired from the way that ant colonies find the shortest route between the food source and their nest. Harmony search algorithm is based on the natural musical performance process that occurs when a musician searches for a better state of harmony. Differential evolution iteratively tries to improve a candidate solution with regards to a given measure of quality.

Among the mathematical programming techniques the sequential quadratic programming method and among the metaheuristic algorithms the differential evolution methods are adopted to obtain the reliability based design optimization of offshore structures in this chapter. Only these 2 methods will be explained in detail in the following sections due to the lack of space.

7.3.1 Sequential Quadratic Programming

Sequential quadratic programming is one of the most effective mathematical programming technique for nonlinearly constrained optimization problems [2]. The method consists of approximating the original nonlinearly constrained problem with a quadratic subproblem and solving the subproblem successively until convergence has been achieved on the original problem [18].

Sequential quadratic programming technique modifies the programming problem given in Eq. (7.1) This is obtained by using the Taylor's expansion [19] as written by,

$$\begin{aligned}
&\text{minimize} && W(\mathbf{d}) + \{\nabla W(\mathbf{d})\}^T \Delta\mathbf{d} + 0.5\Delta\mathbf{d}^T[\nabla^2 W(\mathbf{d})]\Delta\mathbf{d} \\
&\text{subject to} && h_j(\mathbf{d}) + \{\nabla h_j(\mathbf{d})\}^T \Delta\mathbf{d} = 0 \quad j = 1, \ldots, ne \\
& && g_k(\mathbf{d}) + \{\nabla g_k(\mathbf{d})\}^T \Delta\mathbf{d} \le 0 \quad k = 1, \ldots, ni \\
& && \Delta\mathbf{d}_L \le \Delta\mathbf{d} \le \Delta\mathbf{d}_U
\end{aligned} \tag{7.5}$$

where $[\nabla^2 W(\mathbf{d})]$ is the Hessian matrix, which is denoted by $\mathbf{H}$. In actual implementation the real Hessian matrix is not used. Instead, a metric $\mathbf{H}$ is updated in each iteration as it is suggested in variables metric method [20]. In the application of the method $\Delta\mathbf{d}$ is determined after the search direction $\mathbf{s}$ is found by solving the following problem using quadratic programming.

$$\begin{aligned} \text{minimize} \quad & P = \{\nabla W(\mathbf{d})\}^T \mathbf{s} + 0.5\,\mathbf{s}^T\,\mathbf{H}\,\mathbf{s} \\ \text{subject to} \quad & h_j(\mathbf{d}) + \{\nabla h_j(\mathbf{d})\}^T \mathbf{s} = 0 \quad j = 1,\ldots,ne \\ & g_k(\mathbf{d}) + \{\nabla g_k(\mathbf{d})\}^T \mathbf{s} \leq 0 \quad k = 1,\ldots,ni \\ & \mathbf{s}_L \leq \mathbf{s} \leq \mathbf{s}_U \end{aligned} \tag{7.6}$$

The quadratic programming problem posed in Eq. (7.6) finds a feasible direction $\mathbf{s}$ with respect to the current active constraints. After finding the search direction, one has to determine the step size α. The calculation of the step size is based on the criteria that the value of the objective functions should decrease and constraint satisfaction have to improve. There are several ways to achieve this goal. One is to use exterior penalty function method to minimize the following unconstrained function.

$$f(\mathbf{d}^{v+1}) = W(\mathbf{d}^v) + r\sum_{j=1}^{ne} h_j(\mathbf{d}^v)^2 + r\sum_{k=1}^{ni} \max[g_k(\mathbf{d}^v), 0]^2 \tag{7.7}$$

where r is known as penalty constant. Solution of Eq. (7.7) yields the value of g and α. In the case where the minimization problem given by Eq. (7.7) is not desired to solve, then the value of α can be taken as 1 for convenience. Once α is determined, $\Delta\mathbf{d}$ is calculated as $\Delta\mathbf{d} = \alpha\,\mathbf{s}$. The value of new point is then calculated from

$$\mathbf{d}^{v+1} = \mathbf{d}^v + \alpha\,\mathbf{s} \tag{7.8}$$

The steps of sequential quadratic programming method are summarized in the following.

1. Select initial design point $\mathbf{d}^1$, convergence tolerance ε, and maximum number of iterations *maxiter*. Set iteration counter $v = 1$.
2. Using quadratic programming solve the programming problem Eq. (7.6) and find $\mathbf{s}$.
3. Solve unconstrained programming problem Eq. (7.7) and find α or take $\alpha = 1$. Calculate next design point $\mathbf{d}^{v+1}$ from Eq. (7.8). $\Delta\mathbf{d} = \mathbf{d}^{v+1} - \mathbf{d}^v$.
4. Stop the iterations if $\|\Delta\mathbf{d}\| \leq \varepsilon$ or $v = maxiter$. If not, then $v = v+1$, update metric $\mathbf{H}$ and go to step 2.

Further details of the method can be found in [20].

7.3.2 *Differential Evolution Technique*

Differential evolution technique is a stochastic, population-based direct search method that makes use of heuristics to determine the optimum solution in a design domain. Similar to other metaheuristic techniques it does not need gradient computations of the objective function and design constraints of the programming problem. It belongs to the evolutionary optimization algorithms group. It is originated by [21, 22]. It was developed to optimize real parameters of real-valued functions. The stochastic search techniques find the optimum solution of unconstrained functions by searching the design space. Consider the following unconstrained optimization problem.

Find $\mathbf{d}_{\text{opt}}$ such that the objective function $W(\mathbf{d})$ has the minimum value within a region defined as $\mathbf{d}_L \leq \mathbf{d} \leq \mathbf{d}_U$.

Differential evolution algorithm sets up initial population by randomly generating *np* individuals that is expected to cover the entire design space. Uniform probability distribution is used for all random decisions. An individual in a generation represents candidate solution for the optimization problem under consideration which is same as the chromosomes or genomes of genetic algorithm. However, here real numbered representation not binary representation is used for the parameters. The individual is referred as an *agent* and the objective function is called as *fitness function* in differential evolution algorithm. New parameter vectors are generated by adding the weighted difference between 2 population vectors to a third vector. This operation is called *mutation.* The mutated vector's parameters are then mixed with the parameters of another predetermined vector, the target vector, to yield the trial vector. This is referred as *crossover.* If the trial vector yields a lower cost function value than the target vector, the trial vector replaces the target vector in the following generation. This operation is called *selection.* Each population vector has to serve once as a target vector so that *np* competition takes place in one generation. Generations are continued until some stopping criteria such as maximum number of generations is met. The steps of the algorithm are summarized in the following.

1. Set up initial population by generating *np* number of agents $\mathbf{d}$ randomly in the search space.
2. For each agent $\mathbf{d}_j$ where $j = 1,\ldots, np$ carry out the following

   - Select 3 agents $\mathbf{d}_a$, $\mathbf{d}_b$, and $\mathbf{d}_c$ from the population randomly such that they must be distinct from each other and that of $\mathbf{d}_j$.
   - Select a random index k which is between 1 to *np*.
   - Compute the agent's trial vector $\mathbf{d}_t$ by iterating over each $i \in \{1, 2, \ldots\ldots, n\}$ as follows

     - Select a random number $r_i \sim U(0, 1)$.
     - Compute the trial vector as $\mathbf{d}_t = \mathbf{d}_a + F(\mathbf{d}_b - \mathbf{d}_c)$ if $i = k$ or $r_i \leq CR$ otherwise $\mathbf{d}_t = \mathbf{d}_j$ where CR is the crossover rate and F is the scaling (weighting) factor defined by users.

   - Update the trial vector considering the lower and upper bound vectors as $\mathbf{d}_t = \mathbf{d}_L$ if $\mathbf{d}_t < \mathbf{d}_L$, $\mathbf{d}_t = \mathbf{d}_u$ if $\mathbf{d}_t > \mathbf{d}_u$.
   - If $W(\mathbf{d}_t) < W(\mathbf{d}_j)$ then replace the agent $\mathbf{d}_j$ by $\mathbf{d}_t$.

3. The agent $\mathbf{d}_o$ from the population having the lowest fitness $W(\mathbf{d}_o)$ is the best found solution within this generation.
4. Continue the generation until stopping criteria is satisfied.

It is stated that control variables *np*, *F,* and *CR* of the differential evolution algorithm are not difficult to choose in order to obtain good results [21]. It is found reasonable to select the total value of the population between 5 and 10 times of the number of parameters in the optimization problem. 0.5 and 0.1 can be a good initial values for *F* and *CR*. It is best to carry out sensitivity analysis with few values of these parameters in order to find the most appropriate ones for the optimization problem under consideration. In [21] a comparative study is carried out among adaptive simulated annealing, the annealed Nelder and Mead approach, the breeder genetic algorithm, the easy evolution strategy and differential evolution algorithm, and it is affirmed that differential evolution method outperformed all of the above-mentioned minimization techniques in terms of required number of function evaluations necessary to attain the global optimum.

7.4 Mathematical Formulation of the Reliability-Based Design Optimization

The traditional deterministic optimization, that seeks the minimum weight, volume or cost under the specified requirements, has been successfully applied to the engineering designs [23–32]. However, the existence of uncertainties in either engineering simulations or manufacturing processes [33, 34] may affect the obtained result using the deterministic optimization approach. Therefore, the uncertainties associated with the loads, resistances, and structural responses must be included in the optimization process in order to obtain optimal result under realistic conditions. It is possible to represent uncertainties in the design of a structure as random variables with assumed probability distribution functions. The design optimization of a structure with the random variables is called reliability-based design optimization (RBDO). In the RBDO model, an objective function being either structural weight or expected cost of a structural system (i.e. including the initial and failure cost) is minimized under prescribed probabilistic (reliability) constraints. Therefore, it is required that one of the reliability analysis methods, as compared with deterministic optimization, is included in structural optimization process an addition in order to evaluate the reliability constraints, which can be done either by stochastic simulations or by moment methods [33–36]. Thus, it can be realized from above that 3 main components, namely, a structural analysis program, an optimization program, and a reliability analysis program, should be linked together to fulfill RBDO of the structural systems. An optimization program is necessary to evaluate the design variables satisfying all constraints and minimizing the objective function. The reliability analysis is used for the evaluation of the reliability constraints that being the functions of the design variables and the random variables. The structural analysis program is employed to calculate the structural responses. In addition sensitivity analysis, which is responsible for the calculation of the variation of the structural response depending on the random and design variables, is performed both for optimization and for reliability. Due to the integration of components RBDO procedures requires prohibitive computational effort. Depending on the scheme of the integration RBDO formulations can be classified into 3 categories: (a) the 2 level approach, (b) the single loop approach, (c) the decoupled approach [37, 38]. The first one [39–49] considers the probabilistic constraints inside the optimization loop. The RBDO problem is solved in a single loop procedure, where the reliability analysis is avoided, in the second [50–52]. The later [53, 54] consists of separating the reliability analysis from the optimization procedure. These efforts are made to reduce the computation time causing too many repeated searches in the 2 step (level) algorithm. Consequently, a typical structural optimization problem recognizing uncertainties related to loads, geometry, resistance, material, and so on is formulated in terms of random variables vector $\mathbf{X} = \{X_1,..,X_{\text{nrv}}\}^{\text{T}}$, and design variables vector $\mathbf{d} = \{d_1,...d_n\}^{\text{T}}$, where nrv is the number of random variables, and n denates the number of design variables as follows:

$$\begin{aligned}&\text{find } \mathbf{d}, \text{ which minimizes } W(\mathbf{d})\\ &\text{subject to: } P_{fi} = P(G_i(\mathbf{d},\mathbf{X}) \leq 0) \leq P_{fi,\max} \quad i = 1,\ldots,\text{nrc}\end{aligned} \tag{7.9}$$

$W(\mathbf{d})$ is the objective function (e.g. structural mass or volume), $G_i(\mathbf{d}, \mathbf{X})$ is defined as the ith limit state function or performance function, and $G_i(\mathbf{d}, \mathbf{X}) \leq 0$ denotes the failure domain, P(.) is the probability operator, $P_{fi,\max}$ is the admissible failure probability, nrc is the total number of performance functions, or probabilistic constraints. The design variables vector **d** in Eq. (7.9) may be either independent deterministic variables or the mean values of a subset of random variables. Typically, an upper, $\mathbf{d}_U$, and lower, $\mathbf{d}_L$, bounds vector is also defined for the design variables in order to obtain a meaningful result.

In the above model the probabilistic constraints define the feasible region by restricting the probability of violating the limit state function to the admissible probability. The corresponding failure probability for ith limit state function is given by

$$P(G_i(\mathbf{d},\mathbf{X}) \leq 0) = \int_{G_i(\mathbf{d},\mathbf{X}) \leq 0} f_{\mathbf{X}}(\mathbf{x})d\mathbf{x} \tag{7.10}$$

where $f_{\mathbf{X}}(\mathbf{x})$ is the joint probability density function for all random variables involved. Since the exact computation of Eq. (7.10) is impractical, 2 approximate methods are often applied as: (a) stochastic simulations (e.g. crude Monte Carlo [55], importance sampling [35]), (b) moment methods (e.g. first- and second-order reliability methods FORM [33, 34]/SORM [56] to overcome the evaluation of Eq. (7.10). Although the formers are potentially highly accurate, they generally evaluate the limit state function by a requirement for a large number of samples. In the moment methods, the reliability index, β, is calculated as an alternative

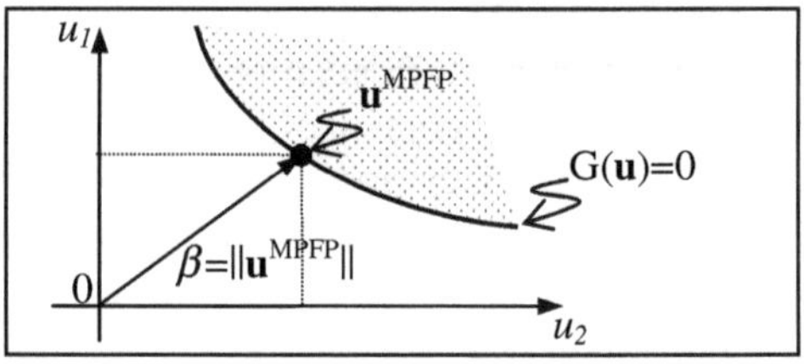

Fig. 7.4 Illustration of reliability index in the standard normal space

measure of failure probability, in which the direct calculation of P_f is avoided. In this case, the probabilistic constraints in Eq. (7.9) are simply replaced by the reliability indices, which is often referred to as RBDO-based on reliability index approach (RIA). An alternative approach, the performance measure approach (PMA), proposed recently for the evaluation of probabilistic constraints [57–60] may be more efficient and stable.

7.4.1 Reliability Index Approach for the RBDO

The formulation where the probabilistic constraints handle with the reliability indices is expressed as:

$$\begin{aligned}&\text{find } \mathbf{d}, \text{ which minimizes } W(\mathbf{d})\\&\text{subject to: } \beta_i \geq \beta_{i,\,\text{target}} \quad i = 1, \ldots, \text{nrc}\end{aligned} \tag{7.11}$$

where β_i and $\beta_{i,\,\text{target}}$ are the structural and the target reliability indices for the ith limit state, respectively. Considering Eq. (7.9) it clearly realized that each $P_{fi} = P(G_i(\mathbf{d},\mathbf{X}) \leq 0)$ is replaced by the reliability index β_i using first-order reliability methods (FORM). In the FORM, a transformation $\mathbf{X} = \mathbf{T}(\mathbf{U})$ is required to map the random variables $\mathbf{X}$ from original space into the U-space of independent, standardized, and normally distributed variables $\mathbf{U}$ (i.e. $\mathbf{u} = \mathbf{T}(\mathbf{x})$) [61–64]. Hence, the definition of the reliability index β associated with the limit state function G_i is defined as the minimum distance from the origin to the point located on the limit state surface and the limit state function where $G_i(\mathbf{u}) = 0$ (see Fig. 7.4). This point is called as the most probable failure point (MPFP) of the failure surface in the standard normal space since the largest contribution to the probability integral Eq. (7.10), comes from the region around that point. Thus, probability of failure is defined as $P_f = \Phi(-\beta)$, in which $\Phi(.)$ is the standard normal cumulative distribution function. According to FORM approximation based on the Hasofer-Lind and Rackwitz-Fiessler (HLRF), an iterative search procedure is used to find the $\mathbf{u}$ vector for a prescribed convergence tolerance (i.e. $\varepsilon = 0.001$). This procedure is formulated in Eq. (7.12)

$$\mathbf{u}^{k+1} = \frac{\nabla G_{\mathbf{u}^k}^T \mathbf{u}^k - G_i(\mathbf{u}^k)}{\nabla G_{\mathbf{u}^k}^T \nabla G_{\mathbf{u}^k}} \nabla G_{\mathbf{u}^k} \tag{7.12}$$

where $\nabla G_{\mathbf{u}^k} = \{\partial G_i/\partial u_1, \partial G_i/\partial u_2, \ldots, \partial G_i/\partial u_{\mathrm{nrv}}\}^{\mathrm{T}}$ is the gradient vector of the ith limit state function with respect to vector $\mathbf{u}^k$ at the kth iteration. Thus, the reliability index is computed as $\beta_i = \|\mathbf{u}\|$ at the end of the iterative search procedure.

From the definition related to β, it is also obtained by solving the constrained optimization problem stated as:

$$\begin{aligned} \underset{\mathbf{u}}{\text{minimum}} \quad & \beta_i = \|\mathbf{u}\| = \sqrt{\mathbf{u}^T\mathbf{u}} \\ \text{subject to} \quad & G_i(\mathbf{u}) = 0 \end{aligned} \tag{7.13}$$

A general optimization method based on the gradient-based, i.e. sequential quadratic programming, or the gradient-free, i.e. simulated annealing, algorithms can be used to solve of Eq. (7.13).

7.4.2 Performance Measure Approach for the RBDO

The approaches formulated above estimate the probability of failure by the reliability index. Tu and Tu et al. [57, 58] offered an alternative means to evaluate the reliability constraints in the RBDO in order to avoid the problems of the RIA concerned with the calculation of the reliability index associated with each reliability constraints during an overall RBDO iteration. In this method, known as the PMA, the reliability constraints are expressed by an inverse formulation as:

$$G_i^p = G_i(\mathbf{u}^*_{=\beta_{i,\text{target}}}, \mathbf{d}) \tag{7.14}$$

where G_i^p is the performance measure corresponding to target reliability of ith reliability constraint evaluated by an inverse reliability analysis, in which minimum distances from the origin in U-space to limit state surfaces are equal to target reliability indices, and subsequently, one of them, at which the limit state function should be minimum, is selected. $\mathbf{u}^*_{=\beta_{i,\text{target}}}$ is the solution to the inverse reliability analysis associated with the optimization problem, which is defined as:

$$\begin{aligned} \underset{\mathbf{u}}{\text{minimum}} \quad & G_i(\mathbf{u}) = 0 \\ \text{subject to} \quad & \|\mathbf{u}\| = \beta_{i,\text{target}} \end{aligned} \tag{7.15}$$

Besides using any optimization algorithms, the inverse reliability analysis based on FORM [57–59] is also used as a tool to calculate $\mathbf{u}$. The updated formula of the developed algorithm based on the advanced mean value approach to solve the problem in Eq. (7.15) is given by

$$\mathbf{u}^{k+1} = -\beta_{i,\text{target}} \frac{\nabla G^T_{\mathbf{u}^k}}{\sqrt{\nabla G^T_{\mathbf{u}^k} \nabla G_{\mathbf{u}^k}}} \tag{7.16}$$

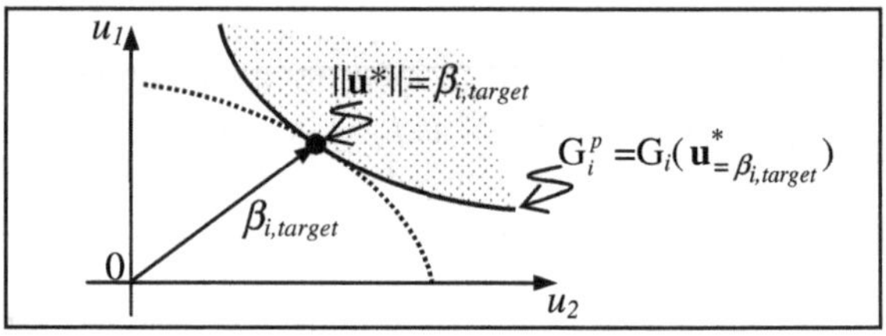

Fig. 7.5 Illustration of PMA in the standard normal space

In addition, some enhanced algorithms developed by [65], i.e. the conjugate mean value and the hybrid mean value algorithms, are employed to solve the problem in Eq. (7.15). Thus PMA for RBDO can be expressed as

$$\begin{aligned} &\text{find } \mathbf{d}, \text{ which minimizes } W(\mathbf{d}) \\ &\text{subject to: } G_i^p \geq 0 \quad i = 1, \ldots, \text{nrc} \end{aligned} \tag{7.17}$$

In contrast to the RIA formulation Eq. (7.13), which is numerically not stable for certain type of distribution, the PMA formulation Eq. (7.15) is usually more efficient and robust because it works on a fixed position, $\|\mathbf{u}\| = \beta_{i,\text{target}}$, in U-space. In other words, the position of $G_i(\mathbf{u}) = 0$ varies with the design point and the search for Eq. (7.13) is performed until reaching the failure surface while the region to be explored by the Eq. (7.15) is the hypersphere having radius equal to the target reliability index [37, 38, 59, 65–70]. A schematic illustration of the solution of Eq. (7.15) in the standard normal space (U-space) is shown in Fig. 7.5.

7.5 Sensitivity Analysis of RBDO of Offshore Structures

Sensitivity analysis quantifies the influence of each parameter on model, function, response, etc. It is crucial integrant both for the reliability analysis and the optimization methods based on the mathematical theory. For the reliability analysis based on FORM, the updated formula given in Eqs. (7.12)–(7.16) needs the gradient information $\nabla G_{\mathbf{u}^k}$ of the limit state function with respect to random variables.

Two distinct ways can be employed to calculate $\nabla G_{\mathbf{u}^k}$. The related gradient information in $\nabla G_{\mathbf{u}^k} = \{\partial G_i/\partial u_1, \partial G_i/\partial u_2, \ldots, \partial G_i/\partial u_{\text{nrv}}\}^{\text{T}}$ can be directly calculated in normalized space in the first way. In the second, applying the chain rule of differentiation, the gradient of the limit state function is calculated in the original space, and then those are multiplied with the derivatives of corresponding random variables calculated in the normalized space as:

$$\nabla G_{\mathbf{u}^k} = \left[\frac{\partial G_i}{\partial x_1}\frac{\partial x_1}{\partial u_1}, \frac{\partial G_i}{\partial x_2}\frac{\partial x_2}{\partial u_2}, \ldots, \frac{\partial G_i}{\partial x_{\text{nrv}}}\frac{\partial x_{\text{nrv}}}{\partial u_{\text{nrv}}}\right]^T \tag{7.18}$$

Since the value of limit state function is generally obtained after performing the structural analysis for a structural engineering problem the second way for obtaining the related gradient information is easily linked to the structural analysis program.

The calculation of the first term in Eq. (7.18) is performed by means of the structural analysis program for the engineering problems in general. Those used for this purpose are generally based on finite element method (FEM). The gradient information is consequently calculated using [71, 72] one of; (1) Finite difference method, (2) Direct differentiation, and (3) Adjoint method.

The second term of Eq. (7.18) ($\partial \mathbf{x}/\partial \mathbf{u}$) can be easily calculated considering $F_{\mathbf{X}}(\mathbf{x}) = \Phi(\mathbf{u}) \Rightarrow \mathbf{u} = \Phi^{-1}(F_{\mathbf{X}}(\mathbf{x}))$, where $F_{\mathbf{X}}(\mathbf{x})$ is the cumulative distribution functions of a continuous random variables and $\Phi(.)$ is the cumulative distribution function for the standard normal distribution, as:

$$\frac{\partial \mathbf{x}}{\partial \mathbf{u}} = \frac{\partial F_{\mathbf{X}}^{-1}(\Phi(\mathbf{u}))}{\partial \mathbf{u}} = \frac{\phi(\mathbf{u})}{f_{\mathbf{X}}(\mathbf{x})} \tag{7.19}$$

in which $\phi(.)$ and $f_{\mathbf{X}}(\mathbf{x})$ are, respectively the probability density function of the standard normal distribution and the corresponding random variable.

The linear elastic static analysis of the structures under the external load can be stated as based on FEM terminology

$$\mathbf{Kq} = \mathbf{F} \tag{7.20}$$

where $\mathbf{K}$ is the structural stiffness matrix, $\mathbf{q}$ is the vector of nodal displacements, and $\mathbf{F}$ is the vector of applied forces. The responses of the structure obtained after performing the linear elastic static analysis are used in the evaluation of the constraints that are generally given by

$$g_i = 1.0 - |\sigma_i|/\sigma_i^a \leq 0 \quad i = 1, 2, \ldots, m \tag{7.21}$$

$$g_{j+m} = 1.0 - |q_j|/q_j^a \leq 0 \quad j = 1, 2, \ldots, r \tag{7.22}$$

where σ_i is the stress in the ith member and σ_i^a is the allowable stress for the same member, q_j is the displacement of the jth node, and q_j^a is its upper bound. Thus, the functions defined for the constraints are implicit functions of the variables, $\mathbf{s}(= \mathbf{d} \cup \mathbf{X})$. The derivatives of the constraint function with respect to $\mathbf{s}$ according to methods mentioned above are calculated as explained in the following sections [66, 71–75].

7.5.1 Finite Difference Method

The value of g is calculated depending on $\mathbf{s}$ at first. Then each variable is perturbed ($\Delta\mathbf{s}$) and the corresponding change in g is computed through multiple deterministic analyses. The derivative of g with respect to $\mathbf{s}$ it can be expressed as if the forward difference approach is used for computation

$$\frac{\partial g}{\partial \mathbf{s}} = \frac{g(\mathbf{s} + \Delta \mathbf{s}) - g(\mathbf{s})}{\Delta \mathbf{s}} \tag{7.23}$$

Since the function g is evaluated $n + 1$ times for n variables, the cost of the gradient increases dramatically with the number of variables. However, it is preferred due to its simplicity. Moreover, it is easily linked with a commercial software program in order to compute the derivatives without making any modification in the software.

7.5.2 Direct Differentiation Method

Using the chain rule of differentiation, the total derivative of g with respect to $\mathbf{s}$ may be calculated as

$$\frac{\mathrm{d}g}{\mathrm{d}\mathbf{s}} = \frac{\partial g}{\partial \mathbf{s}} + \frac{\partial g}{\partial \mathbf{q}} \frac{\mathrm{d}\mathbf{q}}{\mathrm{d}\mathbf{s}} \tag{7.24}$$

Differentiating both sides of Eq. (7.20) with respect to $\mathbf{s}$ $\mathbf{dq/ds}$ can be stated as

$$\frac{\mathrm{d}\mathbf{q}}{\mathrm{d}\mathbf{s}} = \mathbf{K}^{-1} \left[\frac{\partial \mathbf{F}}{\partial \mathbf{s}} - \frac{\partial \mathbf{K}}{\partial \mathbf{s}} \mathbf{q} \right] \tag{7.25}$$

This result is substituted into Eq. (7.24) to obtain

$$\frac{\mathrm{d}g}{\mathrm{d}\mathbf{s}} = \frac{\partial g}{\partial \mathbf{s}} + \frac{\partial g}{\partial \mathbf{q}} \mathbf{K}^{-1} \left[\frac{\partial \mathbf{F}}{\partial \mathbf{s}} - \frac{\partial \mathbf{K}}{\partial \mathbf{s}} \mathbf{q} \right] \tag{7.26}$$

The sensitivity is directly calculated for each variable through Eq. (7.26).

7.5.3 Adjoint Method

An adjoint variables vector λ is introduced as

$$\lambda \cong \left[\frac{\partial g}{\partial \mathbf{q}} \mathbf{K}^{-1} \right]^T = \mathbf{K}^{-1} \frac{\partial g^T}{\partial \mathbf{q}} \tag{7.27}$$

Both sides of Eq. (7.27) is multiplied by the matrix $\mathbf{K}$ to obtain

$$\mathbf{K}\lambda = \frac{\partial g^T}{\partial \mathbf{q}} \tag{7.28}$$

After λ in Eq. (7.28) is solved and substituted into Eq. (7.26), it becomes

$$\frac{\mathrm{d}\mathbf{g}}{\mathrm{d}\mathbf{s}} = \frac{\partial g}{\partial \mathbf{s}} + \lambda^T \left[\frac{\partial \mathbf{F}}{\partial \mathbf{s}} - \frac{\partial \mathbf{K}}{\partial \mathbf{s}} \mathbf{q} \right] \tag{7.29}$$

Although the direct differentiation method and the adjoint method are mathematically identical, their numerical performances might be different. The direct differentiation method may be preferable compared to the adjoint method when the number of variables is larger than the number of constraints and vice versa [66, 71–75].

When a gradient-based algorithms are employed to obtain a solution of problem given in Eqs. (7.11)–(7.17) the sensitivities of objective function and reliability constraints with respect to the design variables must be supplied for the efficient implementation. If the FORM and the inverse FORM approximations summarized above in terms of Eqs. (7.12) and (7.16) are adopted for the reliability analysis for the RIA and PMA, respectively, the sensitivities related to reliability constraints might be computed efficiently depending on 2 type of design variables which are considered in the RBDO application. One is a characteristic value y that is related to the random distribution, such as the mean value of the random variable X. The other is a deterministic parameter z, which is independent of the random variable X. Therefore, the corresponding sensitivities of reliability constraint with respect to design variables vary according to design variable type for the RIA and the PMA [37, 38, 59, 66, 76–78].

Sensitivity for Reliability Index Approach (RIA)

For the RIA, the sensitivity of the ith reliability constraint is obtained from the sensitivity of the reliability index, β. Recalling the definition of the reliability index ($\beta = (\mathbf{u}^{*T}\,\mathbf{u}^*)^{1/2}$) and the most probable point ($\mathbf{u}^* = -\beta \frac{\nabla G_{\mathbf{u}}}{\|\nabla G_{\mathbf{u}}\|}$), the gradient of β with respect to y and z can be expressed as:

$$\frac{\mathrm{d}\beta_i}{\mathrm{d}y} = \frac{1}{\|\nabla G_{\mathbf{u}^*}\|} \frac{\mathrm{d}G_i}{\mathrm{d}y} \quad \text{and} \quad \frac{\mathrm{d}\beta_i}{\mathrm{d}z} = \frac{1}{\beta} \mathbf{u}^* \frac{\mathrm{d}\mathbf{u}^*}{\mathrm{d}z} \tag{7.30}$$

where $\mathbf{u}^*$ is the coordinate of the most probable point, $\mathrm{d}\mathbf{u}^*$ is described by the derivative of the transformation of the associated distribution.

Sensitivity for Performance Measure Approach (PMA)

For the PMA, the sensitivity of the performance measure G_i^p to design variables is expressed as the gradient of the performance function at the minimum performance target point (MPTP $= \mathbf{u}^*_{=\beta_{i,target}}$). Since the most probable failure point (MPFP) and the MPTP are the same if $\beta_i = \beta_{i,\ \mathrm{target}}$, the performance measure does not varies depending on $\mathbf{u}$. Therefore, its sensitivity with respect to y and z can be written more simply than that of the reliability index.

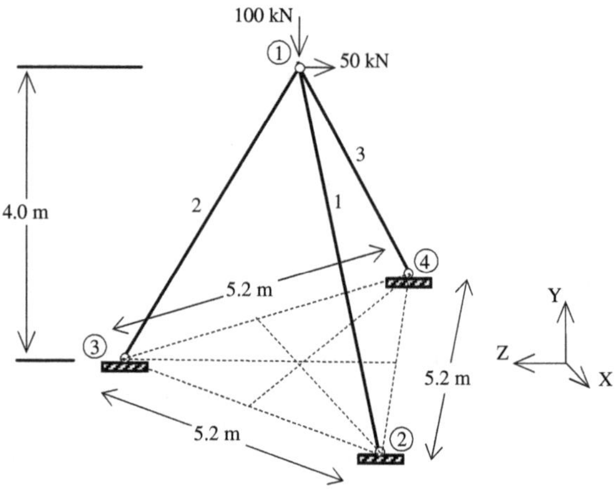

Fig. 7.6 Three-bar space truss

$$\frac{dG_i^p}{dy} = \frac{dG_i(\mathbf{d}, \mathbf{u}^*_{=\beta_{i,target}})}{dy} \quad \text{and} \quad \frac{dG_i^p}{dz} = \frac{dG_i(\mathbf{d}, \mathbf{u}^*_{=\beta_{i,target}})}{dz} \tag{7.31}$$

7.6 Examples

In this section, in the light of the information given in the previous sections, 2 types of design examples are presented. In the first one, 3 bar space truss is considered as an example to demonstrate the mathematical modeling of a deterministic optimum design problem. In the second, 3 numerical examples associated with the monopod, the tripod, and the jacket type offshore towers are presented for reliability-based optimum design.

7.6.1 Deterministic Design Optimization

The 3 bar space truss shown in Fig. 7.6 is subjected to the external loading shown in the figure. It is decided to have the same cross-section for members 2 and 3 while member 1 can have a different section. Modulus of elasticity is taken as 20,000 kN/cm^2. The displacements of joint 1 in X and Y direction are restricted to be not more than 0.25 cm. The compressive stresses in 3 members are required to be not more than 12 kN/cm^2. The optimum design problem is to determine the values of member areas such that the limitations imposed on displacements and stresses are satisfied while the structure has the minimum weight.

The design variables are selected as cross-sectional areas of members. Accordingly the cross-sectional area of member 1 is considered to be A_1 and the

Table 7.1 Direction cosines of members of 3 bar space truss

Member number (i)	$\cos \alpha_i$	$\cos \beta_i$	$\cos \gamma_i$
1	0.60	−0.80	0.00
2	−0.30	−0.80	0.52
3	−0.30	−0.80	−0.52

cross-sectional areas of members 2 and 3 are represented by A_2. In order to express the displacements of joint 1 in terms of design variables, it is necessary to use the matrix displacement method and obtain the stiffness equations which relate the joints displacements to joint loads in global coordinates. The joint displacement vector in global coordinate system is $\mathbf{X} = \{x_1\ y_1\ z_1\}^T$ and corresponding joint load vector in global coordinate system is $\mathbf{P} = \{50, -100,\ 0\}^T$. The joint load vector is related to joint displacement vector as $\mathbf{P} = \mathbf{K}\mathbf{X}$ where $\mathbf{K}$ is the overall stiffness matrix. The overall stiffness matrix of the 3 bar truss can be constructed as

$$\mathbf{K} = \begin{bmatrix} a_1 + a_2 + a_3 & b_1 + b_2 + b_3 & d_1 + d_2 + d_3 \\ b_1 + b_2 + b_3 & c_1 + c_2 + c_3 & e_1 + e_2 + e_3 \\ d_1 + d_2 + d_3 & e_1 + e_2 + e_3 & f_1 + f_2 + f_3 \end{bmatrix} \tag{7.32a}$$

where the parameters in Eq. (7.32a) are as follows:

$$\begin{aligned} a_i &= EA_i \cos^2 \alpha_i/\ell_i, & b_i &= EA_i \cos \alpha_i \cos \beta_i/\ell_i, & c_i &= EA_i \cos^2 \beta_i/\ell_i \\ d_i &= EA_i \cos \alpha_i \cos \gamma_i/\ell_i, & e_i &= EA_i \cos \beta_i \cos \gamma_i/\ell_i, & f_i &= EA_i \cos^2 \gamma_i/\ell_i \end{aligned} \tag{7.32b}$$

in which $(i = 1, 2, 3)$ and $\cos \alpha_i$, $\cos \beta_i$, $\cos \gamma_i$ are the direction cosines of member i. α_i, β_i, γ_i are the angles of member i makes with each global X, Y, and Z axis which are computed from the following expressions.

$$\cos \alpha_i = \frac{X_s - X_f}{\ell_i}, \quad \cos \beta_i = \frac{Y_s - Y_f}{\ell_i}, \quad \cos \gamma_i = \frac{Z_s - Z_f}{\ell_i} \tag{7.33}$$

Where X_f, Y_f, Z_f and X_s, Y_s, Z_s are the coordinates of the first and second end of member i. The direction cosines of the members of 3 bar space truss are given in Table 7.1.

Substituting these into the expression (7.32b) and also noticing that members 2 and 3 are required to have the same cross-section, the stiffness matrix given in Eq. (7.32a) becomes:

$$\mathbf{K} = \begin{bmatrix} 14.4A_1 + 7.2A_2 & -19.2A_1 + 19.2A_2 & 0 \\ -19.2A_1 + 19.2A_2 & 25.6A_1 + 51.2A_2 & 0 \\ 0 & 0 & 21.632A_2 \end{bmatrix} \tag{7.34}$$

where A_1 is the cross-sectional area of member 1 and A_2 is the cross-sectional area of member 2 and 3.

Inverse of the stiffness matrix has the following form.

$$\mathbf{K}^{-1} = \begin{bmatrix} \frac{5A_1 + 10A_2}{324A_1A_2} & \frac{5A_1 - 5A_2}{432A_1A_2} & 0 \\ \frac{5A_1 - 5A_2}{432A_1A_2} & \frac{10A_1 + 5A_2}{1152A_1A_2} & 0 \\ 0 & 0 & \frac{125}{2704A_2} \end{bmatrix} \tag{7.35}$$

Using this matrix the joint displacements can be expressed in terms of cross-sectional areas as written by

$$\begin{Bmatrix} x_1 \\ y_1 \\ z_1 \end{Bmatrix} = \begin{bmatrix} \frac{5A_1 + 10A_2}{324A_1A_2} & \frac{5A_1 - 5A_2}{432A_1A_2} & 0 \\ \frac{5A_1 - 5A_2}{432A_1A_2} & \frac{10A_1 + 5A_2}{1152A_1A_2} & 0 \\ 0 & 0 & \frac{125}{2704A_2} \end{bmatrix} \begin{Bmatrix} 50 \\ -100 \\ 0 \end{Bmatrix} \tag{7.36}$$

The x_1, y_1, z_1 displacements of the joint 1 can be expressed in terms of design variables from Eq. (7.36) as given in the following.

$$x_1 = \frac{-A_1 + 7A_2}{2.592A_1A_2},\ y_1 = \frac{A_1 + 3.5A_2}{3.456A_1A_2},\ z_1 = 0 \tag{7.37}$$

Noticing the fact that the displacement of joint 1 along global Y axis is negative, its absolute value is used in obtaining the constraints related with this displacement so that it can be compared with its upper bound of 0.25 cm which is positive.

Axial stresses at member ends can also be calculated by making use of matrix displacement method. The stress at the first and the second ends of space truss member is computed through the following matrix equation.

$$\begin{Bmatrix} \sigma_{if} \\ \sigma_{is} \end{Bmatrix} = \frac{E}{\ell_i} \begin{bmatrix} \cos\alpha_i & \cos\beta_i & \cos\gamma_i & 0 & 0 & 0 \\ 0 & 0 & 0 & \cos\alpha_i & \cos\beta_i & \cos\gamma_i \end{bmatrix} \begin{Bmatrix} x_1 \\ y_1 \\ z_1 \end{Bmatrix} \tag{7.38}$$

where E is the modulus of elasticity, ℓ_i is the length of member i. Substituting the values of direction cosines of members, the stresses at the first and second end of members are obtained as:

$$\text{in member 1}: \begin{Bmatrix} \sigma_{1f} \\ \sigma_{1s} \end{Bmatrix} = \begin{bmatrix} 24 & -32 & 0 \\ -24 & 32 & 0 \end{bmatrix} \begin{Bmatrix} x_1 \\ y_1 \\ z_1 \end{Bmatrix} \tag{7.39a}$$

$$\text{in member 2}: \begin{Bmatrix} \sigma_{2f} \\ \sigma_{2s} \end{Bmatrix} = \begin{bmatrix} -12 & -32 & 20.8 \\ 12 & 32 & -20.8 \end{bmatrix} \begin{Bmatrix} x_1 \\ y_1 \\ z_1 \end{Bmatrix} \tag{7.39b}$$

$$\text{in member 3}: \begin{Bmatrix} \sigma_{3f} \\ \sigma_{3s} \end{Bmatrix} = \begin{bmatrix} -12 & -32 & -20.8 \\ 12 & 32 & 20.8 \end{bmatrix} \begin{Bmatrix} x_1 \\ y_1 \\ z_1 \end{Bmatrix} \tag{7.39c}$$

Remembering the fact that $z_1 = 0$, the stress expressions for members 2 and 3 become the same. Consequently, the stresses at the first and second end of member 1 and 2 become

$$\begin{Bmatrix} \sigma_{1f} \\ \sigma_{1s} \end{Bmatrix} = \begin{Bmatrix} 24x_1 - 32y_1 \\ -24x_1 + 32y_1 \end{Bmatrix}, \begin{Bmatrix} \sigma_{2f} \\ \sigma_{2s} \end{Bmatrix} = \begin{Bmatrix} -12x_1 - 32y_1 \\ 12x_1 + 32y_1 \end{Bmatrix} \tag{7.40}$$

Substituting Eqs. (7.37) and (7.38) into Eq. (7.40), the stresses are expressed in terms of design variables as in the following.

$$\begin{Bmatrix} \sigma_{1f} \\ \sigma_{1s} \end{Bmatrix} = \begin{Bmatrix} \dfrac{97.222}{A_1} \\ -\dfrac{97.222}{A_1} \end{Bmatrix}, \quad \begin{Bmatrix} \sigma_{2f} \\ \sigma_{2s} \end{Bmatrix} = \begin{Bmatrix} \dfrac{13.89}{A_2} \\ -\dfrac{13.89}{A_2} \end{Bmatrix} \tag{7.41}$$

The design requirements necessitate that displacements of joint 1 should not be more than 0.25 cm and the axial stresses in members should be ≤12 kN/cm^2. Accordingly the displacement constrains become

$$x_1 = \frac{-A_1 + 7A_2}{2.592A_1A_2} \le 0, \qquad y_1 = \frac{A_1 + 3.5A_2}{3.456A_1A_2} \le 0.25 \tag{7.42}$$

Simplification of Eq. (7.42) yield:

$$-A_1 + 7A_2 - 0.648A_1A_2 \le 0, \qquad A_1 + 3.5A_2 - 0.864A_1A_2 \le 0 \tag{7.43}$$

Using the positive values of stresses given in Eq. (7.41) and applying the stress limitation of 12 kN/cm^2, the following stress constraints are obtained.

$$8.102 - A_1 \le 0, \qquad 1.1575 - A_2 \le 0 \tag{7.44}$$

Collecting the constraints, Eqs. (7.43) and (7.44), together with the objective function the optimum design problem of 3 bar space truss has the following form.

$$\begin{aligned} \text{min.} \quad & W = 500(A_1 + 2A_2) \\ \text{subject to} \quad & g_1(A) = -A_1 + 7A_2 - 0.648A_1A_2 \le 0 \\ & g_2(A) = A_1 + 3.5A_2 - 0.864A_1A_2 \le 0 \\ & g_3(A) = 8.102 - A_1 \le 0 \\ & g_4(A) = 1.1575 - A_2 \le 0 \end{aligned} \tag{7.45}$$

Solution by Sequential Quadratic Programming

The solution of the optimum design problem given through Eq. (7.45) is first obtained by the sequential quadratic programming method. This method linearizes the nonlinear constraints at a selected initial design point as shown through Eq.

(7.5) and transforms the nonlinear programming problem into a linear programming problem. Applying this concept to the constraints of the optimum design problem, Eq. (7.45), at a initial design point vector $\mathbf{A}_0$ the following linear programming problem is obtained.

$$[C]\,\{\Delta A\} \le \{b\} \tag{7.46a}$$

where $[C]$ and $\{b\}$ are defined as

$$[C] = \begin{bmatrix} \{\nabla g_1(A_0)\}^T \\ \{\nabla g_2(A_0)\}^T \\ \{\nabla g_3(A_0)\}^T \\ \{\nabla g_4(A_0)\}^T \end{bmatrix} = \begin{bmatrix} -1 - 0.648A_{2,0} & 7 - 0.648A_{1,0} \\ -1 - 0.864A_{2,0} & 3.5 - 0.864A_{1,0} \\ -1 & 0 \\ 0 & -1 \end{bmatrix} \tag{7.46b}$$

$$\{b\} = \begin{Bmatrix} -g_1(A_0) \\ -g_2(A_0) \\ -g_3(A_0) \\ -g_4(A_0) \end{Bmatrix} = \begin{Bmatrix} -A_{1,0} + 7A_{2,0} - 0.648A_{1,0}A_{2,0} \\ A_{1,0} + 3.5A_{2,0} - 0.864A_{1,0}A_{2,0} \\ 8.102 - A_{1,0} \\ 1.1575 - A_{2,0} \end{Bmatrix} \tag{7.46c}$$

Selecting initial design point as $A_0 = \{A_{1,0} = 5, A_{2,0} = 5\}^{\mathrm{T}}$ and substituting the values of $A_{1,0}$ and $A_{2,0}$ into Eq. (7.46a) the following linear programming problem is obtained.

$$\begin{aligned} &\text{min.} \quad f = \Delta A_1 + 2\,\Delta A_2 \\ &\text{subject to} \quad \begin{bmatrix} -4.24 & 3.76 \\ -3.32 & 0.82 \\ -1 & 0 \\ 0 & -1 \end{bmatrix} \begin{Bmatrix} \Delta A_1 \\ \Delta A_2 \end{Bmatrix} \le \begin{Bmatrix} -13.8 \\ -0.90 \\ -3.10 \\ 3.8425 \end{Bmatrix} \end{aligned} \tag{7.47}$$

Solution of the linear programming given by Eq. (7.47) by the Simplex method results in $\Delta A_1 = 3.102$ and $\Delta A_2 = -3.8{,}425$ after 2 simplex iterations. The values of design variables then become $A_{1,1} = 5 + 3.102 = 8.102$ and $A_{2,1} = 5\text{–}3.8425 = 1.1575$. Substituting these new values into Eq. (7.46a) results in the following linear programming problem.

$$\begin{aligned} &\text{min.} \quad f = \Delta A_1 + 2\,\Delta A_2 \\ &\text{subject to} \quad \begin{bmatrix} -1.75 & 1.75 \\ -0.0001 & -3.5 \\ -1 & 0 \\ 0 & -1 \end{bmatrix} \begin{Bmatrix} \Delta A_1 \\ \Delta A_2 \end{Bmatrix} \le \begin{Bmatrix} 6.0765 \\ -4.0506 \\ 0 \\ 0 \end{Bmatrix} \end{aligned} \tag{7.48}$$

Solution of this new linear programming problem gives $\Delta A_1 = 0.0$ and $\Delta A_2 = 1.157$. The new values of the design variables then become $A_{1,2} = 8.102$ and $A_{2,2} = 1.1575 + 1.157 = 2.315$. Carrying out the linearization with these new values yields the following linear programming problem.

Table 7.2 Randomly selected individuals in the initial generation

Individual number	A_1	A_2	$g_1(\mathbf{A})$	$g_2(\mathbf{A})$	$f = A_1 + 2A_2$
1	8.1020	1.3828	−5.6822	3.2620	10.8676
2	8.7711	7.0547	0.5153	−19.9994	22.8805
3	9.9299	8.5701	−5.0840	−33.6010	27.0700
4	8.7385	9.2513	3.6346	−28.7298	27.2411
5	9.6125	8.5206	−3.0426	−31.3310	26.6537
6	8.7569	8.8609	2.9886	−27.2713	26.4788
7	8.2727	9.0122	6.5011	−24.5999	26.2970
8	9.4325	7.6528	−2.6386	−26.1503	24.7380
9	8.6717	1.5971	−6.4665	2.2954	11.8660
10	9.8257	2.0210	−8.5466	−0.2582	13.8678

$$\begin{aligned} \text{min.} \quad & f = \Delta A_1 + 2\,\Delta A_2 \\ \text{subject to} \quad & \begin{bmatrix} -2.5 & 1.75 \\ -1.0 & -3.5 \\ -1 & 0 \\ 0 & -1 \end{bmatrix} \begin{Bmatrix} \Delta A_1 \\ \Delta A_2 \end{Bmatrix} \le \begin{Bmatrix} 4.05 \\ 0.0008 \\ 0 \\ 1.1575 \end{Bmatrix} \end{aligned} \tag{7.49}$$

Solution of this new linear programming problem gives $\Delta A_1 = 0.0$ and $\Delta A_2 = 0.0$ indicating that convergence is obtained after 3 iteration in the sequential quadratic programming. Hence the optimum solution of the design problem is found to be $A_1 = 8.102\ \text{cm}^2$ and $A_2 = 2.315\ \text{cm}^2$. Substitution of these values to the original nonlinear programming problem of Eq. (7.45) shows that all the constraints are satisfied and the objective function has the minimum value which is equal to 6,366 cm^3. This substitution shows that the active constraint in the design problem is the vertical displacement of the joint 1 which dominates the design problem together with the stress constraints. The separate analysis of 3 bar truss under the external loads gives the X and Y displacements of joint 1 as 0.167 cm and 0.25 cm while the compressive stresses in members computed from the members forces obtained as a result of analysis show that they are at their upper bounds of 12 kN/cm^2 which verifies the previous conclusion.

Solution by Differential Evolution Method

The same optimum design problem is also solved by differential evolution method. This method is an evolutionary algorithm similar to genetic algorithms and evolutionary strategies that are population based numerical optimization techniques. The method creates new individuals on a particular manner. A new individual is generated by adding the weighted difference between 2 individuals with a third. If the resulting individual is better than a predetermined individual, the new vector replaces it.

The first step is to select the size of the population which is selected as 10 for the optimum design problem given in Eq. (7.45). Since inequalities $g_3(\mathbf{A})$ and $g_4(\mathbf{A})$ in Eq. (7.45) are lower bounds on design variables they are excluded from the optimum design problem by applying upper and lower bounds on design

Table 7.3 The second generation

Individual number	A_1	A_2	$g_1(\mathbf{A})$	$g_2(\mathbf{A})$	$f = A_1 + 2A_2$
1	8.1020	1.3828	−5.6822	3.2620	10.8676
2	8.7711	7.0547	0.5153	−19.9994	22.8805
3	9.9299	8.5701	−5.0840	−33.6010	27.0700
4	8.7385	3.0256	−4.6919	−28.7298	27.2411
5	9.6125	8.5206	−3.0426	−31.3310	26.6537
6	8.7569	8.8609	2.9886	−27.2713	26.4788
7	10.0000	9.0122	−5.3137	−24.5999	26.2970
8	9.4325	7.6528	−2.6386	−26.1503	24.7380
9	8.6717	1.1575	−7.0735	2.2954	11.8660
10	9.8257	2.0210	−8.5466	−0.2582	13.8678

variables as $8.102 \leq A_1 \leq 10$ and $1.1575 \leq A_2 \leq 10$. The initial population is constructed randomly within these bounds that are given in Table 7.2.

Inspection of the values given in columns belonging to $g_1(\mathbf{A})$ and $g_2(\mathbf{A})$ of Table 7.2 reveals the fact that among these 10 individuals that are selected randomly only the third, fifth, eighth, and tenth individuals satisfy both constraints of $g_1(\mathbf{A})$ and $g_2(\mathbf{A})$ given in Eq. (7.45). Among these the tenth one has the least objective function value as seen from the last column of the Table 7.2 and is considered the best individual in this generation. In order to obtain the next generation 3 individuals are randomly selected from the initial generation. Let these be the r_1, r_2, and r_3 individuals where r_1, r_2, and r_3 are distinct. The donor individual is calculated as

$$\upsilon_i = A_{r1} + F(A_{r2} - A_{r3}) \tag{7.50}$$

where F is mutation factor which is selected as 0.8. The trial individual u_i is developed from the elements of the target individual A_i and elements of the donor individual υ_i with probability CR.

$$u_{j,i} = \left\{ \begin{array}{lll} \upsilon_{j,i} \text{ if } \mathrm{rand}_{j,i} \leq CR & \text{or} & j = I_{\mathrm{rand}} \\ A_{j,i} \text{ if } \mathrm{rand}_{j,i} > CR & \text{and} & j \neq I_{\mathrm{rand}} \end{array} \right\} \tag{7.51}$$

where $i = 1, 2, \ldots, 10$, $j = 1, 2$, $\mathrm{rand}_{j,i}$ is a random number and I_{rand} is a integer 1 or 2 because there are only 2 design variables in the optimum design problem. I_{rand} ensures that $\upsilon_i \neq A_{j,i}$. The target individual A_i is compared with the trial individual υ_i and the one with the lowest objective function value is included in the next generation.

Table 7.4 Design history for 3 bar space truss

Iteration number	A_1	A_2	$f = A_1 + 2A_2$
1	9.8257	2.0211	13.8678
5	9.8257	2.0211	13.8678
10	8.1020	2.3679	12.8378
15	8.1020	2.3679	12.8378
20	8.1020	2.3679	12.8378
25	8.1020	2.3679	12.8378
30	8.1020	2.3679	12.8378
35	8.1034	2.3619	12.8272
40	8.1020	2.3303	12.7625
45	8.1020	2.3210	12.7440
50	8.1028	2.3152	12.7331

$$A_i = \begin{Bmatrix} u_i \text{ if } f(u_i) \leq F(A_i) \\ A_i \text{ otherwise} \end{Bmatrix} \quad i = 1, 2, \ldots, 10 \tag{7.52}$$

The second generation is obtained by applying these rules to the initial generation which is given in Table 7.2. It is apparent from Table 7.3 that 3 new individuals that are fourth, seventh, and ninth are added to the initial generation by replacing their previous counterparts. Inspection of the values of the constraints $g_1(\mathbf{A})$ and $g_2(\mathbf{A})$ in the table shows that third, fourth, fifth, seventh, eighth, and tenth individuals satisfy these constraints. Among these, once again, the tenth individual has the least value for the objective function. Hence the best individual of the initial generation continues to be the best individual in the second generation. The mutation, recombination, and selection are carried in a similar manner explained above until the maximum number of generations is reached. In this example the maximum number of generations is taken as 50. The best individuals attained in every fifth generation are listed in Table 7.4.

It is apparent from the table that differential evolution method also finds the same optimum solution where $A_1 = 8.102$ cm^2 and $A_2 = 2.315$ cm^2. It took 50 iteration with 510 function calls to reach this optimum design whereas sequential quadratic programming found the same result in 3 iterations. Naturally it is clear that metaheuristic methods are computationally expensive. However, they do not need gradient computations of neither the objective function nor the constraints. In some design problem they may be difficult to determine and they may not even exist.

7.6.2 Reliability-Based Design Optimization

In this section, in the light of the information given in the previous sections, 3 numerical examples associated with the monopod, the tripod, and the jacket type

Table 7.5 The stochastic description of the random variables used in the RBDO of the offshore towers

	Description	Mean (μ)	COV (σ/μ)	Distribution
g	gravity	9.81 m/sn^2	–	–
ρ	steel density	7800 kg/m^3	–	–
ρ_w	water density	1024 kg/m^3	–	–
v	Poisson ratio	0.30	–	–
f_y	Yield stress	450.0 MPa.	0.06	Lognormal
E	Young modulus	210.0 GPa.	0.05	Lognormal
M_{deck}	Mass of the deck	problem depended	0.10	Lognormal
H_{max}	Maximum wave height	22.73 m	0.05	Shift Weibull
α_{wave}	Wave steepness	0.06	0.125	Lognormal
c_m	Inertia force coefficient	1.60	0.10	Lognormal

offshore towers are presented. To implement the RBDO of the offshore towers under some uncertainties associated with the loads, the material properties, and environmental data, etc. the developed integrated framework based on the 2 level approaches [42–47] is employed. It has both the RIA and PMA for the RBDO to evaluate the probabilistic constraints. In addition, sequential quadratic programming (SQP) [5, 19, 79, 80] and differential evolution (DE) [21, 22, 81] as optimization methods in order to find the optimum design variables. The mass of the tower is considered as being the objective function; the thickness and diameter of the cross-section of the members of the towers are taken as being design variables of the optimization. The probability distribution types and the characteristics of statistical parameters of the random variables used in the RBDO of the offshore towers are presented in Table 7.5. Three types of limit states as being the functions of design and random variables are used in the RBDO of the tower. These are based on: yielding stress, buckling stress, and natural frequency.

Yield stress function

The probabilistic constraint based on the limit-state-function related to yield stress is defined as:

$$G(\mathbf{d},\mathbf{X}) = f_y - \sigma_{\text{nom}},\ \text{in which}\ \sigma_{\text{nom}} = \frac{N}{A} \mp \frac{M_y}{I_y}\frac{D}{2} \mp \frac{M_z}{I_z}\frac{D}{2} \tag{7.53}$$

where f_y is the yield stress, σ_{nom} is the nominal normal stress, N is the axial force, M_y and M_z are the bending moments about y and z coordinate axes, A is the cross-sectional area, I_y and I_z are the inertia moments, D is the diameter of the member.

Buckling stress function

The reliability constraint based on the limit-state-function related to buckling stress is defined as:

$$G(\mathbf{d},\mathbf{X}) = \sigma_{cr} - \sigma_{\text{nom}} \tag{7.54}$$

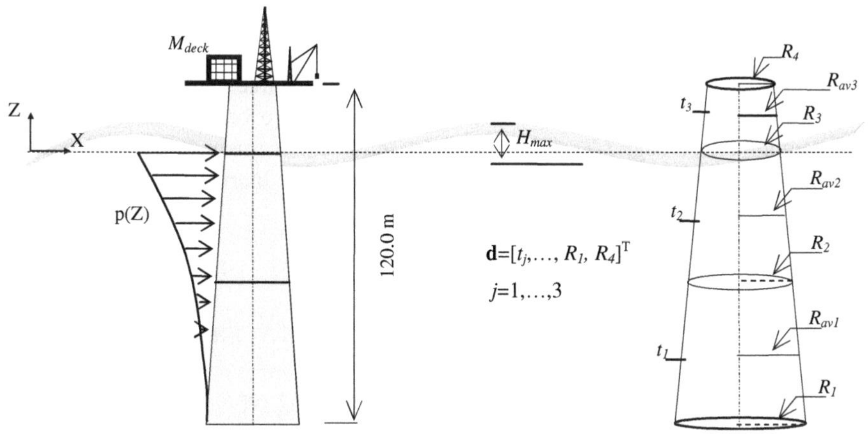

Fig. 7.7 A monopod tower with 3 segments

in which σ_{cr} is the critical buckling stress of the member, which is calculated using the DNV rule [42–47, 82, 83].

Natural frequency function

The limit state function based on the natural frequency is defined as:

$$G(\mathbf{d}, \mathbf{X}) = \omega_n - \omega_{\text{limit}} \tag{7.55}$$

in which, ω_n is the lowest natural frequency of the tower and ω_{limit} is a threshold frequency. The purpose of considering a frequency limit state function is to keep the lowest natural frequency at a reasonable level to reduce dynamic responses and consequently fatigue damages. The threshold frequency is kept far from the peak frequency of the sea spectrum to reduce the dynamic response quantities. A reasonable value of ($\omega_{\text{limit}} = 3.0$ rad/s) is adopted for the threshold frequency.

The following formulation summarizes the RBDO of the offshore towers from the point of view of the aforementioned information.

$$\begin{array}{lll} \text{find} & \mathbf{d} & \\ \text{minimum} & W(\mathbf{d}) = \rho \sum\limits_{j-1}^{ne} A_j L_j & \\ \text{subject to:} & \beta_i \geq \beta_{i,\text{target}} & \text{for RIA} \\ & G_i^p \geq 0 & \text{for PMA} \quad i = 1, \ldots, \text{nrc} \end{array} \tag{7.56}$$

where ρ is density of steel, A_j and L_j are the area and the length of the element j, *ne* is the total number of elements of tower, ndv is the number of design variables adopted for the optimization, nrc is the total number of reliability constraints. A value of $\beta_{i,\text{ target}} = 3.70$ ($i = 1,\ldots,\text{nrc}$) is adopted as target reliability index [84].

7.6.2.1 A Simple Example for Monopod Offshore Tower

For the sake of simplicity, the monopod tower is firstly investigated in order to be followed by the step of RBDO process. It is assumed to be conical and divided by 3 segments (see Fig. 7.7). Parameters related to *j*th segment cross-section such as radius R_{av_j}, area A_{s_j}, diameter D_{av_j}, and moment of inertia I_j, are calculated, respectively as

$$\begin{aligned} R_{av_j} &= \frac{R_j + R_{j+1}}{2}, & D_{av_j} &= 2R_{av_j} \\ A_{s_j} &= 2\pi R_{av_j} t_j, & I_j &= \frac{\pi}{8} D_{av_j}^3 t_j \quad j = 1, \ldots, nseg \end{aligned} \tag{7.57}$$

In Eq. (7.57), t_j is thickness of the *j*th segment and nseg represents the number of segments. Yield stress function, buckling stress, and natural frequency given in Eqs. (7.53)–(7.55) are considered as a service limit state functions being dependent on design and random variables for the RBDO of the tower.

Considering Eq. (7.53), the reliability constraint based on the limit-state-function related to yield stress can be stated as

$$G(d, X) = f_y - \sigma_{\text{nom}} = f_y - \left(\frac{N_j}{A_{s_j}} + \frac{M_j}{I_j} \frac{D_{av_j}}{2} \right) \tag{7.58}$$

in which *j* represents the number of segments, N_j is the normal (axial) force, M_j is the bending moment, I_j and D_{avj} are the inertia moment, and average diameter of *j*th section, respectively, f_y is the yield stress. The maximum normal force *N* acting on the *j*th segment is calculated from

$$N_j = \left[M_{\text{deck}} + \sum_{j=1}^{\text{nseg}=3} \rho A_{s_j} L_{e_j} \right] g \tag{7.59}$$

where M_{deck} denotes the mass of the deck, $L_{e_j}$is the length of the *j*th segment, ρ is density of the steel, and *g* is the gravity acceleration. It is assumed that numbering of segment is started from bottom. The bending moment at the bottom of any segment is calculated from [85]

$$M = \int_{Z=-b_{b_j}}^{0} (b_{b_j} + Z) p(Z) dz \tag{7.60}$$

where b_{b_j} is the bottom boundary of the segment *j*, *p(Z)* is the wave force calculated by using the Morison's equation. For large diameters, the contribution of the drag force term is negligible in comparison with the inertia force term. Hence, the drag force term is ignored and only the inertia force term is considered in the corresponding equation. Besides, since the marine growths are not taken into account the increased diameter for the member is also ignored. Hence, the bending moment can be obtained as expressed by,

$$M_j = \frac{D_{av_j}}{2}\frac{g\rho_w}{m}\hat{\eta}\, C_m \tag{7.61}$$

in which, $\hat{\eta}$ is the wave amplitude and given by $\hat{\eta} = H_{\max}/2$ ($H_{\max}$ is the maximum wave height), ρ_w is the water density, m is the wave number ($m = 2\pi\alpha_{\text{wave}}/H_{\max}$, where $\alpha_{\text{wave}}$is the wave steepness), and C_m is the parameter of the inertia force term defined as

$$C_m = \frac{\pi}{2} c_m D_{av_j}\left(m\, b_{b_j}\tanh(md_w) + \frac{\cosh(m(d_w - b_{b_j}))}{\cosh(md_w)} - 1\right) \tag{7.62}$$

in which d_w is the water depth, c_m is the inertia force coefficient.

Considering Eq. (7.54), the reliability constraint based on the limit-state-function related to buckling stress can be stated as

$$G(d, X) = \sigma_{cr,j} - \sigma_{\text{nom}} = \sigma_{cr,j} - \left(\frac{N_j}{A_{s_j}} + \frac{M_j}{I_j}\frac{D_{av_j}}{2}\right) \tag{7.63}$$

The critical buckling stress $\sigma_{cr,\ j}$ of the jth segment is calculated from the DNV rule [42–47, 82, 83] as given by

$$\sigma_{cr,j} = \frac{f_y}{\sqrt{1 + \lambda_j^4}} \tag{7.64}$$

In Eq. (7.64) f_y is the yield stress, and λ is a dimensionless buckling parameter calculated from

$$\lambda_j^2 = \frac{f_y}{\sigma_{a_j} + \sigma_{b_j}}\left(\frac{\sigma_{a_j}}{\sigma_{Ea_j}} + \frac{\sigma_{b_j}}{\sigma_{Eb_j}}\right) \tag{7.65}$$

where σ_{a_j} and σ_{b_j}denote the stresses due to normal force and bending moment, respectively, the stresses σ_{Ea_j} and σ_{Eb_j} are defined as

$$\begin{aligned}\sigma_{Ea_j} &= (1.5 - 50\chi)C_{a_j}\frac{\pi^2 E}{12(1-\nu^2)}\left(\frac{t_j}{L_{r_j}}\right)^2\\ \sigma_{Eb_j} &= (1.5 - 50\chi)C_{b_j}\frac{\pi^2 E}{12(1-\nu^2)}\left(\frac{t_j}{L_{r_j}}\right)^2\end{aligned} \tag{7.66a}$$

where the corresponded parameters in Eq. (7.66a) are

$$C_{a_j} = \sqrt{1 + (\rho_a \xi)^2}, \qquad C_{b_j} = \sqrt{1 + (\rho_b \xi)^2}$$
$$\rho_a = 0.5\left(1 + \frac{R_{av_j}}{150t_j}\right)^{-0.5}, \quad \rho_b = 0.5\left(1 + \frac{R_{av_j}}{300t_j}\right)^{-0.5} \tag{7.66b}$$
$$\xi = 0.702\mathbb{Z}, \quad \mathbb{Z} = \frac{L_{r_j}}{R_{av_j}t_j}\sqrt{1 - v^2}, \quad L_{r_j} = \frac{L_{e_j}}{nr + 1}$$

in which, v is the Poisson's ratio, E is the Young's modulus, nr is the number of ring-stiffeners (here, $nr = 0$, because ring-stiffener is not considered in the optimization of the tower), χ is a parameter and taken as 0.02. However, since the top segment is in the air, it is not subjected to a bending moment, and therefore, the critical buckling stress is calculated from [87]:

$$\sigma_{cr_{top}} = \frac{E}{\sqrt{3(1 - v^2)}}\left(\frac{t_{top}}{R_{av_{top}}}\right) \tag{7.67}$$

in which, t_{top} and $R_{av_{top}}$ represent thickness and average radius of the top segment.

Recalling Eq. (7.55), the reliability constraint based on the limit-state-function related to natural frequency can be stated as

$$\mathrm{G}(\mathbf{d}, \mathbf{X}) = \omega_n - \omega_{\text{limit}} = \sqrt{\frac{k}{m^*}} - 3.0 \tag{7.68}$$

where k and m^* are the generalized stiffness and mass, respectively. The generalized mass m^* is calculated depending on the deflection shape δ of the structure. Having used the function of δ, which is given approximately by $\delta(z) = \frac{3}{2}\left(\frac{z}{h_s}\right)^2 - \frac{1}{2}\left(\frac{z}{h_s}\right)^3$, where z is measured from the bottom of the tower, the generalized mass for a segment can be stated as

$$m_j^* = \rho A_{s_j} \int_{z_j}^{z_{j+1}} \delta^2(z)dz = \frac{\rho A_{s_j}}{4h_s^4}\left[\frac{z_{j+1}^7 - z_j^7}{7h_s^2} - \frac{z_{j+1}^6 - z_j^6}{h_s} + \frac{9(z_{j+1}^5 - z_j^5)}{5}\right] \tag{7.69}$$

In Eq. (7.69), $j = 1,\ldots,\text{nseg}$, h_s is the height of the tower, z_j and z_{j+1} represent Z coordinates of first and second node of segment j. From this definition, the generalized mass for the tower with segments can be stated as

$$m^* = M_{\text{deck}} + \sum_{j=1}^{\text{nseg}} m_j^* \tag{7.70}$$

where M_{deck} is the mass of the deck and nseg represents the number of segment. Due to the segments having different t and R there are discontinuities along the tower. Therefore, the flexural rigidity formulation $k = \int_0^{h_s} EI(\partial^2\delta/\partial z^2)dz$ is not used directly to calculate the flexural rigidity of the tower. Instead, a segmented integration is carried out. It is obtained as stated by,

$$k = \frac{3EI_{\text{nseg}}}{h_s^3} \frac{1}{\left[\sum_{j=1}^{\text{nseg}-1} \frac{I_{\text{nseg}}}{I_j} \alpha\{3 - (3 + (j-1))\alpha + (3(j^2 - j) + 1)\alpha^2\}\right] + \alpha^3} \tag{7.71}$$

where I_j (j = 1,…,nseg) represents inertia moments of the jth segments, α is equal to 1/nseg.

Up to now, the formulations to employ the limit state functions based on the yield stress, buckling stress, and natural frequency are defined. Now the RBDO process of the tower can be summarized as

Find the design variables vector **d** consisting of thicknesses of segments t_j and radii of bottom and top (R_1 and R_4), which are assumed to be independent, such that the objective function $W(\mathbf{d})$ taken as the mass of the tower has the minimum value within a region defined as 0.010 m $\leq t_j (j = 1,2,3) \leq$ 0.10 m., 5.0 m $\leq R_1 \leq$ 10.50 m., and 2.50 m $\leq R_4 \leq$ 5.25 m. The radii, between the bottom and top are linearly linked to R_1 and R_4. The probability distribution types and characteristic statistical parameters of the random variables are presented in Table 7.5. A shifted Weibull distribution, $F_{H_{\max}}(h) = 1 - \exp[-((h - A)/B)]$, is used for $H_{\max}$ with A = 21.6 m. and B = 1.13 m. [86]. Mass of the deck is taken as 2 × 10^6 kg. For the admissible $\beta_{i,\text{target}}$ values, the minimum value 3.0 is considered for this example, only.

For the calculation convenience, only the point-based algorithm SQP and RIA are employed as the optimization method and the reliability approach. The initial design point is taken into account as $\mathbf{d}^0 = \{t_1 = 1.50$ cm. $t_2 = 1.50$ cm. $t_3 = 1.50$ cm. $R_1 = 6.0$ m. $R_4 = 4.50$ m.$\}^T$. At this point, N_1 = 24.16 MN., N_2 = 22.11 MN., N_3 = 20.28 MN., M_1 = 841.60 MN.m., M_2 = 241.67 MN.m., $M_3 = 0$, $\sigma_{cr,1}$ = 54.27 MPa., $\sigma_{cr,2}$ = 61.26 MPa., $\sigma_{cr,3}$ = 412.05 MPa., ω_n = 1.18 rad/sec. and $W(\mathbf{d}^0)$ = 0.463 megaton(Mt.). Depending on these values, reliability indices for the related limit state functions based on the buckling stresses $\beta_1 = -20.56$, $\beta_2 = -12.60$, $\beta_3 = 19.83$, based on the natural frequency $\beta_4 = -18.27$ and based on the yield stresses $\beta_5 = -2.19$, $\beta_6 = 4.82$, $\beta_7 = 19.76$ are found after the reliability analysis based on the FORM in the RIA is performed. Since the last segment is in the air it is not subject to moment. Therefore, corresponding moment value M_3 is equal to zero. To make the optimization using SQP it needs the gradients related to objective function and constraints in addition to their values. The gradients associated with the design variables are calculated for the objective function as

$\nabla W(\mathbf{d}^0) = \{\,13.937 \quad 12.406 \quad 4.533 \quad 0.044 \quad 0.044\,\}^T$, for the buckling stress constraints as

Table 7.6 Convergences history for the monopod tower with 3 segments

	t_1(cm)	t_2	t_3	R_1(m)	R_4	β_1	β_2	β_3	β_4	β_5	β_6	β_7	W(**d**) Mt.
1	1.50	1.50	1.50	6.00	4.50	−20.56	−12.60	19.82	−18.27	−2.19	4.82	19.76	0.463
2	3.26	2.59	1.00	9.66	2.50	−5.91	−1.80	12.56	−6.01	7.52	8.92	12.85	1.013
3	3.87	2.89	3.13	8.16	3.58	−2.95	−0.06	33.04	−5.57	7.32	9.70	28.85	1.182
4	5.38	3.59	1.00	9.88	3.94	1.88	2.54	12.54	1.28	9.86	11.27	15.93	1.728
5	5.56	3.64	1.59	9.36	4.36	2.21	2.69	20.86	1.48	9.54	11.36	20.53	1.766
6	6.00	3.79	1.00	9.47	4.83	2.95	3.00	12.53	2.87	9.83	11.67	17.32	1.914
7	5.94	3.78	1.09	9.73	4.57	2.96	3.00	14.05	2.89	10.03	11.63	17.67	1.915
8	5.97	3.79	1.12	9.70	4.64	3.00	3.00	14.52	3.00	10.01	11.65	17.99	1.928
9	5.99	3.79	1.08	9.66	4.67	3.00	3.00	13.88	3.00	9.99	11.66	17.75	1.928
10	5.98	3.79	1.07	9.70	4.64	3.00	3.00	13.81	3.00	10.01	11.66	17.66	1.928
11	5.98	3.79	1.07	9.70	4.64	3.00	3.00	13.81	3.00	10.01	11.66	17.66	1.928
12	5.98	3.79	1.07	9.70	4.64	3.00	3.00	13.81	3.00	10.01	11.66	17.66	1.928

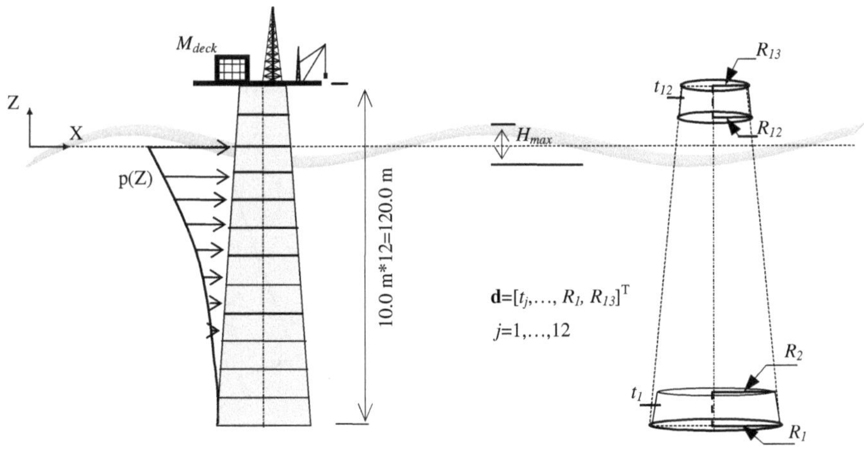

Fig. 7.8 Monopod tower

Table 7.7 Results of the RBDO of monopod tower

d	Deterministic optimization		with RIA: RBDO		with RIA: RBDO (**d** is random)		with PMA: RBDO		with PMA: RBDO (**d** is random)	
	SQP	DE	SQP	DE	SQP	DE	SQP	DE	SQP	DE
t_1 (cm)	4.47	4.41	5.80	5.78	6.35	6.42	5.81	5.79	6.35	6.41
t_2	4.17	4.15	5.43	5.42	5.95	5.97	5.43	5.51	5.95	6.09
t_3	3.89	3.85	5.04	5.05	5.52	5.54	5.04	5.01	5.52	5.60
t_4	3.57	3.54	4.63	4.61	5.06	5.24	4.63	4.78	5.06	5.14
t_5	3.24	3.23	4.19	4.18	4.58	4.65	4.19	4.17	4.58	4.65
t_6	2.89	2.92	3.72	3.73	4.37	4.28	3.72	3.71	4.07	4.33
t_7	2.52	2.54	3.23	3.25	3.73	4.06	3.23	3.28	3.72	3.89
t_8	2.17	2.16	2.71	2.70	3.62	3.45	2.71	2.89	3.54	3.63
t_9	1.81	1.93	2.20	2.44	2.88	3.29	2.20	2.71	2.88	2.98
t_{10}	1.52	1.67	1.82	1.87	2.24	2.42	1.82	1.85	2.75	2.30
t_{11}	1.11	1.39	1.29	1.32	1.74	1.56	1.29	1.44	1.48	1.68
t_{12}	1.00	1.04	1.00	1.05	1.14	1.01	01.00	1.00	1.00	1.13
R_1 (m)	10.50	10.50	10.50	10.49	10.50	10.49	10.50	10.48	10.50	10.49
R_{13}	4.21	4.08	5.10	5.04	5.25	5.25	5.11	4.95	5.25	5.25
W(**d**) Mt.	1.278	1.287	1.708	1.716	1.956	1.987	1.709	1.731	1.947	1.987

$$\nabla G_1(\mathbf{d}^0, \mathbf{X}) = \{\, 1383.352 \quad -30.131 \quad -11.011 \quad -0.719 \quad -1.070 \,\}^T$$

$$\nabla G_2(\mathbf{d}^0, \mathbf{X}) = \{\, 0.0 \quad 1431.143 \quad 13.067 \quad 0.629 \quad -1.082 \,\}^T$$

$\nabla G_3(\mathbf{d}^0, \mathbf{X}) = \{\, 0.0 \quad 0.0 \quad 1196.914 \quad -0.0009 \quad -0.001 \,\}^T$, for the natural frequency constraint as

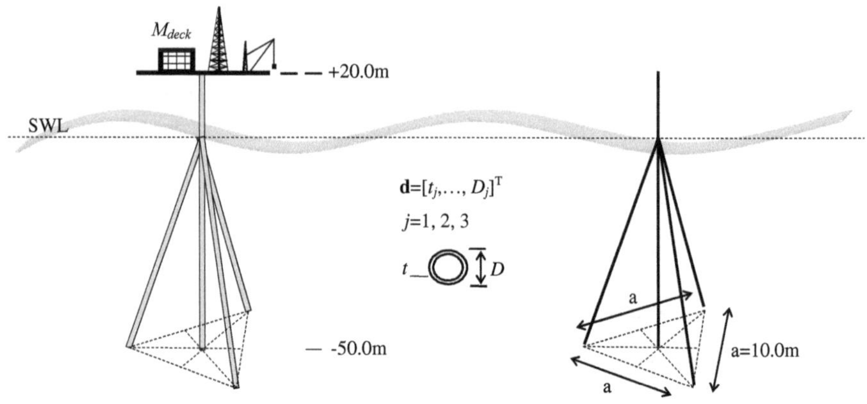

Fig. 7.9 Tripod tower

$\nabla G_4(\mathbf{d}^0, \mathbf{X}) = \{516.128 \quad 116.551 \quad -64.874 \quad 3.757 \quad 1.569\}^T$ and for the yield stress constraints as

$$\nabla G_5(\mathbf{d}^0, \mathbf{X}) = \{538.216 \quad -3.858 \quad -1.410 \quad 0.848 \quad -0.996\}^T$$

$$\nabla G_6(\mathbf{d}^0, \mathbf{X}) = \{0.0 \quad 509.287 \quad -1.828 \quad 0.0063 \quad 0.101\}^T$$

$$\nabla G_7(\mathbf{d}^0, \mathbf{X}) = \{0.0 \quad 0.0 \quad 570.806 \quad 0.155 \quad 1.697\}^T$$

After repeating the RBDO procedure which is illustrated above, the optimum point is obtained using IMSL-Library [80] at the 12 iterations as $\mathbf{d}^{12} = \{t_1 = 5.98 \text{ cm.} \quad t_2 = 3.79 \text{ cm.} \quad t_3 = 1.07 \text{ cm.} \quad R_1 = 9.70 \text{ m.} \quad R_4 = 4.64 \text{ m.}\}^T$, $W(\mathbf{d}^{12}) = 1.928$ Mt.

The iteration history for this problem is presented in Table 7.6, along with reliability indices and mass of the tower.

7.6.2.2 Monopod Offshore Tower

The RBDO of offshore structures is applied to the design of a monopod tower with twelve segments as shown in Fig. 7.8. A total of 14 design variables, which consist of thicknesses of each segment and the radii at the bottom (R_1) and top (R_{13}) segments, and 6 random parameters presented in Table 7.5 are considered in the optimization process. R_1 and R_{13} are assumed to be independent of each other.

Other radii, between the bottom and top are linearly linked to R_1 and R_{13}. The structural elements are made of steel frames with the cross-section of tubular member to be represented by thickness and radii. Three optimum solutions are indicated in Table 7.7 considering different optimization cases including the deterministic optimization performed without consideration of any uncertainties in

Table 7.8 Results of the RBDO of tripod tower

d			with RIA				with PMA			
	Deterministic optimization		RBDO		RBDO (**d** is random)		RBDO		RBDO (**d** is random)	
	SQP	DE	SQP	DE	SQP	DE	SQP	DE	SQP	DE
t_1 (cm)	1.21	1.22	1.78	1.89	1.61	1.67	1.78	1.77	1.58	1.64
t_2	4.29	4.28	6.51	6.47	3.76	3.78	6.51	6.52	3.76	3.78
t_3	1.00	1.00	1.00	1.00	2.42	2.41	1.00	1.00	2.42	2.43
D_1 (m)	10.00	10.00	10.00	10.00	10.00	9.99	10.00	9.99	10.00	9.95
D_2	10.00	10.0	10.00	9.99	10.00	10.00	10.00	10.00	10.00	9.98
D_3	1.00	1.00	1.00	1.00	6.34	6.32	1.00	1.00	6.34	6.31
W(**d**) Mt.	0.618	0.619	0.917	0.918	1.103	1.105	0.916	0.916	1.102	1.107

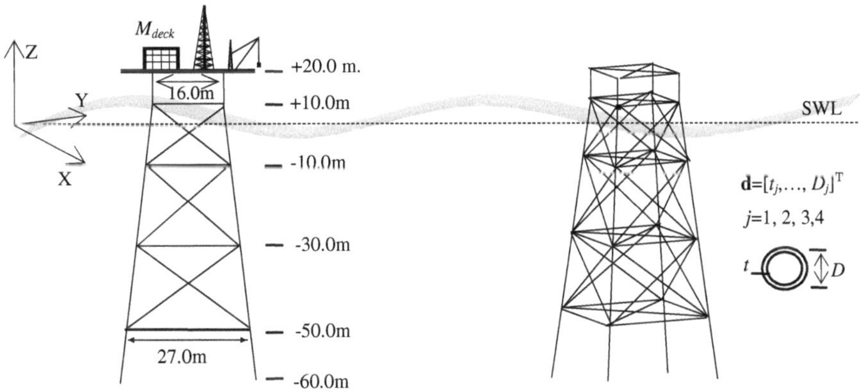

Fig. 7.10 Jacket tower

parameters, the RBDO fulfilled with random parameters given in Table 7.5, and the RBDO with the random design variables in addition to the random variables presented in Table 7.5. For this case, lognormal distributions with (COV = 0.05) are assumed for the probability models of design variables and mean value of the distributions are taken as design variables of the optimization. The adopted lower and upper boundaries for the design variables are $1.0 \leq t_i(\text{cm}) \leq 10.0$, $5.0 \leq R_1(\text{m}) \leq 10.50$, and $2.50 \leq R_{13}(\text{m}) \leq 5.25$ for the thicknesses and the radii, respectively.

For the design variables t_i ($i = 1,\ldots,12$), R_1, and R_{13}, the values of 2.0 cm, 8.50 m, and 3.50 m are assumed as the initial points in the optimization method based on the SQP as it is a point-based algorithm. For DE, the population size is taken as 30; the rates of 0.70 and 0.85 are used for the mutation and crossover. The drag force term of the Morison's equation is ignored in the analysis of the monopod tower since it is negligible compared to the term related to inertia force. Mass of the deck is taken as 2.0 Mt.

Table 7.9 Results of the RBDO of jacket tower

d			with RIA				with PMA			
	Deterministic optimization		RBDO		RBDO (**d** is random)		RBDO		RBDO (**d** is random)	
	SQP	DE	SQP	DE	SQP	DE	SQP	DE	SQP	DE
t_1 (cm)	1.37	1.37	1.77	1.77	1.95	1.95	1.77	1.79	1.95	1.96
t_2	1.00	1.00	1.00	1.00	1.00	1.00	1.00	1.00	1.00	1.00
t_3	1.00	1.00	1.00	1.00	1.00	1.00	1.00	1.00	1.00	1.00
t_4	1.00	1.00	1.00	1.00	1.00	1.00	1.00	1.00	1.00	1.00
D_1 (m)	2.35	2.35	2.79	2.78	2.79	2.79	2.79	2.80	2.80	2.78
D_2	2.17	2.17	2.88	2.90	3.01	2.98	2.88	2.81	3.00	2.99
D_3	1.00	1.00	1.00	1.00	1.00	1.01	1.00	1.00	1.00	1.01
D_4	1.00	1.00	1.00	1.00	1.00	1.00	1.00	1.00	1.00	1.00
W(**d**) Mt.	0.662	0.662	0.836	0.837	0.883	0.883	0.836	0.837	0.882	0.883

7.6.2.3 Tripod Offshore Tower

Figure 7.9 illustrates a tripod tower composed of 3 member groups. The upper and lower boundaries for the design variables are thicknesses and diameters of member groups, which are adopted as $1.0 \leq t_i$(cm, $i = 1,2.3) \leq 10.0$, and $1.0 \leq D_i(\text{m}) \leq 10.0$. In the RBDO of the tower, the diameter of the second member group is taken to be at least equal to that of the first member group ($D_2 \geq D_1$) from the point of view of practical. As the initial points for the t_i and D_i ($i = 1,\ldots,3$) the values of 2.0 cm and 3.50 m are taken in deterministic optimization. However, $t_i(i = 1,2,3) = 2.0$ cm, $D_1 = 5.0$ m, $D_2 = 6.0$ m and $D_3 = 3.50$ m are assumed for the both of the optimizations performed under the uncertainties. To employ the optimization with DE the values of 20, 0.70, and 0.85 are used for the population size, mutation, and crossover rates, respectively. Only the inertia force term is taken into account and mass of the deck is taken as 3.0 Mt.

As similar to Table 7.7, Table 7.8 also presents the results obtained for the different optimization cases explained above. For the RBDO in which **d** is random normal distributions with (COV = 0.05) are assumed for the probabilistic models of design variables and mean values of the distributions are taken as design variables of the optimization.

7.6.2.4 Jacket Type Offshore Tower

The integrated framework is finally applied to solve the RBDO of the jacket tower shown in Fig. 7.10. The jacket consists of 74 elements, which are collected into 4 groups. The first group of the members contains the legs of the structure, horizontal braces, and diagonals locating between the level of +20.0 m and +10.0 m are collected into the second group, vertical diagonals are collected into the third group and finally horizontal braces and diagonals between the level of +20.0 m

Fig. 7.11 A plane truss example

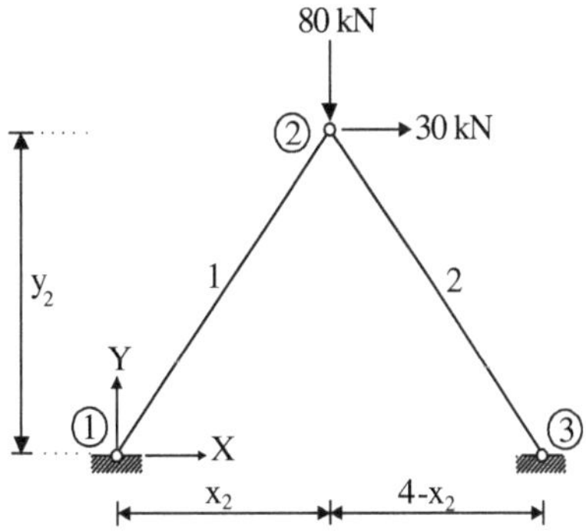

Fig. 7.12 A cantilever beam with rectangular hollow cross-section

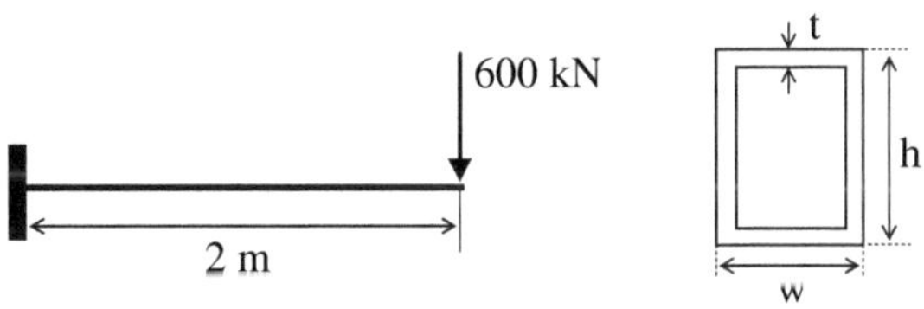

and −50.0 m form the fourth member group, Fig. 7.10. The specified ranges of the design variables consisting of thicknesses and diameters of member groups are given by $1.0 \leq t_i(\text{cm}, i = 1, 2, 3, 4) \leq 5.0$, and $1.0 \leq D_i(\text{m}) \leq 3.50$.

Mass of the deck is taken to be equal to 6.40 Mt. and the drag force coefficient c_d of the wave loading is also taken as being random. A lognormal distribution with ($\mu_{cd} = 1.30$; COV = 0.10) is assumed for the probability model of the c_d. However, in contrast with Table 7.5, in which a reduction factor is used for the coefficient c_m, the mean value of inertia force coefficient c_m is taken as 2.0 for this example. The values of 1.5 cm and 1.75 m are taken as initial values for the t_i and D_i ($i = 1,\ldots,4$) in deterministic optimization whereas the initial values $t_i = 2.0$ cm and $D_i = 1.85$ m are used for RBDO implementations. The values of 30, 0.70, and 0.85 are used for the population size, mutation, and crossover rates, respectively in order to start the optimization with DE.

As well as in Tables 7.7 and 7.8, 3 optimum solutions are indicated in Table 7.9 including the deterministic optimization, the RBDO, and the RBDO with the random design variables in addition to the random variables presented in Table 7.5. For the RBDO, that **d** is considered as being random, lognormal distributions with (COV = 0.06) are assumed for their probability models and mean values of the distributions are taken as design variables of the optimization.

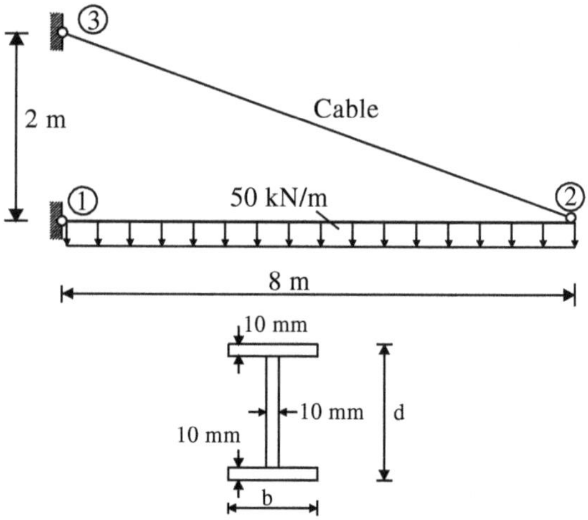

Fig. 7.13 A cable tied I beam

7.6.3 Exercises

7.6.3.1 Exercises for Deterministic Design Optimization

Exercise 1

Redesign 3 bar space truss of Fig. 7.6 by adopting pipe sections for its members having outer diameter of *D* and wall thickness of *t*. Use the Euler critical stress as upper limit for the stress constraints and keep the same displacement limitations.

Exercise 2

A plane truss shown in Fig. 7.11 is required to support the loading which is also shown in Fig. 7.11. Euler critical stress is the upper bound for the axial stresses that develop in members. The displacements of joint 1 along global X and Y axis are restricted to be <10 mm. Modulus of elasticity is 200 kN/mm^2. Formulate an optimum design problem solution of which gives the optimum location of joint 2 as well as cross-sectional areas of members such that the truss has the minimum weight. Treat the coordinates of joint 2 as design variables. Use sequential programming method to determine the optimum solution.

Exercise 3

The rectangular hollow section shown in Fig. 7.12 is used as cantilever beam to carry 600 kN of point load. The allowable bending stress is 265 MPa and the modulus of elasticity is 200 kN/mm^2. The vertical displacement of the tip of the cantilever is required to be <4 mm. Formulate an optimum design problem such that solution of which yield the optimum values of *w*, *h*, and *t* that makes the beam to have the minimum weight.

Exercise 4

In Fig. 7.13, the beam 1–2 is tied to support 3 with cable 2–3. The allowable bending stress is 265 MPa and the modulus of elasticity is 200 kN/mm^2. The beam AB has I-shaped cross-section which will be produced from a steel plate with 10 mm thickness. The allowable stress of the cable material is 400 MPa. Formulate an optimum design problem such that solution of which gives the optimum values for the width b and depth d of I-section as well as the diameter of the cable so that the structure has the minimum weight.

7.6.3.2 Exercises for Reliability-Based Design Optimization

Exercise 1

Repeat the RBDO procedure given above for the monopod tower with 3 segments under the same reliability constraints. However, for this case, consider design variables as being random. Assume normal distributions with (COV = 0.05) for the probability models of design variables and take the mean value of the distributions as design variables of the optimization. $\beta_{i,\mathrm{target}}(i = 1,\ldots,\mathrm{nrc}) = 3.0$

Exercise 2

Find the minimum mass of the monopod tower with 6 segments under the reliability constraints based on the limit-state-function related to yield stress, buckling stress, and natural frequency. Case 1: design variables are not random, Case 2: design variables are random. For their probability models, assume the lognormal distributions with (COV = 0.05) and take the mean value of the distributions as design variables of the optimization. $\beta_{i,\mathrm{target}}(i = 1,\ldots,\mathrm{nrc}) = 3.0$

Exercise 3

For the Exercise 2, show the changes on the design variables and the mass of the monopod tower with six segments considering the $\beta_{i,\mathrm{target}}(i = 1,\ldots,\mathrm{nrc}) = 3.10$, 3.70, 4.20 and 5.0, respectively.

References

1. Horst R, Pardolos PM (1995) Handbook of global optimization. Kluwer Academic Publishers, Dordrecht
2. Nocedal J, Wright JS (1999) Numerical optimization. Springer, New York
3. Chong EKP, Zak SH (2002) Introduction to optimization. Wiley, New York
4. Onwubiko C (2000) Introduction to engineering design optimization. Prentice-Hall, New Jersey
5. Arora J (2004) Introduction to optimum design. Elsevier Academic Press, San Diego
6. Arora J (2011) Optimization of structural and mechanical systems, World scientific Publishing Company, Massachusetts

7. Ravindran A, Ragsdell KM, Relaitis GV (2006) Engineering optimization: methods and applications. Wiley, New Jersey
8. Horst R, Tuy H (1995) Global optimization: deterministic approaches. Springer, Berlin
9. Paton R (1994) Computing with biological metaphors. Chapman & Hall, London
10. Adami C (1998) An introduction to artificial life. Springer, New York
11. Matheck C (1998) Design in nature: learning from trees. Springer, Berlin
12. Mitchell M (1998) An introduction to genetic algorithms. The MIT Press, Cambridge
13. Flake GW (2000) The computational beauty of nature. MIT Press, Cambridge
14. Kennedy J, Eberhart R, Shi Y (2001) Swarm intelligence. Morgan Kaufmann Publishers, San Francisco
15. Glover F, Kochenberger GA (2003) Handbook of metaheuristics. Kluwer Academic Publishers, Dordrecht
16. Dreo J, Petrowski A, Siarry P, Taillard E (2006) Meta-heuristics for hard optimization. Springer, Berlin
17. Sean L (2009) Essentials of metaheuristics, @ http://cs.gmu.edu/~sean/book/metaheuristics/
18. Han SP (1977) Globally convergent method for nonlinear programming. J Optim Theory Appl 22:297–309
19. Venkataraman P (2002) Applied optimization with MATLAB programming. Wiley, New York
20. Davidon WC (1959) Variable metric methods for minimization, U.S. Atomic Energy Commission Research and Development Report No. ANL-5990, Argonne National Laboratory
21. Storn R, Price K (1997) Differential evolution-a simple and efficient heuristics for global optimization over continuous spaces. J Global Optim 11:341–359
22. Price KV, Storn RM, Lampinen JA (2005) Differential evolution: a practical approach to global optimization, Springer, Berlin
23. Lee KS, Geem ZW (2004) A new structural optimization method based on the harmony search algorithm. Comp Struct 82:781–798
24. Rajaev S, Krishnamoorthy CS (1992) Discrete optimization of structures using genetic algorithm. J Struc Eng 118(5):1233–1250
25. Krishnamoorthy CS, Venkatesh PP, Sudarshan R (2002) Object-oriented framework for genetic algorithms with application to space truss optimization. J Comput Civil Eng 16:66–75
26. Camp C, Pezeshk S, Cao G (1998) Optimized design of two-dimensional structure using genetic algorithm. J Struct Eng 124:551–559
27. Erbatur F, Hasançebi Ö, Tütüncü Tütüncü İ, Kılıç H (200) Optimal design of planar and space structures with genetic algorithms. Comput Struct 75:209–224
28. Saka MP (1990) Optimum design of pin-jointed steel structures with practical applications. J Struct Eng 116(10):2599–2620
29. Toğan V, Daloğlu A (2009) Bridge truss optimization under moving load using continuous and discrete design variables in optimization methods. Indian J Eng Mater Sci 16:245–258
30. Toğan V, Daloğlu A (2008) An improved genetic algorithm with initial population and self-adaptive member grouping. Comput Struct 86:1204–1218
31. Toğan V, Daloğlu A (2006) Optimization of 3d trusses with adaptive approach in genetic algorithms. Eng Struct 28:1019–1027
32. Toğan V, Daloğlu A (2006) Shape and size optimization of 3d trusses with genetic algorithm. Tech J Turkish Chamber Civil Eng 17:3809–3826
33. Madsen HO, Krenk S, Lind NC (1986) Methods of structural safety. Prentice-Hall, New Jersey
34. Ditlevsen O, Madsen HO (1996) Structural reliability methods. Wiley, New York
35. Melchers RE (2001) Structural reliability analysis and prediction. Wiley, Chichester
36. Haldar A, Mahadevan S (2000) Reliability assessment using stochastic finite element analysis. Wiley, New York
37. Aoues Y, Chateauneuf A (2009) Benchmark study of numerical methods for reliability based design optimization. Struct Multidisc Optim. doi:10.1007/s00158-009-0412-2

38. Valdebenito MA, Schueller GI (2010) A survey on approaches for reliability-based optimization. Struct Multidisc Optim 42:645–663
39. Enevoldsen I, Sorensen JD (1994) Reliability based optimization in structural engineering. Struct Saf 15:169–196
40. Thanedar PB, Kodiyalam S (1992) Structural optimization using probabilistic constraints. Struct Optim 4:236–240
41. Stocki R, Siemaszko A, Kleiber M (1999) Interactive methodology for reliability-based structural design and optimization. Comput Assisted Mech Eng Sci 6:39–62
42. Toğan V, Karadeniz H, Daloğlu A (2010) An integrated framework including distinct algorithms for optimization of offshore towers under uncertainties. Reliab Eng Syst Saf 95:847–858
43. Karadeniz H, Toğan V, Vrouwenvelder T (2010) Optimization of steel monopod offshore-towers under probabilistic constraints. J Offshore Mech Arct Eng 132: 021605–1, 021605–7
44. Toğan V, Daloğlu A, Karadeniz H (2010) Reliability-based design optimization of structural systems with continuous and discrete design variables. Tech J Turkish Chamber Civil Eng 21:5135–5159
45. Karadeniz H, Daloğlu A, Daloğlu A et al (2010) Reliability based optimization of offshore jacket type structures with an integrated algorithms-system. Ships Offshore Struct 5:67–74
46. Karadeniz H, Toğan V, Vrouwenvelder T, Vrouwenvelder T (2009) An integrated reliability based design optimization of offshore towers. Reliab Eng Syst Saf 94:1510–1516
47. Toğan V, Daloğlu A, Karadeniz H (2011) Optimization of trusses under uncertainties with harmony search. Struct Eng Mech 37:543–560
48. Toğan V(2009) Reliability based design optimization of offshore structures. Dissertation, Karadeniz Technical University
49. Toğan V, Daloğlu A (2006) Reliability and reliability based design optimization. Turk J Eng Environ Sci 30:237–249
50. Kuschel N, Rackwitz R (1997) Two basic problem for reliability-based optimal design. Math Methods Oper Res 46:309–333
51. Kirjner-Neto C, Polak E, Der Kiureghian A (1998) An outer approximations approach to reliability based optimal design of structures. J Optim Theory Appl 98:1–16
52. Kharmanda G, Mohamed A, Lemaire M (2002) Efficient reliability based design optimization using hybrid space with application to finite element analysis. Struct Multidisc Optim 24:233–245
53. Cheng G, Xu L, Jiang L (2006) A sequential approximate programming strategy for reliability based structural optimization. Comput Struct 84:1353–1367
54. Du X, Chen W (2004) Sequential optimization and reliability assessment method for efficient probabilistic design. J Mech Des 126:225–233
55. Marek P, Gustar M, Anagsos T (1996) Simulation-based reliability assessment for structural engineers. CRCPress, Boca Raton
56. Breitung KW (1994) Asymptotic approximations for probability integrals Lect Notes Math. Springer, Berlin
57. Tu J (1999) Design potential concept for reliability based design optimization. PhD Thesis, The University of Iowa, Iowa city
58. Tu J, Choi KK, Park YH (2002) A new study on reliability based design optimization. J Mech Des 121:557–564
59. Lee JO, Yang YS, Ruy WS (2002) A comparative study on reliability index and target-performance based probabilistic structural design optimization. Comp Struct 80:257–269
60. Youn BD, Choi KK, Park YH (2003) Hybrid analysis method for reliability based design optimization. J Eng Mech 125:221–232
61. Hasofer AM, Lind N (1974) An exact and invariant first-order reliability format. J Eng Mech 100:111–121
62. Rackwitz R, Fiessler B (1978) Structural reliability under combined random load sequences. Comput Struct 9:489–494
63. Ditlevsen O (1981) Principle of normal tail approximation. J Eng Mech 107:1191–1209

64. Hohenbichler M, Rackwitz R (1981) Non-normal dependent vectors in structural safety. J Eng Mech 107:1227–1238
65. Youn B, Choi K, Du L (2005) Adaptive probability analysis using an enhanced hybrid mean value method. Struct Multidisc Optim 29:134–148
66. Frangopol DM, Maute K (2005) Reliability based optimization of civil and aerospace structural system, engineering design reliability handbook. CRC Press, Boca Raton
67. Ramu P, Qu X, Youn BD et al (2006) Inverse reliability measures and reliability-based design optimization. Int J Reliab Saf 1:187–205
68. Agarwal H, Mozumder CK, Renaud JE et al (2007) An inverse measure based unilevel architecture for reliability based design optimization. Struct Multidisc Optim 33:217–227
69. Kharmanda G (2007) Numerical and semi-numerical methods for reliability based design optimization. In: Tsompanakis Y, Lagaros ND, Papadrakakis M (eds) Structural design optimization considering uncertainties. Taylor & Francis/Balkema, Leiden
70. Chateauneuf A (2007) Principles of reliability based design optimization. In: Tsompanakis Y, Lagaros ND, Papadrakakis M (eds) Structural design optimization considering uncertainties. Taylor & Francis/Balkema, Leiden
71. Haug EJ, Choi KK, Komkov V (1986) Design sensitivity analysis of structural systems. Academic Press, Orlando
72. Mohamed A, Lemaire M (1999) The use of sensitivity operators in the reliability analysis of structures. In: 3th international conference computer stochastic mechanics, Balkema
73. Choi KK, Kim NH (2005) Structural sensitivity analysis and optimization. Springer, Berlin
74. Kleiber M, Antunez H, Hien TD et al. (1997) Parameter sensitivity in nonlinear mechanics. Wiley, New York
75. Choi SK, Grandhi RV, Canfield RA (2007) Reliability based structural design. Springer, London
76. Yi P, Cheng G, Jiang L (2006) A sequential approximate programming strategy for performance measure based probabilistic structural design optimization. Struct Saf. doi:10.1016/j.strusafe.2006.08.003
77. Enevoldsen I (1994) Sensitivity analysis of a reliability-based optimal solution. J Eng Mech 120:198–205
78. Frangopol D (1985) Sensitivity of reliability-based optimum design. J Struct Eng 111:1703–1721
79. Gill PE, Murray W, Wright HM (1981) Practical optimization. Academic Press, London
80. IMSL Fortran 90 MP Library (1998) Version 4.01, Visual Numeric, Houston
81. Differential Evolution homepage. http://www.icsi.berkeley.edu/~storn/code.html
82. Det Norske Veritas (DNV) (1992) Buckling strength analysis, classification notes. Hovik, Norway
83. Uys PE, Farkas J, Jarmai K et al (2007) Optimization of a steel tower for a wind turbine structure. Eng Struct 29:1337–1342
84. Joint committee on structural safety (JCSS) (2000) Probabilistic model code, Part 1—Basis of Design
85. Karadeniz H (2005) Reliability calculation of RC concrete offshore structures under extreme wave loading. In: Proceeding in 15th ISOPE, June 19–24, Seoul, Korea
86. StuPoc V (1979) Probabilistic reliability analysis for offshore structures. Final report, Netherlands Industrial Council for Oceanology, The Netherlands
87. Lancaster ER, Calladine CR, Palmer SC (2000) Paradoxical buckling behavior of thin cylindrical shell under axial compression. J Mech Sci 42:843–865

MIX
Papier aus verantwortungsvollen Quellen
Paper from responsible sources
FSC® C105338

If you have any concerns about our products,
you can contact us on
ProductSafety@springernature.com

In case Publisher is established outside the EU,
the EU authorized representative is:
Springer Nature Customer Service Center GmbH
Europaplatz 3, 69115 Heidelberg, Germany

Printed by Libri Plureos GmbH
in Hamburg, Germany